BEAM INSTRUMENTATION

Proceedings of the Seventh Workshop

BEAM INSTRUMENTATION

Proceedings of the Seventh Workshop

Argonne, Illinois   May 1996

EDITORS
Alex H. Lumpkin
Catherine E. Eyberger
Advanced Photon Source
Argonne National Laboratory

AIP PRESS

American Institute of Physics

**AIP CONFERENCE
PROCEEDINGS 390**

Woodbury, New York

L.C. Catalog Card No. 97–70568
ISBN 1-56396-612-3
ISSN 0094-243X
DOE CONF- 9605173

Printed in the United States of America

CONTENTS

TUTORIALS

INVITED PAPERS

CONTRIBUTED PAPERS

APPENDICES

PREFACE

The Seventh Beam Instrumentation Workshop (BIW'96) was hosted by Argonne National Laboratory in its 50th anniversary year at the new Advanced Photon Source (APS) Conference Center on May 6–9, 1996. This was the first meeting in the transition of the BIW into the spring and in alternate years from the Particle Accelerator Conference. We had an unusual space-time juxtaposition with the Spring American Physical Society Meeting, which was held later than usual in the first week of May and in Indianapolis, only a few hundred miles away. Some persons were able to participate in both. Our Workshop hotel was at the nearby OakBrook Marriott, situated next to a large shopping mall with several restaurants and only 30 minutes from downtown Chicago.

The Workshop had 140 registrants and 7 vendors. The APS Conference Center auditorium, break-out rooms, atrium area, and lower-level gallery (for posters, vendors, and refreshments) accommodated the crowd comfortably. The initial fraction of abstracts from foreign countries was higher than average. Attendees came from the USA, Canada, United Kingdom, France, Germany, Italy, Japan, Switzerland, and Denmark. The Workshop included 4 invited tutorials, 5 invited papers, 14 contributed oral papers, and 32 contributed posters. In addition, discussion sessions were held on six topics, one of which was diagnostics needs for the fourth-generation light sources. The Workshop welcomes were given by David Moncton, Associate Laboratory Director of the APS, and John Galayda, APS Accelerator Systems Division Director.

The banquet was held during a cruise on Lake Michigan on board *The Spirit of Chicago*. The highlight of the banquet (other than the 1-hour open bar and the Chicago skyline shrouded in fog) was the presentation of the 1996 Faraday Cup Award, sponsored by Bergoz, Inc. This year's winners were Walter Barry of Lawrence Berkeley National Laboratory and Hung-chi Lihn, now of Tencor Instruments, for their contributions to the development and demonstration of a subpicosecond bunch duration measurement technique based on autocorrelation of coherent transition radiation. Their talks describing the technique were presented the following day and are included in these proceedings. The Workshop ended with a closeout session and a tour of the APS storage ring area.

In addition to the BIW'96 Program Committee, many people at Argonne contributed to the success of the Workshop. Special thanks go to Diane Simms as Workshop Secretary, Cathy Eyberger as Workshop Co-Editor, Joan Brunsvold and Jacquie Habenicht as Registrars, Dean Barr for the WWW pages, and Rick Fenner for the poster and audio-visual advice. We also acknowledge Diagnostics Group staff who were at tour stations or led tours, and Susan Barr, Dennis Mills, and Ercan Alp of the APS Experimental Facilities Division who gave us access to their beamline stations for the tour.

We acknowledge staffing support from John Galayda and the Accelerator Systems Division and financial support from Pentland Systems for the afternoon break refreshments.

Finally, we thank the participants for their contributions to the Workshop; this year all tutorial and invited papers are included. We note that having the electronic files of the papers expedited the editing of nontechnical corrections and contributed to the more uniform look of these proceedings. Technical questions were resolved with the aid of the authors. Our thanks to Kelly Jaje of the Accelerator Systems Division who made a major contribution to the electronic editing effort. In addition, the invited speakers' presentations were videotaped and a set of tapes was sent to each of the laboratories with a Program Committee member.

The next workshop in this series will be hosted by the Stanford Synchrotron Radiation Laboratory and PEP-II of the Stanford Linear Accelerator Center and is expected to be held in the Spring of 1998.

Alex Lumpkin
BIW'96 Chairman

ORGANIZING COMMITTEE

Alex Lumpkin, ANL, Chair
Robert Averill, MIT-Bates
Claude Bovet, CERN
Jean-Claude Denard, CEBAF
Robert Hettel, SSRL
Jim Hinkson, LBNL
Heribert Koziol, CERN
George Mackenzie, TRIUMF
Ralph Pasquinelli, FNAL

Mike Plum, LANL
Robert Shafer, LANL
Gary Smith, BNL
Stephen Smith, SLAC
Gregory Stover, LBNL
Robert Webber, FNAL
Richard Witkover, BNL
Jim Zagel, FNAL

LOCAL ARRANGEMENTS COMMITTEE

Dean Barr
Joan Brunsvold
Catherine Eyberger
Richard Fenner

Jacquie Habenicht
Alex Lumpkin
Diane Simms

xi

BEAM INSTRUMENTATION WORKSHOP - May 6-9, 1996

	Monday	Tuesday	Wednesday	Thursday
8:00	Registration at Argonne			
8:30	Welcome/Opening	Tutorial	Tutorial	Tutorial
9:00	Invited Talk	H. Spieler (LBNL)	S. Smith (SLAC)	J. Corlett (LBNL)
9:30	M. Placidi (CERN)			
10:00	Break	Break	Break	Break (15 min)
10:15				Invited Talk (10:15 - 11:15)
10:30	Tutorial	Invited Talk	Invited Talk	Faraday Cup Winners W. Barry and H. Lihn
11:00	R. Siemann (SLAC)	T. Fessenden (LBNL)	A. Lumpkin (ANL)	
11:15				Closeout (11:15 - 11:45)
11:30		Invited Talk	Contributed Talks T. Powers	Tour of APS (11:45-1:15)
12:00	Contributed Talks J. Steimel J. Fox	T. Shintake (KEK)	E. Kahana P. Puzo	
12:30		Lunch	Lunch	
Lunch	Lunch (12:40)			
1:00				
1:30				Lunch (to 2:15)
2:00	Contributed Talks R. Jung R. Anne	Posters/Vendors	Contributed Talks M. Ross C. Fischer	
2:30	X. Wang R. Keller		P. Piot P. Cameron	
3:00	A. Fisher			Hans Bethe Talk
3:30			Discussion Group Summaries (3:40 - 4:15)	
4:00	Break	Break		
4:15			Break (4:15)	
4:30	3 Discussion Groups (to 6:00 pm)	3 Discussion Groups (to 6:00 pm)	Buses to Chicago	
			Banquet (6:45)	

Participants at the 7th Beam Instrumentation Workshop, "bunch compressed" in front of the Advanced Photon Source Conference Center. (Photo by Stan Niehoff.)

Walter Barry (second from left) and Hung-chi Lihn (second from right), recipients of the 1996 Faraday Cup Award, with Julien Bergoz, sponsor of the award (left), and Alex Lumpkin, BIW'96 Chairman (right). (photo by Stan Niehoff)

TUTORIALS

Spectral Analysis of Relativistic Bunched Beams

R. H. Siemann*

Stanford Linear Accelerator Center, Stanford University, Stanford, CA 94309

INTRODUCTION

Particles in a storage ring are oscillating in the longitudinal and transverse dimensions, and therefore, the frequency domain is natural for analyzing many beam-generated signals. Information ranging from oscillation frequencies to beam phase-space distributions can be extracted from the spectral content of these signals.

It is often necessary to switch between time and frequency domains using Fourier transforms. If f(t) is a function of time, $F(\omega)$ given by

$$F(\omega) = \int_{-\infty}^{\infty} f(t)e^{-j\omega t}dt \qquad (1)$$

is its Fourier transform. In this equation $j = \sqrt{-1}$ and ω is the angular frequency which is the appropriate mathematical variable rather than the frequency $\nu = \omega/2\pi$ that is measured by spectrum analyzers. The meaning of the word "frequency" depends on context, but the notation ν for frequency and ω for angular frequency will be rigorous. The inverse transform is

$$f(t) = \frac{1}{2\pi} \int_{-\infty}^{\infty} F(\omega)e^{j\omega t}d\omega . \qquad (2)$$

These notes are restricted to relativistic beams, and for these beams the image current flowing in the walls of an accelerator vacuum chamber, i_w, has a line density equal to the line density of the beam current, i_b,

$$i_w(t) = -i_b(t) \text{ and } I_w(\omega) = -I_b(\omega) \qquad (3)$$

with an overall minus sign since it is an image current. The image current is an ideal current source; nothing placed in the vacuum chamber wall can affect it because the beam would have to be decelerated for that to happen. The image current flowing through a beam detector produces a voltage V_{out} given by

---

* Work supported by the Department of Energy, contract DE-AC03-76SF00515.

$$V_{out}(\omega) = I_w(\omega)S(\omega) \tag{4}$$

where S is the detector longitudinal sensitivity. The sensitivity depends on the detector, on whether it is a cavity, stripline, capacitive pickup, etc., and $S(\omega)$ can vary rapidly or slowly with frequency. I am assuming the variation is slow and that the spectrum of the output voltage is the same as that of the wall and beam currents. Many measurements can be made in a narrow frequency band where the sensitivity can be treated as a constant, so this is a good approximation. If it isn't, corrections must be made to account for frequency dependence of the sensitivity.

If the beam is offset from the vacuum chamber center, the image current is not uniform; instead, it varies around the beam pipe due to the displacement of the beam. A transverse beam detector responds to the dipole moment

$$d(t) = i_b(t)r_\perp(t) \tag{5}$$

where $r_\perp(t)$ is the displacement from the center. The output voltage of a transverse detector is

$$V_{out}(\omega) = D(\omega)S_\Delta(\omega) \tag{6}$$

where S_Δ and D are the transverse sensitivity and the Fourier transform of the dipole moment, respectively. As with the longitudinal, S_Δ is assumed to vary slowly with frequency, and corrections must be made if that isn't a good approximation.

A SINGLE PARTICLE

The spectrum of a single particle is like a Green's function, and it is the key to understanding the spectrum produced by a beam. Three separate cases are considered in an order of increasing complexity: 1) constant revolution frequency, 2) Frequency Modulation introduced by synchrotron oscillations, and 3) Amplitude Modulation introduced by betatron oscillations.

Longitudinal Motion & Constant Revolution Frequency

The current of a single, unit-charge particle with a constant revolution frequency ω_r, Figure 1, is a periodic set of impulses spaced a time $T = 2\pi/\omega_r$ apart

$$i_b(t) = \sum_{n=-\infty}^{\infty} \delta(t - nT). \tag{7}$$

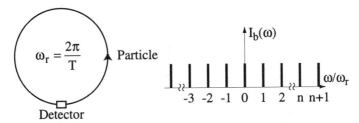

FIGURE 1. A particle with a constant revolution frequency passing a beam detector located at one point in the ring and the spectrum it produces.

Fourier transforming this gives

$$I_b(\omega) = \int_{-\infty}^{\infty} i_b(\omega)e^{-j\omega t}dt = \int_{-\infty}^{\infty} \sum_{n=-\infty}^{\infty} \delta(t-nT)e^{-j\omega t}dt$$

$$= \sum_{n=-\infty}^{\infty} e^{-j\omega nT} = \sum_{n=-\infty}^{\infty} e^{-j2\pi n\omega/\omega_r} . \qquad (8)$$

The last sum is an infinite sum of unit magnitude phasors. If these phasors are at different angles they will add up to zero because there is an infinite number of them, but if they are all at the same angle the sum will be infinite. The latter happens when ω/ω_r equals an integer; every angle is an integer multiple of 2π in that case. This is written formally as

$$I_b(\omega) = \sum_{n=-\infty}^{\infty} e^{-j2\pi n\omega/\omega_r} = \omega_r \sum_{n=-\infty}^{\infty} \delta(\omega - n\omega_r). \qquad (9)$$

The spectrum is a comb of equal amplitude lines spaced at ω_r extending from $\omega = -\infty$ to $\omega = \infty$. It is shown in Figure 1. The spectrum has two properties—the *frequencies* that are given by the argument of the δ-function and the *envelope* that is given by the factor in front. The frequencies are harmonics of ω_r, and the envelope equals ω_r independent of frequency.

There are both positive and negative frequencies in Eq. (9), but spectrum analyzers measure only positive frequencies. How should the negative frequencies be interpreted? Fourier transforming to the time domain

$$i_b(t) = \frac{\omega_r}{2\pi} \sum_{n=-\infty}^{\infty} e^{jn\omega_r t} . \qquad (10)$$

Each term in the sum is a unit magnitude phasor. The angles of the positive frequency phasors increase with time while those of the negative phasors decrease with time. The phasors for the same $|n|$ are shown in Figure 2. The real parts are in phase, and the imaginary parts are 180° out of phase. The physically meaningful quantity is the real part that is given by

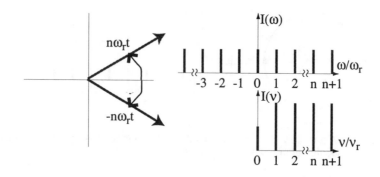

FIGURE 2. Positive and negative frequency phasors for the same |n| and the mathematical and measured, physical spectra.   The line at n = 0 is one-half that for n ≠ 0 because there is no corresponding negative frequency line .

$$\Re(e^{jn\omega_r t} + e^{-jn\omega_r t}) = 2\cos(n\omega_r t). \tag{11}$$

The positive and negative frequency phasors appear in phase and at the same at the same frequency on the spectrum analyzer. The resulting spectrum is shown in Figure 2.

Frequency Modulation Introduced by Synchrotron Oscillations

Synchrotron motion modulates the arrival time by

$$\tau = \tau_a \cos(\omega_s t + \varphi) \tag{12}$$

where τ_a is the synchrotron oscillation amplitude, ω_s is the synchrotron frequency, and φ is the phase.[*] Instead of being a series of impulses spaced exactly T apart, the current is

$$i_b(t) = \sum_{n=-\infty}^{\infty} \delta(t - [nT + \tau_a \cos(\omega_s nT + \varphi)]) , \tag{13}$$

where τ_a is the synchrotron oscillation amplitude, ω_s is the synchrotron frequency, and φ is the phase at n = 0. The synchrotron frequency and synchrotron tune Q_s are related by $\omega_s = Q_s \omega_r$. The Fourier transform is

---

[*]   The synchrotron frequency and synchrotron tune Q_s are related by $\omega_s = Q_s\omega_r$.

$$I_b(\omega) = \sum_{n=-\infty}^{\infty} \exp(-j\omega[nT + \tau_a \cos(\omega_s nT + \varphi)]). \tag{14}$$

In the limit of small amplitude or low frequency, $\omega\tau_a \ll 1$, the second term in the exponent can be approximated with a Taylor expansion

$$I_b(\omega) \approx \sum_{n=-\infty}^{\infty} e^{-jn\omega T} \left(1 - j\omega\tau_a \cos(\omega_s nT + \varphi)\right)$$

$$= \sum_{n=-\infty}^{\infty} e^{-jn\omega T} \left(1 - \frac{j\omega\tau_a}{2} e^{j(\omega_s nT + \varphi)} - \frac{j\omega\tau_a}{2} e^{-j(\omega_s nT + \varphi)}\right). \tag{15}$$

The first of the three sums is the same as for a constant revolution frequency. Take a look at the second one:

$$\sum_{n=-\infty}^{\infty} \exp(-jn\omega T + j\omega_s nT + j\varphi)) = e^{j\varphi} \sum_{n=-\infty}^{\infty} \exp(-j2\pi n(\omega - \omega_s)/\omega_r)$$

$$= e^{j\varphi} \sum_{n=-\infty}^{\infty} \delta(\omega - \omega_s - n\omega_r). \tag{16}$$

The last step follows from the same logic used for a constant revolution frequency—the sum equals zero unless $(\omega - \omega_s)/\omega_r$ is an integer. The frequencies differ from the rotation harmonics by $+\omega_s$. The third term in Eq. (15) leads to frequencies that differ from the rotation harmonics by $-\omega_s$. Synchrotron motion has lead to two new frequency combs displaced from the rotation harmonics. Both of these combs have envelopes that depend linearly on frequency.

The approximation in Eq. (15) is valid for small amplitudes and/or low frequencies. The Taylor series expansion could be continued. The next term would affect the envelope of the rotation harmonics and introduce frequency combs displaced from the rotation harmonics by $\pm 2\omega_s$. However, rather than doing this it is better to perform a Bessel function expansion (1)

$$e^{-jz\cos\theta} = \sum_{k=-\infty}^{\infty} J_k(z) e^{jk(\theta - \pi/2)} \tag{17}$$

where J_k is an ordinary Bessel function of order k. This same expansion leads to Bessel functions in every analysis of Frequency Modulation. Using this in Eq. (14),

$$I_b(\omega) = \sum_{n,k=-\infty}^{\infty} J_k(\omega\tau_a) e^{jk(\omega_s nT + \varphi - \pi/2)} e^{-jn\omega T}$$

$$= \sum_{n,k=-\infty}^{\infty} e^{jk(\varphi - \pi/2)} J_k(\omega\tau_a) e^{-j2\pi n(\omega - k\omega_s)/\omega_r}. \tag{18}$$

As before the summation over n restricts the possible frequencies. In this case it requires that $(\omega - k\omega_s)/\omega_r$ equals an integer. Performing the sum

$$I_b(\omega) = \omega_r \sum_{k=-\infty}^{\infty} e^{jk(\varphi-\pi/2)} J_k(\omega\tau_a) \sum_{n=-\infty}^{\infty} \delta(\omega - k\omega_s - n\omega_r). \qquad (19)$$

Equation (19) gives the general expression for the spectrum of a particle undergoing synchrotron motion. It is an important result that deserves substantial discussion.

1. The approximate treatment based on a Taylor expansion corresponds to three of the terms, $k = -1, 0, 1$, when the lowest order Taylor expansions for the Bessel functions are used.

2. For each rotation harmonic there is an infinite number of sidebands. They are displaced from the rotation harmonic by $k\omega_s$, $k = -\infty,...,\infty$ and have different envelopes. The envelopes illustrated in Figure 3 have the usual properties of ordinary Bessel functions: *i)* they are even (odd) functions of $\omega\tau_a$ if k is even (odd), and *ii)* the first maximum of J_k is at $\omega\tau_a \approx k$. When $|\omega| \ll 1/\tau_a$ only the rotation harmonics, $k = 0$, are present, and as the frequency increases more sidebands appear. This is illustrated in Figure 4.

3. The synchrotron sideband number, k, appears in two different places: *i)* the frequency shift from the rotation harmonic is $k\omega_s$, and *ii)* the envelope is $J_k(\omega\tau_a)$. The best frequency region to observe the k*th* sideband is at $\omega \sim k/\tau_a$. This connection between frequency shift and frequency region for observation will become important when the phase-space structure of a beam is considered.

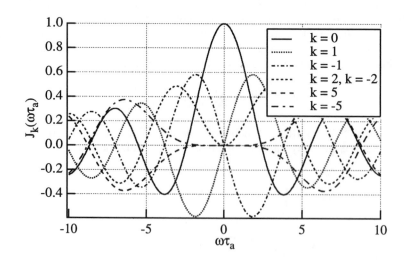

FIGURE 3. The envelopes for different synchrotron sidebands and $\varphi = \pi/2$.

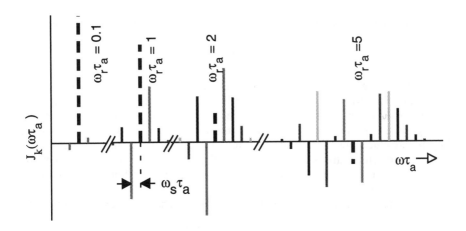

FIGURE 4. The spectra for different $\omega_r \tau_a$ for a synchrotron tune $Q_s = 0.01$ and $\varphi = \pi/2$. The heavy dashed line is $k = 0$ in each case

4. The phase φ is the phase of the synchrotron oscillation when $n = 0$. Since the sum is infinite, $n = 0$ is arbitrary, and, therefore, φ is arbitrary. This situation changes when multiple particles are considered, and the phases of the particles relative to each other has significance.

5. The observed spectrum is obtained by combining the positive and negative frequency components. The signals at $\omega = \omega_{nk} = n\omega_r + k\omega_s$ and $\omega = -\omega_{nk}$ have the same physical frequency, and as the next line shows, they add coherently in the physical spectrum. Fourier transforming back to the time domain and taking the real part gives

$$\Re(i_b(t))$$
$$= \Re\left[e^{jk(\varphi-\pi/2)}J_k(\omega_{nk}\tau_a)e^{j\omega_{nk}t} + e^{-jk(\varphi-\pi/2)}J_{-k}(-\omega_{nk}\tau_a)e^{-j\omega_{nk}t}\right] \quad (20)$$
$$= 2J_k(\omega_{nk}\tau_a)\cos(\omega_{nk}t + k\varphi - k\pi/2) .$$

Amplitude Modulation Introduced by Betatron Oscillations

Constant Revolution Frequency

A particle can be offset from the center of a transverse beam detector due to closed orbit errors, synchrotron oscillations combined with dispersion, or betatron oscillations. A closed orbit offset produces a spectrum identical to that of longitudinal motion alone. Synchrotron oscillations combined with dispersion do not introduce new frequencies although they do affect the envelopes (2). Betatron motion produces Amplitude Modulation that leads to new frequencies and new phenomena in the spectrum. The next two sections concentrate on betatron motion.

Begin with a constant revolution frequency. The displacement is given by

$$r_\perp(t) = A_\beta \cos\omega_\beta t = \frac{A_\beta}{2}(e^{j\omega_\beta t} + e^{-j\omega_\beta t}) \tag{21}$$

where A_β is the betatron amplitude, and ω_β is the betatron frequency that is related to the betatron tune, $Q_{\beta 0}$, by $Q_{\beta 0} = \omega_\beta/\omega_r$. The dipole moment and its Fourier transform are

$$d(t) = i_b(t)r_\perp(t) = \frac{A_\beta}{2}(e^{j\omega_\beta t} + e^{-j\omega_\beta t}) \sum_{n=-\infty}^{\infty} \delta(t - nT), \tag{22}$$

and

$$D(\omega) = \frac{A_\beta}{2}\left[\sum_{n=-\infty}^{\infty} e^{-j(\omega-\omega_\beta)nT} + \sum_{n=-\infty}^{\infty} e^{-j(\omega+\omega_\beta)nT} \right]$$

$$= \frac{A_\beta\omega_r}{2}\left[\sum_{n=-\infty}^{\infty} \delta(\omega - \omega_\beta - n\omega_r) + \sum_{n=-\infty}^{\infty} \delta(\omega + \omega_\beta - n\omega_r) \right]. \tag{23}$$

There are two frequency combs. One is displaced from the rotation harmonics by $+\omega_\beta$ and the other by $-\omega_\beta$. Each has a constant envelope. This is illustrated in Figure 5. The betatron tune is almost always greater than one, and, as consequence of measuring the dipole moment at only one point on the orbit, the integer part of the tune cannot be measured from the spectrum.

There is also an ambiguity in measuring the fractional part of the tune. The spectrum when the fractional part of the tune equals q_β is the same as when the fractional part equals $1 - q_\beta$. Figures 5b and 5d illustrate this. The ambiguity arises because AM produces sidebands above and below the rotation harmonics. One way to resolve it is to increase the tune by strengthening a focusing quadrupole and observing whether the spectral line moves to lower or higher frequency. The frequency of a line in the region $n\omega_r < \omega < (n + 1/2)\omega_r$ will increase if $q_\beta < 0.5$, and it will decrease if $q_\beta > 0.5$.

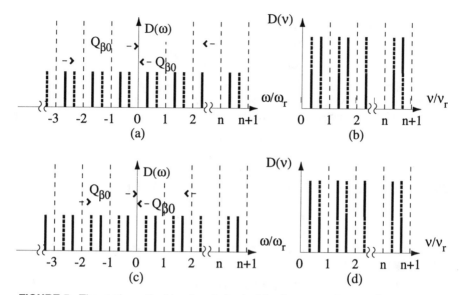

FIGURE 5. The mathematical (a, c) and physical (b, d) spectra given by Eq. (23) when $Q_{\beta 0} \approx 2.33$ (a, b) and $Q_{\beta 0} \approx 1.66$ (c, d). For purposes of illustration the lines from one comb are solid and the lines from the other comb are dashed.

The physical spectra in Figures 5b, 5d could have been derived in a different way. Rather than using Eq. (21) for the transverse displacement, $r_\perp$ could have been written

$$r_\perp(t) = A_\beta e^{j\omega_\beta t} \quad . \tag{24}$$

This would have resulted in a mathematical spectrum with only upper sidebands of amplitude A_β. However, when the real part of the resulting expression was taken, the physical spectra would be the same as in Figures 5b and 5d. A generalization of Eq. (24) is the more convenient form for the transverse displacement when the effects of synchrotron oscillations are included.

Betatron and Synchrotron Motion

The focusing strength of a quadrupole depends on energy, and energy is modulated by synchrotron motion. As a result, the betatron phase does not advance smoothly. The generalization of Eq. (24) in terms of the betatron phase ψ_β is

$$r_\perp(t) = A_\beta e^{j\psi_\beta(t)} \quad . \tag{25}$$

The time rate of change of ψ_β is

$$\frac{d\psi_\beta}{dt} = \omega_\beta(1+\xi\delta) + O(\delta^2) \tag{26}$$

where δ is the fractional deviation from the central energy, and $\omega_\beta = Q_{\beta 0}\omega_r$ is the betatron frequency when $\delta = 0$. The chromaticity ξ measures the variation of betatron tune with energy. It is defined as

$$\xi = \frac{1}{Q_{\beta 0}} \frac{dQ_\beta(\delta)}{d\delta}\bigg|_{\delta=0} . \tag{27}$$

The energy deviation is 90° out of phase with the time deviation, τ

$$\delta = \frac{1}{\alpha}\frac{d\tau}{dt} = -\frac{\omega_s\tau_a}{\alpha}\sin(\omega_s t + \varphi) \tag{28}$$

where α is the momentum compaction. Substituting this into Eq. (26) and integrating gives

$$\psi_\beta(t) = \omega_\beta t + \omega_\xi\tau_a\cos(\omega_s t + \varphi) + \psi . \tag{29}$$

The chromatic frequency is $\omega_\xi = \omega_\beta\varsigma/\alpha$, and ψ is a constant of integration related to the betatron phase at $t = 0$ by

$$\psi = \psi_\beta(t=0) - \omega_\xi\tau_a\cos\varphi = \psi_\beta(t=0) - \omega_\xi\tau(t=0) . \tag{30}$$

The dipole moment is

$$d(t) = i_b(t)r_\perp(t) = A_\beta e^{j\psi_\beta(t)}\sum_{n=-\infty}^{\infty}\delta(t - [nT + \tau_a\cos(\omega_s nT + \varphi)]) . \tag{31}$$

Following the procedure developed above, and identifying the important frequencies gives

$$D(\omega) = \omega_r A_\beta e^{j\psi}\sum_{k=-\infty}^{\infty}e^{jk(\varphi-\pi/2)}J_k((\omega - \omega_\beta - \omega_\xi)\tau_a)$$

$$\times \sum_{n=-\infty}^{\infty}\delta(\omega - (n\omega_r + k\omega_s + \omega_\beta)) . \tag{32}$$

This result gives the general expression for the transverse spectrum of a particle undergoing betatron and synchrotron oscillations and is comparable in importance to Eq. (19).
1. There are an infinite number of synchrotron sidebands centered about the betatron frequencies. These sidebands are displaced from the betatron lines by $k\omega_s$, $k = -\infty,...,\infty$.

2. The envelopes are ordinary Bessel functions with argument $(\omega - \omega_\beta - \omega_\xi)\tau_a$. In contrast to the longitudinal, these envelopes are offset from $\omega = 0$ by $\omega_\beta - \omega_\xi \approx \omega_\xi$ since typically $\omega_\beta \ll \omega_\xi$. A positive chromaticity shifts the envelopes to positive frequencies, etc. This is illustrated in Figure 6.

3. The envelope of sideband k is $J_k((\omega - \omega_\xi)\tau_a)$. The best frequency region to observe the kth sideband is at $\omega \sim \omega_\xi + k/\tau_a$. Depending on the chromatic frequency, high-order sidebands can be seen at low frequencies.

4. The phase of the betatron oscillation when $n = 0$ only has meaning for multiple particles.

5. The observed spectrum is obtained by combining the positive and negative frequencies as has been done above. The mathematical spectrum based on Eq. (25) has only upper sidebands, but the physical spectrum has both upper and lower sidebands. The lower sidebands come from negative frequencies in this mathematics.

MULTIPLE PARTICLES

The expressions in Eqs. (19) and (32) are the basis for understanding beam-generated signals. For example, the longitudinal spectrum of a beam can be derived by convoluting Eq. (19) with the longitudinal phase-space density of the beam $\rho(\tau_a, \varphi)$, where $\rho\tau_a d\tau_a d\varphi$ is the charge in phase space-area $\tau_a d\tau_a d\varphi$;

$$I(\omega) = \omega_r \sum_{k,n=-\infty}^{\infty} \delta(\omega - k\omega_s - n\omega_r) \int_0^\infty \tau_a d\tau_a \int_0^{2\pi} d\varphi \rho(\tau_a, \varphi) e^{jk\varphi} J_k(\omega\tau_a) . \quad (33)$$

The envelope has changed, but the frequencies haven't.

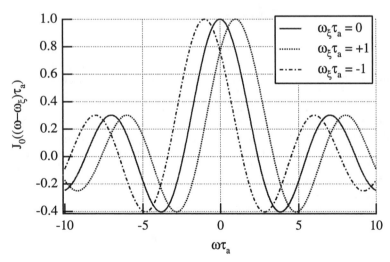

FIGURE 6. Envelopes for the betatron line (k = 0) for three different values of chromatic frequency.

The examples below will show how information about the beam can be extracted from spectra. These examples are intended to illustrate methods that can be applied to many problems.

Longitudinal Phase-Space Structure

When beam intensity is low, particle motion is determined by magnet and rf cavity fields. Beam-generated fields can be neglected and particles move independently of each other. The longitudinal phase-space density can depend on τ_a, but it can't depend on φ. If it did, particles would not be independent.[*] The phase-space density is $\rho(\tau_a, \varphi) = \rho_0(\tau_a)/2\pi$, and the beam-generated signal is

$$I(\omega) = \omega_r \sum_{k,n=-\infty}^{\infty} \delta(\omega - k\omega_s - n\omega_r) \int_0^\infty \tau_a d\tau_a \frac{1}{2\pi} \int_0^{2\pi} d\varphi \rho_0(\tau_a) e^{jk\varphi} J_k(\omega\tau_a)$$

$$= \omega_r \sum_{n=-\infty}^{\infty} \delta(\omega - n\omega_r) \int_0^\infty \tau_a d\tau_a \rho_0(\tau_a) J_0(\omega\tau_a) \ .$$

(34)

Only the $k = 0$ term is not equal to zero once the φ integral is performed. There are rotation harmonics but no synchrotron sidebands.

If the beam has charge Q and is Gaussian in τ with rms bunch length σ_τ,

$$\rho_0(\tau_a) = \frac{Q}{\sigma_\tau^2} \exp(-\tau_a^2/2\sigma_\tau^2)$$

(35)

and (3)

$$I(\omega) = Q\omega_r \exp(-\omega^2\sigma_\tau^2/2) \sum_{n=-\infty}^{\infty} \delta(\omega - n\omega_r) \ .$$

(36)

The spectrum is a comb of rotation harmonics with a Gaussian envelope with rms width of $1/\sigma_\tau$. A detector with a flat sensitivity up to $\omega \sim 1/\sigma_\tau$ can be used to measure the bunch length. Alternatively, an amplitude measurement at a fixed frequency of $\omega \sim 1/\sigma_\tau$ could be used to monitor changes in bunch length.

The appearance of synchrotron sidebands and azimuthal structure in longitudinal phase space are directly related. Observation of synchrotron sidebands implies azimuthal phase-space structure, and azimuthal phase-space structure leads to synchrotron sidebands. Phase-space structure arises from interaction of the beam with its own fields. Since these fields depend on the bunch distribution, that distribution is the solution of a self-consistency problem formulated using the Vlasov equation (4).

The phase-space density can be written as a Fourier expansion

$$\rho(\tau_a, \varphi) = \frac{1}{2\pi} \sum_{m=-\infty}^{\infty} \rho_m(\tau_a) e^{im\varphi} \ .$$

(37)

---

[*]   This holds down to the Schottky noise level of the beam.

Substituting into Eq. (33),

$$I(\omega) = \frac{\omega_r}{2\pi} \sum_{k,n,m=-\infty}^{\infty} \delta(\omega - k\omega_s - n\omega_r) \int_0^\infty \tau_a d\tau_a \int_0^{2\pi} d\varphi \rho_m(\tau_a) e^{j(k+m)\varphi} J_k(\omega\tau_a)$$

$$= \omega_r \sum_{m,n=-\infty}^{\infty} \delta(\omega + m\omega_s - n\omega_r) \int_0^\infty \tau_a d\tau_a \rho_m(\tau_a) J_k(\omega\tau_a).$$

(38)

There is a direct relation between each sideband and a specific harmonic of the phase-space structure; the m*th* harmonic of the phase-space structure produces a signal at the -k*th* sideband.

As a specific example, suppose that the beam has a quadrupole perturbation illustrated in Figure 7

$$\rho(\tau_a, \varphi) = \frac{\rho_0(\tau_a)}{2\pi} \left(1 + A_2 \frac{\tau_a^2}{\sigma_\tau^2} \cos(2\varphi) \right)$$

(39)

where ρ_0 is given by Eq. (35). The τ_a dependence of the perturbation was chosen for the purpose of this example. Substituting Eq. (39) into (38) and performing the integrals (3),

$$I(\omega) = Q\omega_r \exp(-\omega^2\sigma_\tau^2/2) \left[\sum_{n=-\infty}^{\infty} \delta(\omega - n\omega_r) \right.$$

$$\left. + A_2 \frac{\omega^2\sigma_\tau^2}{2} \left(\sum_{n=-\infty}^{\infty} \delta(\omega - n\omega_r + 2\omega_s) + \sum_{n=-\infty}^{\infty} \delta(\omega - n\omega_r - 2\omega_s) \right) \right].$$

(40)

There are rotation harmonics and sidebands at $\pm 2\omega_s$. The envelopes are shown in Figure 8. The sideband signal is a maximum at $\omega \sim 1.5/\sigma_\tau$ and is about a factor of 100 less than the rotation harmonics in this frequency range for $A_2 = 0.01$.

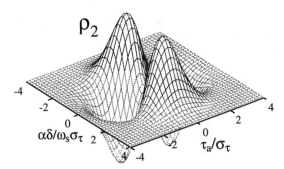

FIGURE 7. Phase space for the quadrupole perturbation given by Eq. (39).

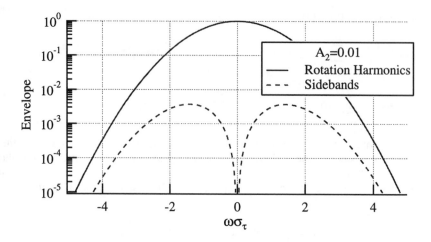

FIGURE 8. The envelopes for the rotation harmonics and second synchrotron sidebands for the example in Eq. (39).

Longitudinal Schottky Noise

Particles move independently of each other when beam-generated fields are negligible, and it was argued in the previous section that a consequence is that the longitudinal phase-space density $\rho(\tau_a, \varphi)$ cannot depend on φ. However, there is a limit to this because a beam consists of individual particles not a smooth distribution.   The particles cannot arrange themselves to remove all φ dependence. They wouldn't be independent if they could. There is some residual φ dependence and some signal, called Schottky noise, from it.

Taking account of individual particles, the phase-space density is

$$\rho(\tau_a, \varphi) = q \sum_{p=1}^{P} \delta(\tau_a - \tau_{ap}) \delta(\varphi - \varphi_p) \tag{41}$$

where q is the particle charge, τ_{ap} and φ_p are the amplitude and phase of the *p*th particle, and there are P particles in the beam. Using this distribution, Eq. (33) becomes

$$I(\omega) = \sum_{k,n=-\infty}^{\infty} \delta(\omega - k\omega_s - n\omega_r) I_{nk}(\omega) ;$$

$$I_{nk}(\omega) = q\omega_r \sum_{p=1}^{P} e^{jk\varphi_p} J_k(\omega_{nk}\tau_{ap}); \ \omega_{nk} = k\omega_s + n\omega_r . \tag{42}$$

The current depends on the phase-space coordinates of P particles, and it is different for every ensemble of particles.

This is just like a random walk problem. The result of a random walk is unknowable, but statistical quantities based on an ensemble of random walks have meaning. The relationship between the phase-space coordinates of different particles changes with time due to nonlinearities and random processes such as emission of synchrotron radiation photons. Therefore, while the current at a particular time is unknowable, the mean and rms currents are meaningful statistical quantities. They can be determined by averaging over all possible samples of P particles. This averaging is denoted by angular brackets: $< >$.

Each of the terms in the expression for I_{nk} (Eq. (42)) is a phasor with magnitude given by the Bessel function and direction given by the argument of the exponential. When $k \neq 0$, the phasors point in all directions, and when averaged over all possible samples $< I_{nk} > = 0$. The phase which had no meaning for a single particle is critically important for a beam. When $k = 0$ all the phasors point in the same direction, and the sum does not vanish. The discrete particle nature of the beam is critical for evaluating the phasor sum, but once that is done the sum over particles can be evaluated with an integral using the phase-space density $\rho_0(\tau_a)$. The mean current $< I_{n0} >$ is given Eq. (34).

The square of the rms current is given by

$$\left\langle I_{nk}^2(\omega_{nk}) \right\rangle = \frac{1}{2}\left\langle I_{nk}(\omega_{nk})I_{nk}^*(\omega_{nk}) \right\rangle$$

$$= \frac{q^2\omega_r^2}{2}\left\langle \sum_{p=1}^{P} e^{jk\varphi_p} J_k(\omega_{nk}\tau_{ap}) \sum_{s=1}^{P} e^{-jk\varphi_s} J_k(\omega_{nk}\tau_{as}) \right\rangle \quad (43)$$

$$= \frac{q^2\omega_r^2}{2}\left\langle \sum_{p,s=1}^{P} e^{jk(\varphi_p - \varphi_s)} J_k(\omega_{nk}\tau_{ap}) J_k(\omega_{nk}\tau_{as}) \right\rangle .$$

The terms in the double sum are phasors. In general they point in arbitrary directions making the sum equal to zero when the average is taken. However, for any sample of P particles the phasors line up and add coherently when $\varphi_p = \varphi_s$; i.e., when particle p is the same as particle s. This removes one of the sums, and

$$\left\langle I_{nk}^2(\omega_{nk}) \right\rangle = \frac{q^2\omega_r^2}{2}\left\langle \sum_{p=1}^{P} J_k^2(\omega_{nk}\tau_{ap}) \right\rangle$$

$$= \frac{q\omega_r^2}{2} \int_0^{\infty} \tau_a d\tau_a \rho_0(\tau_a) J_k^2(\omega_{nk}\tau_a) . \quad (44)$$

The sum over particles has been replaced by an integral over τ_a in the last step. When the beam is Gaussian and ρ_0 is given by Eq. (35) (5),

$$\left\langle I_{nk}^2(\omega_{nk}) \right\rangle = \frac{qQ\omega_r^2}{2} \exp\left(-\omega_{nk}^2\sigma_\tau^2\right) I_k\left(\omega_{nk}^2\sigma_\tau^2\right). \qquad (45)$$

Equations 44 and 45 have several interesting features. First, since $Q = qP$, the rms current is proportional to $\sqrt{P}$ as expected from shot noise. In contrast to Eq. (38) that depends on unknown constants (e.g., A_2 in Eq. (39)), the noise calculation is absolute. Equation 44 gives the noise current, and currents above this value are due to phase-space structure. This is illustrated in Figure 9 where the current from a quadrupole structure with $A_2 = 10^{-4}$ is larger than Schottky noise for $\omega\sigma_\tau < 3.5$. If $A_2 = 10^{-5}$ the cross over point would move to $\omega\sigma_\tau \sim 2.0$.

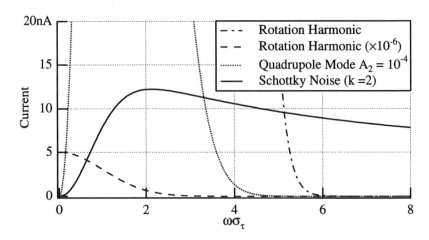

FIGURE 9. Envelopes for the rotation harmonics, a coherent structure with $A_2 = 10^{-4}$ (Eq. (40)), and the $k = 2$ rms Schottky current for 10^{11} particles and $v_r = 50$ kHz.

Transverse Phase-Space Structure

Analyses of Schottky noise and signals associated with phase-space structure can be performed for transverse motion also. The Schottky analysis is an extension of the previous section and is not repeated here. It can be found in Ref. 2. The signals from transverse phase-space structure introduce new ideas and are treated in this section.

In general, a four-dimensional phase-space density κ must be used to calculate the transverse signal because Eq. (32) depends on the betatron amplitude and phase and the synchrotron amplitude and phase

$$D(\omega) = \omega_r \sum_{n,k=-\infty}^{\infty} \delta(\omega - (n\omega_r + k\omega_s + \omega_\beta)) \int A_\beta dA_\beta d\psi \tau_a d\tau_a d\varphi$$

$$\times \kappa(A_\beta, \psi, \tau_a, \varphi) A_\beta e^{j\psi + jk\varphi} J_k((\omega - \omega_\beta - \omega_\xi)\tau_a) \,.$$

(46)

The phase-space density can be expanded in a Fourier series in ψ and φ. The only component of the ψ expansion with a non-zero dipole moment is the one with symmetry $\exp(-j\psi)$. All of the others equal zero when the ψ integral in Eq. (46) is performed. Each of the components of the φ expansion can produce a signal, and, as in Eq. (38), there is a relation between phase-space structure and sidebands.

The new aspect is that the betatron amplitude can depend on the synchrotron amplitude, $A_\beta = A_\beta(\tau_a)$. In general, coherent transverse modes have such dependencies (4). As a specific example, consider all of the particles oscillating with the same amplitude A_0 and a Gaussian distribution in τ_a (Eq. (35)) with no φ dependence. The phase-space density is

$$\kappa = \frac{1}{(2\pi)^2} \rho_0(\tau_a) \delta(A_\beta - A_0) e^{-j\psi}$$

(47)

and

$$D(\omega) = \omega_r A_0 Q \sum_{n=-\infty}^{\infty} \delta(\omega - (n\omega_r + \omega_\beta)) \exp\left(-\frac{(\omega - \omega_\beta - \omega_\xi)^2 \sigma_\tau^2}{2} \right) . \quad (48)$$

The envelope is a Gaussian centered at $\omega = \omega_\beta + \omega_\xi \approx \omega_\xi$ and shifts as the chromaticity changes.

The transverse displacement at $t = 0$ is (Eqs. (25) and (30))

$$r_\perp(t=0) = A_\beta e^{j\psi_\beta(t=0)} = A_\beta e^{j(\psi + \omega_\xi \tau_a \cos\varphi)} = A_\beta e^{j(\psi + \omega_\xi \tau)} . \quad (49)$$

The connection between the transverse displacement and the shift of envelope with chromaticity can be understood by plotting the dipole moment, $\rho_0 r_\perp$, for different chromaticities and at different times. This is done in Figure 10. When $\xi = 0$, the transverse motions of all the particles are in phase. The signal has an average value, i.e., a component at $\omega = 0$; the time scale of the dipole moment is the bunch length σ_τ and the characteristic frequency is $\omega \sim 1/\sigma_\tau$. As the chromaticity increases a head-to-tail phase shift is introduced, and the head and tail are out of phase. The average value of the signal decreases thereby decreasing the signal at $\omega = 0$. In addition, the dipole moment varies more rapidly along the bunch. This shortens the time scale and increases the characteristic frequency; the envelope has shifted to higher frequency as given by Eq. (48).

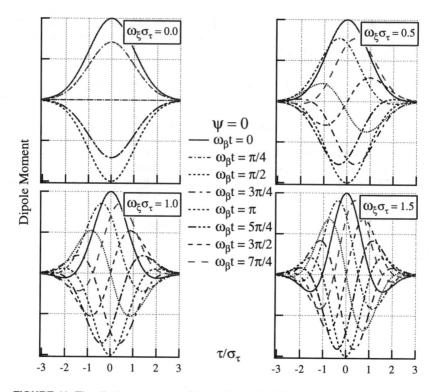

FIGURE 10. The dipole moment at different times for different values of chromaticity.

Coupled-Bunch Signals

The examples above have all involved convolution of single particle signals with the beam phase-space density. The final one goes back to the derivation leading to Eq. (19). Assume that the beam has two equally spaced bunches each of which can be treated as a macroparticle and that these bunches are coupled together by a long-range wake field. The phase shift between bunches $\Delta\varphi$ can have only discrete values from the following argument. If the phase of the first bunch is φ, the phase of the second bunch is $\varphi + \Delta\varphi$, and the phase of the third bunch is $\varphi + 2\Delta\varphi$. The third bunch is the same as the first bunch, so the possible values for $\Delta\varphi$ are $\Delta\varphi = 0, \pi$. The first bunch current is given by Eq. (19). Assuming the second bunch has the same amplitude, its current is given by Eq. (18) with a phase factor $\exp[j(k\Delta\varphi + jn\omega T/2)]$ multiplying it. The first term is the bunch-to-bunch phase shift, and the second is due to the arrival time delay of the second bunch. Adding these two currents together gives

$$I_b(\omega) = \omega_r \sum_{k,n=-\infty}^{\infty} e^{jk(\varphi-\pi/2)} J_k(\omega\tau_a)(1 + e^{j(k\Delta\varphi+\omega T/2)})\delta(\omega - k\omega_s - n\omega_r) \,. \quad (50)$$

For rotation harmonics the phase difference between bunches doesn't matter; the current equals zero when n is odd, and it is double the single bunch current when n is even. The latter two follow from $n\omega_r T/2 = n\pi$. When $k = \pm 1$, the sidebands of the odd rotation harmonics are present if $\Delta\varphi = \pi$ and are missing (neglecting a factor $O(\omega_s T/2)$) if $\Delta\varphi = 0$. The sidebands of the even rotation harmonics are missing if $\Delta\varphi = \pi$ and present if $\Delta\varphi = 0$. This is illustrated in Figure 11. The presence or absence of particular sidebands tells the relative motion of the bunches.

This analysis can be generalized to many bunches. For B bunches the possible values for the phase shift between bunches are $\Delta\varphi = 2\pi m/B$, where m = 0,..., B − 1 is called the coupled-bunch mode number. Only every B*th* rotation harmonic appears, and if one sees synchrotron sidebands of the n*th* rotation harmonic it is from a coupled-bunch mode with mode number n = mod(n, B). This follows because the phasors representing the current of the individual bunches have the same phase and add up coherently only when

$$\Delta\varphi + \frac{n\omega_r T}{B} = \frac{2\pi m}{B} + \frac{2\pi n}{B} = 2\pi p; \quad p = \text{integer} \,. \quad (51)$$

If bunches are unequally spaced or have different charges or amplitudes, these simple results may not hold, but the same phasor addition can be used to identify dominant lines for different coupled-bunch modes.

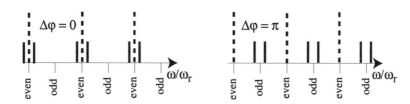

FIGURE 11. The two bunch spectra for coupled-bunch modes with 0 and π phase shift between bunches.

CONCLUDING REMARK

There is a wealth of information in the beam spectrum.   The techniques used in the examples above can be applied and/or extended to chromaticity measurements, coherent frequency shifts, etc.   In addition, there is a close relationship between beam-generated signals and beam stability because it is the current and dipole moment that drive accelerator impedances and produce forces that act on the beam.

This paper was intended to introduce ideas and methods for understanding beam spectra. I hope it serves that purpose.

REFERENCES

1.   I. S. Gradshteyn and I. M. Ryshik, *Tables of Integrals, Series, and Products* (New York: Academic Press, 1965), eq. 8.511.4.

2.   R. H. Siemann, AIP Conference Proceedings **184** (New York, 1989, edited by M. Month and M. Dienes), p. 430.

3.   Reference 1, eq. 6.631.4

4.   J. L. Laclare, *11th International Conference of High-Energy Accelerators* (Basel: Birkhauser Verlag, 1980, edited by W. S. Newman), p. 526.

5.   Reference 1, eq. 6.633.2

Introduction to Radiation-Resistant Semiconductor Devices and Circuits[*]

Helmuth Spieler

Ernest Orlando Lawrence Berkeley National Laboratory, Physics Division, 1 Cyclotron Road, Berkeley, CA 94720, USA

Abstract. This tutorial paper provides an overview of design considerations for semiconductor radiation detectors and electronics in high-radiation environments. Problems specific to particle accelerator applications are emphasized and many of the presented results originated in extensive studies of radiation effects in large-scale particle detectors for the SSC and LHC. Basic radiation damage mechanisms in semiconductor devices are described and specifically linked to electronic parameter changes in detectors, transistors, and integrated circuits. Mitigation techniques are discussed and examples presented to illustrate how the choice of system architecture, circuit topology, and device technology, can extend the range of operation to particle fluences $>10^{14}$ cm^{-2} and ionizing doses >100 Mrad.

INTRODUCTION

Radiation-resistant electronics have been integral to the aerospace, nuclear reactor, and weapons communities for many years, but only rather recently have they become important for particle accelerators and accelerator-based experiments. The Superconducting Super Collider (SSC) made the design of radiation-resistant detectors and electronic read-out systems a key design consideration for high-energy physics experimentalists. The energy frontier has now shifted to the Large Hadron Collider (LHC), which requires even higher luminosities to achieve its physics goals. Even at existing machines, for example the Tevatron at Fermi National Accelerator Laboratory (FNAL), radiation-hard electronics are required in the innermost tracking systems. The vertex detector for BaBar at the SLAC B-Factory requires radiation-hard electronics. On the accelerator side, higher beam currents and the increased sophistication of monitoring and diagnostic systems are bringing the need for radiation-resistant electronics to the forefront of designers' concerns.

Although one can argue that vacuum tubes are extremely radiation hard, the complexity of today's electronics systems restricts our focus to semiconductor devices. For all practical purposes this leaves us with silicon and gallium-arsenide technology. For a variety of reasons silicon transistors and integrated circuits

---

[*] This work was supported by the U.S. Department of Energy, Office of High Energy and Nuclear Physics.

23

comprise the bulk of radiation-hard electronics. In designing SSC and LHC detectors, we have found no compelling justification for GaAs electronics in any radiation-sensitive application. Indeed, in some areas silicon technology provides critical performance advantages. For these reasons, despite the fascinating physics of compound semiconductors, this tutorial will emphasize silicon technology.

The study of radiation effects in semiconductor electronics and the development of radiation-resistant integrated circuits have formed an active scientific community that has produced a wealth of data and conceptual understanding. Although access to some of these results and techniques is restricted, most of the data and papers are in the public domain and readily accessible, and they provide a valuable resource (please refer to the Bibliography). Much has been published on basic damage mechanisms and on device properties for specific applications. Nevertheless, when attempting to apply this information to an area outside the traditional purview of the radiation effects community, key pieces of information needed to link basic damage mechanisms to usable design guidelines are often missing. This became very clear in the development of detectors for the SSC and LHC, where both the application of detectors with deep depletion regions and novel circuit designs combining low noise, high speed, and low power pushed developments into uncharted territory.

This is a very complicated field and developing a general road map is not easy, but one can apply a few fundamental considerations to understanding the effects of radiation on device types in specific circuit topologies and narrow the range of options that must be studied in detail. That is the thrust of this tutorial. Due to space limitations, some treatments are sketchier than desirable and the reader should consult the references and the bibliography. A more detailed version of this paper is on the World Wide Web and will be expanded and modified in response to criticisms and new developments (1).

RADIATION DAMAGE MECHANISMS

Semiconductor devices are affected by two basic radiation damage mechanisms:

- **Displacement damage**: Incident radiation displaces silicon atoms from their lattice sites. The resulting defects alter the electronic characteristics of the crystal.
- **Ionization damage**: Energy absorbed by electronic ionization in insulating layers, predominantly SiO_2, liberates charge carriers that diffuse or drift to other locations where they are trapped, leading to unintended concentrations of charge and, as a consequence, parasitic fields.

Both mechanisms are important in detectors, transistors, and integrated circuits. Some devices are more sensitive to ionization effects, some are dominated by displacement damage. Hardly a system is immune to either one phenomena and most are sensitive to both.

Ionization effects depend primarily on the absorbed energy, independent of the type of radiation. At typical incident energies ionization is the dominant absorption mechanism, so that ionization damage can be measured in terms of energy absorption per unit volume, usually expressed in rad or gray (1 rad = 100 erg/g, 1 Gy= 1 J/kg = 100 rad). Since the charge liberated by a given dose depends on the absorber material, the ionizing dose must be referred to a specific absorber, e.g., 1 rad(Si), 1 rad(SiO_2), 1 rad(GaAs) [or in SI units 1 Gy(Si), etc.].

Displacement damage depends on the non-ionizing energy loss, i.e., energy and momentum transfer to lattice atoms, which depends on the mass and energy of the incident quanta. A simple measure as for ionizing radiation is not possible, so that displacement damage must be specified for a specific particle type and energy.

In general, radiation effects must be measured for both damage mechanisms, although one may choose to combine both, for example by using protons, if one has sufficient understanding to unravel the effects of the two mechanisms by electrical measurements. Even non-ionizing particles can deposit some ionization dose via recoils, but this contribution tends to be very small: 2×10^{-13} rad per 1 MeV neutron/cm^2, for example (2).

To set the scale, consider a tracking detector operating at the LHC with a luminosity of 10^{34} $cm^{-2}s^{-1}$. In the innermost volume of a tracker the particle flux from collisions is $n' \approx 2 \times 10^9/r_\perp^2$ $cm^{-2}s^{-1}$, increasing roughly twofold in the outer layers due to secondary interactions and loopers. At $r_\perp = 30$ cm the particle fluence after one year of operation (10^7 s) is about 2×10^{13} cm^{-2}. A fluence of 3×10^{13} cm^{-2} of minimum ionizing particles corresponds to an ionization dose of 1 Mrad, obtained after 1.5 years of operation. Albedo neutrons from a calorimeter could add a yearly fluence of 10^{12} to 10^{13} cm^{-2}.

Displacement Damage

An incident particle or photon capable of imparting an energy of about 20 eV to a silicon atom can dislodge it from its lattice site. Displacement damage creates defect clusters. For example, a 1-MeV neutron transfers about 60 to 70 keV to the Si recoil atom, which in turn displaces roughly 1000 additional atoms in a region of about 0.1 μm size. Displacement damage is proportional to non-ionizing energy loss (3), which is not proportional to the total energy absorbed, but depends on the particle type and energy. Non-ionizing energy loss for a variety of particles has been calculated over a large energy range (4). Although not verified quantitatively, these curves can be used to estimate relative effects. X rays do not cause direct displacement damage, since momentum conservation sets a threshold energy of 250 keV for photons. ^{60}Co γ rays cause displacement damage primarily through Compton electrons and are about three orders of magnitude less damaging per photon than a 1-MeV neutron (5). Table 1 gives a rough comparison of displacement damage for several types of radiation.

TABLE 1. Relative Displacement Damage for Various Particles and Energies

Particle	proton	proton	neutron	electron	electron
Energy	1 GeV	50 MeV	1 MeV	1 MeV	1 GeV
Relative Damage	1	2	2	0.01	0.1

Displacement damage manifests itself in three important ways:

- Formation of mid-gap states, which facilitate the transition of electrons from the valence to the conduction band. In depletion regions this leads to a generation current, i.e., an increase in the current of reverse-biased *pn*-diodes. In forward-biased junctions or non-depleted regions, mid-gap states facilitate recombination, i.e., charge loss.
- States close to the band edges facilitate trapping, where charge is captured and released after a certain time.
- A change in doping characteristics (donor or acceptor density).

The role of mid-gap states is illustrated in Fig. 1. Because interband transitions in Si require momentum transfer ("indirect band-gap"), direct transitions between the conduction and valence bands are extremely improbable (unlike GaAs, for example). The introduction of intermediate states in the forbidden gap provides "stepping stones" for emission and capture processes. The individual steps, emission of holes or electrons and capture of electrons or holes, are illustrated in Fig. 1. As shown in Fig. 1(a), the process of hole emission from a defect can also be viewed as promoting an electron from the valence band to the defect level. In a second step, (b), this electron can proceed to the conduction band and contribute to current flow, generation current. Conversely, a defect state can capture an electron from the conduction band, (c), which in turn can capture a hole, (d). This "recombination" process reduces current flowing in the conduction band.

Since the transition probabilities are exponential functions of the energy differences, all processes that involve transitions between both bands require mid-gap states, to proceed at an appreciable rate. Given a distribution of states, these processes will "seek out" the mid-gap states. Since the distribution of states is not necessarily symmetric, one cannot simply calculate recombination lifetimes from generation currents and vice versa (as is possible for a single mid-gap state, as assumed in textbooks). Whether generation or recombination dominates depends on the relative concentration of carriers and empty defect states. In a depletion region the conduction band is underpopulated, so generation prevails. In a forward-biased junction carriers flood the conduction band, so recombination dominates. Figure 1(e) shows a third phenomenon: defect levels close to a band edge will capture charge and release it after some time, a process called "trapping."

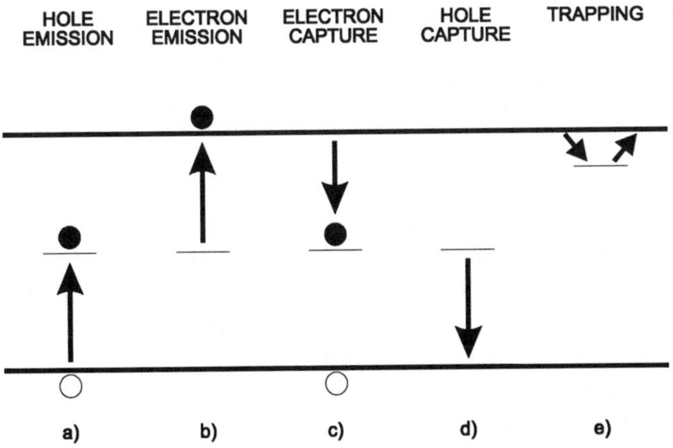

FIGURE 1. Emission and capture processes through intermediate states. The arrows show the direction of electron transitions.

In a radiation detector or photodiode system the increased reverse-bias current increases the electronic shot noise. The change in doping level affects the width of the depletion region (or the voltage required for full depletion). The decrease in carrier lifetime incurs a loss of signal as carriers recombine while traversing the depletion region. As will be shown later, the same phenomena occur in transistors, but are less pronounced, depending on device type and structure. Displacement damage effects will be discussed in more detail in the section on diodes.

Ionization Damage

As in the detector bulk, electron-hole pairs are created in the oxide. The electrons are quite mobile and move to the most positive electrode. Holes move by a rather complex and slow hopping mechanism, which promotes the probability of trapping in the oxide volume and an associated fixed positive charge. Holes that make it to the oxide-silicon interface can be captured by interface traps. This is illustrated in Fig. 2, which shows a schematic cross section of an n-channel metal-oxide silicon field effect transistor (MOSFET). A positive voltage applied to the gate electrode attracts electrons to the surface of the silicon beneath the gate. This "inversion" charge forms a conductive channel between the n+ doped source and drain electrodes. The substrate is biased negative relative to the MOSFET to form a depletion region for isolation. Holes freed by radiation accumulate at the oxide-silicon interface. The positive charge buildup at the silicon interface requires that the gate voltage be adjusted to more negative values to maintain the negative charge in the channel.

Trapped oxide charge can also be mobile, so that the charge distribution generally depends on time, and more specifically, how the electric field in the oxide changes with time. The charge state of a trap depends on the local quasi Fermi level, so the concentration of trapped charge will vary with changes in the applied voltage and state-specific relaxation times. As charge states also anneal, ionization effects depend not only on the dose, but also on the dose rate. Figure 2 also shows a thick field oxide, which serves to control the silicon surface charge adjacent to the field effect transistor (FET) and prevent parasitic channels to adjacent devices. The same positive charge buildup as in the gate oxide also occurs here, indeed it can be exacerbated because the field oxide is quite thick. For more details, see ref. (6), currently the authoritative text on ionization effects.

In summary, ionization effects are determined by
- Interface trapped charge
- Oxide trapped charge
- The mobility of trapped charge
- The time and voltage dependence of charge states

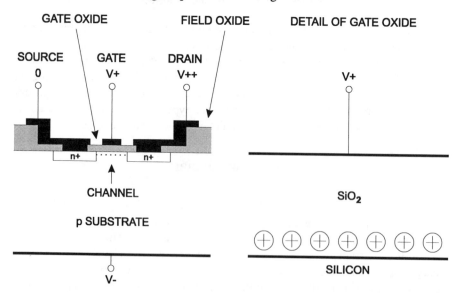

FIGURE 2. Schematic cross section of an n-channel MOSFET (left). A detail of the gate oxide shows the trapped holes at the oxide-silicon interface (right).

Although the primary radiation damage depends only on the absorbed ionizing energy, the resulting effects of this dose depend on the rate of irradiation, the applied voltages and their time variation, the temperature, and the time variation of the radiation itself. Ionization damage manifests itself most clearly in MOS field effect transistors, so it will be discussed in more detail in that section.

EFFECTS ON DEVICE CHARACTERISTICS

Radiation Damage in Diodes

Diode structures are basic components of more complex devices, for example bipolar transistors, junction FETs, and integrated circuits. Since the properties of diode depletion regions are determined primarily by bulk properties, measurements on diodes will serve to illustrate the effects of displacement damage. Reverse biased diodes with large depletion depths are used as radiation detectors and photodiodes. Because of their large depletion depths, typically hundreds of microns, detector diodes are very sensitive to bulk damage, and extensive work by the SSC/LHC community has produced many insights into bulk radiation effects. Affected are the detector leakage current, the doping characteristics, and charge collection.

A theoretical analysis from first principles is quite complex, due to the many phenomena involved. Take doping changes as an example. Si interstitials are quite active and displace either P donors or B acceptors from substitutional sites and render them electrically inactive. These interstitial dopants together with oxygen, commonly present in the lattice as an impurity, react in very different ways with vacancies to form complexes with a variety of electronic characteristics (see ref. (7) and references therein). Fortuitously, although a multitude of competing effects can be invoked to predict and interpret experimental results, the data can be described by rather simple parameterizations.

The increase in reverse bias current (leakage current) is linked to the creation of mid-gap states. Experimental data are consistent with a uniform distribution of active defects in the detector volume. The bias current after irradiation

$$I_{\text{det}} = I_0 + \alpha \cdot \Phi \cdot Ad \,, \tag{1}$$

where I_0 is the bias current before irradiation, α is a damage coefficient dependent on particle type and fluence, Φ is the particle fluence, and the product of detector area and thickness Ad is the detector volume. For 650-MeV protons $\alpha \approx 3 \times 10^{-17}$ A/cm (8,9) and for 1-MeV neutrons (characteristic of the albedo emanating from a calorimeter) $\alpha \approx 2 \times 10^{-17}$ A/cm (9). The parameterization used in Eq. (1)

is quite general, as it merely assumes a spatially uniform formation of electrically active defects in the detector volume, without depending on the details of energy levels or charge states.

The coefficients given above apply to room temperature operation. The reverse bias current of silicon detectors depends strongly on temperature

$$I_R(T) \propto T^2 e^{-E/2k_B T} \tag{2}$$

if the generation current dominates (10), as is the case for substantial radiation damage. The effective activation energy $E = 1.2\,\text{eV}$ for radiation-damaged samples (8,11,12), whereas unirradiated samples usually exhibit $E = 1.15\,\text{eV}$. The ratio of currents at two temperatures T_1 and T_2 is

$$\frac{I_R(T_2)}{I_R(T_1)} = \left(\frac{T_2}{T_1}\right)^2 \exp\left[-\frac{E}{2k_B}\left(\frac{T_1 - T_2}{T_1 T_2}\right)\right]. \tag{3}$$

After irradiation the leakage current initially decreases with time. Pronounced short-term and long-term annealing components are observed and precise fits to the annealing curve require a sum of exponentials (9). Experimentally, decreases by factors of 2 to 3 have been observed with no further improvement after five months or so (8,5). In practice, the variation of leakage current with temperature is very reproducible from device to device, even after substantial doping changes due to radiation damage. The leakage current can be used for dosimetry and diodes are offered commercially specifically for this purpose.

The effect of displacement damage on doping characteristics has been investigated in the course of detector studies for the SSC and LHC and is still the subject of ongoing study. Measurements on a variety of strip detectors and photodiodes by groups in the U.S., Japan, and Europe have shown that the effective doping of n-type silicon initially decreases, becomes intrinsic (i.e., undoped) and then turns p-like, with the doping density increasing with fluence. This phenomenon is consistent with the notion that acceptor sites are formed by the irradiation, although this does not mean that mobile holes are created (13). Initially, the effective doping level N_d-N_a decreases as new acceptor states neutralize original donor states. At some fluence the two balance, creating "intrinsic" material, and beyond this fluence the acceptor states dominate. In addition, there is evidence for a concurrent process of donor removal (14,15). Since the probability of donor removal is proportional to the initial donor concentration N_{d0}, whereas the formation of defects leading to acceptor states is proportional to fluence, the effective doping concentration N_{eff} of n-type starting material after exposure to a particle fluence Φ is described by (16)

$$N_{eff}(\Phi) = -N_{d0}e^{-c\Phi} + g_c\Phi + g_s\Phi \cdot e^{-t/\tau(T)} + N_Y(\Phi,t,T), \qquad (4)$$

where a negative or positive sign of N_{eff} denotes whether the effective doping is n- or p-like. The first term describes the removal of donors and the second the creation of acceptors. c and g_c are constants for a given particle type and energy that describe the stable component of radiation damage. The third and fourth terms describe the time- and temperature-dependent changes in the effective doping concentration and will be discussed later. For high-energy protons the average from many measurements is $c = (0.96 \pm 0.19) \times 10^{-13}$ cm^2 and $g_c = (1.15 \pm 0.09) \times 10^{-2}$ cm^{-1}. Type inversion from n- to p-type silicon occurs at a fluence of about 10^{13} cm^{-2}. Data for 1-MeV-equivalent neutrons yield $c = (2.29 \pm 0.63) \times 10^{-13}$ cm^2 and $g_c = (1.77 \pm 0.07) \times 10^{-2}$ cm^{-1} (9).

After a proton fluence $\Phi = 10^{14}$ cm^{-2} the acceptor concentration before annealing is 10^{12} cm^{-3}, which requires a bias voltage of 165 V for full depletion of a 300-μm-thick detector. At first glance, it would seem that beginning with a higher n doping level N_{d0} (lower resistivity) would increase overall detector lifetime. Although the inversion fluence increases with larger values of N_{d0}, the difference in doping concentration is negligible at larger fluences since the exponential term quickly becomes insignificant (15). For example, as shown in Fig. 3, materials with initial doping densities of 10^{12} cm^{-3} and 10^{13} cm^{-3} lie within 15% at $\Phi = 5 \times 10^{13}$ cm^{-2}.

Very high resistivity silicon ($\rho > 10$ kΩcm or $N_d < 4 \times 10^{11}$ cm^{-3}) is often highly compensated, $N_{eff} = N_d - N_a$ with $N_d \sim N_a >> N_{eff}$, so that minute changes to either donors or acceptors can alter the net doping concentration significantly, and the above equations must be modified accordingly. Moderate resistivity n-type material ($\rho = 1$ to 5 kΩcm) used in large area tracking detectors is usually dominated by donors.

Annealing of Ionized Acceptor States

After defect states are formed by irradiation, their electronic activity changes with time. A multitude of processes contribute, some leading to beneficial annealing, i.e., a reduction in acceptor-like states, and some increasing the acceptor concentration. The third term in Eq. (4) describes the beneficial annealing (17), where $g_s = 1.93 \times 10^{-2}$ cm^{-1} and $\tau(T) = (6 \times 10^6) \times \exp[-0.175(T - 273.2)]$ s (to set the scale, $\tau(0°C) = 70$ d). The fourth term in Eq. (4)

$$N_Y(\Phi, t_{1/2}, T) = g_Y\Phi \cdot \left[1 - \frac{1}{1 + g_Y\Phi \cdot k(T) \cdot t}\right], \qquad (5)$$

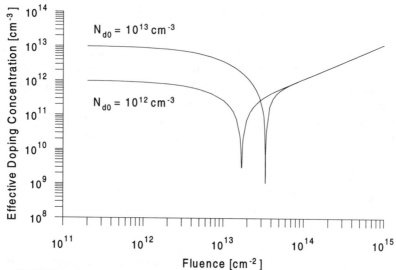

FIGURE 3. Calculated effective doping concentration vs. high-energy proton fluence for silicon with initial donor concentrations N_{do} of 10^{12} and 10^{13} cm^{-3}.

where for 1-MeV neutrons $g_Y = (4.6 \pm 0.3) \times 10^{-2}$ cm^{-1} and for 1-GeV protons values of $g_Y = (4.97 \pm 0.23) \times 10^{-2}$ cm^{-1} (16) and $(5.8 \pm 0.3) \times 10^{-2}$ cm^{-1} (9) have been found. The temperature-dependent evolution is determined by

$$k(T) = k_0 \, e^{-E_a / k_B T}. \tag{6}$$

Typical parameter sets are $k_0 = (0.85 + 25$-$0.82)$ cm^3/s and $E_a = 1.16 \pm 0.08$ eV (16), and $k_0 = (520 + 1590 - 392)$ cm^3/s and $E_a = 1.31 \pm 0.04$ eV (9).

Anti-annealing is a concern because of its effect on detector depletion voltage, i.e., the voltage required to collect charge from the complete thickness of the silicon detector. Since this voltage increases with the square of the doping concentration, anti-annealing of 20 to 40% can easily exceed the safe operating range, especially at high fluences. The relative effect of anti-annealing increases strongly with fluence and temperature, as illustrated in Table 2, which shows the relative increase in doping and depletion voltage. Clearly, low temperature operation is beneficial. Nevertheless, even a low-temperature system will require maintenance at room temperature and warm-up periods must be controlled very carefully (9,16).

TABLE 2. Relative Anti-annealing After 100 h vs. Fluence and Temperature

Fluence [cm^{-2}]	10^{13}	10^{13}	10^{13}	10^{14}	10^{14}	10^{14}
Temperature [°C]	0	20	40	0	20	40
$N_a(t=100h)/N_a(0)$	1.00	1.02	1.39	1.01	1.21	4.71
$V_{depl}(t=100h)/V_{depl}(0)$	1.00	1.04	1.92	1.02	1.46	22.2

Data on charge collection efficiency are still rather sketchy. The primary mechanism is expected to be trapping of signal charge at defect sites, i.e., a decrease in carrier lifetime τ. Since the loss in signal charge is proportional to $\exp(-t_c/\tau)$, reducing the collection time t_c mitigates the effect. Since either the operating voltage is increased or depletion widths are reduced at damage levels where charge trapping is appreciable, fields tend to be higher and collection times decrease automatically with radiation damage, provided the detector can sustain the higher fields.

Typical measurements have determined the signal charge vs. bias voltage and have taken the plateau value (or the maximum signal charge just below break-down). Lemeilleur et al. (18) find $\Delta Q/Q_0 = \gamma\Phi$, where $\gamma = (0.024 \pm 0.004) \times 10^{-13}$ cm^2 for 1-MeV-equivalent neutrons. Fretwurst et al. (19) find similar results, with a dependence $1/\tau = \gamma\Phi$, where for holes $\gamma_p = 2.7 \times 10^{-7}$ cm^{2}s and for electrons $\gamma_e = 1.2 \times 10^{-6}$ cm^{2}s for $\Phi > 10^{13}$ cm^{-2} of 1-MeV-equivalent neutrons. For a fluence $\Phi = 5 \times 10^{13}$ cm^{-2}s^{-1}, a 400-μm-thick detector with a depletion voltage of 130 V operated at a bias voltage of 200 V would show a decrease in signal charge of 12%. Ohsugi et al. (20) have demonstrated the operation of strip detectors to neutron fluences beyond 10^{14} cm^{-2}, with signal losses of about 10%. Similar results have been obtained on fully irradiated strip detectors read out by LHC-compatible electronics (21).

The basic detector is insensitive to ionization effects. In the bulk, ionizing radiation creates electrons and holes that are swept from the sensitive volume; charge can flow freely through the external circuitry to restore equilibrium. The problem lies in the peripheral structures, the oxide layers that are essential for controlling leakage paths at the edge of the diode and preserving inter-electrode isolation in segmented detectors.

The positive space charge due to hole trapping in the oxide and at the interface (see Fig. 2) attracts electrons in the silicon bulk to the interface. These accumulation layers can exhibit high local electron densities and form conducting channels, for example between the detector electrodes. This is especially critical at the "ohmic" electrodes in double-sided detectors, where the absence of *pn* junctions makes operation rely on full depletion of the silicon surface. (Even without radiation, the silicon surface tends to be n-type, so the ohmic side of n-type detectors is inherently more difficult to control.) (22,23)

Some detectors include integrated coupling capacitors and biasing networks. Biasing structures such as punch-through resistors and MOSFET structures are subject to ionization damage. Although these devices can remain functional, substantial changes in voltage drop have been reported for punch-through and accumulation-layer devices, whereas measurements on polysilicon resistors irradiated to 4 Mrad (65 MeV p) show no effect (24).

Radiation Damage in Transistors and Integrated Circuits

In principle, the same phenomena discussed for detectors also occur in transistors, except that the geometries of transistors are much smaller (depletion widths <1 μm) and the typical doping levels are higher (>10^{15} cm^{-3}).

Bipolar Transistors

The most important damage mechanism in bipolar transistors is the degradation of DC current gain at low currents. The damage mechanism is the same that causes increased leakage current in detectors: formation of mid-gap states by displacement damage. The difference is that the base-emitter junction is forward biased, so the high carrier concentration in the conduction band tips the balance from generation to recombination (see Fig. 1). The fractional carrier loss depends on the relative concentrations of injected carriers and defects. Consequently, the reduction of DC current gain due to radiation damage depends on current density. For a given collector current a small device will suffer less degradation in DC current gain than a large one.

Since the probability of recombination depends on the transit time through the junction region, reduced base width will also improve the radiation resistance. Base width is strongly linked with device speed, so that the reduction in DC current gain β_{DC} scales inversely with a transistor's unity gain frequency f_T (25)

$$\Delta\beta_{DC} \equiv \frac{1}{\beta} - \frac{1}{\beta_0} \propto \frac{\Phi}{f_T}. \qquad (7)$$

Since integrated circuit (IC) technology is driven strongly by device speed, mainstream market forces will indirectly improve the radiation resistance of bipolar transistor processes. Mid-gap states also limit the low current performance before irradiation. Over the past decade, evolutionary improvements in contamination control and process technology have also yielded substantially better low-current performance. Measurements on bipolar transistors from several vendors have shown that processes not specifically designed for radiation resistance are indeed quite usable in severe radiation environments, even at low currents (26,27,28).

Changes in doping levels have little effect in bipolar transistors. Typical doping levels in the base and emitter are $N_B = 10^{18}$ and $N_E = 10^{20}$ cm^{-3}. In the collector depletion region doping levels are smaller, typically 10^{16}, rising to 10^{18} or 10^{19} cm^{-3} at the collector contact. At these levels the change in doping level due to displacement damage ($\Delta N_A \approx 10^{12}$ cm^{-3} at $\Phi = 10^{14}$ cm^{-2}) is negligible, although local device temperatures may be sufficiently high that anti-annealing leads to noticeable effects.

Figure 4 shows measured DC current gain for *npn* and *pnp* bipolar transistors irradiated to a fluence of 1.2×10^{14} cm^{-2} (800-MeV protons) (26). These devices, fabricated in AT&T's CBIC-V2 high-density complementary *npn-pnp* IC process, exhibit $f_T = 10$ GHz for the *npn* and 4.5 GHz for the *pnp* transistors. In the CAFE chip designed for the ATLAS silicon tracker (29), the *npn* input device is operated at a current density of about 2 μA/(μm)^2 where the post-rad current gain decreases to about 60% of its initial value. Although a smaller transistor would deteriorate less, the thermal noise contribution of the parasitic base resistance would be excessive, so a compromise is necessary.   No measurable changes in transconductance were measured, as expected. The output resistance of these devices decreased by <10% after irradiation. Similar results have been measured on comparable devices fabricated by Maxim (Tektronix) (27,28) and Westinghouse (30).

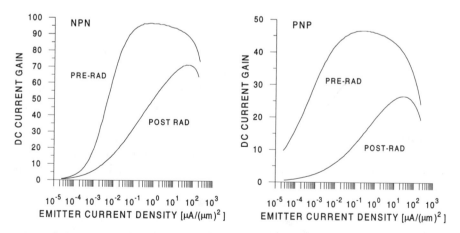

FIGURE 4. DC current gain of npn and pnp transistors before and after irradiation to a fluence of 1.2×10^{14} cm^{-2} (800-MeV protons).

Noise degradation has been measured on individual transistors and complete preamplifier circuits. The results are consistent with the measured degradation in DC current gain and no change in transconductance or parasitic resistances, as expected. Figure 5 shows the measured spectral noise density of a monolithically integrated preamplifier before and after irradiation to 1.2×10^{14} cm^{-2} (800-MeV protons) (26). The gain increased by a few percent after irradiation, so the input noise increase is somewhat smaller than shown.

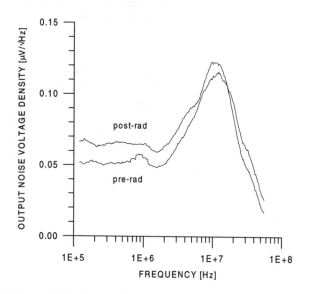

FIGURE 5. Noise of a bipolar transistor preamplifier before and after irradiation to a fluence of 1.2×10^{14} cm^{-2} (800-MeV protons).

Junction Field Effect Transistors (JFETs)

JFETs (either silicon or GaAs) can be quite insensitive to both ionization and displacement effects. In these devices a conducting channel from the source to the drain is formed by appropriate doping, typically *n* type. The gate electrode is doped *p* type so that applying a reverse bias voltage relative to the channel will form a depletion region that changes the cross section of the conducting channel (31). At low values of gate and drain voltages the channel is contiguous and resistive. At higher voltage levels the channel becomes fully depleted near the drain, but the current flow is still determined by the conducting channel near the source. Since the gate voltage now controls both the geometry and potential distribution, voltage-current characteristics become more complex and the device acts much like a controlled current source, i.e., it exhibits a high output resistance.

With respect to radiation effects, the important fact is that device characteristics are determined essentially by the geometry and doping level of the channel. Typical doping levels are 10^{15} to 10^{18} cm^{-3}, so the effect of radiation-induced acceptor states is small. Silicon JFETs exhibit very good radiation resistance. Measurements on both standard commercial devices and custom-designed integrated circuits have shown minimal changes in gain at fluences $>10^{14}$ neutrons/cm^2 and ionization doses up to 100 Mrad (32,33,34). Low-frequency noise ($f < 100$ kHz) may increase by an order of magnitude, but at high frequencies very little change in noise is observed. Measurements of Si JFETs at 90K also exhibit excellent radiation characteristics (33).

In some applications, e.g., analog storage circuitry, gate leakage current is important. Generation current in the gate depletion region due to displacement damage can affect the gate current strongly. Measurements on commercial JFETs irradiated by high-energy electrons to 100 Mrad ($\Phi \approx 10^{15}$ cm^{-2}) show the gate reverse current increasing 100-fold from an initial value of 70 pA (35). Here one should choose the smallest geometry device commensurate with other requirements.

At this point it is worth noting that the superior radiation resistance claimed for GaAs ICs has more to do with the use of JFETs or metal semiconductor FETs (Schottky barrier JFETs) than the properties of the semiconductor. These devices are more radiation resistant than silicon MOSFETs (discussed below) but suffer from a much lower circuit density.

Metal-Oxide Silicon Field Effect Transistors (MOSFETs)

Within the FET family, MOSFETs present the most pronounced ionization effects, as the key to their operation lies in the oxide that couples the gate to the channel. As described above and illustrated in Fig. 2, positive charge buildup due to hole trapping in the oxide and at the interface shifts the gate voltage required for a given operating point to more negative values. This shift affects the operating points in analog circuitry and switching times in digital circuitry. Reducing the thickness of the gate oxide t_{ox} greatly improves the radiation resistance; gate voltage shifts scaling with t_{ox}^2 to t_{ox}^3 for a given dose have been observed (6). Thinner gate oxides are required for small channel lengths, so higher density processes tend to improve the radiation resistance even without special hardening techniques. The gate voltage shift is typically expressed in terms of threshold voltage V_T, which roughly marks the onset of appreciable current flow.

Typical threshold shifts for a 1.2-µm radiation-hardened complementary metal oxide semiconductor (CMOS) IC process with a 20-nm-thick gate oxide are shown in Fig. 6 (36). After exposure to 5 Mrad(Si) of ^{60}Co irradiation, N-channel MOS (NMOS) thresholds shift by 200 mV and P-channel MOS (PMOS) levels change by 150 mV. For both NMOS and PMOS devices the threshold voltage shifts to more negative values as expected from positive charge buildup in the oxide. The slight upturn above 2 Mrad in the NMOS curve is typical and reflects the buildup of interface states (6). About 70% of the threshold shifts occur during the first 250 krad, also a typical phenomenon. Measurements to 125 Mrad on a similar process show a total threshold shift of 400 mV for NMOS and 100 mV for PMOS with little increase beyond 10 Mrad (37).

Figure 7 shows the normalized transconductance g_m/I_d vs. I_d/W before and after irradiation (36). For the selected channel length this representation allows direct scaling to any device width at a given current density. For example, to operate a 1.2-µm NMOS transistor in moderate inversion one might choose a normalized drain current $I_d/W = 0.3$ A/m, yielding $I_d = 0.3$ mA for a 1-mm-wide transistor. The normalized transconductance $g_m/I_d = 15.4$ V^{-1} or $g_m = 4.6$ mS. After exposure to 5 Mrad, $g_m/I_d = 11.8$ V^{-1} or $g_m = 3.5$ mS. Typically, the NMOS devices suffer a 20 to 30% degradation, whereas the PMOS devices are quite insensitive to radiation, with only a few percent decrease in transconductance at 5 Mrad. About half of the observed change at 5 Mrad occurred before attaining a dose of 1 Mrad.

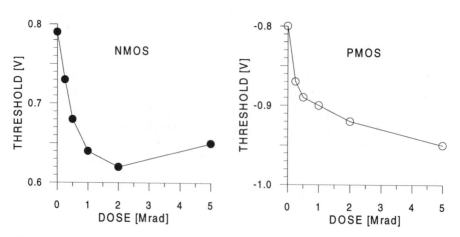

FIGURE 6. Threshold voltage shifts for radiation-hardened NMOS and PMOS transistors vs. ^{60}Co radiation dose.

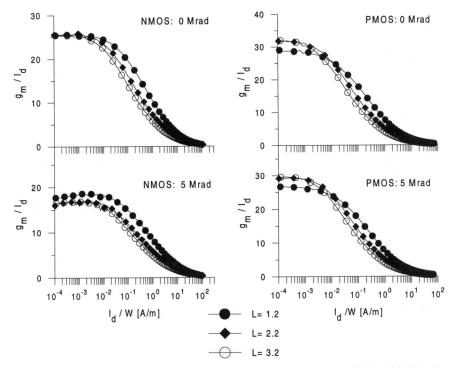

FIGURE 7. Normalized transconductance g_m/I_d vs. drain current I_d/W for NMOS and PMOS transistors with channel lengths of 1.2, 2.2 and 3.2 μm before and after ^{60}Co irradiation to 5 Mrad (Si).

Extensive noise measurements have been performed at the University of Pennsylvania (38) and by a UCSC/LBNL group (36). In the latter, spectral noise density was measured over a frequency range of 10 kHz to 10 MHz before and after ^{60}Co irradiation to a dose of 5 Mrad(Si). The noise was measured at three representative drain current densities I_d/W. Again, these data can be scaled to any device width, where the noise scales with $W^{-1/2}$. The difference between the NMOS and PMOS results is striking. The NMOS devices show a much greater degradation and the PMOS devices also exhibit substantially less low-frequency noise. The low-frequency noise spectral density of the NMOS devices can be described by $v_n^2 = A_f/q + B$, where q ranges from 0.8 to 1.0 and is constant for all currents for the same geometry. The changes in q after a dose of 5 Mrad are of order 0.1. The noise coefficient A_f is about 1.0 to 1.5 × 10[-30] V² pre-rad and 5 to 10 × 10[-30] V² post-rad. Before irradiation, A_f scales well with inverse gate area, but no clear pattern is observed after irradiation. The low-frequency noise behavior of the PMOS devices is more complex and cannot be parameterized in this simple manner, but the devices exhibit substantially better noise than the NMOS transistors.

White noise was evaluated at high frequencies and is characterized by the noise coefficient $\gamma_n = v_n^2 \cdot g_m/4kT$ to assess the inherent noise properties independently of transconductance. Results for various device geometries and current densities are shown in Table 3. For these measurements the substrate was biased at the source potential.

TABLE 3. Noise coefficients $\gamma_n = v_n^2 \cdot g_m/4kT$ for NMOS and PMOS transistors of various widths and lengths, operated at current densities $I_d/W = 0.03$, 0.1 and 0.3 A/m, before and after ^{60}Co irradiation to 5 Mrad(Si). Widths and lengths are given in μm.

Type	NM	PMOS	NMOS	PMOS	NMOS	PMOS	NMOS	PMOS
Width	75	75	1332	1332	888	888	1332	1332
Length	1.2	1.2	1.2	1.2	2.2	2.2	3.2	3.2
$I_d/W = 0.03$								
0 Mrad			0.81	0.61	0.64	0.59	0.66	0.50
5 Mrad			2.17	0.84	1.00	0.58	1.50	0.69
$I_d/W = 0.1$								
0 Mrad	1.10	0.70	1.20	1.10	0.80	0.80	0.80	0.60
5 Mrad	3.80	1.10	3.40	1.60	1.30	0.90	1.70	0.70
$I_d/W = 0.3$								
0 Mrad	1.60	1.30	2.00	1.70	1.10	1.00	1.10	0.77
5 Mrad	5.00	2.90	4.80	2.70	1.60	1.40	1.20	0.81

Again, we see substantially less post-radiation degradation in the PMOS devices. One can also observe the higher intrinsic noise of NMOS short channel devices. Although the observed degradation is quite small in some cases, typically it is quite substantial and would need to be compensated for by a considerably higher operating current. Seller et al. have exposed low-noise preamplifiers fabricated in a rad-hard 1.2-μm bulk CMOS process to a dose of 100 Mrad and measured noise and gain (37). Gain decreased by no more than 7%, but the increase in equivalent input noise at high frequencies ranged from 20 to 75%. This process is only specified to 5 Mrad, so these results indicate that circuits are still quite usable at much higher doses, if one can accommodate the increase in noise.

Due to the presence of mobile trapped charge, threshold behavior can become quite difficult to predict when the gate voltage changes appreciably with varying duty cycles, as in logic circuitry. Detectors and analog circuitry are simpler by comparison, since the voltage levels are either static or change with a fixed period, as in analog pipelines, for example. In general, when performing ionization damage tests devices must be operated at typical operating voltages and digital circuitry must be clocked at frequencies and patterns approximating typical operation.

Generally speaking, both bulk and silicon on insulator (SOI) CMOS are subject to the effects described above. SOI is often cited as a specifically radiation-hard technology because of its resistance to transient radiation effects, primarily latchup due to photocurrents developed at high intensity bursts of radiation ($>10^6$-10^7 rad/s) typical of nuclear detonations. Although SOI can provide superior device speed because of reduced stray capacitance, this technology is not inherently more resistant to radiation in our applications. If anything, the additional oxide interfaces tend to complicate matters and at this time most radiation-resistant CMOS processes are on bulk silicon.

Radiation Effects in Integrated Circuit Structures

The preceding discussion has emphasized the properties of individual devices. In integrated circuits many devices are placed close together. As mentioned above, the silicon surface is naturally n-type, so isolation structures are required to preclude unwanted cross-coupling between devices. Two basic techniques are used:

- Junction isolation, where reverse-biased pn junctions provide both ohmic and capacitive isolation
- Oxide isolation, where oxide layers with carefully controlled interface properties deplete the adjacent silicon of mobile charge

More detailed information on these processes can be found in texts on IC technology, for example (39).

Junction isolation is very robust, but requires substantial additional space. Oxide isolation allows higher packing densities and is used by most high-density IC processes. All "standard" CMOS processes utilize some form of oxide isolation, whereas bipolar transistor processes can be found with both junction and oxide isolation. Under irradiation the oxide layers used for isolation suffer from the same phenomena described for the gate oxide of MOSFETs (see field oxide in Fig. 2). Since isolation oxides are thicker than gate oxides, more electron-hole pairs are formed by incident radiation. Furthermore, the fields in the isolation oxide tend to be much lower, so charge trapping in the oxide will be exacerbated. Developing radiation-hard isolation oxides (field oxides) was a major challenge in the development of high-density, radiation-hard CMOS and remains one of the few "secret" process ingredients (for a basic discussion see (6)).

Problems can occur when inherently radiation-hard devices, notably JFETs and bipolar transistors, are used in non-hardened, oxide-isolated processes. Here radiation effects in the isolation structures can severely affect the radiation resistance of the devices. Clues to the importance of such parasitic ionization effects can be gleaned from a comparison of neutron and photon irradiations. Conventional (non-hardened) processes using oxide isolation have yielded good results in measurements to proton fluences $>10^{14}$ cm^{-2} or >4 Mrad (27,28), demonstrating that oxide isolation can be acceptable and that the suitability of these processes must be determined case by case.

IC processes also use special device structures to facilitate the integration of different device types. A prime example is the lateral *pnp* transistor, a structure more compatible with a standard CMOS process than "classic" vertical bipolar transistors. In a lateral transistor the emitter, base, and collector are arranged along the surface of the silicon with large-area exposure to oxide interfaces. Unlike vertical bipolar transistors, lateral devices are very susceptible to ionizing radiation, as surface leakage causes severe degradation of DC current gain. Lateral *pnp* transistors can be used as current sources or high impedance loads if the biasing circuitry is designed to accommodate substantial increases in base currents.

MITIGATION TECHNIQUES

Although little can be done to reduce radiation damage in a given device, many techniques can be applied to reduce the effects of radiation damage to an overall system. The goal of radiation-hard design is not so much to obtain a system whose characteristics do not change under irradiation, but rather to maintain the required performance characteristics over the lifetime of the system. The former approach tends to utilize mediocre to poor technologies that remain so over the course of operation. The latter starts out with superior characteristics, which gradually deteriorate under irradiation. Depending on the specific system, these designs may die gradually, although at some fluence or dose a specific circuit, typically digital, may cease to function at all. Clearly, the best mitigation technique is to avoid the problem, either by shielding or by reducing the electronics in the radiation environment to the minimum required to do the job. The latter runs counter to prevailing trends, which favor digitizing as close to the front-end as possible and tend to implement even simple control functions with digital circuitry.

Detectors

Increased detector leakage current has several undesirable consequences.

1. The integrated current over typical signal processing times can greatly exceed the signal.
2. Shot noise increases.
3. The power dissipated in the detectors increases ($I_{det} \cdot V_{det}$).

Since the leakage current decreases exponentially with temperature, cooling is the simplest technique to reduce diode leakage current. For example, reducing the detector temperature from room temperature to 0 °C reduces the bias current to about 1/6 of its original value.

Detector power dissipation is a concern in large-area silicon detectors for the LHC, where the power dissipation in the detector diode itself can be of order 1 to 10 mW/cm^2. Since the leakage current is an exponential function of temperature, local heating will increase the leakage current, which will increase the local heating, and so on, ultimately taking the device into thermal runaway. To avoid this potentially catastrophic failure mode, the cooling system must be designed to provide sufficient cooling of the detector, a challenging (but apparently doable) task in a system that is to have minimal mass.

Reducing the integration time reduces both baseline changes due to integrated detector current and shot noise. Clearly, this is limited by the duration of the signal to be measured. To some degree, circuitry can be designed to accommodate large baseline shifts due to detector current, but at the expense of power. AC-coupled detectors eliminate this problem. In instrumentation systems that require DC coupling, correlated double sampling techniques can be used to sample the baseline before the signal occurs and then subtract from the signal measurement.

One of the most powerful measures against detector leakage current is segmentation. For a given damage level, the detector leakage current *per signal channel* can be reduced by segmentation. If a diode with a leakage current of 10 μA is subdivided into 100 subelectrodes, each with its own signal processing channel, the DC current in each channel will be 100 nA and shot noise reduced by a factor of 10. Smaller electrodes also reduce the equivalent noise charge of the amplifier, allowing greater signal loss with radiation damage. This is why large area silicon tracking detectors can survive in the LHC environment. Fortuitously, increased segmentation is also required to deal with the high event rate. Pixel detectors with small electrode areas offer great advantages in this regard.

The most severe restriction on radiation resistance is imposed by type inversion, where the net acceptor concentration at some fluence becomes so large that the detector will no longer sustain the required voltage for full depletion. This is especially critical for position-sensing detectors with electrodes on both sides (double-sided detectors), for which full depletion is essential.

One can circumvent the type-inversion limit by using back-to-back single-sided detectors. The initial configuration uses n type segmented strip electrodes on n bulk, with a contiguous p electrode on the backside. Initially, the pn junction is at the backside. This does require full depletion in initial operation, which is no problem for the non-irradiated device and becomes easier to maintain as increasing fluence moves the bulk towards type inversion. After type inversion the bulk becomes p type and the junction shifts to the n electrodes, so that the bulk around the electrodes will be depleted and maintain inter-electrode isolation even in partial depletion.

Electronics

The design of the electronic systems is governed by changes in transistor parameters under irradiation, but circuit design and, at a higher level, architecture are equally important. Amplifiers are sensitive to changes in gain, bandwidth, and noise, so that effects on transconductance and noise parameters are important. Comparators used for threshold determination and timing rely critically on threshold shifts. Analog storage cells and switched capacitor systems tend to be sensitive to leakage currents. Digital circuitry is affected by threshold shifts that affect propagation delays and device transconductance, which determines switching speed.

In analog circuits, shorter shaping times improve tolerance to leakage currents. High rate systems require fast response time anyway, so experimental desires and engineering considerations interfere constructively. Since the system must be designed to tolerate a substantial shot noise current, utilization of bipolar junction transistors becomes very attractive, since the base shot noise becomes an acceptable contribution (in contrast to systems that emphasize noise minimization, as in x-ray spectrometry or liquid argon calorimetry).

In general, for use in amplifiers bipolar transistor circuitry is superior to CMOS. In logic circuitry, especially at low overall switching rates, CMOS is advantageous both because of power consumption and circuit density. For example, the on-detector silicon tracker front-end under development for the ATLAS experiment at the LHC uses bipolar transistor technology for the amplifier - pulse shaper - comparator IC and radiation-hard CMOS for a clock-driven digital pipeline buffer and data readout.

In amplifiers, bipolar transistors offer higher bandwidth for a given power in addition to superior device matching, which is a prime consideration in highly segmented systems with a correspondingly large number of channels. Threshold shifts in bipolar transistors are quite small with excellent matching between devices. JFETs yield excellent noise performance in applications where power consumption and circuit density are not prime considerations. Even when a

CMOS front end is chosen because of the use of a switched capacitor analog memory or the desire to combine the analog and digital circuitry on the same chip, amplifiers can be made quite radiation resistant, since the circuitry can be made to adjust for shifts in threshold voltage (1). In general, the use of fully differential circuitry and current mirrors yields circuitry whose operating point relies primarily on relative device matching (40,26,28). Changes in threshold voltages or current gain in adjacent devices tend to track after radiation damage, so the circuit will maintain its operating point. Circuitry should also be designed to minimize single-point failure modes. Failure of common bias networks will cause all associated circuitry to fail. Local biasing with highly parallel architectures reduces these problems (see ref. (26,29), for example).

The design of CMOS logic circuitry does not offer this flexibility. Since the threshold shifts of n and p MOSFETs are not complementary, circuit switching thresholds change. At high damage levels the device transconductance also suffers due to buildup of interface charge and increased scattering of charge carriers in the channel. Both effects change propagation delays, which can lead to race conditions (mismatches in propagation delays of streams whose results are combined) that cause circuit failure. These problems can be mitigated somewhat by careful design, but they point out a qualitative criterion for radiation-resistant system design: complexity. As a general rule, simple logic circuitry can be made more radiation resistant than complex circuitry that requires relative control of many mixed serial and parallel paths. Fully clocked systems avoid this problem, but at substantial penalties in power, speed, and area. Careful consideration should be given before incorporating complete wish-lists of circuitry (on-chip digitization, digital signal processing, microprocessor controlled readout, etc.) in a severe radiation environment (apart from common-sensical considerations such as reliability and maintenance of components that are not accessible without major disassembly). Simplest tends to be best.

IMPLEMENTATIONS

Several ICs for high-energy physics using the radiation-hard CMOS are installed in running experiments. Clock-driven pipelines designed for ZEUS and SDC have been fabricated and tested and are operating successfully. The SVX IC designed for the CDF silicon vertex detector has been transferred to the rad-hard UTMC process. SVX-H ICs are installed in both CDF and L3 (41) and are providing excellent results. All of these are full-custom designs, which allow control over device and process selection. Otherwise, the use of a non-hardened bipolar transistor IC process (26,28) would be extremely risky. However, full custom technology may not be required in all applications.

In many instrumentation applications discrete designs are suitable. As shown above, bipolar transistors and JFETs can provide very high radiation resistance without resorting to qualified radiation-hard devices. The same typically holds for

ECL logic ICs. If the ionizing dose does not exceed 100 or 200 krad, standard sub-micron CMOS may be adequate, because the thinner gate oxides (~ 20 nm) required in short channel devices provide a significant improvement in threshold shift with respect to the 50-nm oxides of earlier 3-μm devices. One caveat is in order, however. The radiation characteristics of standard (non rad-hard) CMOS processes are inherently unpredictable from lot to lot. If devices from a given production run are tested and found satisfactory (including a substantial performance margin), devices from the same lot should be used in the final system. This practice should be followed with any "off the shelf" IC that is not radiation-qualified. Especially if the system is readily accessible for maintenance or replacement, this course may be quite acceptable.

A more reliable approach is to use radiation-qualified transistors and ICs available commercially as standard parts. Power MOSFETs are offered with full specifications to 1 Mrad and limited use to 3 Mrad. Displacement damage is specified to 10^{14} n/cm^2. Operational amplifiers are available with guaranteed specifications to 1 Mrad(Si). CMOS logic ICs (inverters, gates, flip-flops, shift registers) are also specified to 1 Mrad. As mentioned above, the circuit design must accommodate increased propagation delay and reduced clock rates. Devices with higher integration levels are also available, for example 32K × 8 SRAMs specified to 300 krad. Twenty MSPS 8-bit flash ADCs implemented in 1.25-μm, junction-isolated, rad-hard CMOS have been tested to 81 Mrad ^{60}Co with no loss in performance (42).

The last example is also a reminder of a phenomenon that has been illustrated above (37) and observed repeatedly (43). The typical pattern is that parameters change most up to 1 Mrad and then plateau. Modern radiation-hard CMOS devices perform well at doses well beyond their rated maximum dose. The reason for this is the expense of fully qualifying a radiation-hard process in accordance with the requirements of the military and aerospace agencies, so devices are guaranteed only to the required specification, rather than the capabilities of the fabrication process.

CONCLUSION

Judicious evaluation of the radiation fields coupled with a stringent analysis of application requirements can yield electronic systems capable of performing well to ionizing doses of 100 Mrad and particle fluences of 10^{14} and probably 10^{15} cm^{-2}. Developing radiation-resistant systems does require great attention to detail and substantially more testing effort than conventional designs, but the effort is necessary if we are to exploit the high-luminosity accelerators on the horizon. For many applications we are limited less by technology than by ingenuity.

BIBLIOGRAPHY

Semiconductor Devices and Integrated Circuit Technology

1. Grove, A.S., *Physics and Technology of Semiconductor Devices* (John Wiley & Sons, New York, 1967).
2. Sze, S.M., *Physics of Semiconductor Devices* (John Wiley & Sons, New York, 1981) TK 7871.85.S988, ISBN 0-471-05661-8.
3. Nicollian, E.H. and Brews, J.R., *MOS (Metal Oxide Semiconductor) Physics and Technology* (John Wiley & Sons, New York, 1982) TK7871.99.M44N52, ISBN 0-471-08500-6.
4. Wolf, S., *Silicon Processing for the VLSI Era, Volume 2 - Process Integration* (Lattice Press, Sunset Beach, 1990) ISBN 0-961672-4-5.

Radiation Effects

1. Ma, T.P. and Dressendorfer, P.V., *Ionizing Radiation Effects in MOS Devices and Circuits* (John Wiley & Sons, New York, 1989) TK7871.99.M44I56, ISBN 0-471-84893-X.
2. Messenger, G.C. and Ash, M.S., *The Effects of Radiation on Electronic Systems* (van Nostrand Reinhold, New York, 1986) TK7870.M4425, ISBN 0-442-25417-2.
3. Srour, J.R. et al., *Radiation Effects on and Dose Enhancement of Electronic Materials* (Noyes Publications, Park Ridge, 1984) TK7870.R318, ISBN 0-8155-1007-1.
4. van Lint, V.A.J. et al., *Mechanisms of Radiation Effects in Electronic Materials* (John Wiley & Sons, New York, 1980) TK7871.M44, ISBN 0-471-04106-8.

Journals

Most papers on radiation effects in semiconductor devices are presented at the IEEE Nuclear and Space Radiation Effects Conference and published in the annual conference issue (usually December) of the IEEE Transactions on Nuclear Science. Additional papers, primarily from the high energy physics community, are published in the Conference Record of the IEEE Nuclear Science Symposium and in the conference issue of the IEEE Transactions on Nuclear Science. Other conferences on detector instrumentation tend to publish their proceedings in Nuclear Instruments and Methods. Many of the new results on detectors (Si and GaAs) and low-noise front ends appear as internal notes of the ATLAS and CMS collaborations in preparation for the LHC.

REFERENCES

1. http://www-atlas.lbl.gov/strips/doc/tutorials.html.

2. Messenger, G.C. and Ash, M.S., *The Effects of Radiation on Electronic Systems* (van Nostrand Reinhold, New York, 1986), p. 166.

3. Burke E.A., "Energy Dependence of Proton-Induced Displacement Damage in Silicon," *IEEE Trans. Nucl. Sci.* **NS-33/6**, 1276 (1986).

4. Van Ginneken A., "Non Ionizing Energy Deposition in Silicon for Radiation Damage Studies," FNAL report FN 522 (October 1989).

5. Srour, J.R. et al., "Radiation Damage Coefficients for Silicon Depletion Regions," *IEEE Trans. Nucl. Sci.* **NS26/6**, 4784 (1979).

6. Ma, T.P. and Dressendorfer, P.V., *Ionizing Radiation Effects in MOS Devices and Circuits* (John Wiley & Sons, New York, 1989).

7. Tsveybak, I. et al., "Fast Neutron-Induced Changes in Net Impurity Concentration of High-Resistivity Silicon," *IEEE Trans. Nucl. Sci.* **NS-39/6**, 1720-1729 (1992).

8. Barberis, E. et al., "Temperature effects on radiation damage to silicon detectors," *Nucl. Instr. and Meth.* **A326**, 373-380 (1993).

9. Chilingarov, A. et al., "Radiation Hardness of the Silicon Counter Tracker (SCT) for ATLAS," *Proceedings of the 27th International Conference on High Energy Physics*, Glasgow, July 20-27, 1994, contribution 0943.

10. Sze, S.M., *Physics of Semiconductor Devices* (John Wiley & Sons, New York, 1981), p. 90.

11. Gill, K. et al., "Radiation damage by neutrons and photons to silicon detectors," *Nucl. Instr. and Meth.* **A322**, 177 (1992).

12. Ohsugi, T. et al., "Radiation Damage in Silicon Microstrip Detectors," *Nucl. Instr. and Meth.* **A265**, 105-111 (1988).

13. Li, Z., "Modeling and simulation of neutron induced changes and temperature annealing of N_{eff} and changes in resistivity in high resistivity silicon detectors," *Nucl. Instr. and Meth.* **A342**, 105-118 (1994).

14. Pitzl, D. et al., "Type Inversion in Silicon Detectors," *Nucl. Instr. and Meth.* **A311**, 98 (1992).

15. Giubellino, P. et al., "Study of the Effects of Neutron Irradiation on Silicon Strip Detectors," *Nucl. Instr. and Meth.* **A315**, 156 (1992).

16. Matthews, J.A.J. et al., "Bulk Radiation Damage in Silicon Detectors and Implications for ATLAS SCT," New Mexico Center for Particle Physics Note and ATLAS INDET-NO-118, 1995.

17. Ziock, H.-J. et al., "Temperature dependence of the radiation induced change of depletion voltage in silicon PIN detectors," *Nucl. Instr. and Meth.* **A342**, 96-104 (1994).

18. Lemeilleur, F. et al., "Neutron-induced radiation damage in silicon detectors," *IEEE Trans. Nucl. Sci.* **NS-39/4**, 551-557 (1992).

19. Fretwurst, E. et al., "Radiation Hardness of Silicon Detectors for Future Colliders," *Nucl. Instr. and Meth.* **A326**, 357-364 (1993).

20. Tamura, N. et al., "Radiation effects of double-sided silicon strip detectors," *Nucl. Instr. and Meth.* **A342**, 131-136 (1994).

21. Unno, Y. et al., to be published in *IEEE Trans. Nucl. Sci.*

22. Barberis, E. et al., "Capacitances in silicon microstrip detectors," *Nucl. Instr. and Meth.* **A342**, 90-95 (1994).

23. Wheadon, R. et al., "Radiation tolerance studies of silicon microstrip detectors for the LHC," *Nucl. Instr. and Meth.* **A342**, 126-130 (1994).

24. Kubota, M. et al., "Radiation Damage of Double-Sided Silicon Strip Detectors," *Conference Record of the 1991 IEEE Nuclear Science Symposium and Medical Imaging Conference*, Nov. 2 - 9, 1991, Santa Fe, New Mexico, Vol. 1, p. 246.

25. Messenger, G.C. and Ash, M.S., *The Effects of Radiation on Electronic Systems* (van Nostrand Reinhold, New York, 1986), p. 183.
26. I. Kipnis et al., "An Analog Front-End Bipolar-Transistor IC for the SDC Silicon Tracker," *IEEE Trans. Nucl. Sci.* **NS-41/4**, 1095-1103 (1994).
27. Cartiglia, N. et al., "Radiation Hardness Measurements on Bipolar Test Structures and an Amplifier-Comparator Circuit," *Conference Record of the IEEE Nuclear Science Symposium*, Oct. 25-31, 1992, Orlando, Florida (ISBN 0-7803-0883-2), Vol. 2, pp. 819-821, 1992.
28. Spencer, E. et al., "A Fast Shaping Low Power Amplifier-Comparator Integrated Circuit for Silicon Strip Detectors," *IEEE Trans. Nucl. Sci.* **NS-42/4**, 796-802 (1995).
29. Kipnis, I., "CAFE: A Complementary Bipolar Front-End Integrated Circuit for the ATLAS SCT," on WWW at http://www-atlas.lbl.gov/strips/doc/ reports.html.
30. Kipnis, I., private communication.
31. Sze, S.M., *Physics of Semiconductor Devices* (John Wiley & Sons, New York, 1981), pp. 314-322.
32. Citterio, M. et al., "A Study of Low Noise JFETs Exposed to Large Doses of Gamma Rays and Neutrons," *Conference Record of the IEEE Nuclear Science Symposium*, Oct. 25-31, 1992, Orlando, Florida (ISBN 0-7803-0883-2), Vol. 2, pp. 794-796, 1992.
33. Citterio, M. et al., "Radiation Effects at Cryogenic Temperatures in Si-JFET, GaAs MESFET, and MOSFET Devices," *IEEE Trans. Nucl. Sci.* **NS-42/6**, 2266-2270 (1995).
34. Radeka, V. et al., "JFET Monolithic Preamplifier with Outstanding Noise Behavior and Radiation Hardness Characteristics," *IEEE Trans. Nucl. Sci.* **NS-40/4**, 744-749 (1993).
35. Stephen, J.H., "Low noise field effect transistors exposed to intense ionizing radiation," *IEEE Trans. Nucl. Sci.* **NS-33/6**, 1465-1470 (1986).
36. Dabrowski, W. et al., "Noise Measurements on Radiation-Hardened CMOS Transistors," *Conference Record of the 1991 IEEE Nuclear Science Symposium and Medical Imaging Conference*, Nov. 2 - 9, 1991, Santa Fe, New Mexico, IEEE catalog no. 91CH3100-5, Vol. 3, pp. 1536-1540, 1991.
37. Seller, P. et al., RAL-TR-95-055 and "ATLAS SCT Technical Proposal Backup Document," ATLAS INDET-NO-085, 1995 (CERN).
38. Tedja, S. et al., "Noise Spectral Density Measurements of a Radiation-Hardened CMOS Process in the Weak and Moderate Inversion," *IEEE Trans. Nucl. Sci.* **NS-39/4**, 804-808 (1992).
39. Wolf, S., *Silicon Processing for the VLSI Era, Volume 2 - Process Integration* (Lattice Press, Sunset Beach, 1990) ISBN 0-961672-4-5.
40. Spieler, H., "Analog Front-End Electronics for the SDC Silicon Tracker," *Nucl. Instr. and Meth.* **A342**, 205-213 (1994).
41. Acciarri, M. et al., "The L3 silicon microvertex detector," CERN-PPE/94-122, 1994.
42. Nutter, S. et al., "Results of Radiation Hardness Tests and Performance Tests of the HS9008RH Flash ADC," *IEEE Trans. Nucl. Sci.* **NS-41/4**, 1197-1202 (1994).
43. "ATLAS SCT Technical Proposal Backup Document," ATLAS INDET-NO-085, 1995 (CERN).

Beam Position Monitor Engineering

Stephen R. Smith

Stanford Linear Accelerator Center, Stanford, California 94309

Abstract. The design of beam position monitors often involves challenging system design choices. Position transducers must be robust, accurate, and generate adequate position signal without unduly disturbing the beam. Electronics must be reliable and affordable, usually while meeting tough requirements on precision, accuracy, and dynamic range. These requirements may be difficult to achieve simultaneously, leading the designer into interesting opportunities for optimization or compromise. Some useful techniques and tools are shown. Both finite element analysis and analytic techniques will be used to investigate quasi-static aspects of electromagnetic fields such as the impedance of and the coupling of beam to striplines or buttons. Finite-element tools will be used to understand dynamic aspects of the electromagnetic fields of beams, such as wake fields and transmission-line and cavity effects in vacuum-to-air feedthroughs. Mathematical modeling of electrical signals through a processing chain will be demonstrated, in particular to illuminate areas where neither a pure time-domain nor a pure frequency-domain analysis is obviously advantageous. Emphasis will be on calculational techniques, in particular on using both time domain and frequency domain approaches to the applicable parts of interesting problems.

INTRODUCTION

We will work through a beam position monitor (BPM) system from transducer to digitization; starting with a simple case, expressed in a simplified model. Then we will analyze a tougher problem, again using simple mathematical models. Finally we will apply more complicated analysis to understand some details of the system.

SIGNAL MODELING

A common difficulty is choosing a particular approach to the problem at hand: Is a frequency domain or a time domain approach more suitable? Table 1 contains some suggestions of when to best use either the time domain or frequency domain.

The approach used here is to describe time-domain phenomena as discrete quantities sampled at a uniform rate in time and then to transform back and forth from time to frequency domain as needed to take advantage of the best features of each representation. For example, let a voltage $V(t)$ be represented by N samples

V_j sampled at times $t_j = t_{max} \dfrac{j}{N}$. The discrete Fourier transform and its inverse are given by

$$fV_k = \frac{1}{N} \sum_j V_j e^{-2\pi i \frac{jk}{N}} \quad \text{(FFT)}, \qquad V_j = \sum_k fV_k e^{2\pi i \frac{jk}{N}} \quad \text{(FFT}^{-1}\text{)}.$$

TABLE 1. Time Domain versus Frequency Domain

Frequency Domain appropriate when:	Time Domain appropriate when:
Periodic processes	Single shot
High Q, low bandwidth	Low Q, high bandwidth
Frequency-dependent parameters	Amplitude-dependent parameters, e.g. limits of linear range of components
• Complex impedances • Filter response	
	• Saturation (P_{1dB}) • Slew rates • Damage thresholds (V_{max})
Linear phenomena	Non-linear phenomena
	• mixers • diodes
Discrete frequency phenomena	Discrete-time operations
• oscillators	• Sample & Hold • Digitization

ANALYZE A SIMPLE BPM

Let's analyze a button BPM in some ring with stored beam, bunched at frequency f_b. We wish to calculate the signal, noise, and position sensitivities. With this information we can establish a noise figure budget needed to achieve some required resolution. This is an obvious case for frequency-domain analysis (1,2).

We estimate the intrinsic resolution from the ratio of signal to thermal noise. First we calculate the signal. The image charge on a button is given by

$$Q(\omega) = \frac{ButtonArea}{DuctCircumference} \cdot \rho(\omega),$$

where $\rho(\omega)$ is the linear charge density, which we have assumed varies slowly on the scale of the button size. Then the image current out of the button is given by

$$I_{img} = \frac{dQ}{dt} = \frac{ButtonArea}{DuctCircumference} \cdot \frac{d\rho}{dt} .$$

We have made the (usually excellent) approximation here that the beam acts as a perfect current source in generating image currents. Expressing the linear charge density in terms of the beam current,

$$\rho(\omega) = \frac{I(\omega)}{\beta c} \quad so \quad \frac{d\rho}{dt} = \frac{1}{\beta c} \cdot \frac{dI}{dt} = \frac{i\omega}{\beta c} \cdot I(\omega)$$

$$I_{img} = \frac{\pi a^2}{2\pi b} \frac{i\omega}{\beta c} I(\omega) .$$

The button voltage is the product of the current out of the button and the impedance seen by this current. The dominant pieces of this impedance are the impedance of the cable shunted by the reactance of some parasitic capacitance C_b between the button and the walls of the beam duct:

$$Z = Z_{coax} \| \frac{1}{i\omega C_b} \qquad Z \to Z(\omega)$$

$$V_b = Z \cdot I_{img} = \frac{\pi a^2}{2\pi b} \frac{\omega}{\beta c} Z(\omega) I(\omega) .$$

Later we will find it useful to rearrange this so the button voltage appears as the product of beam current times something that looks like an impedance, which we will call a "transfer impedance" Z_t.

$$V_b = Z_t(\omega) I(\omega) \qquad \text{where in this case} \quad Z_t = \frac{\pi a^2}{2\pi b} \frac{\omega}{\beta c} Z(\omega) .$$

For an average current I_{avg} circulating with angular frequency $\omega_b = 2\pi f_b$ the beam current is given by

$$I(t) = I_{avg} \cdot \left[1 + 2 \sum_m A_m \cos(m\omega_b t) \right] .$$

So in frequency space the current consists of a line spectrum where the amplitude of the mth harmonic is given by

$$I_m = I_{avg} \cdot \begin{cases} 1 & for \quad m = 0 \\ 2A_m & for \quad m > 0 \end{cases}.$$

The coefficients A_m are determined by the shape of the bunch and are near unity for frequencies well below the inverse of the bunch length. Now we pick a processing frequency $f_0 = m\omega_b / 2\pi$ corresponding to the mth revolution harmonic and find the signal voltage:

$$V_b = \frac{\pi a^2 Z}{2\pi b \beta c} \cdot \frac{dI}{dt} = \frac{\pi a^2 Z}{b \beta c} 2A_m f_0 I_{avg}.$$

To make our example more concrete, we specify a few of the parameters in Table 2.

TABLE 2. Example Parameters for a Narrow-Band Beam Position Monitor

Parameter	Symbol	Value
Duct radius	b	3 cm
Beam current	I_{avg}	10 mA
Bunch frequency	f_b	500 MHz
Button radius	a	5 mm
Coax impedance	Z	50 Ω
Beam velocity	β	1
Measurement bandwidth	B	1 MHz
Required resolution	σ_x	10 μm
Processing harmonic	m	1

The signal power is

$$P_s = \frac{1}{2} \frac{V_b^2}{Z} = \frac{2\pi^2 a^4}{b^2 \beta^2 c^2} ZA_m^2 f_0^2 I_{avg}^2 = 0.19 \mu W = -37 dBm.$$

The noise power in this bandwidth is:

$$P_n = k_B TZB = -114 dBm \qquad (\text{T=300 K,} \quad B = 1 \text{ MHz})$$

$$SNR = \frac{P_s}{P_n} = 77 dB.$$

For small displacements from the center of the beam duct, the beam position in terms of voltages on hypothetical left and right buttons is given by

$$X = \frac{b}{2} \cdot \frac{V_L - V_R}{V_L + V_R} \qquad \text{(Difference-over-Sum algorithm).}$$

Translating the voltage noise into a position error,

$$\sigma_X = \frac{b}{2} \cdot \frac{\sqrt{2}\sigma_V}{2V} = \frac{b}{2\sqrt{2}} \cdot \frac{1}{\sqrt{SNR}} = 1.5 \mu m.$$

This assumes the parameters in Table 2 and noiseless, lossless processing. Assuming a required resolution of $\sigma_X = 10 \mu m$, we have a noise and loss budget

$$F_n = \frac{10 \mu m}{1.5 \mu m} = 6.7 = 16 dB,$$

which we can allocate to losses and electronics noise.

A MORE INTERESTING CASE...

The first example was readily handled in the frequency domain. Now let's look at a system which must respond to a single bunch, a few bunches, or a continuous train of beam bunches. In this case we'll need to choose the most convenient point of view for each aspect of the problem. We'll follow the signal from the beam pipe to the ADC. We'll start by analyzing the response of a button BPM to a single beam bunch. Essential parameters for this example are listed in Table 3. The tool we use to do all of the calculations, transformations, and plotting is Mathcad.

TABLE 3. Example Parameters for a Wide-Band Beam Position Monitor, Taken (loosely) from the PEP-II Straight Section BPMs

Parameter	Symbol	Value
Duct radius	b	4.4 cm
Bunch charge	Q	8×10^8 electrons
Bunch frequency	f_b	238 MHz
Button radius	a	7.5 mm
Coax cable impedance	Z	50 Ω
Beam velocity	β	1
Bunch shape	Gaussian	
Bunch length (rms)	σ_z	1 cm
Processing harmonic	m	4

Again we start at the button, but this time we calculate the image charge as a function of time. At any given time, the image charge on the button is found by integrating the button angular coverage over the longitudinal beam charge distribution

$$Q_{img}(t) = \int \rho(z) \cdot \frac{width(z)}{2\pi b} \cdot dz \ ,$$

where z is the distance coordinate in the beam direction and the button width as a function of z is given by

$$width(z) = 2\sqrt{a^2 - z^2}$$

for a round button of radius a. Since the beam charge density propagates down the beam pipe at $v = c$, the time-dependent image charge

$$Q_{img}(t) = \int \rho(z - ct) \cdot \frac{width(z)}{2\pi b} \cdot dz \ .$$

This is just the convolution of the charge density and the button shape so we can evaluate by multiplying the Fourier transforms and transforming back:

$$Q_{img}(t) = FFT^{-1}(FFT(\frac{width(ct)}{2\pi b}) \cdot FFT(\rho(ct))).$$

Image current is the time derivative of the image charge, which is easy to do by transforming to the frequency domain, multiplying by $i\omega$, and transforming back to the time domain (see Fig. 1):

$$I_{img} = \frac{dQ_{img}}{dt} = FFT^{-1}(i\omega \cdot FFT(Q_{img})).$$

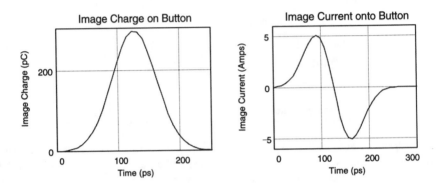

FIGURE 1. Image charge and current vs. time.

The impedance seen by the button is that of the coax shunted by the button capacitance. Expressed as a function of frequency the impedance is given by

$$Z_b = \left(Z_0^{-1} + i\omega C_b\right)^{-1} \quad \text{therefore} \quad V_b = FFT^{-1}(Z_b \cdot FFT(I_{img})),$$

as shown in Fig. 2.

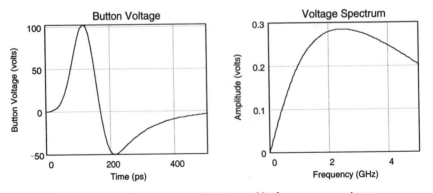

FIGURE 2. Signal voltage on a button and its frequency spectrum.

A coaxial cable with skin effect losses has a frequency response given by (3)

$$fcoax(f) = e^{-(1+i)\sqrt{\frac{f}{f_e}}}.$$

Here f_e is the frequency at which the amplitude is attenuated by a factor of e. For example, a cable of length 40 m with a loss of 6 dB/100ft will have f_e = 1.2 GHz. Convoluting this response with the button voltage yields the voltage at the other end of the cable. The modeled frequency response of the cable and its convolution with the position monitor signal are shown in Fig. 3.

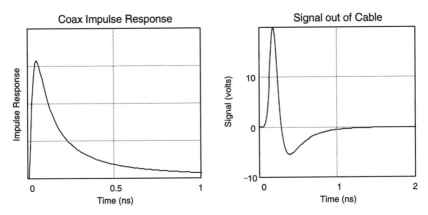

FIGURE 3. Impulse response of a coaxial cable and the beam signal at the far end.

This signal is passed through a two-pole Bessel bandpass filter which is specified by its Laplace transform (conveniently supplied by MATLAB), a polynomial in $s = i\omega$ in the Laplace transform sense

$$Bessel(s) = \frac{0.102 \cdot s^2}{s^4 + 0.554 \cdot s^3 + 71.61 \cdot s + 1278} \qquad fBessel(f) = Bessel(2\pi i f).$$

The impulse response of this filter is shown in Fig. 4.

The ringing signal can now be demodulated. We choose synchronous detection. This is an obvious place to model via a time-domain approach, since demodulation is an inherently nonlinear phenomenon. We'll do it by multiplying the ringing signal with a pure sine function whose phase is chosen to maximize the demodulated signal. Then we apply a low-pass filter, again with a Bessel characteristic, which leaves only the baseband signal as shown in Fig. 5.

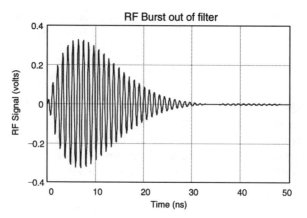

FIGURE 4. An rf burst out of the bandpass filter.

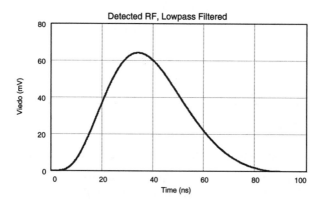

FIGURE 5. Demodulated rf after 3-pole Bessel low-pass filter.

But what happens if these bunches occur every $t_b = 4.2$ ns? We send a finite train through our simulation by convoluting the button signals with a finite-comb response. In the frequency domain, a sixteen-element comb of delta functions separated by time δ looks like

$$fComb(\omega) = \sum_{n=1}^{16} e^{-i\omega n\delta} \ .$$

The convolution of the comb response with the button signal before and after the bandpass filter is shown in Fig. 6. The demodulated rf bursts from both the single-bunch and 16-bunch cases are shown in Fig. 7.

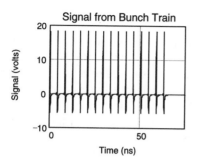
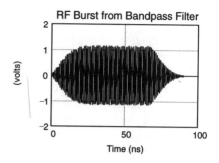

FIGURE 6. Button and bandpass filter responses to a train of 16 beam bunches.

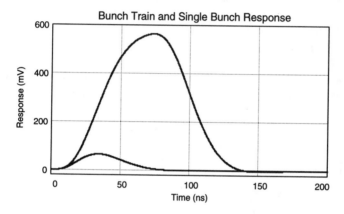

FIGURE 7. Demodulated rf for 16 bunch train (upper trace) and single bunch (lower trace).

Now we calculate position resolution from the peak signal and noise. Using the previous formula to convert signal amplitude to position resolution, we get for the single bunch case

$$\sigma_x = \frac{b}{2} \cdot \frac{\sqrt{2}\sigma_v}{2V} = \frac{b}{2\sqrt{2}} \cdot \frac{2\mu V}{64mV} = 0.5\mu m.$$

The intrinsic resolution is ten times better for the multibunch case. This is for thermal noise only; add electronic noise figure and system losses to this to get the real resolution.

We've generated a reasonable first approximation to a BPM system. There are limitations to the accuracy of this model: we've assumed a round beam pipe, taken a low-frequency approximation to the button response (good to a few GHz), and of course, we only get out what we've thought to put in the model.

DUCT-BUTTON COUPLING

Determination of signal amplitude and position sensitivity of a position monitor requires knowledge of the coupling of the beam to the transducer, whether button or stripline, as a function of beam position. For relativistic beams, two-dimensional electrostatic calculations give sufficient estimates of coupling for most cases. Analytic calculations are good for simple cases, such as round pipes. Usually beam ducts are more complicated structures, often requiring numerical techniques to estimate coupling. Typical tools are POISSON, ANSYS, Electro, and even spreadsheets like Excel using the relaxation method to solve Poisson's equation.

We have used conformal mapping to solve for the field in non-trivial beam ducts, in particular elliptical and octagonal ducts. This gives (formally) analytic solutions, although one must still evaluate the resulting expressions by numerical techniques. Figure 8 shows the results of the calculation of electric field and equipotential contours for a beam displaced by 1 cm in the PEP-II high energy ring arc beam duct. We have used this technique to find the optimal location for the BPM buttons, the beam coupling, and its dependence on beam position.

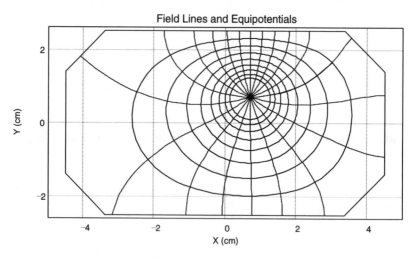

FIGURE 8. Field lines and equipotential contours found by conformal mapping.

THREE-DIMENSIONAL FIELD CALCULATION

Now we analyze the consequences of the real three-dimensional geometry of the beam duct, the buttons, and the vacuum feedthroughs. The goal is twofold: to look for the effects of the actual geometry on the coupling of the beam to the button, which we have so far modeled only in two dimensions, and to estimate the wake field impedances presented by the position monitor to the circulating beam.

The tool used is MAFIA, a finite-difference, three-dimensional electromagnetic field solver. A detailed accounting of this analysis is presented in ref. (4). The beam duct for the PEP-II high-energy ring arcs, whose cross section through the BPM buttons is shown in Fig. 9, is modeled in three dimensions as shown in Fig. 10. Since we are interested in wake fields for a few consecutive beam bunches separated by 2.1 ns, a section of pipe up to 5 meters long is modeled.

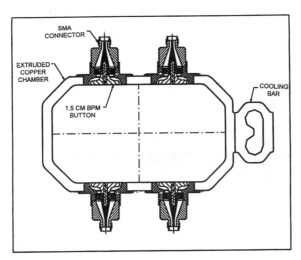

FIGURE 9. PEP-II high-energy ring arc vacuum chamber cross section taken through the BPM buttons.

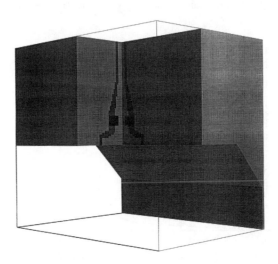

FIGURE 10. Geometry of beam duct and BPM buttons as described in MAFIA.

Wake Field Impedances

We also need to know the effect of the buttons on the beam. The beam induces fields on the position monitor buttons; that's how we deduce beam position. These fields act back on the beam, in particular on subsequent bunches. Narrow-band resonances in this response can lead to coupled-bunch instabilities. The longitudinal electric field left in the PEP-II beam pipe after passage of a single beam bunch is shown in Fig. 11. Its frequency spectrum is shown in Fig. 12.

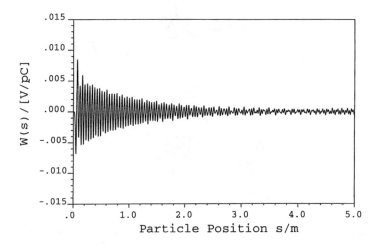

FIGURE 11. Longitudinal wake field E_z vs. distance behind bunch.

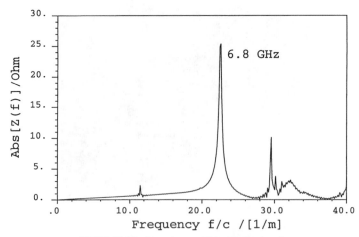

FIGURE 12. Spectrum of longitudinal wake.

A resonance is apparent at 6.8 GHz. This is due to a TE_{11}-like resonance localized in the gap between the edge of the button and the beam duct wall. This gap, running around the edge of the button, is like a slotline waveguide, so that modes propagating around the button are in resonance when their wavelength is approximately the button circumference. Figure 13 shows a snapshot of the electric field around the edge of the button after passage of a beam bunch, as calculated by MAFIA. The response shown in Fig. 12 meets the PEP-II impedance budget. However, the initial design for the PEP-II BPM button called for buttons with a 2-cm diameter. This analysis showed that the TE_{11} resonance would have been intolerable. Reducing the button diameter to 1.5 cm eliminated the problem; the coupling of the beam to the resonance is reduced by roughly the area of the button, and the resonance is moved up to the 6.8 GHz shown in Fig. 12, by which frequency the bunch power spectrum has fallen drastically. Of course the position signal amplitude is also reduced along with the button area, hence the change could only be made along with a comparable reduction of the noise budget for the processing system, made possible by the introduction of a low-noise preamplifier in the electronics.

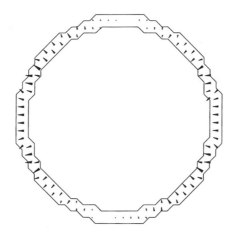

FIGURE 13. Calculated electric field around the edge of the BPM button.

Transfer Impedance

We have previously calculated the coupling from beam to button assuming cylindrical symmetry for the beam pipe, plus ideal transmission lines from the buttons out through the vacuum wall. We then used MAFIA to calculate the fields induced by the beam in the coaxial cable leading to the electronics, incorporating the full three-dimensional geometry of the beam pipe, buttons, vacuum feedthrough, and transition to the coaxial cable. In particular we want to know how the beam couples to TEM modes propagating up the coaxial cable to the processing electronics. The three-dimensional model includes a short stub of coax attached to the button, properly terminated at the far end. MAFIA projects the propagating TEM modes out of the calculated fields. Figure 14 shows the coax voltage versus time calculated in this manner. The frequency spectrum is shown in Fig. 15. We extract the transfer impedance from this plot; at the signal processing frequency of 952 MHz, the transfer impedance is 0.65 Ω, in good agreement with the estimates based on two-dimensional approximations.

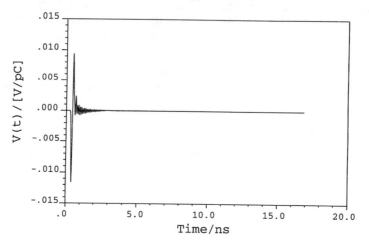

FIGURE 14. Calculated beam signal fom button.

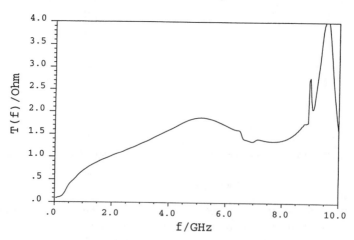

FIGURE 15. Transfer impedance of BPM.

CONCLUSIONS

We have shown a few tools which allow calculation of the performance of a beam position monitor system. We started with simple ways of estimating signals in simplified cases. Then we addressed ways of incorporating more detail in the models. We concluded with a full three-dimensional analysis of time-dependent electromagnetic fields.

REFERENCES

1. R. Shafer, "Beam Position Monitoring," in *AIP Conf. Proc.* **212**, 26-58 (1990).
2. R.H. Siemann, "Spectral Analysis of Relativistic Bunched Beams," these proceedings.
3. R.L. Wiginton and N.S. Nahman, "Transient Analysis of Coaxial Cables Considering Skin Effect," in *Proc. IRE* **45**, 166-174 (1957).
4. C.K. Ng, T. Weiland, D. Martin, S. Smith, N. Kurita, "Simulation of PEP-II Beam Position Monitors," in *Proc. of 16th IEEE Particle Accelerator Conference (PAC 95) and International Conference on High Energy Accelerators*, Dallas, Texas, 1-5 May 1995, 2485-2487 (1996).

Impedance of Accelerator Components*

John N. Corlett

*Center for Beam Physics, Lawrence Berkeley National Laboratory,
1 Cyclotron Road, Berkeley, California 94720*

Abstract. As demands for high luminosity and low emittance particle beams increase, an understanding of the electromagnetic interaction of these beams with their vacuum chamber environment becomes more important in order to maintain the quality of the beam. This interaction is described in terms of the wake field in time domain, and the beam impedance in frequency domain. These concepts are introduced, and related quantities such as the loss factor are presented. The broadband Q=1 resonator impedance model is discussed. Perturbation and coaxial wire methods of measurement of real components are reviewed.

INTRODUCTION

At low beam currents the motion of a charged particle beam can be described by the optics in the accelerator—the beam experiences accelerating and focusing forces due to the external magnetic and electric fields purposely applied and controlled. In addition to this interaction with external fields, the beam also communicates with its surrounding vacuum chamber through electromagnetic fields generated by the beam itself. When the beam current is sufficiently large, the effects of the beam-induced fields become more important and can limit the performance of an accelerator. The various accelerator components, such as rf cavities, bellows, injection septa, dielectric walls, and even a smooth pipe of finite conductivity, result in scattering or trapping of the beam-induced fields. These fields can last long enough to be experienced by a charge following the exciting charge, causing perturbations to the energy or angle of the following particle's orbit. The dynamics of bunches of particles due to their interaction with the environment of the accelerator through the beam-induced electromagnetic field are generally described in terms of "collective effects" which may be highly disruptive to the beam. Problems may also be encountered in heating of accelerator components as a result of these scattered or trapped fields.

Prediction and measurement of the effects of the beam-induced fields, in terms of beam impedance or wake function, is necessary for accurate assessment of machine performance, and a considerable formalism has been developed to

---

*This work was supported by the U.S. Department of Energy, under Contract DE-AC03-76SF00098.

describe the beam interaction with electromagnetic fields. Analytical estimates, computer simulation, test laboratory measurements, and measurements with beam in an accelerator are used to determine the beam impedance (1 - 4).

The effects of the self-fields of the beam may be analyzed in either time domain or frequency domain, and each has its advantages and drawbacks. For circular accelerators, the periodic nature of the beam signals makes the frequency-domain approach generally more useful, whereas the time-domain approach is more often applied to linear machines. In the time domain the beam-induced electromagnetic field in an accelerator component may be described by *wake function* and in the frequency domain by the *beam impedance* (sometimes known as the *coupling impedance*). The beam impedance is a complex quantity: the real part is associated with extraction of energy from the beam and the imaginary part with deformation of the beam profile. The wake function and impedance are equivalent in the sense that the impedance is the Fourier transform of the wake function.

BASIC CONCEPTS

The electric field vector for a charged particle moving in free space, in the laboratory frame, may be written as (5)

$$\vec{E} = \frac{q\vec{r_0}}{4\pi\varepsilon_0\gamma^2 r^2} \frac{1}{\left(1 - \frac{v^2}{c^2}\sin^2\Psi\right)^{3/2}},$$

where Ψ is the angle between the observer at $\mathbf{r_0}$ and the particle velocity $\mathbf{v}$.

The opening angle of the radial E-field is of the order $1/\gamma$. For ultra-relativistic particles the fields resemble plane waves—$\mathbf{E}$ and $\mathbf{B}$ are transverse to each other and lie in a disk transverse to the particle velocity. We have only radial E field and azimuthal H field, confined to a disk perpendicular to the direction of motion of the charge, producing a δ-function distribution in the direction of motion. Outside this disk there are no fields, and consequently there would be no forces acting on a charge ahead of or following the particle. The situation remains the same for charges moving along the axis of an infinitely conducting smooth cylindrical pipe. The electric field lines are then terminated with surface charges on the inside wall.

For non-relativistic particles the situation is more complex. γ is determined by the energy of the particle beam divided by the rest energy of the constituent particles. For low-γ beams the space-charge force which cancels out with the magnetic forces of ultra-relativistic charges cannot be neglected, and the fields associated with a charge are not confined to a disk around the charge. Here, we will deal mainly with the simpler case of ultra-relativistic charges and ignore these latter complications.

Longitudinal Wake Fields

The beam-induced electromagnetic fields are called the *wake fields* since, in the limiting case of charges moving at the speed of light, causality requires that the fields exist only behind the charge.

Consider a point charge q_1 traveling with velocity $v = \beta c$ at position z_1 and followed by another point "test" charge q traveling with equal velocity parallel to q_1 and at position z. The time delay between the charges is τ, and their longitudinal coordinates are given by $z_1(t) = vt$, $z(t) = v(t - \tau)$. Figure 1 shows the coordinate system.

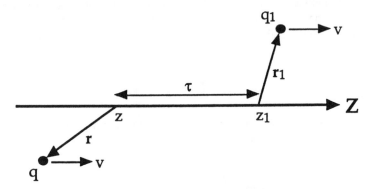

FIGURE 1. Coordinates of the point charges.

The Lorentz force experienced by the test charge q due to fields created by the exciting charge q_1 is given by

$$F = q\,[E + v \times B]$$

In general, the wake fields and resulting force will have transverse and longitudinal components. The energy lost from the leading charge q_1 is given by the work done against the electromagnetic fields

$$\Delta U_{11} = -\int_{-\infty}^{\infty} F(r_1, z, t) \cdot dz; \quad t = \frac{z_1}{v},$$

and for the longitudinal component we find

$$\Delta U_{11} = -q_1 \int_{-\infty}^{\infty} E_z(r_1, z_1, t) \cdot dz; \quad t = \frac{z_1}{v}.$$

This accounts for the energy loss to resistive heating in the vacuum chamber walls, to fields scattered at discontinuities in the pipe, and to fields stored in irregular regions of the pipe.

The *loss factor* $k(\mathbf{r}_1)$ is defined as the energy loss to the self-field per unit charge squared

$$k(\mathbf{r}_1) = \frac{\Delta U_{11}(\mathbf{r}_1)}{q_1^2}$$

and has units of volt per Coulomb.

The test charge experiences an energy change due to the fields produced by the leading charge

$$\Delta U_{21} = -q \int_{-\infty}^{\infty} E_z(\mathbf{r}, z, \mathbf{r}_1, z_1, t)dz; \quad t = \frac{z_1}{v} + \tau.$$

The *longitudinal wake function* $w_z(\mathbf{r},\mathbf{r}_1,t)$ may be defined as the energy lost by the trailing charge q per unit charge of both q_1 and q:

$$w_z(\mathbf{r}, \mathbf{r}_1, \tau) = \frac{\Delta U_{21}(\mathbf{r}, \mathbf{r}_1, \tau)}{q_1 q},$$

or equivalently in terms of the longitudinal electric field

$$w_z(\mathbf{r}, \mathbf{r}_1, \tau) = \frac{\int_{-\infty}^{\infty} E_z(\mathbf{r}, z, \mathbf{r}_1, z_1, t)dz}{q_1}; \quad t = \frac{z_1}{v} + \tau.$$

Like the loss factor, the wake function has units of volt per Coulomb.

We need to know the wake function for a distribution of particles in a bunch. By using the wake function as defined above for a point charge, also known as a *Green function*, we apply linear superposition to add the effects of a bunch of particles. Thus the wake field of an arbitrary charge distribution $i_b(t)$, where

$$q_1 = \int_{-\infty}^{\infty} i_b d\tau,$$

is obtained by the convolution of the δ-function wake function with the bunch distribution. We omit the transverse position dependence and write the integral over the bunch coordinate τ'

$$W_z(\tau) = \frac{\int_{-\infty}^{\infty} i_b(\tau')w_z(\tau - \tau')d\tau'}{q_1}. \qquad (1)$$

Note that the wake function is zero for time $\tau' > \tau$, by causality—the distant tails of the exciting bunch cannot influence a test particle closer to the center of the bunch. For a unit test charge $q_1 = 1$, the wake function is known as the *wake potential* $V(\tau)$.

The loss factor for a bunch may now be defined as

$$K = \frac{\int_{-\infty}^{\infty} W_z(\tau)i_b(\tau)d\tau}{q_1}. \qquad (2)$$

Longitudinal Beam Impedance

The frequency-domain or impedance representation may be related to the time-domain wake function by Fourier transform. For a point charge the *beam impedance* or *coupling impedance* is defined as the Fourier transform of the wake function

$$Z(\omega) = \int_{-\infty}^{\infty} w_z(\tau)e^{-j\omega\tau}d\tau$$

and has units of Ohms.

The Fourier spectrum of the charge distribution, $I(\omega)$, is

$$I(\omega) = \int_{-\infty}^{\infty} i_b(\tau)e^{-j\omega\tau}d\tau.$$

By transforming Eqs. (1) and (2), we find the wake function in terms of the impedance and the frequency-domain representation of the loss factor

$$W_z(\tau) = \frac{\int_{-\infty}^{\infty} Z(\omega)I(\omega)e^{j\omega\tau}d\omega}{2\pi q_1},$$

$$K = \frac{\int_0^{\infty} Z_{real}(\omega)|I(\omega)|^2 d\omega}{\pi q_1^2},$$

where we have taken the real part of the impedance in calculating the loss factor.
For the case of a Gaussian charge distribution

$$I(\omega) = q_1 e^{-f\frac{(\omega\sigma_t)^2}{2}},$$

the loss factor for a bunch is given by

$$K = \frac{\int_0^{\infty} Z_{real}(\omega)e^{-(\omega\sigma_t)^2}d\omega}{\pi},$$

and it is apparent that the loss depends on the bunch length.
We may also write the wake potential as

$$V_z(\omega)=I(\omega)Z(\omega),$$

and we see that a convolution integral in time domain has become a simple product in frequency domain.

Longitudinal Wake of a Resonant Cavity

The wake function of a cavity resonant at frequency ω_r may be found by modeling the cavity as a parallel resistance-inductance-capacitance (RLC) circuit and calculating the response of the circuit to a δ-function pulse. The impedance of the RLC circuit is given by

$$\frac{1}{Z_{\parallel}} = \frac{1}{R_s} + \frac{1}{j\omega L} + j\omega C,$$

which gives

$$Z_{\parallel} = \frac{R_s}{1 - jQ\left(\dfrac{\omega_r}{\omega} - \dfrac{\omega}{\omega_r}\right)} \qquad (3)$$

where

$$\omega_r = \frac{1}{\sqrt{CL}}, \qquad Q = R_s\sqrt{\frac{C}{L}}.$$

Charge q induces a voltage given by

$$\frac{q}{C} = q\frac{\omega_r R_s}{Q}.$$

The energy stored in the capacitor can be related to the loss factor by

$$\Delta U = \frac{q^2}{2C} = \frac{\omega_r R_s}{2Q}q^2 = kq^2,$$

where we have introduced the loss factor for a point charge, or the *parasitic mode loss factor* k for a given cavity mode

$$k = \frac{\omega_r R_s}{2Q}.$$

The wake function is given by the inverse Fourier transform of Eq. (3)

$$w(t) = 2q\alpha R_s e^{\alpha t}\left(\cos\bar{\omega}t + \frac{\alpha}{\bar{\omega}}\sin\bar{\omega}t\right),$$

where

$$\alpha = \frac{\omega_r}{2Q}, \quad \bar{\omega} = \sqrt{\omega_r^2 - \alpha^2}$$

and for high Q-value

$$w(t) \approx q \frac{\omega_r R_s}{Q} e^{\omega_r t \left(j - \frac{1}{2Q} \right)} .$$

Transverse Wake Fields

Consider the leading charge q_1 transversely displaced from the axis. The charge can excite electromagnetic fields which can be expanded in multipole components (dipole, quadrupole, etc.); for small displacements the dipole component is usually dominant. The test charge q receives a momentum change (a kick) from the fields

$$\Delta \mathbf{p}_{21} = q \int_{-\infty}^{\infty} (\mathbf{E} + \mathbf{v} \times \mathbf{B})_{\perp} dz; \quad t = \frac{z_1}{v} + \tau,$$

which, in general, depends on the positions of both the exciting and trailing charges and is not in the direction of the displacement of the leading charge. The *transverse wake function* $w_{\perp}(\tau)$, defined as the kick per unit of both charges, is given by

$$w_{\perp}(\tau) = \frac{\Delta p_{21}(\tau)}{q_1 q}$$

and has units of volt per Coulomb. Here, the symbol $\perp$ represents either the x or y direction. Analogous to the longitudinal case, we define the *transverse loss factor* $k_{\perp}$ as the transverse kick given to the charge by its own wake per unit charge squared,

$$k_{\perp} = \frac{\Delta p_{11}}{q_1^2},$$

and again has units of volt per Coulomb.

Usually the dipole component dominates, and this term is proportional to the transverse displacement of the exciting charge q_1. The *transverse dipole wake function* $w'(\tau)$ is defined as the transverse wake function per unit transverse displacement

$$w'_\perp (\tau) = \frac{w_\perp(\tau)}{r_1}$$

and has units of volt per Coulomb per meter. Similarly, the *transverse dipole loss factor* is given by

$$k'_\perp = \frac{\Delta p_{11}}{q_1^2 r_1}.$$

Transverse Beam Impedance

The *transverse beam impedance* $Z_\perp$ may be found from the Fourier transform of the transverse wake function

$$Z_\perp(\omega) = j \int_{-\infty}^{\infty} w_\perp (\tau) e^{-j\omega\tau} d\tau;$$

the units are Ohms.

Since the transverse dynamics is dominated by the dipole wake, we define the dipole transverse impedance as

$$Z'_\perp (\omega) = \frac{Z_\perp(\omega)}{r_1},$$

and the units are Ohms per meter. If the impedance is known, the transverse wake may be calculated from the inverse Fourier transform

$$w_\perp (\tau) = \frac{j}{2\pi} \int_{-\infty}^{\infty} Z_\perp(\omega) e^{j\omega\tau} d\tau.$$

Relationship Between Transverse and Longitudinal Wake Fields and Impedance

If we differentiate the momentum kick experienced by a charge q with respect to time, we obtain

$$\frac{\partial \Delta \mathbf{p}}{\partial t} = q \int_{t_a}^{t_b} \left(\frac{\partial \mathbf{E}}{\partial t} dt + \frac{\partial (\mathbf{v} \times \mathbf{B})}{\partial t} dt \right),$$

$$\frac{\partial \Delta \mathbf{p}}{\partial t} = q \int_{t_a}^{t_b} \left(\frac{\partial \mathbf{E}}{\partial t} dt + \mathbf{v} \times \frac{\partial \mathbf{B}}{\partial t} dt + \mathbf{B} \times \frac{\partial \mathbf{v}}{\partial t} dt \right).$$

For relativistic particles of constant velocity

$$d\mathbf{z} = \mathbf{v} \ dt,$$

we have

$$\frac{\partial \Delta \mathbf{p}}{\partial t} = q \cdot \int_{t_a}^{t_b} \frac{\partial \mathbf{E}}{\partial t} dt + \int_a^b d\mathbf{z} \times \frac{\partial \mathbf{B}}{\partial t} dt.$$

Using Maxwell's equation

$$\frac{\partial \mathbf{B}}{\partial t} = -\nabla \times \mathbf{E}$$

and the identity

$$d\mathbf{z} \times \nabla \times \mathbf{E} = \nabla(d\mathbf{z} \cdot \mathbf{E}) - (d\mathbf{z} \cdot \nabla)\mathbf{E} = \nabla(d\mathbf{z} \cdot \mathbf{E}) - \frac{\partial \mathbf{E}}{\partial z} dz,$$

we then find

$$\frac{\partial \Delta \mathbf{p}}{\partial t} = q \int_{t_a}^{t_b} \frac{\partial \mathbf{E}}{\partial t} dt - \int_a^b \left(\nabla(d\mathbf{z} \cdot \mathbf{E}) - \frac{\partial \mathbf{E}}{\partial z} dz \right),$$

$$\frac{\partial \Delta \mathbf{p}}{\partial t} = q \int_a^b (\nabla(d\mathbf{z} \cdot \mathbf{E}) + 2d\mathbf{E}).$$

The transverse components are

$$\frac{\partial}{\partial t}(\Delta \mathbf{p}_\perp) = q\int_a^b [-\nabla_\perp (dz \cdot \mathbf{E}) + 2d\mathbf{E}_\perp],$$

and noting that

$$\Delta E = q\int_a^b dz \cdot \mathbf{E},$$

we find

$$\frac{\partial}{\partial t}(\Delta \mathbf{p}_\perp) = -\nabla_\perp (\Delta E) + 2q[\mathbf{E}_\perp(b) - \mathbf{E}_\perp(a)].$$

We choose to extend the bracketed term over the region where the entry and exit fields $\mathbf{E}_\perp(a)$ and $\mathbf{E}_\perp(b)$ are zero. Then for fields with sinusoidal time variation we have the *Panofsky-Wenzel* theorem

$$\frac{j\omega\Delta \mathbf{p}_\perp}{q} = -\frac{1}{q}\nabla_\perp (\Delta E) = -\nabla_\perp V.$$

This theorem tells us that the transverse kick can be described purely in terms of the longitudinal electric field. There must be a longitudinal electric field component in order to produce a transverse momentum change in a particle traveling through a structure. The frequency dependence shows that the higher the frequency at which the deflecting fields are encountered, the less of a kick they impart.

IMPEDANCE MEASUREMENTS

Perturbation Measurements

For narrow-band impedances, e.g., cavity resonances, perturbation measurements are an accurate method for mapping fields and determining the beam impedance. The method involves the introduction of a small perturbing object into the fields of the resonator (6).

Upon introducing the object, the change in resonant frequency is proportional to the relative change in electric and magnetic stored energy

$$\frac{\Delta\omega}{\omega} = \frac{\Delta U_E - \Delta U_M}{U}.$$

For a small object (i.e., one for which the unperturbed field is roughly constant over the volume of the perturbing object), the perturbed energy is generally expressible as the product of a) the stored energy of the unperturbed field integrated over the volume of the perturbing object and b) a form factor which depends on the shape, orientation, and electromagnetic properties of the perturbing object. For a small sphere of radius r we find

$$\frac{\Delta\omega}{\omega} = \frac{\Delta U}{U} = -\frac{\pi r^3}{U}\left[\varepsilon_0\frac{\varepsilon_r - 1}{\varepsilon_r + 2}E_0^2 + \mu_0\frac{\mu_r - 1}{\mu_r + 2}H_0^2\right].$$

For a dielectric bead with $\mu_r = 1$,

$$\frac{\Delta\omega}{\omega} = -\frac{\pi r^3}{U}\left[\varepsilon_0\frac{\varepsilon_r - 1}{\varepsilon_r + 2}E_0^2\right].$$

For a metal bead with $\mu_r \to 0$ and $\varepsilon_r \to \infty$,

$$\frac{\Delta\omega}{\omega} = -\frac{\pi r^3}{U}\left[\varepsilon_0 E_0^2 - \frac{\mu_0}{2}H_0^2\right].$$

For the case of a monopole mode, with zero magnetic field on axis, the electric field is

$$|E| = \sqrt{\frac{\Delta\omega}{\omega_r}\frac{U}{\pi r^3\varepsilon_0}}.$$

A metallic bead gives a large frequency shift, but the perturbation is sensitive to both E and H fields. A dielectric bead will only perturb the E field. Shaped perturbing objects, such as needles, can enhance the perturbation and offer directional sensitivity. For an ellipsoid, the enhanced perturbation can be calculated analytically; other objects can be calibrated in a known field.

The absolute fields may be calculated if the power and Q value are known; however, the geometrical factor R/Q can be found from the longitudinal field distribution. By mapping the longitudinal distribution of $E_\parallel$ and integrating, the R/Q of the cavity mode can be measured as

$$\frac{R}{Q} = \frac{V^2}{2\omega_r U} = \frac{\left(\int E_\parallel dz\right)^2}{2\omega_r U} = \frac{\left(\int \sqrt{\frac{\Delta\omega}{\omega_r}\frac{U}{\pi r^3 \varepsilon_0}}\,dz\right)^2}{2\omega_r U} = \frac{\left(\int \sqrt{\Delta\omega}\,dz\right)^2}{2\pi\omega_r^2 r^3 \varepsilon_0}.$$

The *transit-time factor* T is defined as the ratio of energy received by a charge passing through the time-varying fields to that which would be received if the field everywhere along the path were at its time-maximum value:

$$T = \frac{\left|\int E_\parallel e^{j\frac{\omega}{c}z}\,dz\right|}{\int E_\parallel dz}.$$

Consider a charge traveling at speed $v = z/t$ through a field extending a length $\pm g$, where the field is

$$E_\parallel = \frac{V_0}{g}\cos\omega t.$$

Then the energy gain is given by

$$\Delta E = \int_{-g/2}^{g/2} qE_\parallel dz = \int_{-g/2}^{g/2} \frac{eV_0}{g}\cos\frac{\omega z}{v}dz,$$

$$\Delta E = 2qV_0\frac{v}{\omega g}\sin\frac{\omega g}{2v},$$

$$\Delta E = qV_0\frac{\sin\theta}{\theta}.$$

The transit-time factor is given by

$$T = \frac{\sin\theta}{\theta},$$

and the transit angle θ is

$$\theta = \frac{\omega g}{2v}.$$

The transit-time-corrected shunt impedance RT^2 of an even mode (symmetric in z about the center of the cavity) is obtained from

$$\frac{RT^2}{Q} = \frac{\left(\int \sqrt{\Delta\omega}\left(\cos\frac{\omega_r z}{c}\right)dz\right)^2}{2\pi\omega_r^2 r^3 \varepsilon_o}.$$

Transverse Dipole Mode Impedance

To determine the transverse impedance of a dipole mode we measure the longitudinal impedance of the dipole mode, at a radial offset r, and use the Panofsky-Wenzel theorem to calculate the transverse impedance.

The transverse energy change is related to the transverse voltage by

$$\Delta E_\perp = \beta c \Delta p_\perp = e V_\perp ,$$

and the dipole transverse impedance is

$$R_\perp = j\frac{V_\perp}{Ir}.$$

Then, for ultra-relativistic particles

$$R_\perp = j\frac{c\Delta p_\perp}{eIr} = -\nabla_\perp (V_\parallel)\frac{c}{\omega Ir}.$$

For small radial displacements in dipole modes, $E_\parallel$ is proportional to the radial offset r, and we may write

$$\nabla_\perp (V_\parallel) = \frac{V_\parallel (r)}{r},$$

which gives

$$R_\perp = \frac{V_\parallel(r)c}{\omega I r^2} = \frac{R_\parallel(r)c}{\omega r^2} = \frac{R_\parallel(r)}{k r^2}.$$

Coaxial Wire Measurements

The coaxial wire impedance measurement uses a conducting rod placed along the beam axis in the vacuum chamber, forming the center conductor in a coaxial line system (7,8). Tapers at either end of this section allow for smooth impedance transformation from the 50-Ω lines used in common microwave measurement equipment to the characteristic impedance of the vacuum chamber and center conductor, on the order of hundreds of Ohms.

A smooth vessel of the same entrance/exit cross section and length as those of the device under test is used in a reference measurement. Resonances within the apparatus are difficult to avoid completely and require careful placing of absorptive material, manufacture of test and reference chambers, and assembly of apparatus.

Frequency Domain

Figure 2 shows a schematic of the measurement apparatus and the currents in the apparatus. I_0 is the incident current applied upstream of the impedance to be measured, and I_r is the upstream current induced by the voltage V generated across the impedance Z. The coaxial wire forms a line of characteristic impedance R with the vacuum chamber. A voltage V is generated at the impedance, inducing currents V/2R traveling equally upstream and downstream.

For a localized impedance (small in extent compared with the wavelength of the applied current), the current that excites the voltage V in the impedance Z is

$$I_e = I_0 - I_r.$$

The perturbation in wire current is

$$\Delta I = I_0 - I_e = \frac{V}{2R} = \frac{I_e Z}{2R}$$

and

$$Z = \frac{2R(I_0 - I_e)}{I_e} = 2R\left(\frac{I_0}{I_e} - 1\right).$$

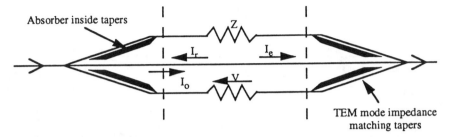

FIGURE 2. Coaxial wire impedance measurement apparatus. Absorptive material is placed in the impedance-matching transformers at the ends of the device under test to avoid resonances of traveling waveguide modes within the apparatus. A reference vessel replaces the device under test between the vertical dashed lines, to normalize to the losses in the tapers, cables etc.

S_{21} measurements without the impedance Z (reference measurement) and with the impedance Z (object measurement) give

$$Z = 2R\left(\frac{S_{21}^{\text{reference}}}{S_{21}^{\text{object}}} - 1\right).$$

Time Domain

The time-domain measurement gives the wake potential and the loss factor for a bunch simulated by a current pulse from a pulse generator. Figure 3 shows a schematic of the measurement apparatus. The time-domain measurements are generally made sequentially—through the reference line with characteristic impedance R to determine the "unperturbed" current pulse i_o and through the device under test to measure the perturbed pulse i_m.

The energy in the unperturbed current pulse at the output of the reference line is

$$U_o = \int R i_o^2(t)dt.$$

For the perturbed pulse, at the output of the device under test line we have

$$i_m(t) = i_o(t) + \Delta i(t),$$

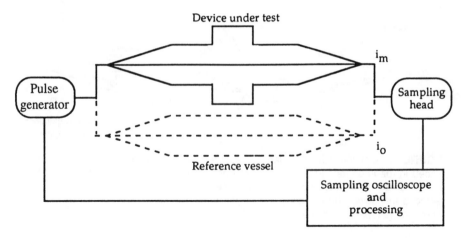

FIGURE 3. Coaxial wire impedance measurement apparatus, time domain.

and the energy in the pulse at the output of the device under test line is

$$U_m \approx \int R i_o(t)[i_o(t) + 2\Delta i(t)]dt.$$

The energy difference $U_m - U_o$ can be interpreted as the energy lost by the pulse traveling through the device under test. Then the loss from a bunch is given by

$$\Delta U = kq^2 = k\left(\int i_o dt\right)^2 = 2R\int i_o(i_m - i_o)dt,$$

the loss factor k is

$$k = 2R\frac{\int i_o(i_m - i_o)dt}{\left(\int i_o dt\right)^2},$$

and the wake potential is given by

$$w(t) = 2R(i_o - i_m).$$

Transverse Impedance Measurements

Transverse impedance measurements are made with two off-axis wires driven differentially with a hybrid or by using a single wire and a ground-plane. The technique is basically the same as the longitudinal measurements described above. The transverse impedance is given by

$$Z_\perp = -j\frac{\Delta V_\perp}{I\Delta x} = \frac{1}{I\Delta x}\frac{c}{\omega}\frac{\partial V_\parallel}{\partial x} = \frac{1}{k\Delta x}\frac{\partial}{\partial x}(Z_\parallel),$$

$$Z_\perp = \frac{2R_w}{k(\Delta x)^2}\left(\frac{S_{21}^{reference}}{S_{21}^{object}} - 1\right),$$

where Δx is one half of the separation of the effective electrical centers (i.e., the location of the "point" wire which would give a cylindrical equipotential at the surface of the actual finite wire), and R_w is the impedance of the twin-wire line.

EFFECTS OF WAKES OR IMPEDANCE

The fields of a passing bunch, or train of bunches, induce image charges on the walls which terminate the field lines of the free charges in the beam. The associated image currents can cause heating on the surface of vacuum chamber components, which may present problems for certain structures. In particular, if the beam current spectrum can excite resonances in vacuum chamber components, large wall currents can be induced and significant damage to the accelerator may result.

The type of impedance creating the wake fields may be useful in identifying potential problems. A broad-band, or low-Q, impedance will decay rapidly, and the heating will be well characterized by the single-pass calculation. Narrow-band, or high-Q, impedance will have a memory which may enhance the heating if the resonant frequency is close to a beam harmonic.

Resistive heating due to image currents in the resistive walls of the vacuum chamber may be calculated from the Fourier series of the beam current and the wall resistance, taking into account the skin-depth penetration of the fields. The Fourier coefficients of the beam current for bunches of length σ_t spaced by time T_b, $\omega_b = 2\pi/T_b$ are

$$C_n = I_o e^{-\frac{(n\omega_b\sigma_t)^2}{2}}.$$

For a circular pipe of length L, radius r, and conductivity σ_{dc}, the resistance at the frequency corresponding to the nth harmonic of the bunch frequency is given by

$$R = \frac{L}{2\pi r}\sqrt{\frac{n\omega_b\mu_o}{2\sigma_{dc}}}.$$

Then the total power is

$$P = \sum_{n=-\infty}^{\infty} C_n^2 R_n = \frac{L}{2\pi r}\sqrt{\frac{\mu_o\omega_b}{2\sigma_{dc}}}\, I_o^2 \sum_{n=1}^{\infty} e^{-(n\omega_b\sigma_t)^2\sqrt{n}}.$$

The summation is approximated by an integral evaluated using the definition of the gamma function; the final result is

$$P = \frac{L}{8\pi^2 r}I_o^2\Gamma\!\left(\frac{3}{4}\right)\sqrt{\frac{\mu_o}{2\sigma_{dc}}}\,\frac{T_b}{\sigma_t^{(3/2)}}.$$

Single-pass power loss due to short-range wake fields (or broad-band, low-Q impedance) that decay before the arrival of the next bunch can be calculated from the loss factor

$$P = k\frac{q_b^2}{T_b} = kT_b I_o^2,$$

where q_b is the charge in each of N_b equally spaced bunches. This calculation assumes that the wake field has decayed in the time interval between bunches.

For wakes that persist until the arrival of the next bunch, the situation may be quite different. Depending on the phase of the wake field at the time of passage of the following bunch, there may be energy imparted to that bunch or extracted from the bunch. The multibunch losses may be more or less than the losses calculated from the loss factor for single-pass effects.

For the narrow-band case, an analytic expression may be derived in time domain (9), or the power lost from the beam can be calculated from the product of the beam current squared and the impedance, summing over all beam spectral lines. For a uniformly bunched beam, the current spectrum is given by

$$I_b(\omega) = I_o + 2I_o e^{-\frac{(\omega\sigma_t)^2}{2}} \sum_{n=0}^{\infty} \delta(\omega - n\omega_b).$$

Broadband Q=1 Impedance Model

When considering single-bunch collective effects we are concerned with the wake fields over the bunch length. If we Fourier analyze these short-range wake fields, we find that the effective impedance sampled by a single bunch does not show the detail that may exist in the actual impedance of a structure; the short-range wake "smoothes out" the impedance. The actual impedance of the many and varied components of an accelerator is often replaced with an effective impedance described by a low-Q resonator of the form of Eq. (3). The resonant frequency ω_r is generally taken to be the TM-mode cut-off of the vacuum chamber, and the Q-value is taken to be unity. The shunt impedance may be estimated by making the energy loss to the low-Q resonator equal to the total loss of the individual component resonances in the accelerator:

$$k_{Q=1} = \sum k_{components}.$$

For resonant modes, and Gaussian bunches

$$k_{Q=1} = \sum_{n \text{ modes}} \frac{\omega_n R}{2Q} e^{-(\omega_n \sigma_t)^2}.$$

In the case of components with mostly inductive impedance, such as shallow cavities, the situation may be better modeled by equating the low-frequency inductance of the resonator to the calculated inductance of the components. The imaginary part of the resonant impedance at low-frequencies is given by

$$Z_i = j \frac{\omega R}{\omega_r Q}.$$

A convenient measure of broad-band impedance is $|Z/n|$, the magnitude of the resonant impedance divided by the normalized frequency n (where $n = \omega/\omega_0$, the frequency divided by the revolution frequency of the ring). $|Z/n|$ is approximately constant below the resonant frequency, and this makes the model attractive when calculating single-bunch collective effects in some cases. Approximate expressions dependent upon $|Z/n|$ are often used to estimate single-bunch effects.

The Q = 1 resonator has limitations, particularly for very short bunches. In this case the beam power spectrum extends beyond the beam-pipe cut-off frequency, and the bunch resolves more of the detail of the beam impedance.

ACKNOWLEDGEMENTS

The author wishes to acknowledge the assistance of D. Goldberg in reading the manuscript and suggesting improvements.

REFERENCES

1. Palumbo, L., Vaccaro, V. G., and Zobov, M., "Wake Fields and Impedance," in *Proceedings of the Fifth Advanced Accelerator Physics Course*, CERN 95-06, pp. 331-390, 1995.
2. Weiland, T., and Wanzenberg, R., "Wake Fields and Impedances," unpublished DESY Note DESY 91-06, 1991.
3. Karantzoulis, E., "An Overview on Impedances and Impedance Measuring Methods for Accelerators," unpublished Sincrotrone Trieste Note ST/M-91/1, 1991.
4. Palumbo, L., and Vaccaro, V., "Wake Fields Measurements," unpublished INFN note LNF-89/035(P), 1989.
5. Edwards, D. A., and Syphers, M. J., *An Introduction to the Physics of High Energy Accelerators*, (New York, J. Wiley, 1993), Chap. 6, pp. 187-189.
6. Maier, L. C. Jr., and Slater, J. C., "Field Strength Measurements in Resonant Cavities," *J. Appl. Phys.* **23**, 68-77 (1952).
7. Caspers, F., "Beam Impedance Measurements using the Coaxial Wire Method," in *Proceedings of the Workshop on Impedance and Current Limitations,* unpublished ESRF Report, October 1988.
8. Sands, M., and Rees, J., "A Bench Measurement of the Energy Loss of a Stored Beam to a Cavity," unpublished SLAC Note PEP-95, 1974.
9. Sands, M., "Energy Loss to Parasitic Modes of the Accelerating Cavities," unpublished SLAC Note PEP-90, 1974.

INVITED PAPERS

Absolute Beam Energy Measurements in e⁺e⁻ Storage Rings

M. Placidi

CERN, Geneva, Switzerland

Abstract. The CERN Large Electron Positron collider (LEP) was dedicated to the measurement of the mass M_Z and the width Γ_Z of the Z^0 resonance during the LEP1 phase which terminated in September 1995. The Storage Ring operated in Energy Scan mode during the 1993 and 1995 physics runs by choosing the beam energy E_{beam} to correspond to a center-of-mass (CM) energy at the interaction points (IPs) $E_{CM}^{peak} \pm 1762$ MeV. After a short review of the techniques usually adopted to set and control the beam energy, this paper describes in more detail two methods adopted at LEP for precise beam energy determination that are essential to reduce the contribution to the systematic error on M_Z and Γ_Z. The positron beam momentum was initially determined at the 20-GeV injection energy by measuring the speed of a less relativistic proton beam circulating on the same orbit, taking advantage of the unique opportunity to inject two beams into the LEP at short time intervals. The positron energy at the Z^0 peak was in this case derived by extrapolation. Once transverse polarization became reproducible, the Resonant Depolarization (RD) technique was implemented at the Z^0 operating energies, providing a $\leq 2 \times 10^{-5}$ instantaneous accuracy. RD Beam Energy Calibration has been adopted during the LEP Energy Scan campaigns as well as in Accelerator Physics runs for accurate measurement of machine parameters.

INTRODUCTION

This paper is intended to give an overview of the several techniques which can provide information on the beam energy in storage rings and to account for the associated precisions.

Accurate knowledge of the beam energy is of relevant interest in storage rings dedicated to the measurement of the mass and the width of resonances. Reducing the uncertainty on the energy scale improves the quality of the determination of the resonance parameters since the energy information level sets the standard for the systematic contribution to the global error. Besides this specific issue on the energy scale calibration, the possibility of setting, monitoring, and controlling the nominal energy of the accelerator with a high degree of reliability greatly improves the performance of the operation.

After a brief review of standard methods, which provide an energy information essentially derived from the measurement of the magnetic field in reference magnets or in the ring dipoles, the paper describes in detail more sophisticated techniques based on direct measurements of specific properties of beams, such as particle **velocity** and **polarization**.

The first method is based on the determination of the velocity of protons on a specific and reproducible orbit through measurement of their revolution frequency. This provides the momentum per unit charge associated with that particular orbit which is the same for protons and leptons. A second method, (RD), by far the most accurate, measures the spin precession frequency of a vertically polarized beam, directly related to the average beam energy, by means of a controlled depolarizing resonance. Other techniques involving special detectors or dedicated experiments are also described, to complete the review.

Experimental results from the 1995 LEP beam energy calibration campaign with RD are shown and effects responsible for changes in the CM energy at the IPs during the physics runs are described.

STANDARD TECHNIQUES FOR ENERGY MONITORING

The Integrating Coil

The usual method adopted to set the energy of a storage ring to some desired level consists of measuring the *integrated magnetic field* $\int B ds$ in a reference magnet powered in series with the main ring dipoles. The information from a digital integrator connected to a long rotating coil properly positioned in the gap provides a continuous measurement of $\int B ds$ and is used as a reference for the current-regulated control of the main power converters. This is particularly important when the physics runs take place at energies different from the injection energy and a magnet-cycling procedure is applied to compensate for hysteresis effects. The $\int B ds$ information is also part of the log-in data set provided to the experiments at every fill.

The reproducibility and the resolution of this method, referred to as *field display* (FD) (1), are in the range of $\pm$ (20-30) ppm.

For the field display to reproduce the situation represented by the dipoles in the accelerator tunnel, the reference magnet should obey two basic rules: **be structurally identical to the ring magnets and undergo the same environmental history.*** Temperature changes affect the nominal value of $\int B ds$ as they modify both the gap height and the length of the dipole cores: the FD technique provides information on the global effect.

---

*   This includes primarily *temperature* changes although in the case of the LEP dipoles, where the yokes are made of a mixture of steel laminations and concrete, it is important to monitor and control other parameters such as the *local humidity*.

The presence of a long coil and associated instrumentation in the magnet gap together with the need for maintenance access discourage installation of the reference magnet in the tunnel itself; consequently, the FD information requires that other calibration techniques be used as an absolute energy monitor.

The Nuclear Magnetic Resonance Probes

Nuclear magnetic resonance probes provide a very precise measurement of the *local magnetic field* down to an $\sim 10^{-6}$ accuracy. Their compact size allows allows them to be installed directly in the gap of the magnets with reduced interference with the vacuum chamber. These probes provide on-line monitoring of the time evolution of the dipolar field. NMR probes do not provide information on the variations of the magnet length with temperature; so their use in a reference magnet should be associated with adequate temperature measurements in the accelerator tunnel.

The accuracy of this information is limited by the small number of sampled dipoles. The overall homogeneity of both the magnetic properties of the cores and their thermal behavior is required.

The LEP Flux-Loop System

The LEP reference magnet, made from a stack of standard dipole laminations, is installed in a temperature-controlled environment and series-connected with the main dipoles. Measurements of the integrated magnetic field are carried out with a flip coil mounted in the magnet gap along the position of the central orbit.

The LEP dipoles have iron-concrete cores which undergo aging and are sensitive to changes in both temperature and humidity. The field-display information from the reference magnet is calibrated periodically by a direct measurement of the flux variations in an 8-fold loop consisting of electric wires mounted in the lower pole of each of the bending magnets and connected in series throughout each LEP octant (2). When the whole dipole system undergoes a magnetic cycle of given excursion, the induced voltage from the flux variation in the loops is measured by eight digital integrators in the even underground areas of the machine and provides direct information on the $\int B ds$ along the accelerator from the dipoles in the actual environmental conditions. Polarity reversal permits a measurement of the remanent field which is of particular importance for the LEP low-field dipoles.

The attainable accuracy of the beam momentum is on the order of 5×10^{-4} since the method does not account for additional dipolar bending for orbits off-axis in the quadrupoles.

THE NOMINAL ENERGY

For a given magnetic structure the **nominal energy** E_0 is defined for a beam circulating on the **central orbit** C_0 going on average through the center of the quadrupoles so that the bending strength experienced by the beam over a machine revolution comes only from the dipoles:

$$\frac{\beta E_0}{ec} = \frac{p_0}{e} = \frac{1}{2\pi} \oint_{C_0} B(s) \, ds \; . \tag{1}$$

The integration is extended over the central orbit C_0 defined by the central revolution frequency f_{rev}^0:

$$f_{rev}^0 = \frac{\beta c}{C_0} = \frac{c}{h\lambda_{RF}^0} \; , \tag{2}$$

where λ_{RF}^0 is the wavelength of the **nominal rf frequency** f_{RF}^0 and h is the harmonic number.

Measuring the Central Orbit

The nominal momentum of a beam off the central orbit in a FODO structure is modified by the additional integrated dipolar field in the quadrupoles according to Eq. (1):

$$\Delta p = \frac{p_0 \Delta C}{\alpha_c C_0} = \frac{e}{2\pi} \oint_C G(s)x(s) \, ds \; , \tag{3}$$

where α_c is the momentum compaction for the used optics, $G(s)$ is the field gradient, and $C = C_0 + \Delta C$ is the length of the actual reference orbit. This effect modifies the betatron tunes (chromaticity) and generates closed orbit distortions induced by angular kicks from the (de)focusing strengths, which are minimum when the beam is centered.

In principle a direct method to define the central orbit would consist of looking for minimum orbit distortion (orbit differences) as a function of the rf frequency for different settings in the strength of the arc quadrupoles, but the attainable accuracy is essentially limited by the associated tune shift.

The method commonly adopted in a regular FODO-lattice magnetic structure exploits the fact that in each magnetic cell the sextupoles are installed on the same girder as the quadrupoles and precisely aligned with respect to them. In this assumption a beam off-axis in the quadrupoles receives additional focusing from the same misalignment it has in the center of the sextupoles. This makes the betatron tunes $Q_{x,y}$ depend on the sextupole excitation until the orbit is *on axis* in the quadrupole-sextupole complex.

One would then measure the dependence of the betatron tunes over the radial position of the orbit in the arcs by changing the rf frequency for different settings of the sextupole families, i.e., for different chromaticities $Q' = \Delta Q/\Delta p/p$. For energy changes that are small compared to the acceptance of the machine, the chromaticity varies linearly with the sextupole excitation and the lines $Q_{x,y}(f_{RF})$ will cross at one point defining the central rf frequency f_{RF}^0. If the particle velocity is known, as in the case of ultra-relativistic leptons, the method provides a measurement of the central orbit length C_0 (3).

Figure 1 shows an example of the use of positrons ($1 - n\beta_{e+} \sim 3 \times 10^{-10}$) to determine the length of the LEP **actual circumference**[†] from

$$C_0 = \frac{c\beta_{e+}h_{e+}}{f_{RF,\,e+}^0} \, . \tag{4}$$

The harmonic number $h_{e+} = 31324$ gives $C_0^{LEP} = 26.658873$ km. The error in the circumference measurement is of the order of 0.3 mm corresponding to a relative error of about 1×10^{-8}. The accuracy on the frequency measurement is more than one order of magnitude better.

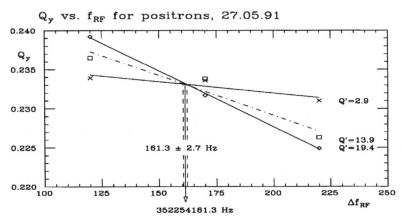

FIGURE 1. *Positron* chromaticity measurements to determine the central revolution frequency and the length of the *central orbit*, ref. (3).

[†]   As stated in (4), the concept of *terra firma* has to be reconsidered when dealing with alignment stability in accelerators: the length of the central orbit is time-dependent due to seasonal and periodic ground motion as discussed in the last chapter.

THE CENTRAL MOMENTUM

Determinting the momentum on the central orbit requires the use of non-ultra-relativistic particles. Protons have been proposed (5) as their velocity $c\beta_p$ is measurably different from the speed of light.

After measuring the central orbit with positrons using the method discussed above, protons are injected keeping the **same magnetic settings** and trapped by the rf system on a different harmonic number h_p associated with the new velocity, as determined from the knowledge of the nominal energy inferred from magnetic measurements. Measurement of the chromaticity Q'_p for different sextupole settings defines the central rf frequency $f^0_{RF, p}$ for protons and hence their velocity:

$$\beta_p = \frac{C_0 f^0_{rev, p}}{c} = \beta_{e^+} \frac{h_{e^+}}{h_p} \frac{f^0_{RF, p}}{f^0_{RF, e^+}} . \tag{5}$$

The momentum on the **same central orbit** is common to positrons and protons

$$P_{e^+} \equiv P_p = c\beta_p m_p \gamma_p = \frac{\beta_{e^+} E_{e^+}}{c} \tag{6}$$

within an accuracy

$$\frac{\Delta p}{p} = \gamma_p^2 \left(\frac{\Delta \beta}{\beta}\right)_p . \tag{7}$$

Figure 2 shows the application of this method to the determination of the central momentum for LEP at the nominal injection energy of 20 GeV (3). Due to the accuracy of the measurements the lines associated with the different chromaticities do not cross exactly at one point and an error ellipse for one standard deviation is shown on the figure. As it can be gathered from the error on the determination of the central frequency ($\Delta\beta/\beta \sim 3 \times 10^{-8}$), a relative precision of $\sim 10^{-5}$ is attainable but it deteriorates to $\sim 10^{-4}$ when extrapolated to higher energies.

The dependence on γ^2 in Eq. (7) suggests the possibility of accelerating protons or heavy ions to the final energies (6) to improve the accuracy; however, the technical problems involved are quite considerable.

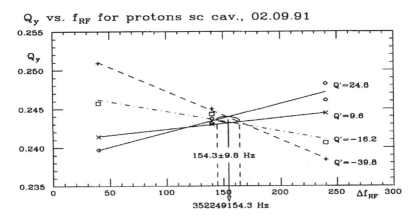

FIGURE 2. *Proton* chromaticity measurements to determine the velocity and the *momentum* on the central orbit, ref. (3).

SYNCHROTRON RADIATION IN THE VISIBLE RANGE

This method, based on the angular and energy dependence of synchrotron radiation distribution in storage rings (7), makes use of a special collimator housing two variable-amplitude slits to intercept portions of the distribution at two different positions in the vertical plane. The central beam energy can be determined when the same photon flux is transmitted by the two slits. The slits must be constructed and controlled to better than 0.1 μm.

The method, mainly applicable to beam transfer lines where special magnets can be inserted, also requires knowledge of the dipolar field at the point of emission to an ~ 5 × 10^{-6} level. While the proposed use of the visible part of the emitted radiation (and hence visible optics) instead of x rays to separate the narrow emission angles is elegant on paper, this is difficult to realize.

ELECTRON SCATTERING

Scattering of the bound electrons of hydrogen atoms in a gas target from *electrons* (Møller e⁻e⁻ scattering (8)) or *positrons* (Bhabha e⁺e⁻ scattering) offers another possibility of measuring the beam energy in a storage ring.

Basic Møller Method

From the two-body kinematic relation for elastic scattering on a target electron at rest, the electron beam energy in units of the electron mass m_e

$$\frac{E_{beam}}{m_e} = \left(\frac{8}{\theta^2 1 - \kappa^2} - 1\right) \approx \left(\frac{8}{\theta^2(1 - \kappa^2)}\right) \tag{8}$$

is determined from the momentum balance $\kappa = \cos\theta^*$ between the energies $E_{1,2}$ of the scattered electrons and the opening angle

$$\theta = \tan\theta_1 + \tan\theta_2 . \tag{9}$$

The scattered electrons are detected in a near-symmetric configuration ($E_{1,2} \cong E_{beam}/2$, $\theta_1 \cong \theta_2$) and the minimum opening angle ($\theta_1 = \theta_2$, $\kappa = 0$)

$$\theta_{min} = \sqrt{\frac{8}{1 + E_{beam}/m_e}} \approx \sqrt{\frac{8m_e}{E_{beam}}} \tag{10}$$

typically ranges from 9.53 to 6.73 mrad for $E_{beam} \in [45\text{--}90]$ GeV. Calorimetric and/or geometric determinations of E_{beam} are possible depending on how the parameter κ in Eq. (8) is measured:

$$\kappa_A = \frac{E_{beam} + m_e}{E_{beam} - m_e} \frac{E_1 - E_2}{E_1 + E_2} , \tag{11}$$

$$\kappa_B = \frac{\tan\theta_1 - \tan\theta_2}{\tan\theta_1 + \tan\theta_2} . \tag{12}$$

A typical detector considered (9) for LEP2 would consist of:
- a hydrogen gas jet target (GJT),
- a position-sensitive silicon micro-strip detector,
- a high-resolution electro-magnetic calorimeter, and
- a small silicon micro-strip detector in vacuum, close to the GJT, detecting recoil protons from elastic e-p scattering to monitor the spatial stability of the gas-$e^\pm$ beam interaction area.

With a luminosity at the target electrons (8) of 4×10^{31} cm^{-2} s^{-1}, the Møller rate is $\sim 2.2 \times 10^6$ events/hr and the statistical accuracy estimated for the LEP2 energies is of the order of 2 MeV in about 30 min. The Bhabha scattering of the cross section is about 1/4 of the Møller one and the associated statistical error is two times larger for the same measuring time.

The scattering method provides on-line information on the beam energy **at the target location**[‡]. The extrapolation to the IPs requires careful monitoring of the horizontal orbit and the rf parameters.

CALIBRATION BY RESONANT DEPOLARIZATION

Transverse Polarization in e⁺e⁻ Storage Rings

Lepton beams circulating in storage rings become vertically polarized via the Sokolov-Ternov radiative polarization process (11): a small spin-flip probability associated with the quantum emission of synchrotron radiation has a large asymmetry in orienting the e⁺e⁻ magnetic moments along the guiding magnetic field in parallel/anti-parallel directions.

In an ideal accelerator the self-polarization mechanism builds up towards a theoretical asymptotic level $\hat{P} = 8/5\sqrt{3} = 92.4\%$ with a rise time

$$\tau_{ST} = c_{ST}\frac{\rho_{eff}^3}{\gamma^5} \quad , \quad c_{ST} = \frac{8}{5\sqrt{3}} \frac{m_e}{\hbar r_e} = 2.832 \times 10^{18} \text{ s m}^{-3} , \quad (13)$$

where $r_e = e^2/m_e c^2$ is the electron classical radius and ρ_{eff} is defined as (12)

$$(\rho_{eff}^3)^{-1} = \frac{1}{C} \oint \frac{ds}{|\rho(s)|^3} . \quad (14)$$

Typical τ_{ST} values are:

$$\tau_{ST} = 320 \text{ min /LEP1 at the } Z^0 \text{ energy} \quad (15)$$
$$= 260 \ (220) \text{ min /PEP II, LER(HER)} .$$

The dynamics of the spin vector $\vec{S}$ is determined by the particle magnetic history along the ring and described by the Thomas-BMT equation (13,14):

[‡]   The beam energy varies along the beam path due to the localized compensation of synchrotron radiation losses in the rf stations (10) (energy sawtooth).

$$\frac{d\vec{S}}{dt} = \vec{\Omega}_s \times \vec{S} \ , \quad \vec{\Omega}_s = -\frac{e}{\gamma_e m_e}[(1 + a_e\gamma_e)\vec{B}_\perp + (1 + a_e)\vec{B}_{||}] \ , \tag{16}$$

where $a_e = (g-2)/2 = 1.159\ 652\ 188 \times 10^{-3}$ is the e^+e^- gyro-magnetic anomaly and $\vec{B}_\perp$, $\vec{B}_{||}$ are the field components transverse and parallel to the trajectory, respectively.

The effect of longitudinal fields on the spin precession frequency $\vec{\Omega}_s$ is γ_e times weaker than that from transverse ones and, in an ideal case where only vertical fields $\vec{B}_y$ are present, the spin vector will precess around the stable solution $\vec{n}_0$ of Eq. (16) parallel to $\vec{B}_y$ with a frequency $(a_e\gamma_e)$ times the cyclotron frequency $\Omega_c = -eB_y/m_e\gamma_e$.

The quantity $\nu = a_e\gamma_e$ measures the number of spin precessions in a machine revolution and, analogous with the terminology adopted for the betatron and synchrotron motion, is called **spin tune**.

In a real machine *non-vertical* magnetic fields on the particle trajectory originate from vertical misalignments in the quadrupoles and in the corrector dipoles used to compensate them, causing unwanted tilts to $\vec{n}_0$ and generating spurious vertical dispersion. Consequently, the orientation of the $\vec{n}_0$ vector becomes sensitive to energy fluctuations and dependent on the position of the particle along the ring due to quantum emission of radiation, thus reducing the degree of polarization (*spin-orbit coupling* (15)).

Refined orbit correction techniques known as *Harmonic Spin Matching* (HSM) have been conceived to compensate the tilt of the $\vec{n}_0$ vector by controlled spin rotations (16) generated by appropriate combinations of orbit bumps. The magnetic alignment of the machine elements and the quality of the orbit measurement are essential ingredients for a successful implementation of the HSM method. Beam-based techniques (17) to determine the relative offsets between quadrupoles and beam position monitors, associated with refined optical survey (18) and reliable beam orbit acquisition (19), proved to be extremely useful in implementing the HSM method at LEP (20) and improving the initial 10% polarization level (21) up to the 57% best result in 1993 (22).

Polarimetry in Storage Rings

Laser polarimeters based on spin-dependent Compton scattering of circularly polarized photons on polarized electrons or positrons (23) have become standard equipment in most e^+e^- storage rings.

The differential Compton scattering cross section is (24,25)

$$\frac{d\sigma}{d\Omega} = \frac{1}{2}\left(r_e\frac{k}{k_0}\right)^2[\Phi_0 + \Phi_1(\vec{\xi}) + \Phi_2(\vec{\xi}, \vec{P})] , \tag{17}$$

where k, k_0 are the scattered and incident photon momenta in the e^+e^- CM system, respectively, $\vec{p}$ is the electron polarization vector, and $\vec{\xi} \equiv (\xi_1, \xi_2, \xi_3)$ is the photon polarization vector in terms of the normalized Stokes parameters ($\sum_{i=1}^{3} \xi_i = 1$).

When illuminating the beam with circularly polarized light [$\vec{\xi} \equiv (0, 0, \pm 1)$], the term Φ_1 vanishes and Eq. (17) shows an asymmetric behavior with respect to the helicity of the incoming photons via the term

$$\Phi_2(\vec{\xi}, \vec{P}) = -\xi_3[P_\parallel F_1(k_0, k, \theta) + P_\perp F_2(k, \theta)\sin\phi] , \tag{18}$$

where θ, ϕ are the photon scattering angles.

Both recoil photons and/or leptons can be detected to measure the degree of polarization. When dealing with *longitudinally* polarized e^+e^- beams the measured effect from a light helicity flipping $\xi_3 = \pm 1$ consists of a *backscattered rate* dependence. With *transverse* beam polarization, the azimuthal asymmetry

$$A(\vec{P}, \vec{\xi}) = \frac{\Phi_2}{\Phi_0} = P_\perp\xi_3 F(k_0, k, \theta)\sin\phi \tag{19}$$

produces an *up-down shift* in the mean $\tilde{y}$ of the vertical distribution of the high energy back-scattered photons directly proportional to the degree of the polarization of both the photon and the e^+e^- beams:

$$\Delta\tilde{y}(P_\perp\xi_3, t) = \left(\kappa_{pol} \xi_3 \hat{P} \frac{\tau_{eff}}{\tau_{ST}}\right)[1 - \exp(-t/\tau_{eff})] . \tag{20}$$

$$(\Delta\tilde{y})_\infty$$

The analyzing power κ_{pol} is calibrated by accurate determination of the effective rise time τ_{eff} and the asymptotic mean-shift $(\Delta\tilde{y})_\infty$ from a fit to the time evolution (Eq. (20)) normalized to the actual light polarization state ξ_3.

The analyzing power of the LEP polarimeter (26,27) is

$$\kappa_{pol}^{LEP} = (4.4 \pm 0.3) \ \mu m/\%$$

and its global accuracy is ~ 1% in 1 minute of data collection with 100% circularly polarized light. Extreme care in producing and controlling the appropriate light polarization states is required to monitor very low e^+e^- polarization levels.

Resonant Depolarization (RD)

The **mean beam energy** of a storage ring can be determined to great accuracy by measuring the spin tune $\nu = a_e\gamma_e$ of a polarized e^+e^- beam

$$\nu = N + \delta\nu = \left(\frac{a_e}{m_ec^2}\right)E_{beam} = \frac{E_{beam}}{0.4406486(1)} \tag{21}$$

where $m_ec^2/a_e = 440.6$ MeV is the energy change for a unitary spin tune change and the integer N is known from magnetic measurements.

Typical spin tune values are

$$\nu = 103.469 \text{ at the } Z^0 \text{ energy} \tag{22}$$
$$= 20.455 \text{ at PEP II / HER } (E_{beam} = 9.014 \text{ GeV})$$
$$= 7.046 \text{ at PEP II / LER } (E_{beam} = 3.105 \text{ GeV}).$$

A frequency-controlled radial rf magnetic field makes the particle spin to precess away from the vertical axis, and a depolarizing resonance occurs when the perturbing field oscillates at the spin precession frequency

$$\omega_{dep}^{res} \equiv \Omega_s = 2\pi\ \delta\nu\ f_{rev} = 2\pi(1 - \delta\nu)\ f_{rev}\ . \tag{23}$$

The depolarizer frequency f_{dep}^{res} at the resonance gives the fractional spin tune, and the beam energy at the revolution frequency f_{rev} (**actual** orbit)

$$E_{beam} = \nu\left(\frac{m_ec^2}{a_e}\right) = \left(N + \frac{f_{dep}^{res}}{f_{rev}}\right) \times 0.4406486 \text{ GeV}\ . \tag{24}$$

The mirror ambiguity (*aliasing*) $f_{dep}^{res} = \delta\nu\ f_{rev} = (1 - \delta\nu)\ f_{rev}$, typical of sampling data at equally spaced intervals (Fig. 3) is solved by determining the change in f_{dep}^{res} when varying the beam energy through a known change in the rf frequency, i.e., *without modifying the magnetic setup of the machine*.

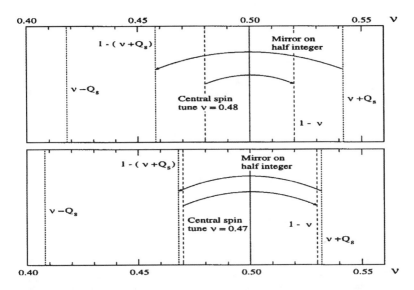

FIGURE 3. Synchrotron satellites and mirror ambiguity (aliasing), ref. (20).

Resonant depolarization also occurs on synchrotron satellites $\delta v \pm Q_s$ of the main resonance. They can be separated from the latter as this is not shifted by a change in the synchrotron tune Q_s.

In practice, the frequency of the depolarizer is slowly varied with time in steps over a given range until the polarization level drops to zero. The frequency step for which depolarization occurs defines the value of the spin tune interval, and the step amplitude δf_{dep} gives the resolution of the spin tune and beam energy measurement. By properly gating the excitation of the perturbing field, individual bunches can be selectively depolarized and the strategies meant to identify the true resonance can be applied while the re-polarization process of the previously depolarized bunches takes place. An example of resonant depolarization is shown in Fig. 4.

Beam energy calibration by resonant depolarization was successfully implemented in LEP (28) and first applied in 1992 to reduce the systematic error on the mass of the Z^0 boson (29)**. In order to implement the RD method under accelerator operation conditions as close as possible to those used for data collection, the spin rotation induced by the experimental solenoids has to be compensated. The adopted scheme (30), first suggested by Steffen (31) and Rossmanith (32), was experimentally verified (33) in 1992. Following these results, the RD method became the LEP standard energy calibration procedure (34) and was routinely adopted at the end of physics fills during the three-point energy scan campaigns in 1993 (35) and 1995.

** From (29): Mz =(91.187 ± 0.007) GeV.

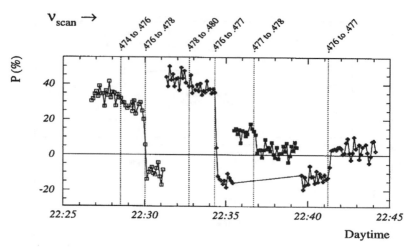

FIGURE 4. Example of energy calibration with resonant depolarization. The upper scale shows the non--integer part of the spin tune corresponding to the sharp drop in the polarization level at the depolarizing resonance. Selective bunch depolarization is indicated by the different symbols. A partial spin–flip was checked to be real by flipping it back at the same spin tune (0.476–0.477), ref. (34).

The 1995 calibration results (36) collected in Fig. 5 show the accuracies attained in measurements of the beam energy and residual effects on the time evolution of the beam energy once the known perturbing mechanisms (see next section) are taken into account.

MECHANISMS CAUSING ENERGY CHANGES

Alignment tolerances in modern and future accelerators have become more and more critical with the introduction of strong focusing magnetic elements to contain beam phase space and with increasing beam currents implying precise positional requirements to reduce interactions of wake fields with the beam environment (37,38). Besides occasional motion from seasonal variations of water content in the soil (39), natural micro-seismic disturbances, and other effects, earth tides are the major example of **periodic ground motion**.

Terrestrial Tides

The equilibrium between the gravitational attraction of the moon and sun on the earth and the centrifugal force between them all· distorts the "spheroidal" sur-

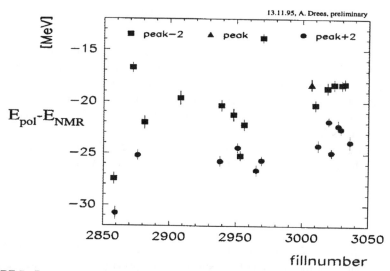

FIGURE 5. Resonant depolarization results from the 1995 LEP run. The systematic shift E_{pol}-E_{NMR} between the beam energy from resonant depolarization and that estimated from NMR information shows a ~ 9 GeV drift for both $E_{e^\pm}^{peak} \pm 2$ GeV beam energies over the 26 calibrated fills in the two-month scan campaign, ref. (36).

faces of constant gravity solutions to Poisson's equation at the earth's surface. The result is a quadrupolar deformation of the earth's crust that produces two daily bulges with asymmetric amplitudes (for an observer far from the equator) due to the inclinations of both the earth's rotational axis ($\varepsilon_E = 23°26'$) and the lunar orbital plane ($\varepsilon_M = 5°08'$) to the ecliptic. As the moon transit time is longer than an earth day, the lunar and sun tide periods differ by 48 min 38 s; the global effect is additive in the full/new moon configurations and subtractive in quadrature. The resulting earth tides appear as a surface deformation with a period of 12 h 22 min, the amplitude of which depends on the observer latitude and on the earth-moon relative position. Maximum tides occur twice a month corresponding to *full* or *new moon* phases.

A wide spectrum of periodicities, including ellipticity and oscillations of the earth and moon orbits, equinox precession from Earth oblateness, and other components, makes the real picture much more complex (40,41).

Two important observables are associated to the above phenomena, i.e., the *strain* and the *local gravity variations*. The strain variation is a six-component, second-order tensor describing the **lateral motion** on the earth's crust according to the local geological structure. Gravity variations can be measured and predicted. Models using a Cartwright-Tayler-Edden (CTE) potential, including up to 505 harmonic components, are used in geophysics and available in Centers for Earth Tides (42).

Effects on Accelerators

Due to the crust strain the horizontal position of the magnetic elements in an accelerator changes periodically with time, proportional to the amplitude of the crust motion in the vertical direction (43). As a consequence, the energy of particles circulating on the orbit defined by the *operational* rf frequency becomes *time-dependent* due to an additional bending strength from periodical off-axis passage in the quadrupoles.

Relative beam energy changes are related to strain-driven differences $\Delta C(t)$ between the lengths of the two orbits by the momentum compaction α_c

$$\frac{\Delta E(t)}{E_0} = -\frac{1}{\alpha_c} \frac{\Delta C(t)}{C_0} = \kappa_{tide} \, \Delta g(t) , \tag{25}$$

where $\Delta g(t)$ is the time-dependent local gravity excursion, and the tide coefficient κ_{tide} accounts for the coupling between strain and gravity changes. The sign convention in Eq. (25) indicates that a positive strain (*expanding ring*) induces a *reduction* of the beam energy on the operational orbit with the usual notations above transition.

The LEP TidExperiment

In a dedicated experiment (44) the evolution of the relative beam energy variation due to terrestrial tides has been measured at the LEP accelerator. A gravity excursion Δg exceeding 140 μGal[††] occurs in the Geneva area during high tides and about 15% couples into transverse crust motion causing a variation of ± 0.5 mm in the LEP circumference.

As shown in Fig. 6, a ± 100 ppm relative energy excursion was measured in agreement with the behavior predicted from Eq. (25) for the momentum compaction factor associated with the actual beam optics and was also measured with the RD technique. A tide coefficient

$$\kappa_{tide} = (-0.86 \pm 0.08) \; ppm/\mu Gal \tag{26}$$

was determined from Eq. (25) in good agreement with predictions (42).

---

†† $1 Gal = 1 \; cm \; s^{-2}$ is the gravity acceleration unit adopted in geophysics (after G. Galilei).

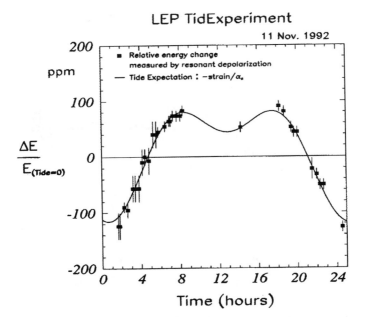

FIGURE 6. Time evolution of the relative LEP beam energy variations driven by terrestrial tides measured over 24 hours. The solid line is calculated using the CTE tide model (41) with the tide coefficient from Eq. (26), ref. (44).

Environmental Perturbations (Earth Currents)

Variations of the LEP beam energy during a physics fill have been observed and traced back to changes in the bending field experienced by the beams as monitored by the two NMR probes available in 1995 in the tunnel (45). Both exhibit drifts, steps, and saturation as shown in the complex time behavior of Fig. 7 where the ordinate axis represents the *equivalent* beam energy as deduced from the measured field.

Short-term perturbations of the order of 15 MeV together with long-term shifts up to 10 MeV are likely to be driven by earth currents circulating along the LEP beam pipe which appear to be strongly correlated with electric current leaking into earth from the nearby railway system. A clear correlation is shown in Fig. 8 between the time behavior of the rail voltage and the voltage on the vacuum pipe, and the resulting NMR-measured main bending field on the pole tips. Short-term effects should not be seen by the beam due to the screening effect from the aluminum vacuum chamber.

Behaviour of LEP NMR

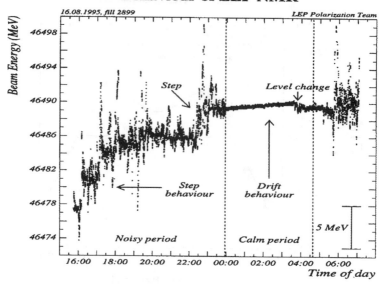

FIGURE 7. Various changes in the LEP main bending field registered during a physics fill as detected by an NMR probe installed in one of the tunnel dipoles, ref. (45).

The present understanding of the effect claims that parasitic currents are not directly responsible for the field changes but rather modulate the field in the magnets through mini-hysteresis loops.

A 1.7×10^{-4} modulation of the main bending field applied in a few steps at the beginning of the fills seems to reduce the overall drift to about 3 MeV. This *conditioning* procedure has been adopted during the 1995 physics run as a partial cure to the effect.

SUMMARY

An overview has been presented of methods adopted or suggested to perform absolute beam energy measurements in storage rings. Various techniques have been illustrated to try to give as complete a picture as possible of both the achievable performance and the technical difficulties associated with them. Experimental results obtained at the LEP Storage Ring to determine the momentum on the central orbit using protons and the time evolution of the beam absolute energy during the 1995 Scan Campaign have been presented. In addition, some interesting

Correlation between trains and LEP

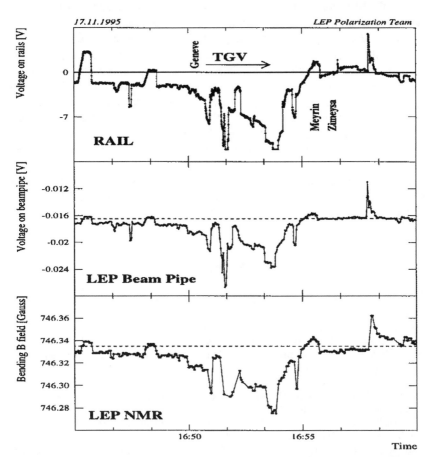

FIGURE 8. Time correlation between voltage on rails, parasitic voltage on LEP vacuum chamber, and the resulting main bending field variations, ref. (45).

phenomena which affect the CM energy at the IPs by amounts well above the accuracies attainable with the adopted calibration methods were discussed.

ACKNOWLEDGMENTS

This review is the result of the collaboration with several working groups dealing with the problem of the beam energy monitoring and calibration.

The experimental results presented here are based on the work of many colleagues in the LEP Polarization Team, carried out over an extended period of time, during long MD shifts and endless nights in the control room.

It is a pleasure to give credit to the members of the Working Group on LEP Energy for their extensive work performed in the last years analyzing the large quantity of information collected in connection with the running of LEP for physics with the aim of understanding the several phenomena affecting the energy stability of the beams and producing a LEP Energy Model for use in evaluating energy changes from logged variations in the main parameters of the accelerator.

Stimulating input from the Working Group on Energy Calibration as a subgroup of the 1995 Workshop on Physics at LEP2 are accounted for in the chapter on electron scattering.

Finally, the importance of the yearly Chamonix Workshop on LEP Performance as a forum for the continuous progress and evolution of the LEP Storage Ring must be stressed and acknowledged.

REFERENCES

1. J. Billan et al., Proc. XIV Int. Conf. on High Energy Accel., Tsukuba, Japan, August 1989, and *Particle Acc.* **29**, 215 (1990).
2. J. Billan et al., Proc. 1991 PAC, San Francisco, CA, May 1991.
3. R. Bailey et al., "LEP Energy Calibration," Proc. 2nd EPAC, Nice, France, June 12-16, 1990.
4. G.E. Fischer, CERN Accelerator School on Magnetic Measurements and Alignment, Montreux, Sept. 1992, CERN 92-05.
5. A. Hofmann and T. Risselada, CERN LEP Note 383, 1982.
6. D. Plane, Contribution to the WG4/Energy Calibration Subgroup, PHYSICS AT LEP2, CERN 96-01, Vol. 1, (Feb. 1996).
7. I.P. Karabekov, CEBAF PR-92-004 (March 1992) and CEBAF PR-93-009 (April 1993).
8. P. Galumian et al., *NIM* A **327**, 269-276 (1993).
9. C. Cecchi, J.H. Field, T. Kawamoto and D. Schaile, Contribution to the WG4/Energy Calibration-Subgroup, PHYSICS AT LEP2, CERN 96-01, Vol. 1, (Feb. 1996).
10. M. Bassetti, Proc. 11th Int. Conf. on High Energy Accelerators, Geneva, Switzerland, July 7-11, 1980.
11. A.A. Sokolov and I.M. Ternov, *Sov. Phys. Dokl.* **8**, 1203 (1964).
12. M. Böge, private communication.
13. L.H. Thomas, *Philos. Mag.* **3**, 1 (1927).
14. V. Bargmann, L. Michel and V.L. Telegdi, *Phys. Rev. Lett.* **10**, 435 (1959).
15. Ya.S. Derbenev and A.M. Kondratenko, *Sov. Phys. JETP* **37**, 968 (1973).
16. R. Rossmanith and R. Schmidt, *NIM* A **236**, 231 (1985).
17. I. Barnett et al., "Dynamic Beam Based Calibration of Orbit Monitors at LEP," Proc. 4th Int. Workshop on Accelerator Alignment, Tsukuba, Japan, Dec. 1995, KEK Proc. 95-12 and CERN-SL/95-97 (BI).
18. M. Mayoud and A. Verdier, "Survey and correction of LEP elements," Proc. 3rd Workshop on LEP Performance, J. Poole, Ed., CERN-SL/93-19-DI, (Feb. 1993).
19. J. Borer, "BOM system hardware status," Proc. 3rd Workshop on LEP Performance, J. Poole, Ed., CERN-SL/93-19-DI, (Feb. 1993).

20. R. Assmann, "Transversale Spin-Polarisation und ihre Anwendung für Präzisionmessungen bei LEP," PhD Thesis, 1994.

21. L. Knudsen et al., "First Evidence of Transverse Polarization In LEP," Proc. Symp. High Energy Spin Physics, Bonn, Germany, Sept. 1990.

22. M. Placidi et al., "Lepton Beam Polarization at LEP," Proc. Int. Symposia on H. E. Spin Physics and Polarization Phenomena in Nuclear Physics, (SPIN '94), Bloomington, Indiana, U.S.A., Sept. 15-22, 1994.

23. V.N. Baier and V.A. Khoze, Sov. J. Nucl. Phys. 9, 238 (1969).

24. U. Fano, J. Opt. Soc. Am 39, 859 (1949).

25. F.W. Lipps and H.A. Toelhoek, Physica, XX, 85 and 395 (1954).

26. M. Placidi and R. Rossmanith, NIM A 274, 79 (1989).

27. B. Dehning, "Elektronen- und Positronen-Polarisation im LEP-Speicher ring und Präzisionsbestimmung der Masse des Z-Teilchens," PhD Thesis, 1996.

28. L. Arnaudon et al., "Accurate determination of the LEP beam energy by resonant depolarization," Z. Phys. C 66, 45-62 (1995).

29. L. Arnaudon et al., Phys. Lett. B 284, 431 (1992).

30. The Working Group on LEP Energy and the LEP Collaborations ALEPH, DELPHI, L3 and OPAL, Phys. Lett. B 307, 187 (1993).

31. A. Blondel, LEP-Note 629 (1990).

32. K. Steffen, Int. Report DESY PET-82 (1982), unpublished.

33. R. Rossmanith, LEP-Note 525 (1985).

34. M. Placidi, "Polarization Results and Future Perspectives," Proc. 3rd Workshop on LEP Performance, J. Poole, Ed., CERN-SL/93-19-DI, (Feb. 1993).

35. The Working Group on LEP Energy, "The Energy Calibration of LEP in the 1993 scan," Z. Phys. C 66, 567-582 (1995).

36. A. Drees, "LEP Beam Energy Calibration in 1995," Proc. XXXIst Rencontres de Moriond, "Electroweak Interactions and Unified Theories," Les Arcs, France, March 16-23, 1996.

37. G.E. Fischer, SLAC-PUB-3392 Rev., July 1985.

38. G.E. Fischer, Proc. First Int. Workshop on Accelerator Alignment, SLAC, July 1989 and SLAC-375.

39. J. Wenninger, "Study of the LEP Beam Energy with Beam Orbits and Tunes," CERN-SL/94-14 (BI), 1994.

40. K.E. Lang, Astrophysical Formulae, 2nd Edition, Springer Verlag.

41. P. Melchior, The Tides of the Planet Earth, 2nd Edition, Pergamon Press, 1983.

42. Code provided by Prof. P. Melchior from International Center for Earth Tides, Bruxelles, Belgium.

43. G. Fischer and A. Hofmann, Proc. 2nd Workshop on LEP Performance, J. Poole, Ed., CERN-SL/92-29-DI, (Feb. 1992).

44. L. Arnaudon et al., "Effects of terrestrial tides on the LEP beam energy," NIM A 357, 249-252 (1995).

45. M. Geitz, "Investigations on environmental effects on the beam energy—magnetic field perturbations," Diploma Thesis, 1996.

Diagnostics for Induction Accelerators*

Thomas J. Fessenden

Lawrence Berkeley National Laboratory
Berkeley, CA 94720

Abstract. The induction accelerator was conceived by N. C. Christofilos and first realized as the Astron accelerator that operated at Lawrence Livermore National Laboratory (LLNL) from the early 1960s to the end of 1975. This accelerator generated electron beams at energies near 6 MeV with typical currents of 600 Amperes in 400-ns pulses. The Advanced Test Accelerator (ATA) built at Livermore's Site 300 produced 10,000-Ampere beams with pulse widths of 70 ns at energies approaching 50 MeV. Several other electron and ion induction accelerators have been fabricated at LLNL and Lawrence Berkeley National Laboratory (LBNL). This paper reviews the principal diagnostics developed through efforts by scientists at both laboratories for measuring the current, position, energy, and emittance of beams generated by these high-current, short-pulse accelerators. Many of these diagnostics are closely related to those developed for other accelerators. However, the very fast and intense current pulses often require special diagnostic techniques and considerations. The physics and design of the more unique diagnostics developed for electron induction accelerators are presented and discussed in detail.

INTRODUCTION

Induction accelerators are low impedance devices that are able to accelerate beams at currents much greater than those from rf-based accelerators. Electron current pulses are typically measured in kiloamps with pulse times measured in nanoseconds and contain no microstructure. The accelerating voltage is isolated from the outside world by inductors that must support the total accelerating voltage for the duration of each pulse. Diagnostics for induction accelerators must reproduce the waveforms of the accelerator parameters with great fidelity. Fortunately, the large currents and voltages typical of an induction accelerator are easily detected and small signal levels are not usually a concern.

This paper focuses on the very fast beam monitors developed for measuring the current and position of high-current electron beams in induction accelerators. Also described are two systems for measuring the beam energy in the Astron and ATA accelerators, x-ray wire scanning techniques for determining beam profiles, and two methods of measuring pulsed magnetic fields in a harsh environment. Most of this work dates back to the 1970s and early 1980s when electron induction linacs and their diagnostics were under intensive development. Many of these diagnostics are also summarized in a paper presented by K. W. Struve (1) at a 1986 conference at the Bureau of Standards in Gaithersburg, Maryland.

---

*Preparation of this report was supported by the Director, Office of Energy Research, Office of Fusion Energy, U.S. Dept. of Energy, under Contract No. DE-AC03-76SF00098.

INDUCTION ACCELERATORS

The linear induction accelerator was invented by N. C. Christofilos in about 1960 for the purpose of generating the "E-layer" for the Astron thermonuclear fusion experiment. Even in the proof-of-principle experiment then under development at the Lawrence Livermore Laboratory, the Astron concept (2) required relativistic electrons at currents far beyond those available from conventional accelerators. The Astron accelerator (3) was developed as a part of this program and, near the end of its operational life, produced 400-ns pulses of 6-MeV electrons at currents up to 800 Amperes. The accelerator could be operated at many repetition rates including a burst mode of 100 pulses at rep-rates up to 1440 Hz intended to build and maintain the E-layer. The accelerator normally operated at an average repetition rate of 2 Hz, set mainly by consideration of x-radiation levels at neighboring laboratory experiments. The accelerator used cable pulse-forming lines switched by hydrogen thyratrons to generate square current pulses with a rise time near 10 ns. The Astron accelerator was in operation from 1961 through 1975. The Astron accelerator technology was further advanced by a development (4) under J. Leiss at the National Bureau of Standards during the mid 1970s aimed at extending the pulse width and reducing the cost of the expensive induction cores that are a part of this accelerator concept. This machine was transferred to the Naval Research Laboratory in the early 1980s for use as the driver for a free-electron laser experiment.

The Astron accelerator was followed by the ERA accelerator (5) at Lawrence Berkeley Laboratory in 1972. This machine used oil Blumlein pulse-forming lines switched by spark gaps to generate 30-ns pulses of electrons at currents up to 3 kA and energies near 4 MeV. The accelerator was used for research on the electron ring collective accelerator concept at the Lawrence Berkeley Laboratory. This accelerator was followed by three related accelerators at the Lawrence Livermore Laboratory: the ETA accelerator (6) in 1979 *(10 kA, 4.5 MeV, 40 ns, 1 Hz)*; the FXR accelerator (7) in 1982 *(4 kA, 20 MeV, 60 ns, 1/3 Hz)*; and the ATA accelerator (8) in 1984 *(10 kA, 50 MeV, 70 ns, 1 Hz average or 10 pulse burst at 1 kHz)*. These three accelerators used water-filled Blumlein pulse-forming lines and spark gap switches. The ETA and ATA accelerators were part of the Strategic Defense Initiative and the FXR accelerator is used for flash radiography. An ETA II (9) accelerator that used magnetic modulator technology operated briefly in the late 1980s.

A four-beam ion induction linac named MBE-4 (10) was built and operated from 1986 to 1991. This machine accelerated four parallel beams of cesium[+] to nearly 1 MeV while increasing the current from 4×10 mA to nearly 4×90 mA. This proof-of-principal experiment was part of the Heavy Ion Fusion Accelerator Research program at the Lawrence Berkeley Laboratory. A recirculating ion induction accelerator (11) is currently being fabricated at LLNL to demonstrate the physics of this approach to a more economical heavy ion fusion accelerator/driver.

Diagnostics developed for this accelerator were discussed by S. Eylon (12) and D. Berners with L. Reginato (13) at previous Beam Instrumentation conferences. Of all these induction accelerators, only the FXR machine is currently in operation.

CURRENT AND POSITION MEASUREMENTS—BEAM BUGS

Introduction

Perhaps the most important diagnostics developed for induction accelerators is the device that monitors the beam current and position in the accelerator and associated beam transport lines. These instruments commonly called "beam bugs" are capable of measuring the kiloampere currents and beam position of the ETA and ATA accelerators with rise times of less than 0.2 ns and relative position resolutions less than 100 μm. They operate by measuring the currents induced in the wall by the passage of the beam.

One of the first to use this technique for monitoring beam current and position was a team (14) at the Lawrence Berkeley Laboratory. They used a band of resistors to interrupt the wall current and generate a current and position signal from the beam in the ERA accelerator. The idea was rapidly adapted for use on the Astron accelerator (15) at Livermore. Fifty one-ohm carbon resistors were placed in the tube wall at nearly 30 locations along the Astron accelerator and transport section.

For the ETA and ATA accelerators, the carbon resistors were replaced by a 1-mil-thick (27 μm) foil or band of stainless steel that is laser or E-beam welded across an insulated break in the beam tube wall. The diameter of the band equals the inner diameter of the beam tube so that the beam sees no abrupt steps or extraneous capacitance during its passage through the bug. A small overlap is formed as the band encircles the inside diameter of the tube. As we will show later, this overlap limits the absolute accuracy of the beam position measurements to a few tenths of a millimeter. Figure 1 shows the design (16) of the beam "bug" used with the ATA accelerator. The resistance of the resistor foil was approximately 2mΩ with a resulting sensitivity of the instrument of 1 kA/V. Also shown are measurements of the response of the bug to a 10 amp fast rising current pulse from a test pulser. A ferrite inductor is placed behind the resistor foil to prevent the beam current from flowing around the resistor. Because of the very low foil resistance, the inductance/resistance (L/R) time for the current to decay is very much longer than the beam pulse. A polyimide ring is used to form an insulating vacuum seal. To achieve the very high frequency response of this instrument we found it necessary to reduce the capacitance of the insulating gap by using a thick polyimide ring. Eight pickoffs as shown in Fig. 1 are used to develop the current and position signals. These are brought out through 50-Ω resistors as shown to keep the cables matched.

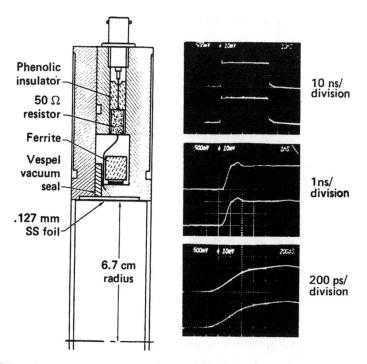

FIGURE 1. Drawing of an ATA beam bug. Also shown is the response of the bug (lower trace) to a 10-Amp pulse (upper trace) along the bug axis from a mercury pulser.

Theory

R. C. Weber (17) has described some of the physics of this type of monitor in the first of these beam instrumentation meetings. He describes the wall currents flowing near a beam bug or wall current monitor but considers the performance of a monitor with a much larger resistance than is of interest here. Let us extend his arguments with a more quantitative analysis of the physics of a beam in a conducting pipe.

It is possible to show (by the method of images, for example) that a current I flowing within a pipe of radius a at a distance ΔR off axis causes a surface current $K(\theta)$ to flow on the inner pipe wall that is given by

$$K(\theta) = (I/2\pi a)\left[(\rho^2 - 1)/(1 + \rho^2 - 2\rho\cos\theta)\right] ,$$

(1)

where $\rho = \Delta R/a$. The angle θ is defined by Fig. 2. Consider interrupting the beam pipe with a resistive band or ring placed at the inner circumference of the pipe with total resistance R. The surface current $K(\theta)$ passing through the resistor that initially or for a short time develops a voltage around the pipe $V(\theta)$ given by

$$V(\theta) = K(\theta) \, 2\pi a \, R = RI \, \frac{\rho^2 - 1}{1 + \rho^2 - 2\rho \cos\theta} \, .$$

(2)

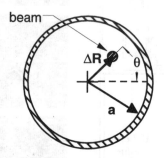

FIGURE 2. Sketch of a beam off axis in a conducting pipe.

However, the voltage variation around the pipe will drive a current that flows through the gap to the outside of the pipe, which will equalize this voltage. After equalization, the surface current pattern on the outside of the pipe may look something like that sketched in Fig. 3, and the voltage across the gap will be independent of θ and given by $V(\theta) = I \, R$. However, until the current leaks through the gap, both beam current and position measurements are possible.

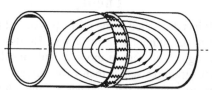

FIGURE 3. Sketch of the current pattern eventually established on the outside of the pipe as a result of the resistive band.

As we will show, the leakage field pattern is approached approximately exponentially with a time constant that varies directly as the area of the pipe and inversely with resistor value R. For parameters used in the Astron accelerator (10-20 cm diameter, 20 mΩ shunt resistance), this time constant is a few tenths of a microsecond. Note that the lower shunt resistance of the ATA bug increases this time by a factor of about 10 but the ferrite does not slow down the establishment of this leakage current. We will return to the issue of this leakage current in a later section. Let us first assume that the beam pulse is short compared to the time for the leakage to occur. For this time the voltage around the gap is correctly given by Eq. (2).

The beam current signal is developed by adding the voltages from four pickoffs located 90° apart with a resistive summing circuit in which great care was taken to avoid reflections in the cabling. From Eq. (2) we find

$$V_I = 1/4 \ \left[V\,(0) + V\,(90°) + (180°) + V\,(270°) \right]. \tag{3}$$

A little algebra yields

$$V_I = IR\, \frac{\left(\rho^4 - 1\right)\left(\rho^4 + 1\right)}{\left(\rho^4 + 1\right)^2 - 2\rho^4\left(1 + \cos 4\theta\right)}. \tag{4}$$

Similarly, the position signals are obtained by adding a negative signal (pickoff goes the other way around the ferrite) at 180° + θ to a positive signal at θ. Again we find

$$V_\rho = RI\, \frac{2\rho \cos \theta\left(1 - \rho^2\right)}{\left(1 + \rho^2\right)^2 - 2\rho^2\left(1 + \cos 2\theta\right)}. \tag{5}$$

In the limit of small beam displacement from the axis ($\rho \ll 1$)

$$V_I = RI \tag{6}$$

and

$$V_\rho = 2\rho RI \cos \theta \tag{7}$$

or

$$\frac{x}{a} = \rho_x = \frac{V_x}{2V_I}; \quad \frac{y}{a} = \rho_y = \frac{V_y}{2V_I}. \tag{8}$$

These are the expressions used for finding the beam position within the transport system of the accelerator and beam transport line. The x or y position signal is digitized and divided by twice the current signal. Obviously, signal timing is extremely important and large errors can be expected at the beginning and end of the pulse.

Let us consider in detail the error in the current signal resulting from the assumption of small beam displacement ρ.   Figure 4 shows plots of the hypothetical current measured by the bug divided by the beam current as a

function of the normalized beam displacement ρ. These curves were obtained
from Eq. (5). The two curves show the bounds on the measurements to be
expected for the case of the beam angle near a pickoff point and for the case of a
beam half-way between two pickoff points. At a normalized displacement of $\rho =$
1/2, the error is less than or equal to $\pm$ 1/8. For displacements greater than $\rho = 1/2$
the error becomes very large. Therefore as a practical limit, current measurements
are of little value if the beam centroid is more than one-half the distance to the
tube wall.

A similar analysis of beam position measurements shows that position errors
as large as $\pm 30\%$ can be expected at normalized displacements $\rho = 1/2$.

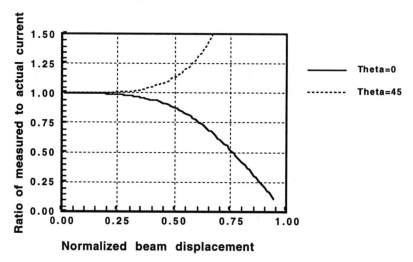

FIGURE 4. Plot of the ratio of the measured to actual beam current as a function of the
normalized beam displacement from axis. For arbitrary angle, the point will be bounded by
the two curves.

Response at Large Time (Asymptotic Response)

Let us return to the issue of the leakage current and the way this current shorts
out the position signal. This problem was considered by Cooper and Neil (18)
and by Fessenden (19). Consider the sketch showing a transverse view of a beam
bug as presented in Fig. 5. The beam axis and symmetry axis are vertical. The
signal voltage $V(\theta)$ is isolated by the ferrite inductor from the foil of thickness δ
and length ℓ_f. The ferrite is connected to the foil by two conducting rings
separated by a gap ℓ_{gap}.

Assume at $t = 0$ that a beam induces a voltage across the foil given by Eq. (1).
Because of the azimuthal voltage $V(\theta)$, a leakage current $I(\theta)$ will flow around the
gap. This current flows on the area on either side of the gap at the foil and on the
surfaces of the pipe. This current generates a magnetic field that flows in and out

of the ferrite, through the foil, and into and out of the interior region of the pipe. Associated with this field is an inductance per unit length L. The determination of this inductance requires a solution of Maxwell's equations for the magnetic fields and surface current distributions in the region near the foils.

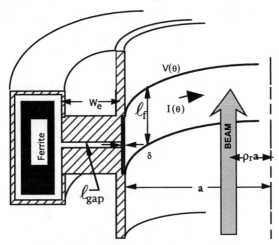

FIGURE 5. Cutaway sketch of an idealized beam bug.

As a very crude estimate let us assume that this azimuthal surface current only flows on both sides of the gap. With this approximation the inductance is found to be

$$L = \frac{\mu \ell_{gap}}{W_e} \quad . \tag{9}$$

In any case let us assume that Eq. (9) defines an effective width W_e such that L is the right value obtained from a proper solution to Maxwell's equations. For a gap $\ell_{gap} \ll a$, this value will be close to that shown in Fig. 1. We can now write a differential equation for $V(\theta)$ and $I(\theta)$, the voltage and current around the gap:

$$V(\theta_2) = V(\theta_1) - L\frac{dI(\theta)}{dt} a\Delta\theta \quad , \tag{10}$$

$$I(\theta_2) = I(\theta_1) - G a\Delta\theta \quad . \tag{11}$$

Here L is the inductance/m and G is the conductance/m around the resistive foil and is given by

$$G = \frac{\sigma\delta}{\ell_f} \quad, \tag{12}$$

where σ is the conductivity of the foil material, δ is the foil thickness, and ℓ_f is the length of the foil. Let us also define the fundamental time of the problem as

$$\tau = LGa^2 = \mu\sigma a^2 \frac{\delta}{W_e} \frac{\ell_{gap}}{\ell_f} \quad. \tag{13}$$

This time constant plays an important role in the work that follows. An estimate of its magnitude based on the dimensions given in Fig. 1 is near 1 μs.

We now have a differential equation for the voltage around the circumference of the gap given by

$$\frac{d^2V}{d\theta^2} = \tau \frac{dV}{dt} \quad. \tag{14}$$

The solution to this equation for the initial distribution given by Eq. (2) can be shown to be

$$V(\theta,t) = I_b R \left[1 + \sum_{n=1}^{\infty} \rho^n e^{-2nt/\tau} \cos n\theta\right] \quad. \tag{15}$$

Here I_b is the beam current. This equation has the required asymptotic time dependence since only the first term survives at large t. Substituting this into the x portion of Eq. (5) yields the expression for the displacement signal as a function of time

$$V(x,t) = 2I_b R \left[\rho e^{-t/\tau} \cos\theta + \sum_{m=1}^{\infty} \rho^{2m+1} e^{-(2m+1)^2 t/\tau} \cos(2m+1)\theta\right]. \tag{16}$$

Note that the first term decays with the time constant τ given by Eq. (13) and the higher-order terms vanish rapidly. The y–position signal is described by a similar equation.

These relations show that, for times short compared to the time τ (≈ 1 μs), beam position measurements using beam bugs are possible with no further signal conditioning. The ETA and ATA accelerators had a pulse width of 40 to 70 ns. As a consequence, no conditioning of the position signal from the bugs used on these accelerators was necessary. For longer pulse accelerators such as the Astron accelerator, the decay of the position signal had to be extended. This was accomplished with a circuit of the type shown in Fig. 6.

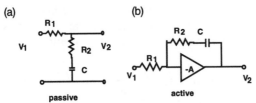

(a)   (b)

passive   active

FIGURE 6. Simple compensation circuits for the beam bug position signals.

The passive circuit has the transfer function

$$V_1 / V_2 = \frac{R_2}{R_1 + R_2} \frac{(s + 1/R_2 C)}{(s + 1/(R_1 + R_2) C)}. \tag{17}$$

If $R_2 C$ is made equal to τ, the effective droop time is increased by the ratio $(R_1 + R_2)/R_2$ whereas the rise time is unchanged. The signal level is, of course, reduced by the same factor. In principle, perfect compensation can be obtained through the use of an operational amplifier as suggested in Fig. 6.

The beam current signal is generated by adding the voltage $V(\theta, t)$ at four equidistant points around the foil. Using Eq. (18) and proceeding as before gives

$$V_I(t) = 4 I_b R \left[1 + \sum_{n=1}^{\infty} \rho^{4n} e^{-(4n)^2 t / \tau} \cos(4n\theta) \right]. \tag{18}$$

Here the higher-order terms vanish even faster than in Eq. (15) and no compensation is required.

Position Errors Resulting from the Overlap

As mentioned before, beam bugs are fabricated using a strip of resistive material that is welded around the circumference with a small overlap. The overlap is necessary to prevent the large beam return current from generating noise signals within the bug. However, for the length of the overlap Δ, the azimuthal voltage $V(\theta)$ is one-half what it should be.

An analysis similar to the previous reveals that for a beam on axis the overlap will generate a voltage around the foil given by

$$V(r,t) = IR \left[1 - \frac{\Delta}{4 \pi a} - \sum_{n=1}^{\infty} \frac{1}{2 \pi n} \sin \frac{n \Delta}{2a} e^{-n^2 t / \tau} \cos(n\phi) \right], \tag{19}$$

where ϕ is the angle to the overlap. Thus we see that the beam current measurement will be in error by a constant term plus a term that decays rapidly in time. This current measurement error is not as serious as the position error resulting from the foil overlap. For a small overlap Δ/a that is away from any pickoff, we find the major contribution to the x or position signal of

$$V_e\ (r,t) = - I_b\ \frac{\Delta}{4\pi r}\left(1 + e^{-t/\tau}\ \cos\phi\right)\ .$$
(20)

The time dependence of this term is identical to that of the x or y position signal and will be misinterpreted. For example, suppose that we superimpose this effect with that produced by an offset beam. We find terms like

$$V_m(t) = 2I_b\ e^{-t/\tau}\left(\rho\cos\theta - \frac{\Delta}{4\pi r}\cos\left(\theta - \phi\right)\right)\ .$$
(21)

Here as before the beam is located at the angle θ. Thus the smallest beam offset one can expect to reliably measure is bounded by $\Delta/4\pi$. Therefore, it is important to minimize the foil overlap and to avoid locating it near the pickoff points.

On the ATA and ETA II accelerators, these beam bugs were integrated into computerized data collection systems capable of displaying the beam current and position along the length of these systems. For further descriptions and examples of collected data see reference (8).

BEAM SIZE AND POSITION MEASUREMENTS

Although the beam bugs are useful for finding the centroid of the beam in the accelerator or transport system, they give no information about the size or current density profile of the beam. A common and very useful diagnostic (20) for determining these parameters was to observe the light emitted with a gated TV system as the beam struck the dump or passed through a metal or carbon foil. This diagnostic was in common use after approximately 1970 on all the Livermore electron accelerators. A digital frame grabber/scanning system was use to generate beam profiles or contour plots of the beam current density. For a long time, the physics behind the light emission was not understood. In 1983, not coincidentally during the time R. Fiorito of the Naval Surface Weapons Laboratory was visiting LLNL, the group began to realize that the source of the light was transition radiation emitted as the relativistic beam struck the material. Since then Fiorito and his colleagues have intensively investigated the physics of transition radiation generated by relativistic beams striking foils. The 1993 Faraday Cup prize was awarded to R. Fiorito and D. Rule for the development of diagnostics (21) based on transition radiation.

Another technique developed by Beam Research personnel for determining the profile and position of the electron beam depends on the x-rays generated as a wire is scanned through the beam. The x-rays from the wire are detected by a fast scintillator and photo-multiplier tube. The detector is collimated by lead shielding to minimize x-ray noise from other sources. At the higher ATA beam energies as much as 0.3 m of lead was required to reduce the x-ray noise to acceptable levels. This diagnostic requires many highly repeatable beam pulses as the wire is scanned across the beam. Because of the excellent time response of the x-ray detector, spatial beam profiles can be obtained at any time during the pulse. X-rays are generated by all the electrons that strike along the wire. To obtain true profiles, the data must be unfolded (Abel inversion). This was seldom performed in the actual experiments. A refinement was to use a tungsten or tantalum dot as the x-ray scatterer. The dot was much smaller than the beam and consisted of a small carbon container filled with tungsten or tantalum powder that was suspended by carbon fibers. The profiles obtained by scanning the dot through the beam require no unfolding but the signal-to-noise ratio tended to be smaller.

ENERGY MEASUREMENTS

Energy analyzers for induction linacs are similar to those developed for other types of accelerators. I will describe two analyzers developed at LLNL, the first was used with the Astron accelerator with some service on the ETA accelerator, and the second was developed for use at energies near 50 MeV on the ATA accelerator.

The Astron On-line Beam Energy Analyzer

This energy analyzer (22) was a spectrometer and as such is similar to several developed by others (23). The analyzer was used in three separate modes: the absolute energy of the beam was determined to within 2%; relative beam energy variations smaller than 0.1% occurring in times less than 10 ns could be measured; and the beam current averaged over a pulse as a function of energy was measured.

A diagram of the analyzer is presented in Fig. 7. A fine wire (originally 38 μm tungsten but later 50 μm carbon) was placed in the path of the primary beam. The wire served two purposes: (a) as a scattering center for the primary beam; and (b) as one part of the collimator for the secondary beam passed through the analyzing magnet. The current scattered from the primary beam is completely negligible ($\approx$ 0.1%), and thus the analyzer causes no interference with the accelerator operation.

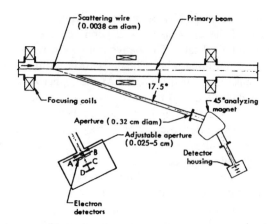

FIGURE 7. Diagram of the Astron on-line beam energy analyzer.

The secondary beam originating at the scattering wire is collimated by a 0.32-cm-diameter hole just before the analyzing magnet. Thus the collimating system consists of the tungsten scattering wire and the 0.32-cm bore. The secondary beam is bent 45° and focused by the analyzing magnet. The magnetic field of the analyzing magnet is measured to 0.01% by a rotating coil Gauss meter. At the detector slit, the magnetic field focuses the beam to a vertical line. That is, a horizontal focus exists at the plane of the A and B detectors. Detectors A, B, C, and D are $1 \times 0.5 \times 0.05$-cm chips of doped silicon which provide approximately 10^5 secondary electrons for each energetic electron striking their surfaces.

An absolute calibration of the system was obtained by allowing x-rays produced by the primary beam to generate neutrons in a block of beryllium. The threshold energy of 1.66 MeV for neutron emission was detected and used to calibrate the analyzer. The calibration thus obtained is believed accurate to 2%.

To determine the absolute energy of the beam to within 2%, the outputs of detectors A and B are electronically subtracted and displayed on an oscilloscope. The signal from detector C is also displayed on the oscilloscope. The absolute energy measurement is obtained by maximizing the signal from detector C and then measuring the magnetic field of the analyzing magnet. The waveform displayed by the A-B signal is then a crude display of the beam energy as a function of time. An example of these signals is given in Fig. 8.

Measurements of relative beam energy variations smaller than 0.1% occurring in times less than 10 ns can be obtained by increasing the gain on the A-B signal and recording the time(s) of the zero crossing(s) of the oscilloscope trace as a function of the magnetic field of the analyzing magnet. The percentage change in the magnetic field of the analyzing magnet is then plotted as a function of the recorded zero crossing times resulting in a curve such as that shown in Fig. 9.

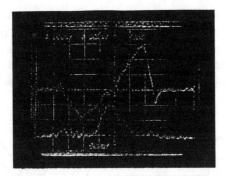

FIGURE 8. Analysis of an Astron beam with a large energy spread. Top trace—A-B signal at 0.5 %/div.; bottom trace is C signal; sweep speed 50 ns/div. Beam energy is 5.13 MeV.

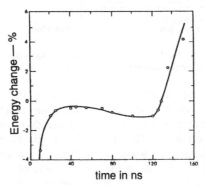

FIGURE 9. Change in the Astron beam energy as a function of time.

For more powerful beams at higher beam energies the scattering wire will not survive, and the range of the beam electrons in material becomes impractically long. Sharp beam images cannot be produced by aperturing the beam with slits or pinholes because the electrons scatter back into the aperture. As a consequence, standard pepper-pot emittance diagnostic techniques cannot be used and other methods must be found to diagnose the beam. Moreover, at high beam powers the design of a beam dump that will survive the beam becomes an issue. A multi-purpose device was developed for diagnosing the 50-MeV, 10-kAmp ATA beam.

The ATA Diagnostic Box

The beam dump/diagnostic system (24) developed for the ATA accelerator consisted of a series of carbon plates whose collective thickness totaled approximately 1.5 ranges at 50 MeV. The dump was designed to dissipate up to

175 kW of average power. The beam power absorbed by the plates was radiated to a water-cooled wall. A small hole along the axis of the plates formed a collimator that permitted a low current beamlet to pass through an energy analyzer. The analyzer consists of a 60° bending magnet and two high sensitivity (high resistance) beam bugs. Figure 10 is a schematic of the assembly.

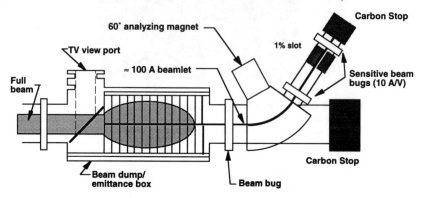

FIGURE 10. Sketch of the high-power beam dump and diagnostic assembly developed for ATA.

The beam current and position at the entrance to the dump/diagnostic assembly are measured with a beam bug of the type described above. After passing through the beam bug, the beam enters the dump which consists of a series of range–thin carbon disks. The first disk was placed at an angle of 45° to the beam and used to develop an image of the beam for a gated TV system. A small hole of diameter D (1/4") along the axis through all of the carbon disks is used to form a collimator that produces a beamlet of current determined by the beam emittance ε and the acceptance α of the collimator according to $i = I_b(\alpha /\varepsilon)^2$. The collimator acceptance $\alpha = D^2/4L$ where L is the collimator length.

After leaving the dump, an analyzing magnet bends the beamlet through an angle of 60° into a secondary beamline. The beamlet is detected by a pair of high-sensitivity beam bugs. The first is located close to the exit of the analyzing magnet and measures the total current of the beamlet passing through the collimator. The ratio of the beamlet current to the primary current is used to obtain an estimate of the beam emittance of the on-axis component of the primary beam. The beamlet continues through the secondary beamline and then through a slot designed to accept an energy variation of up to 1%. The second sensitive current monitor is used to measure the time variation of the beamlet within the 1% energy acceptance set by the slot. The beamlet is finally stopped by a six-inch-thick (~1.25 ranges) solid carbon dump. X-ray-induced noise in the signal cables was a problem near this diagnostic. The x-ray noise was greatly alleviated (25) by adding two inverted negative bug pickoffs to two positive bug pickoffs away from the radiation field.

MAGNETIC FIELD MEASUREMENTS

The measurement of fast magnetic fields is often of interest. Often these are less than ideal, particularly at high frequencies, because the magnetic pickups are also sensitive to electric fields that may also be present. There are at least two possible ways of avoiding the electrical field pickup: Birx loops and shield-integrated probes.

Birx Loops

D. Birx solved the problem (26) with the design shown in Fig. 11. The pickup is shielded from electric fields by equal lengths of metal on each side of the loop. The magnetic field is admitted to the loop interior by the small gap leading to the precisely known area. It is important to locate the gap so that the shield lengths on each side are exactly equal, and the electric field pickup of one side is exactly canceled by pickup from the other. Obtaining the required mechanical precision is a relatively easy task for a machinist.

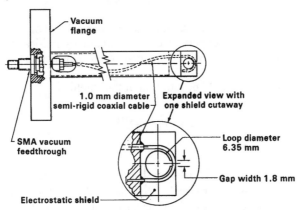

FIGURE 11. Drawing of a shielded rf pickup loop.

Shield-Integrated Probes

Shield-integrated loops (27) were developed for moderately fast magnetic field measurements in particularly harsh environments (i.e., in the presence of strong electric fields and/or energetic electrons). The spurious electric field can be eliminated by placing the magnetic pickup coil within an electrostatic shield. The shield impedes the penetration of the external magnetic field in such a way that for certain periods the field inside the shield is proportional to the integral of the field outside. Since the coil effectively differentiates the field, the detected signal is approximately proportional to the external magnetic field.

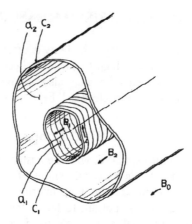

FIGURE 12. Generic shield-integrated magnetic probe. B_0 is the externally applied field, B_1 is the field within the coil of area a_1, B_2 is the field in the area a_2 between the pickup coil and the shield.

For simplicity consider the case of a long cylindrical coil of N turns and length ℓ enclosed by a good, but not perfect, conducting thin cylindrical shell as shown in Fig. 12. At $t = 0$ an external uniform magnetic field B_0 is applied to the outside of the coil.

The time constant for field penetration of the shield is

$$\tau_s = L_s / R_s = \frac{\sigma \delta \mu_o}{C_2} \left(a_1 + a_2 \right) , \qquad (22)$$

where C_2 is the circumference of the shield, δ is the thickness, and σ is the conductivity of the shield material. Similarly, the coil has a time constant determined by its inductance and the resistance in the circuit of

$$\tau_c = L / R = \frac{\mu_o N^2 a_1}{R\ell}. \qquad (23)$$

Here L is the self-inductance of the coil, ℓ is the length of the shield, and R is the total of the coil resistance R_c plus the resistance R_{ext} of the external circuit (generally 50 Ω).

Defining α as the ratio of the area inside the coil to the total area ($\alpha = a_1/(a_1 + a_2)$), one can obtain equations for the change in field across the shield and coil

$$B_o - B_2 = \tau_s \left[(1 - \alpha) \frac{\partial B_2}{\partial t} + \alpha \frac{\partial B_1}{\partial t} \right], \qquad (24)$$

$$B_2 - B_1 = \tau_c \frac{\partial B_1}{\partial t}. \tag{25}$$

These equations are easily solved for the output voltage in terms of the externally applied field B_o by Laplace transformation. For the signal voltage V_o one obtains

$$V_o = -\frac{R_{ext}}{R} \frac{sNa_1 B_o(s)}{\tau_1 \tau_2 (s + 1/\tau_1)(s + 1/\tau_2)}, \tag{26}$$

where τ_1 and τ_2 are the two solutions of the equation

$$(\tau_{1,2})^2 - \tau_{1,2}(\tau_c + \tau_s) + \tau_c \tau_s (1-\alpha) = 0. \tag{27}$$

If, as is generally true, time constants τ_1 and τ_2 are not of comparable magnitude, one finds in terms of the smaller and larger of the time constants τ_s, τ_c:

$$\tau_1 \cong (1-\alpha) \frac{\tau_s \tau_c}{\tau_s + \tau_c} \cong (1-\alpha) \tau_{smaller}, \tag{28}$$

$$\tau_2 \cong \tau_{larger}. \tag{29}$$

Let us suppose that the external field B_o is switched on at $t = 0$. In this case $B_o(s) = B_o/s$. By Laplace inverting Eq. (13) one obtains

$$V_o(t) \cong -\frac{R_{ext}}{R} \frac{N_a B_o}{\tau_2} \left(e^{-t/\tau_1} - e^{-t/\tau_2} \right). \tag{30}$$

The rise time of the signal is approximately the smaller of the shield and coil time constants reduced by the factor $(1 - \alpha)$. The fall time is approximately the longer time constant. Note that as the space between the coil and the shield becomes small ($\alpha \to 1$), the rise time improves.

Notice that there is complete duality between the roles of the shield and coil time constants. The shield and coil time constants affect the system response in an entirely symmetrical way. Depending on relative magnitudes, either could establish the rise time and/or fall time of the probe. This means that there are two methods the experimenter can use to make a self integrating probe. The shield time constant can be made much longer than the L/R time constant of the coil producing a shield-integrated probe. Conversely, as is more commonly done, one can make the coil L/R time constant much longer than the shield time constant, producing a coil-integrated probe.

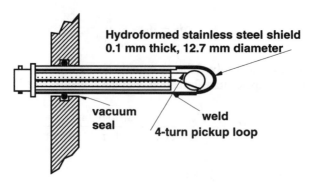

FIGURE13. A shield-integrated magnetic field probe.

The design of a shield-integrated magnetic probe is shown in Fig. 13. This probe was used for measuring the fields created around a 20-ns, 20-kiloamp electron beam. The sensitivity of the probe was 14 Gauss/Volt. The rise time was 7 ns and the fall time set by the shield was near 433 ns. This was extended to approximately 10 μs with a compensating circuit of the type shown in Fig. 6a with a corresponding loss of sensitivity.

ACKNOWLEDGMENT

This paper summarizes many contributions made by the experimental physicists that worked in the Astron and Beam Research Programs at the Lawrence Livermore Laboratory from the period 1970 to 1985. It is a pleasure for me to acknowledge the many interactions and exchanges of ideas with my colleagues and the particularly significant contributions of B. Stallard, E. Lauer, L. Reginato, J.Clark, D. Birx, K. Struve, Y.P. Chong, and F. Chambers to the diagnostic systems developed at LLNL.

REFERENCES

1. K. Struve, "Electrical Measurement Techniques for Pulsed High Current Electron Beams," submitted to the Measurement of Electrical Quantities in Pulse Power Systems-II, National Bureau of Standards, Gaithersburg, Maryland (1986), also LLNL Report UCRL-93261 (1986).
2. N. C. Christofolis, Proc. U.N. Conf. on Peaceful Uses of Atomic Energy 2nd Geneva, Vol. **32**, 279 (1958).
3. N. C. Christofolis, R.E. Hester, W.A.S. Lamb, D.D. Reagan, W.A. Sherwood, and R.E. Wright, *Rev. Sci. Instrum.* **35**, 886 (1964).
4. J.E. Leiss, N.J. Norris, and M.A. Wilson, *Part. Accel.* **10**, 223 (1980).
5. R. Avery et. al., "The ERA 4 MeV Injector," Proc. of the 1971 PAC, Chicago, IL (1971) also LBL Report UCRL-20174 (1971).
6. R.E. Hester, et al., *IEEE Trans. Nucl. Sci.* **NS-26,** 4180 (1979); T.J. Fessenden, et al., *IEEE Trans. Nucl. Sci.* **NS-28** (1981).
7. B. Kulke, T.G. Innes, R. Kihara, and R.D. Scarpetti, "Initial Performance Parameters on FXR," Proc. 5th IEEE Power Modulator Symposium, 307 (1982).

8. L. Reginato, *IEEE Trans. Nucl. Sci.* **NS-30**, 2970 (1983).
9. F.W. Chambers et. al., "Diagnostics and Data Analysis for the ETA II Linear Induction Accelerator," Proc. of the 1991 PAC, San Fransisco, CA., 3085 (1971) and subsequent papers.
10. R.T. Avery et. al., *IEEE Trans. Nucl. Sci.* **NS-32**, 3187 (1985).
11. A. Friedman et. al., "Recirculating Induction Accelerators for Inertial Fusion: Prospects and Status," Proceedings of the International Symposium on Heavy Ion Inertial Fusion, to be published by *Fusion Engineering and Design* (1996).
12. S. Eylon, "The Diagnostics System for MBE-4," AIP Conference Proceedings No. 252: Beam Instrumentation Workshop, 225 (1992).
13. D. Berners and L. Reginato, "Beam Position and Total Current Monitor for Heavy Ion Fusion Beams," AIP Conference Proceedings No. 281: Beam Instrumentation Workshop, 168 (1992).
14. R.T. Avery, A. Faltens, and E.C. Hartwig, "Non-Intercepting Monitor of Beam Current and Position," LBL Report UCRL-20166 (1971).
15. T.J. Fessenden, B. W. Stallard, and G.G. Berg, "Beam Current and Position Monitor for the Astron Accelerator," *Rev. of Sci. Instr.* **43**, 1789 (1972).
16. K. Struv, "The ATA Beam Bug," informal communication, LLNL.
17. R.C. Weber, "Longitudinal Emittance—An Introduction to the Concept and Survey of Measurement Techniques Including Design of a Wall Current Monitor," AIP Conference Proceedings No. 212: Beam Instrumentation Workshop, 85-110 (1989).
18. R.K. Cooper and V.K. Neil, "Resistor Beam Bugs," LLNL Report UCID-16057 (1972).
19. T.J. Fessenden, "Beam Bugs--Asymptotic Response," informal communication, LLNL.
20. Y.P. Chong, R. Kalibjian, J.P.Cornish, J.S. Kallman, and D. Donnelly, "Beam Profile Measurements on the Advanced Test Accelerator using Optical Techniques," Proc. of the "Beams 86" Conference, Kobe, Japan, 738 (1986).
21. R.B. Fiorito and D.W. Rule, "Optical Transition Radiation Beam Emittance Diagnostics," AIP Conference Proceedings No. 319: Beam Instrumentation Workshop, 21 (1994) and references therein.
22. T.J. Fessenden, "The Astron On-Line Beam Energy Analyzer," *Rev. of Sci. Instr.* **43**, 1090 (1972).
23. See for instance Ref. (17) p. 96.
24. J.M. White et. al., "Beam Dump/Diagnostic Box for a 10 kA, 50 MeV, 50 ns Electron Beam," *IEEE Trans. Nucl. Sci.* **NS-30**, 2207 (1983).
25. K. Struve, "Radiation Induced Noise Signals in Diagnostic Cabling of the Advanced Test Accelerator," Proc. of the "Beams 86" Conference, Kobe, Japan, 742 (1986).
26. D. Birx, private communication.
27. T.J. Fessenden and B.W. Stallard, "Shielded Magnetic Probes for Pulsed Magnetic Field Measurements," LLNL Report UCID-16115 (1972).

Beam Profile Monitors for Very Small Transverse and Longitudinal Dimensions Using Laser Interferometer and Heterodyne Techniques

Tsumoru Shintake

KEK: National Laboratory for High Energy Physics
Tsukuba, Ibaraki 305 Japan

Abstract. In future e^+e^- linear colliders of 0.5-1 TeV center-of-mass energy region, very small beam sizes will be used at the interaction region in order to get enough beam-beam luminosity. Typical design parameters assume a cross section of 3-5 nm by 300 nm in vertical and horizontal, respectively, and a bunch length of 0.1 mm. In order to measure the nanometer vertical size, we use laser interference fringes. By scanning the electron beam over the fringe pattern and monitoring the high-energy gamma rays due to Compton scattering, beam information can be obtained. The modulation depth in the gamma-ray flux gives the spot size information. This method was used in the Final Focus Test Beam (FFTB) test at SLAC, and a 70-nm spot was measured. To measure the longitudinal bunch length, a new method, utilizing a laser-heterodyne technique, will be proposed in this paper.

INTRODUCTION

In future linear colliders in the TeV-region, a typical spot size of a few hundred nanometers by a few nanometers and the bunch length of 0.1 mm at the interaction point is required to get enough luminosity (1,2). This spot size is much smaller than that in any conventional colliding-beam machines; therefore one of the most important tasks is to develop new technologies for beam diagnostics, especially a reliable spot size and bunch length measurement systems.

A conventional beam size monitor employing a carbon wire (3), such as that used in the Stanford Linear Collider (SLC), is limited to beam dimensions around one micron, since a wire with diameter smaller than a few µm is not available, and if it were it would not be able to withstand the destructive energy of the high intensity electron beam (4). Therefore, several new schemes have been proposed, some of which utilize the interaction of high-energy electrons with ions in a gas (5,6). The beam sizes are determined by monitoring the ion velocity and its number, or bremsstrahlung. However, the space-charge fields are so high that the gas will be ionized by the tunneling ionization in a wide area around the beam. This tunneling ionization limits the resolution.

In 1990 I proposed a new method (7) which uses Compton-scattered gamma rays from a photon target of intense laser light and determines the spot size by using the laser interference fringe as a physical scale. This method has no limitation due to the high space-charge field of the electron beam. The observed data are directly related to the electron beam size by a simple equation without any calibrations. This monitor was designed (8) and installed in the Final Focus Test Beam (FFTB) (9,10) at the Stanford Linear Accelerator Center (SLAC) in 1993. In the first FFTB beam test in April-May 1994, an average spot size of 75 nm was achieved and these data were successfully measured by this system (11,12).

For the bunch length measurement, H. Lihn et al. proposed a method that uses a far-infrared Michelson interferometer to measure coherent transition radiation emitted at wavelengths longer than the bunch length via optical autocorrelation. They experimentally demonstrated subpicosecond resolving power at the Stanford SUNSHINE facility in 1995 (13). However, in the linear collider very high intensity beam has to be measured; that is, each bunch contains 1×10^{10} electrons, and there are 100 bunches in one train. Therefore, the foil would be damaged due to pulsive thermal deposition from the beam. To overcome this problem, a new method has been proposed here, which utilizes the laser-heterodyne technique. The laser beam is not subject to damage from the electron beam, and the measurement is non-destructive. It has a single-bunch selection capability from the multi-bunch train and its physical scale is very reliable since it is defined by the laser wavelength.

In this paper, Part I describes the idea of the spot size measurement and its experimental results, and Part II describes the proposal of the laser-heterodyne technique for bunch-length measurement.

PART I: TRANSVERSE SPOT SIZE MEASUREMENT

System Diagram

Figure 1 is a schematic showing the setup for spot size measurement. The laser beam is split into two beams by a half mirror, then focused into common foci, creating an interference fringe. First, the electron beam is injected on the focused laser beam from the normal direction. Next, the high-energy gamma ray is measured downstream, after the bending magnet that sweeps out the electrons to the beam dump. Finally, the electron beam is slowly scanned by means of a weak steering magnet upstream, and the modulation depth in the gamma-ray flux is measured to determine the spot size.

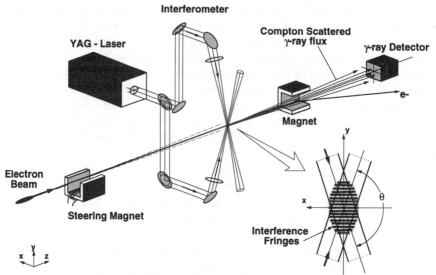

FIGURE 1. System diagram of the spot size monitor.

Interaction of the Electron Beam with the Interference Fringe

Here we consider the electromagnetic field in the interference fringe and calculate the beam trajectory and its radiation. Consider two laser beams $\mathbf{k}_1$ and $\mathbf{k}_2$ (see Fig. 2):

$$\mathbf{k}_1 = \begin{bmatrix} k_0 \cos\alpha \\ k_0 \sin\alpha \\ 0 \end{bmatrix}, \quad \mathbf{k}_2 = \begin{bmatrix} k_0 \cos\alpha \\ -k_0 \sin\alpha \\ 0 \end{bmatrix} \tag{1}$$

where $\theta = 2\alpha$. The vector components are represented as $\begin{bmatrix} k_x, k_y, k_z \end{bmatrix}$. Suppose the laser beam has S-polarization on the interferometer table, that is, the electric field is along the z-axis (electron beam direction).

The electron will be periodically accelerated in the transverse direction by the Lorentz force due to the magnetic field. The magnetic field in a two-wave interference fringe is given by

$$\mathbf{B}_1 + \mathbf{B}_2 = 2B_0 \begin{bmatrix} \sin\alpha \cdot \sin k_y y \cdot \sin(\omega t - k_x x) \\ -\cos\alpha \cdot \cos k_y y \cdot \cos(\omega t - k_x x) \\ 0 \end{bmatrix} \tag{2}$$

where

$$k_x = k_0 \cos\alpha \quad \text{and} \quad k_y = k_0 \sin\alpha. \tag{3}$$

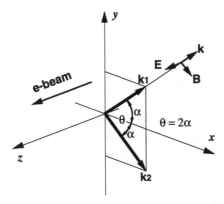

FIGURE 2. Definition of wave vectors and coordinates. Here we assumed S-polarization, whose electric field is parallel to the mirror surface on the interferometer table (*xy*-plane).

Equation (2) shows that the interference pattern has a traveling-wave component in the x direction, but it is a standing wave in the y direction.

When a point charge is running along the z-axis, it will be accelerated in the transverse direction by the magnetic field:

$$\mathbf{a}_\perp = q(\mathbf{E} + \mathbf{u} \times \mathbf{B})_\perp = qc(\mathbf{e}_z \times \mathbf{B})_\perp = qc \begin{bmatrix} -B_y \\ B_x \end{bmatrix}$$

$$= 2qB_0c \begin{bmatrix} \cos\alpha \cdot \cos k_y y \cdot \cos(\omega t - k_x x) \\ \sin\alpha \cdot \sin k_y y \cdot \sin(\omega t - k_x x) \end{bmatrix}. \tag{4}$$

Since there is a 90° phase difference between the horizontal and vertical accelerations, the electron will travel along a spiral trajectory. The amplitudes are functions of the crossing angle of the lasers and the vertical position y. When the crossing angle θ is 90°, the amplitude is the same in each direction, and the trajectory becomes perfectly circular or linear according to the beam position y. Therefore, by choosing the electron beam position, we can make the effective laser polarization linear, clockwise circular or anti-clockwise circular. This interesting feature can be used in some applications in the future.

In our spot size monitor, we have no interest in the polarization, but we measure the total number of gamma rays from the radiation. The radiation power and the number of gamma rays are given by the time average of the square of the transverse acceleration:

$$N_\gamma \propto P \propto \langle \mathbf{a}_\perp{}^2 \rangle \propto \langle (B_x^2 + B_y^2) \rangle \propto 2B_0^2 (\sin^2\alpha \cdot \sin^2 k_y y + \cos^2\alpha \cdot \cos^2 k_y y)$$

$$\propto B_0^2 (1 + \cos\theta \cdot \cos 2k_y y). \tag{5}$$

This equation represents the interference fringe pattern. We note that the modulation amplitude, or the contrast of the fringe, looking from a charged particle depends on the laser crossing angle θ. We must take this effect into account in a formula for calculating the spot size. In the special case of perpendicularly crossing laser beams ($\theta = 90°$), the modulation will disappear. Therefore, for the spot size monitor application, we cannot use a 90° crossing angle.

An actual electron beam has a Gaussian distribution. The variation of the gamma-ray number is given by integrating the Gaussian distribution function of the electron beam multiplied by Eq. (5), as follows:

$$N_\gamma(y_0) = \int_{-\infty}^{\infty} \frac{1}{\sqrt{2\pi}\sigma_y} \cdot e^{-(y-y_0)^2 / 2\sigma_y^2} \cdot (1 + \cos\theta \cdot \cos 2k_y y) \cdot dy, \qquad (6)$$

where $N_\gamma(y_0)$ is arbitrary normalized, y_0 is the vertical position of the electron beam in the interaction region, and k_0 is the wavenumber of the laser light $k_0 = 2\pi / \lambda_0$. Here we define the modulation depth to describe the degree of variation of the gamma-ray number $M = \Delta N / N_0$, where ΔN is the difference between the average and the maximum gamma-ray number and N_0 is the average value. Integrating Eq. (6) and comparing the maximum and the average values, we have the modulation depth

$$M = |\cos\theta| \cdot \exp\left(-\frac{1}{2}(2k_y\sigma_y^*)^2\right). \qquad (7)$$

The fringe pitch is obtained from the wave number as follows:

$$2k_y p = 2\pi, \qquad (8)$$

$$p = \frac{\lambda_0}{2\sin(\theta/2)}. \qquad (9)$$

For example, if we use a 1064-nm wavelength laser and choose a crossing angle of 6°, we have 10.2-μm pitch spacing, which has been used to measure a 1-μm horizontal spot size at the FFTB test.

Correction Factors

In the previous section, we discussed the interaction of an electron beam with the interference fringe generated by two plane waves. In a practical situation, the electron beam is not a parallel beam but it has a finite emittance, and the actual laser has an intensity distribution inside the beam.

We have to take the following factors into account.

1. Laser Beam Profile
    We assume a Gaussian laser beam profile, whose power density distribution is formalized by

    $$I(r) = I_0 \exp\left(-r^2 / 2\sigma_L^2\right).$$ (10)

2. Power Imbalance of Laser Beams
    In a practical spot size monitor, the two laser beams do not always have the same intensity because of imperfections in the mirror coating. This imbalance causes the contrast of the interference fringe to drop below 100%. The fringe contrast is given by

    $$C_p = \frac{2\sqrt{P_2 / P_1}}{1 + P_2 / P_1},$$ (11)

    where P_2 / P_1 is the power ratio of the two laser beams. Equation (11) indicates that the fringe contrast maintains an almost perfect value (=1.0) over a wide range of power imbalance. This is an important feature of the spot size monitor. Thanks to this phenomenon, the spot size monitor can create a high contrast fringe and measure the spot size with high accuracy.

3. Beta-Function of the Electron Beam
    The electron beam has finite emittance, and its transverse beam size changes around the focus point as follows:

    $$\sigma_y = \sigma_y^* \sqrt{1 + \left(z / \beta_y^*\right)^2},$$ (12)

    where σ_y^* and β_y^* are the minimum spot size and the minimum beta function at the focus point, respectively. Since the laser beam has also a finite thickness in the z direction, the measured spot size becomes an average value within the laser beam, which is slightly larger than the minimum spot size σ_y^* at the focus point.

With all these factors taken into account, the modulation depth becomes

$$M = |\cos\theta| \cdot C_p \frac{1}{\sqrt{1 + (2k_y\sigma_y^*)^2 (\sigma_L / \beta_y^*)^2}} \cdot \exp\left(-\frac{1}{2}(2k_y\sigma_y^*)^2\right).$$ (13)

To measure the minimum spot size precisely, at least the laser beam size must be smaller than the beta function. In the FFTB experiment, since $\beta_y^* = 100$ μm, we chose the laser size as $\sigma_L = 50$ μm. With this condition, the correction factor for a 60-nm beam becomes

$$\frac{1}{\sqrt{1+(2k_y\sigma_y^*)^2(\sigma_L/\beta_y^*)^2}} = 0.94. \tag{14}$$

This makes the spot size 7 nm larger, that is, the raw data become 67 nm for an actual spot size of $\sigma_y^* = 60$ nm.

Gamma Ray Generation

Here we calculate the number of available gamma rays and their energy using the FFTB experiment as an example. We assume use of a neodymium:yttrium-aluminum-garnet (Nd:YAG) laser with 1064-nm wavelength and an output energy of a few hundred mJ per pulse within a 10-ns pulse. These parameters are typical in Q-switched YAG lasers of medium power. In the actual spot size monitor, the laser beam is split into six beamlines with the energy in one beamline at the 10-mJ level, and the peak power at the 10-MW level.

As discussed in the previous section, the laser beam size should be small enough to measure precisely the electron beam size at the focus point. A smaller laser beam size is also desirable in order to provide a higher number of gamma rays. However, a smaller beam size makes alignment of the laser beams more difficult. Therefore, we chose the laser beam size $\sigma_L = 50$ μm. The diameter of the laser beam is $D \approx 2\sigma_L = 100$ μm. The power density of the laser beam at the focus point becomes

$$\frac{P_L}{S} = \frac{10 \text{ MW}}{\pi(0.05 \text{ mm})^2} = 1.3 \times 10^{15} \text{ W/m}^2. \tag{15}$$

The average photon density in one traveling wave of the laser beam is given by

$$n = \frac{1}{h\nu} \cdot \frac{P_L}{cS} = 2.3 \times 10^{19} \text{ photons/cm}^3, \tag{16}$$

where we used $h\nu = 1.17$ eV as the dominant frequency of the Nd:YAG laser. Then we calculate the number of gamma rays by Compton scattering. Since the well-known Klein-Nishina equation for Compton scattering cross section is formulated in the electron-rest-frame, we need to transform the laser photons into the electron-rest-frame by the Lorentz transformation. This process is discussed in the appendix of ref. (7). The average number of gamma rays generated by Compton scattering from one laser beam is given by

$$<N_\gamma> = \sigma_c n D N_e. \tag{17}$$

Here N_e is the number of electrons in one bunch and σ_c is the total cross section of the Compton scattering, which is given by (14,15)

$$\frac{\sigma_c}{\sigma_0} = \frac{3}{4}\left[\frac{1+\varepsilon_1}{\varepsilon_1^3}\left\{\frac{2\varepsilon_1(1+\varepsilon_1)}{1+2\varepsilon_1} - \log(1+2\varepsilon_1)\right\} + \frac{1}{2\varepsilon_1}\log(1+2\varepsilon_1) - \frac{1+3\varepsilon_1}{(1+2\varepsilon_1)^2}\right] \tag{18}$$

where σ_0 is the Thomson scattering cross-section $\sigma_0 = 8\pi r_0^2/3 = 6.65 \times 10^{-25}$ cm^2 and ε_1 is the dimensionless photon energy in the electron rest frame

$$\varepsilon_1 = \gamma\frac{hv_0}{m_0c^2}. \tag{19}$$

In the non-relativistic case $\varepsilon_1 \ll 1$, Eq. (18) gives the cross section equal to the Thomson scattering: $\sigma_c = \sigma_0$. On the other hand, for the extreme relativistic case $\varepsilon_1 \gg 1$, the cross section becomes very small. In the case of our interesting electron energy range in the FFTB, ε_1 is 0.22 and the cross section becomes $0.72\sigma_0$. Therefore, the reduction of cross section due to the relativistic effect is not so severe. The gamma ray takes a wide spectrum (7); its maximum is 16 GeV in the FFTB test (we use the first harmonic of the YAG laser of 1064-nm wavelength and 50-GeV electron beam).

Using the electron number of $N_e = 1\times10^{10}$, we have the number of gamma rays per bunch: $<N_\gamma>$ = 1000 gamma rays/bunch. In the FFTB experiment, with careful control of the trajectory of the incoming electron beam, it was possible to reduce the background noise of bremsstrahlung due to spent electrons, and the Compton-scattered gamma-ray signal was measured with enough signal-to-noise ratio.

Practical Design

System Overview

Figure 3 shows the interferometer table installed in the FFTB beamline. Optical components are mounted on a 1.6 × 1.5-m optical table, 110 mm thick with a honeycomb structure, which is mounted on a final Q-magnet table. The laser system is in a clean room outside the accelerator tunnel, and the laser beam is transported through a vacuum pipe for 20 m. At the entrance of the interferometer the position of the laser beam is monitored, and the position signal is sent back to the laser system. This is used as a feedback signal to control the mirror angle at the exit of the YAG laser to keep the incoming laser position at the injected into a vacuum interaction chamber through glass windows, and focused at the interaction point. The generated gamma ray is monitored by a gas Cherenkov light detector located downstream of the beamline, which is also used to detect gamma rays from a wire scanner near the laser interferometer.

FIGURE 3. Photograph of the installed spot size monitor at the focal point in the FFTB beamline. The electron beam runs from right to left through the interferometer table which is mounted on a massive girder together with the final focusing Q-magnet. To protect optical components from dust, the interferometer table is covered with an aluminum box.

Interferometer System Design

The designed system can set up three different pitches of interference fringes, as shown in Fig. 4. The beam size responses for these modes are plotted in Fig. 5. The system sensitivity is 1 μm down to 40 nm for vertical, and 4 μm down to 0.7 μm for horizontal.

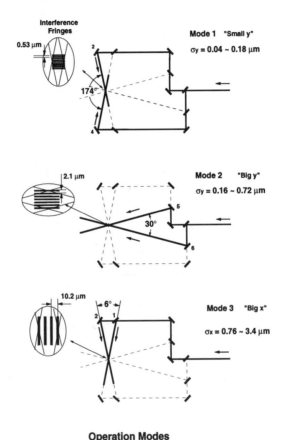

**Operation Modes
of Laser-Compton Spot Size Monitor**

FIGURE 4. Three modes of operation for generating three different fringes.

YAG Laser

We use a Nd:YAG laser, model GCR-3 produced by Spectra Physics Inc.; its specifications are summarized in Table 1. Generally speaking, the power density distribution (spatial beam profile) of the output beam from a high power

Nd:YAG laser shows large spatial distortion of high spatial frequencies. Such high frequency (or short wavelength) components are generated by interference of parasitic waves with the main beam. They are diffraction waves mainly generated by the Gaussian tail intercepting the edge of the YAG laser rod. In our spot size monitor, a smooth beam profile is required for accurate measurement. To improve the spatial profile, reflection mirrors of smaller curvature were used in the oscillator section, which made the beam diameter smaller and reduced the beam interception.

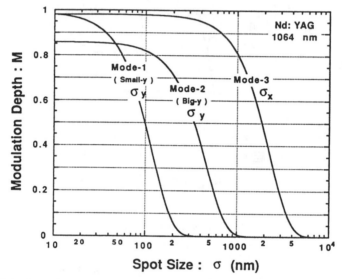

FIGURE 5. Spot size response curve for three measurement modes calculated by Eq. (7).

TABLE 1. Specifications of the Nd:YAG Laser System

Wavelength	1064	nm
Pulse Energy*		
without amplifier	200	mJ
with amplifier	600	mJ
Pulse Width	8 - 9	ns
Energy Stability	1	%
Repetition Frequency (optimum)	10	pps
Spatial Beam Profile (Gaussian fit)	> 85	%
Line Width (with injection seeding)	< 0.003	cm^{-1}
Beam Pointing Stability	< 100	μrad/hour

*Note: We mostly used the oscillator section without the amplifier in the FFTB experiment.

Damage Threshold of Optical Components

To protect the optical components from high power damage, we use long focus lenses of f 500 mm in the last section. The laser beam sizes on every optical component are kept to 3-7 mm in diameter, and the power density to 0.1-0.5 GW/cm^2, which is well below the damage threshold of 5 GW/cm^2 for the dielectric optical coating on the mirrors.

Coherency of the YAG Laser

In principle, in the interferometer scheme, it is possible to choose the path difference between two laser beams to be zero. However, in the actual system there are path differences of 5 to 10 cm. The natural width of the spectral line of a Nd:YAG laser is about 1 cm^{-1}. The path difference is longer than the coherent length, and the interference fringe pattern will fade out. The line width required to get enough fringe contrast (visibility > 95%) is 0.01 cm^{-1}. The seeding injection laser technique can reduce the line width of a YAG laser below 0.003 cm^{-1}.

The seeding laser also improves the temporal waveform of the output pulse. The natural waveform of the output power from a Q-switched YAG laser has high frequency spikes due to overlapping of many longitudinal modes inside the laser resonator. The seeding laser limits the longitudinal mode to a single mode and makes the temporal profile smooth (Fourier-limited waveform), and the pulse width becomes 8-9 ns. This is an important point in reducing the gamma-ray fluctuation due to timing jitter and improving data statistics.

Background-Noise Subtraction

Since the spot size is determined by the modulation depth in the Compton-scattered gamma-ray data, i.e., the ratio of AC to DC components, for accurate measurements it is very important to eliminate baseline shifts due to background noise. To do this, a synchronous detection technique is used. When the electron beam runs at 30 Hz, the laser is fired at 10 Hz. That is, for each electron beam pulse, the laser beam is fired alternately: ON_OFF_OFF_ON_OFF_OFF... The laser-ON gamma-ray data are averaged for six pulses, the background-noise (laser-OFF) gamma-ray data are averaged for twelve pulses, and then the background-noise data are subtracted from the laser-ON data.

Beam Operation

Operating the spot size monitor involved two key issues: background-noise control and laser beam alignment.

Since the active length of the laser beam spot along the electron beam trajectory is only 0.1 mm, the number of gamma rays generated is relatively low, as discussed previously. Therefore, it is vital to reduce the background gamma-ray noise in the beam experiments. By tuning the beam trajectory and adjusting a variable mask, we can reduce the background to a level five or ten times smaller than the Compton-scattered gamma-ray signal. However, this process usually takes a fairly long time.

Concerning laser beam alignment, we have to align the laser beams at the common focal point with an accuracy better than ± 20 μm to make the fringe contrast better than 98% (corresponding to an accuracy of ± 3 nm at a 60-nm spot size). We did this alignment in two steps as follows.

Laser Beam Slit Scan (z-axis Alignment)

For z-axis alignment of the laser beams, we prepare a slit scanner. We insert the slit at the focal point and scan the z position of the laser beam with the mirror mover. Then we measure the transmitted laser power through the slit using a photo-detector on the opposite side, find the offset of the laser beam, and move it to the slit center.

Electron Beam Finder (xy-plane Alignment)

For xy-plane alignment, we use the gamma-ray signal from the electron beam. By scanning the electron beam trajectory along the x or y axis, we look for a gamma-ray peak of the laser beam. By adjusting the x or y axis of the mirror, we can align each laser beam to the center (on the electron beam axis).

Spot Size Measurement

After the laser beam alignment, we set up the desired fringe mode. If we scan the electron beam trajectory with fine steps, we observe a periodic intensity modulation in the gamma-ray data, as shown in Fig. 6(a). The solid curve is a least-mean-square fit of the analytical function

$$Y = A + B\sin(\frac{2\pi}{p}y + C), \tag{20}$$

where p is the period of the fringe pattern (dark-to-dark distance). We use a theoretical value of $p = \lambda_0 /[2\sin(\theta/2)]$, where λ_0 is the wavelength of the laser, θ is the crossing angle of the laser beams (in this case, $\lambda_0 = 1064$ nm, $\theta = 174°$, and $p = 533$ nm), and A, B, and C are unknown free parameters which must be determined by data fitting. In this case $A = 106.7$, $B = 72.6$, and the modulation depth is $B/A = 0.68$.

Using Eq. (13), we have the spot size $\sigma_y^* = 66$ nm, where we used $|\cos\theta| = 0.9945$, $P_2/P_1 = 1.26$, $C_P = 0.993$, $\beta_y^* = 100$ μm.

Figure 6(b) shows the measured spot size where the correction term due to the finite thickness of the laser beam is neglected. If we include this effect, the actual spot size is about 6-7 nm smaller than in the figure.

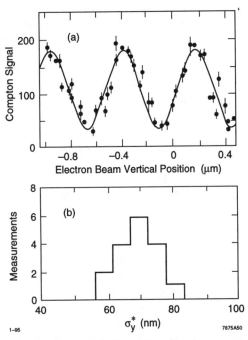

FIGURE 6. (a) Measured gamma-ray data. The solid curve is the least-square-fit of the sine-function. (b) Spot size distribution for 3 hours measurement.

Discussion

Features of Laser Interferometry

The features of this laser interferometer method are summarized below.

1. The measurement mechanism is simple and easy to understand.
2. High-power beam capability. The usual wire scanner using carbon fiber is subject to being broken by a high-power electron beam, but the laser beam is not.

3.   No question about the measurement scale. The fringe pitch is defined
     by the wavelength of the laser beam (well known) and the crossing angle
     (precisely controlled by mechanical dimension).
4.   The fringe is quite stable. Thanks to the interference effect, the
     interference fringe is quite stable. Fringe contrast is always kept close to
     100%, and it is quite insensitive to power imbalance or misalignment.
5.   The reading errors are always on the safe side. Every source of
     measurement error, except errors in the noise subtraction circuit (this
     can be calibrated), produce positive errors on the measured spot size.
     (Every error tends to fade out the fringe contrast.) Therefore, the raw
     data on measured spot sizes are always bigger than the actual spot size.
6.   Relatively wide range measurement. By changing the crossing angle
     (from narrow angle to head-on crossing), we can choose a fringe pitch in
     a wide range from millimeter to nanometer.

Linear Collider Application

In an actual linear collider we use a short-wavelength laser such as the 5th
harmonic radiation of an Nd:YAG laser at 213 nm. Fully utilizing the synchronous
background-noise subtraction technique and using a stable laser oscillator, it will
be possible to utilize the modulation amplitude up to 95% or more. Therefore the
spot size of 3-4 nm, which is usually assumed at the interaction point in linear
colliders, can be measured with enough accuracy by this method.

However, it is hard to fit this monitor inside the detector at the interaction
point. We may use this monitor for machine tuning before the detector is rolled in
to the interaction point. During the physics run time, we use this monitor at the
image focus point upstream of the final Q-magnet where the spot size is 100
nanometers. For the final beam tuning, the beam-beam interaction can be used to
monitor the spot size during beam collision (16).

PART II: BUNCH-LENGTH MEASUREMENT

Principle of Operation

Figure 7 is a schematic diagram of the bunch-length measurement system.
By mixing two laser beams of different frequency, an intensity modulation with a
frequency equal to the frequency difference between the two laser beams (beat
frequency) is created. We inject electron bunches into the laser beam from the
normal direction and measure the Compton scattered gamma rays downstream.
Since the amplitude of the electric field in the beat wave is periodically modulated,
the number of the gamma-ray flux will change according to the injection phase of
the electron bunch.

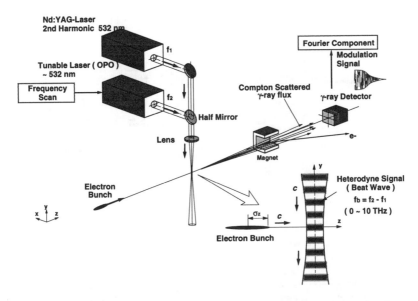

FIGURE 7. Schematic diagram of the bunch-length measurement system using the laser-heterodyne technique.

Figure 8 shows the operating mechanism. If the pitch of the beat wave is longer than the bunch length (Fig. 8(a)), the electron bunch hits bright or dark zones, and the generated gamma-ray flux will show intensity variation according to its injection phase. Since it is very hard to control the beam injection phase with the beat wave (the injection phase is random), the gamma-ray number shows intensity fluctuation pulse-to-pulse. Next we tune the laser wavelength to increase the beat frequency. If the pitch of the beat wave becomes much smaller than the electron bunch length (Fig. 8(b)), the electron bunch will collide with bright and dark zones during its passage. The total gamma-ray number is averaged for high and low values and becomes almost constant.

Using the techniques available with a tunable laser, we can scan the wavelength of one laser precisely. We scan the beat-wave pitch from longer pitch to shorter pitch, for example, and measure the histogram of the gamma-ray flux. When the pitch becomes shorter than the bunch length, the fluctuation in the gamma ray will disappear. From this threshold, we can determine the bunch length.

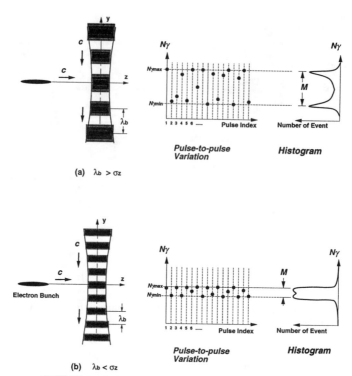

(a) $\lambda_b > \sigma_z$

(b) $\lambda_b < \sigma_z$

FIGURE 8. Pulse-to-pulse fluctuation in gamma-ray flux.

Bunch Spectrum Measurement

The combined laser power is amplitude-modulated with the difference frequency

$$P(y,t) = P_1 + P_2 + 2\sqrt{P_1 P_2}\, \cos f_b(t - y/c), \qquad (21)$$

where $f_b = f_2 - f_1 = f_0 \dfrac{\Delta \lambda}{\lambda_0}$ and $\Delta \lambda = \lambda_1 - \lambda_2$ is the wavelength difference between two lasers. If $P_1 = P_2 = P_0$, the resulting signal is 100% modulated. Assuming the laser has a Gaussian beam profile, the power density becomes

$$p_L(x,y,z,t) = \frac{2P_0}{2\pi\sigma_{Lx}\sigma_{Lz}}\, \exp\!\left(-\frac{x^2}{2\sigma_{Lx}^2} - \frac{z^2}{2\sigma_{Lz}^2}\right)\!\cdot\!\left[\, 1 + \cos\{f_b(t - y/c) + \phi_b\}\,\right], \quad (22)$$

where σ_{Lx} and σ_{Lz} are the laser radii along the x and z axes, respectively, at which point the intensity has decreased to $e^{-1/2}$ or 0.607.

Let's assume the electron population in one bunch by the Gaussian profile in the x and y direction:

$$dN_e = \frac{N_e}{2\pi\sigma_x\sigma_y}\exp\left(-\frac{x^2}{2\sigma_x^2}-\frac{y^2}{2\sigma_y^2}\right)\cdot f(t-z/c)\cdot dxdydz, \qquad (23)$$

where $f(t-z/c)$ represents the longitudinal bunch-shape which is running along the z axis with speed of the light.

Since the number of gamma rays generated by the Compton scattering is proportional to the photon density in the laser, or power density of the laser, the gamma-ray flux is given by the following integration:

$$N_\gamma \propto \int P_L(x,y,z,t)dN_e dt. \qquad (24)$$

Assuming the laser width is much larger than the horizontal size of the electron beam (that is, $\sigma_{Lx} \gg \sigma_x$), integrating Eq. (24) we have

$$N_\gamma \propto 2N_e P_0 \left[F_0 + \exp\left(-\frac{(k_b\sigma_\Sigma)^2}{2}\right)\left\{A(f_b)\cos\phi_b + B(f_b)\sin\phi_b\right\} \right]$$

$$= 2N_e P_0 \left[F_0 + \exp\left(-\frac{(k_b\sigma_\Sigma)^2}{2}\right) F_\omega \sin(\phi_b + \phi_f) \right], \qquad (25)$$

where $\sigma_\Sigma^2 = \sigma_y^2 + \sigma_{Lz}^2$. A, B, and F represent the Fourier spectrum of the electron bunch and are defined as follows:

$$F_\omega^2 = A(f_b)^2 + B(f_b)^2,$$

$$A(f_b) = \frac{1}{\pi}\int_{-\infty}^\infty f(t)\cos f_b t\cdot dt,$$

$$B(f_b) = \frac{1}{\pi}\int_{-\infty}^\infty f(t)\sin f_b t\cdot dt, \qquad (26)$$

$$F_0 = \frac{1}{\pi}\int_{-\infty}^\infty f(t)\cdot dt.$$

In a practical system, since the injection-phase ϕ_e is random in bunch-by-bunch, what we can detect is the maximum and the minimum limits of the fluctuation. From Eq. (25), we have

$$M = \frac{N_{\gamma\,max} - N_{\gamma\,min}}{N_{\gamma\,max} + N_{\gamma\,min}} = \exp\left(-\frac{(k_b\sigma_\Sigma)^2}{2}\right)\cdot\frac{F_\omega}{F_0}. \qquad (27)$$

As seen in Eq. (27), by measuring the modulation depth in the gamma-ray flux we can determine the relative Fourier spectrum of the electron bunch. The exponential term in Eq. (27) is a correction term to compensate for the effect due to the laser radius and the electron bunch height.

Figure 9 shows a calculated bunch spectrum for a Gaussian bunch at a typical bunch length of 0.1 mm. As seen in the figure, the spectral intensity extends up to 1 THz. Using the laser-heterodyne technique, it is easy to generate a beat wave at 1 THz by de-tuning 1 nm from 532 nm on one laser beam. The modern pulsed high-power laser can do this precisely (described later).

Practical Example Parameter

Electron Beam Parameter

Here we apply the laser-heterodyne technique to measure a short bunch in a future linear collider. We measure the bunch length at the exit of the bunch compressor in the Japanese Linear Collider (JLC). Typical parameters of the electron beam are listed in Table 2.

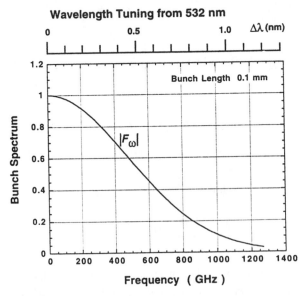

FIGURE 9. Frequency spectrum for a 0.1-mm-length bunch.

TABLE 2. Beam Parameters at the Bunch Compressor in the JLC.

Normalized Emittance	$\gamma \varepsilon_x$	3×10^{-6}	m·rad
	$\gamma \varepsilon_y$	3×10^{-8}	m·rad
Beam Energy	E	1.75	GeV
Beta Function	β_x	16	m
	β_y	41	m
Beam Size	σ_x	118	μm
	σ_y	19	μm
	σ_z	85	μm

Here we assume the laser width in the z direction σ_{Lz} to be 30 μm. As seen in Table 2, the horizontal width of the electron beam is larger than 100 μm. In order to cover this size of electron, we assume the laser beam cross section to be a long elliptic shape, whose long radius is $\sigma_{Lx} = 200$ μm. Applying a cylindrical lens in the laser focusing system, an elliptic spot can be easily produced. (This technique is widely used in matching sections between a laser diode and a fiber optic system.)

In this condition, a correction factor of 0.94 (due to the finite laser depth and the electron beam height) can be calculated using $\sigma_\Sigma = \sqrt{\sigma_y^2 + \sigma_{Lz}^2} = 34$ μm and substituted into the expression $\exp\{-(k_b\sigma_\Sigma)^2/2\}$. Here we assumed a typical laser beating frequency at 500 GHz. The measured intensity modulation will be 6% lower than the expected depth defined by the bunch-shape Fourier spectrum. Correction to this factor will not cause a substantial error in the measurements.

Laser System

Modern technology in laser physics allows users to tune the wavelengths of high-power lasers, such as the Optical Parametric Oscillator (OPO) (17), which can provide coherent radiation ranging from the ultraviolet to the far infrared wavelengths at power levels of 10 to 100 mJ energy per pulse. Recently a new model has been developed (18) which uses the nonlinear crystal β-Barium Borate (BBO) pumped by the 355-nm, third-harmonic output of a Nd:YAG laser. Using an injection-seeded Nd:YAG laser and a tunable seed beam from the master oscillator, a system output linewidth of less than 0.2 cm⁻¹ can be achieved. At 532 nm, the linewidth is only 0.006 nm. The system can generate 100 mJ energy per pulse at 50 pps maximum, 3- to 8-ns pulse width. This amount of power is quite enough for the present application. The expected gamma-ray number from the

Compton scattering is almost ten times higher than that in the spot size experiment at FFTB.

Accuracy and Practical Problems

There are two issues for accurate measurement of the bunch spectrum. One is the frequency accuracy, and the other is the amplitude accuracy.

The wavelength accuracy in the OPO laser mentioned above is equal to its line width of 0.006 nm. The frequency accuracy of the heterodyne signal becomes 6 GHz, which is 1% of the typical bunch spectrum at 500 GHz. Using a spectrometer (resolution 0.2 cm^{-1}), we can always monitor the wavelength separation of two laser beams to confirm this accuracy.

The amplitude accuracy is limited by the statistical error in the gamma-ray measurement, which depends on the total flux of the gamma ray per bunch. In the present example, the average number per bunch is expected to be a few thousand; therefore the statistical error will be less than 1%. The laser power fluctuation can be normalized by a photodiode signal in pulse-to-pulse.

However, there is another error source from background noise which will dominate the measurement error. To reduce the background noise due to spent electrons or electron-to-gas collisions, we have to choose the best location to install this monitor in a beamline and design appropriate masks in the electron beamline and in the gamma-ray extraction line.

Discussions

We summarize the features of this method.

1. Non-destructive monitor. No foil or wire is needed. The laser beam does not destroy the electron beam.
2. The laser beam performance parameters (intensity, spot shape, and beat frequency, etc.) are well isolated and independent of the electron beam.
3. Bunch selection capability. The YAG laser and OPO laser can generate a short laser pulse at a few ns, which enables one bunch from the multi-bunch train to be hit.
4. Accurate time base at THz frequency. The beat-wave frequency is defined by the laser wavelength, which is well controlled in modern OPO lasers and can be calibrated by a grating spectrometer.

ACKNOWLEDGMENT

The spot size monitor experiment was done by a number of people at Kawasaki Heavy Industries and the High Energy Physics (KEK) and also by the

FFTB group at the Stanford Linear Accelerator Center (SLAC). The author would like to thank all the people who contributed to the work.

REFERENCES

1. R. H. Siemann, "Linear Collider Research and Development," Proc. of the 17th Int'l. Linear Accelerator Conference, August 21-26, 1994, Tsukuba Japan, 24-28 (1994); also published as SLAC-PUB-6646, August 1994.
2. G. A. Loew, "Review of Studies on Conventional Linear Colliders in the S- and X-band Regime," Proc. XVth Int. Conf. on High Energy Accelerators, Hamburg Germany, July 20-24, 1992, 777-783 (1992).
3. G. Bowden, D. Burke, C. Field and W. Koska, "Retractable Carbon Fibers Targets for Measuring Beam Profiles at the SLC Collision Point," SLAC-PUB-4744, January 1989.
4. Roger A. Erickson, "Monitoring in Future e^+e^- Colliders," SLAC-PUB-4974, May 1989.
5. Jean Buon, "Possibility to measure very small spot sizes using gas ionization at future linear colliders," LAL/RT89-05, August 1989, Proc. of the XIV International Conference on High-Energy Accelerators, August 22-26, 1989, Tsukuba, Japan, *Particle Accelerators* **31**, 39-46 (1990).
6. The gas ionization monitor has been installed in FFTB, and successfully operated. P. Puzo, "First Results of the Orsay Beam Size Monitor for Final Focus Test Beam," Proc. European Accelerator Conf., EPAC 94, London, UK, 1994, 31-33 (1994).
7. T. Shintake, "Proposal of a nanometer beam size monitor for e^+e^- linear colliders," *Nucl. Instrum. Methods A* **311**, 453 (1992).
8. T. Shintake et. al., "Design of Laser-Compton Spot Size Monitor," Proc. XVth Int. Conf. on High Energy Accelerators, Hamburg, Germany July 20-24, 1992, 215-218 (1992).
9. M. Berndt et al., "Final Focus Test Beam Design Report," SLAC-REF-376 (1991).
10. B. Schwarzschild, "New Stanford Facility Squeezes High-Energy Electron Beams," *Physics Today*, 22, July 1994.
11. T. Shintake et. al., "Experiments of Nanometer Spot Size Monitor at FFTB Using Laser Interferometry," Proc. of the 1995 Particle Accelerator Conf. and Int. Conf. on High Energy Accelerators, May 1-5, 1995, Dallas, Texas, U.S.A. 2444-2446 (1996); also KEK Preprint 95-46, May 1995.
12. V. Balakin et al., "Focusing of Submicron Beams for TeV-Scale e^+e^- Linear Colliders," *Phys. Rev. Let.* **74**, 2479 (1995).
13. Hung-chi Lihn et. al., "Measurement of Subpicosecond Electron Pulse," SLAC-PUB-95-6958, August 1995.
14. I. F. Ginzburg, G. L. Kotkin, V. G. Serbo, and V. I. Tel'nov, "Colliding γe and $\gamma\gamma$ beams from single-pass e^+e^- accelerators," *Sov. J. Nucl. Phys.* **38** (2), Aug. 1983.
15. V. I. Telnov, "Problems of Obtaining $\gamma\gamma$ and γe Colliding Beams at Linear Colliders," Proc. of the International Workshop on the Next Generation of Linear Colliders, Stanford, California (1988).
16. T. Tauchi, "Beam Size Monitor by Using Pairs Created in Beam-Beam Interaction at IP," Proc. of the Fifth Int'l. Workshop on Next Generation Linear Colliders, LC93, SLAC, Oct. 13-21, 1993, SLAC-436, 1993.
17. S. E. Harris, "Tunable Optical Parameteric Oscillators," *Proc. IEEE* **57** (12), 2096-2113, Dec. 1969.
18. Spectra-Physics Lasers, 1330 Terra Bella Avenue, Mountain View, CA, 94039-7013 USA.

Commissioning Results of the APS
Storage Ring Diagnostics Systems*

Alex H. Lumpkin

Advanced Photon Source, Argonne National Laboratory
9700 South Cass Avenue, Argonne, Illinois 60439 USA

Abstract. Initial commissionings of the Advanced Photon Source (APS) 7-GeV storage ring and its diagnostics systems have been done. Early studies involved single-bunch measurements for beam transverse size ($\sigma_x \approx 150$ µm, $\sigma_y \approx 50$ µm), current, injection losses, and bunch length. The diagnostics have been used in studies related to the detection of an extra contribution to beam jitter at ~ 6.5 Hz frequency; observation of bunch lengthening ($\sigma \approx 30$ to 60 ps) with single-bunch current; observation of an induced vertical, head-tail instability; and detection of a small orbit change with insertion device gap position. More recently, operations at 100-mA stored-beam current, the baseline design goal, have been achieved with the support of beam characterizations.

INTRODUCTION

The commissioning of the Advanced Photon Source (APS) 7-GeV storage ring (SR) was supported by a full complement of diagnostics subsystems (1-3). Both the machine and, in some cases, specific features of the diagnostics systems were evaluated and demonstrated in a phased manner. Essential steps of first injected beam, first single turns, first multiple turns, stored beam, and increased stored beam were tracked with generally standard measurement techniques. However, assistance from the single-turn rf beam position monitor (BPM) capability and the functioning photon monitor for optical synchrotron radiation (OSR) imaging for transverse beam size measurements and early bunch length measurements were very useful, particularly in the single-bunch, first-turn measurements. Early studies included single-bunch measurements for beam transverse size, orbit stability, current, injection losses, and bunch length. The detection and resolution of an unexpected horizontal beam motion contribution at ~ 6.5 Hz frequency as revealed through analysis of rf BPM data (4) is addressed as well as bunch lengthening effects with single-bunch current and an induced vertical, head-tail instability. The path traveled to get both the machine and the diagnostics functioning at the baseline, 100-mA stored-beam current is also discussed.

---

* Work supported by U. S. Department of Energy, Office of Basic Energy Sciences under Contract No. W-31-109-ENG-38.

EXPERIMENTAL BACKGROUND AND PROCEDURES

The APS storage ring diagnostics include a full complement of subsystems. As shown in Fig. 1, there are 360 rf BPMs, one direct-current current transformer (DCCT), ten intercepting profiling screens with cameras, two striplines for tune measurements, five scrapers, ten loss rate monitors, and an installed OSR imaging station on a bending magnet source. As a comment in passing, a subset of this figure was shown in the 1992 BIW (2), but at that point they were only planned; now all are installed and functional except the undulator radiation source and beamline. Basic accelerator parameters for diagnostics are listed in Table 1. The fundamental rf frequency is 351.9 MHz with a harmonic number of 1296 in the 1104-m-circumference ring. The horizontal and vertical damping times are 9.4 ms with horizontal and vertical tunes being 35.22 and about 14.30, respectively. The longitudinal damping time is 4.7 ms at 7 GeV and the synchrotron frequency is about 1.9 kHz, depending on the rf gap voltage. The baseline design was for 5 mA (18 nC) in a single bunch with 20 bunches to make the 100-mA design goal. However, we typically find better lifetimes with a fill of at least 36 bunches and often use a few hundred. There are 40 sectors in the ring with the injection point in sector 40 and the first insertion device in sector 1.

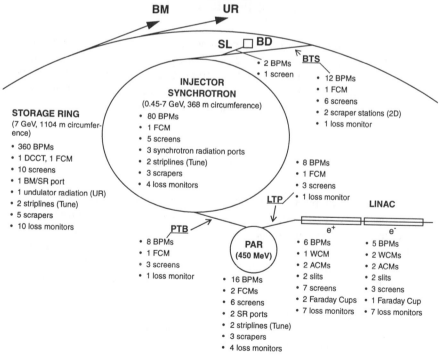

FIGURE 1. Schematic of beam diagnostics in the APS.

TABLE 1. Accelerator Parameters for Diagnostics

Parameter	Storage Ring	Inj. Synch.
Energy (GeV)	7	0.45 - 7
rf Freq. (MHz)	351.93	351.93
Harmonic No.	1296	432
Min. Bunch Spacing (ns)	2.8	1228
Rev. Period (µs)	3.68	1.228
No. of Bunches	1-1296	1
Design Max. Single Bunch Current (mA)	5	4.7
Bunch Length (2σ) (ps)	35-100	61-122
Damping Times $\tau_{h, v}$ (ms)	9.46	2.7 @ 7 GeV
Tunes ν_h, ν_v	35.22, 14.30	11.76, 9.80
Damping Time τ_s (ms)	4.73	1.35 @ 7 GeV
Synch. Freq. f_s (kHz)	1.96	21.2

Beam Profile Monitors

Beam profile monitors are based on intercepting, chromox screens viewed by standard Vicon charge-coupled device (CCD) cameras. For initial injection, camera stations in sectors 40, 1, 7, 14, and 39 supported transport through the first third of the ring and then the S39 flag supported evidence for the first full turn. (In fact, the OSR port in sector 35 was used to monitor the first turn.)

Beam Position Monitors

Beam position monitors are located in all 40 sectors around the ring. There are 360 stations using rf BPM button pickups and their associated electronics. The storage ring standard vacuum chamber uses 10-mm-diameter button feedthroughs, and the insertion device vacuum chambers (IDVC) use 4-mm-diameter buttons. The BPM electronics have been designed for single-turn capability using a monopulse receiver that involves the amplitude to phase conversion technique. In stored beam mode, a separate module with a boxcar averager can process values from the nine signal conditioning and digitizing units (SCDUs) in each VXI crate for each sector. Very good position resolution has been achieved which will be briefly discussed in the next section and more completely in Ref. 5.

Beam Current Monitors

Monitoring of the current/charge in the transport lines and injector rings is generally based on the use of fast current transformers (FCTs) manufactured by Bergoz, shielded housings, and in-house electronics (6). In the storage ring a Bergoz parametric current transformer (PCT) Model PCT-S-175 is the basis of the DC current measurement. Additionally, an integrating current transformer (ICT) is mounted in the same housing as the DCCT and allows single-bunch evaluation of charge in the SR.

Beam Loss Rate Monitors

The beam loss rate monitors (LRMs) cover the transport line from the booster to storage ring (BTS) as well as the circumference of the ring in 10 sections. A gas-filled coaxial cable acting as an ionization chamber is used to detect bremsstrahlung radiation. In practice, it is also sensitive to the hard storage ring x-rays; so for stored beam conditions, specific losses are more difficult to identify. This system is discussed in more detail elsewhere in these proceedings (7).

Synchrotron Radiation Monitors

Transverse beam profiles and bunch lengths have been measured using OSR and x-ray synchrotron radiation (XSR) imaging techniques. A bending magnet source is dedicated to particle beam characterization. The early commissioning data were taken using an OSR pick-off mirror based on a water-cooled Moly assembly. A standard Vicon CCD camera linked to a Questar telemicroscope was initially used within the accelerator tunnel (8). Subsequently, an in-air transport line with a series of mirrors was used to bring the OSR out of the tunnel and onto an optical diagnostics table for measuring with a CCD camera and the Hamamatsu C5680 dual-sweep streak camera.

The main features of the dual-sweep streak system include a synchroscan sweep unit phase-locked to the 117.3-MHz source from the rf system oscillator. On its fastest range it has a resolution $\sigma_{res} \sim 0.6$ ps with time jitter expected to be less than that. The four selectable ranges have a time axis span of 0.15 to 1.5 ns. The slow sweep axis can be triggered at up to 10 Hz with a selection of time spans from 100 ns to 100 ms. For single-bunch experiments with a ring circulation time of 3.68 μs, the 10-, 20- and 100-μs ranges were often used. The streak images were read-out at the standard 30-Hz video rate using a Peltier-cooled CCD chip camera. Controls were done through a GPIB interface to the local computer. The video images were also shipped via the video mux to the main control room for on-line display on TV monitors.

By the time of the high current runs, a second split mirror was installed for the OSR pick-off to the streak camera, and an x-ray pinhole imaging station was operational within the tunnel as described in an accompanying paper in these proceedings (9).

PRIMARY DIAGNOSTICS AND INITIAL COMMISSIONING OF THE STORAGE RING

As mentioned earlier, the primary diagnostics were used for initial injection and first turns. The DCCT, five intercepting screens, the loss rate monitors, over 300 rf BPMs with single-turn capability, and a single-turn OSR monitor were available. Early in machine commissioning we also addressed key issues of rf BPM resolution, beam stability, transverse size, and bunch length which will be discussed in more detail below.

rf BPM Resolution Tests

In stored-beam mode the resolution of the rf BPM monitors can be effectively improved by averaging over many turns or increasing the beam charge in the ~ 50-ns-wide integration window of the electronics. The details of the electronics are covered elsewhere and a more complete report is given in these proceedings (5). The assessment of contributions from electronics noise versus actual beam motion was facilitated by deliberately cross-connecting selected cables to the front-end rf BPM buttons at a location. By crossing the top two buttons in S6B:P1, rotating by 90° the connections in S6A:P1, and crossing the inboard buttons in positions S9A:P1 and S9B:P1, the 4-button set's sensitivity to beam motion in the horizontal plane is nulled, the vertical goes to horizontal in the electronics, and the vertical plane is nulled, respectively. The sector 7 rf BPMs had all normal connections. The BPM resolution was then evaluated at fixed current and N turns or with fixed average and varied beam current. Examples of the results are shown in Fig. 2. In Fig. 2(a) the reduction of the variance in the averages of measurements Δx, Δy are shown versus sampling number N. An approximate $1/\sqrt{N}$ slope is seen down to ~ 8 µm at 2048-turn average for the S6B:P1:x data and the S9B:P1:y data with their insensitivities to x and y motion, respectively. The sector 7 x data do not follow this but flatten out at about 30 µm. This was attributed to actual horizontal beam motion of about 30 µm rms. In Fig. 2(b), the averaging at N = 2 turns is fixed and the approximate inverse dependence on beam current (I_b) is exhibited. Again the sector 7 x and y data do not follow the trend like the S6B:P1:x and the

S9B:P1:y data. Actual beam motion in the two planes adds to the variance in the x position. This 30-μm rms horizontal motion was actually beyond the orbit stability specification and will be addressed in the next section. The summary results of the rf BPM resolution are that the electronic contributions to resolution are about

$$\Delta x_e \sim \left\{ \begin{array}{ll} 0.16 & \text{regular} \\ 0.1 & \text{ID} \end{array} \left[\frac{\mu m \cdot mA}{\sqrt{Hz}} \right] \right\} \times \frac{\sqrt{BW}}{I_b[mA]}, \tag{1}$$

where BW is the measurement bandwidth in Hz. That is, for regular rf BPM buttons in a normal gap vacuum chamber for 2-mA beam and 100-Hz bandwidth, a resolution of 0.8 μm is expected. It is 60% lower for the miniature rf BPM locations. This result can be compared to the baseline orbit stability specification of 17 μm horizontally and 4 μm vertically (rms). Additional issues of offset long-term drift, bunch pattern dependency, and beam intensity dependence are being evaluated (5).

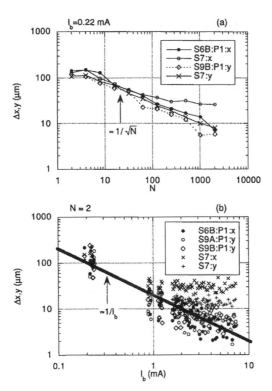

FIGURE 2. Results of rf BPM resolution tests for (a) fixed current and various averages of multiple turns and (b) fixed average and varying current.

Beam Stability

As noted in the previous section, evidence for rms-horizontal beam motion of about 30 μm was detectable in these resolution tests. It had been detected earlier in commissioning by examining the frequencies of beam motion contributions using a vector signal analyzer (1). The rf BPM system is able to detect rms beam motion with resolution better than 4 μm in the 1-20 Hz band. Figure 3 shows the frequency analysis of this motion by plotting the power log10(V) versus delta frequency with 135.7 kHz the center (carrier) frequency. Noticeable broad peaks at ± 6.5 Hz are seen as well as at about 11 Hz, the girder vibration contribution. A search for the source including the BPM electronics was undertaken. By sampling the performance of the various classes of power supplies for the magnet sets in the ring, it was noticed that the sextupoles (family 1) had a strong correlation to the problem. Figure 4 shows the decrease in the observed beam motion by turning off the power supplies. In fact, with them off the rms motion was below 20 μm. A representative sextupole power supply's noise was analyzed for frequency content, and a spike at 6.5 Hz was found. The sextupole regulators were modified with an additional filter, and in October of 1995 the upgrades were completed. Figure 5 shows an example of the motion frequency contents before and after the upgrade. As shown in Table 2, the rms motion in the 1→15 Hz band dropped from 27 to 15.5 μm. This rather unanticipated effect was related to current ripple of the S1A sextupoles at 5→8 Hz. The eddy current induced by the ripple field in the sextupole produced a vertical dipole field to drive the beam horizontally. It is noted the extruded vacuum chamber is horizontally nonsymmetric (4).

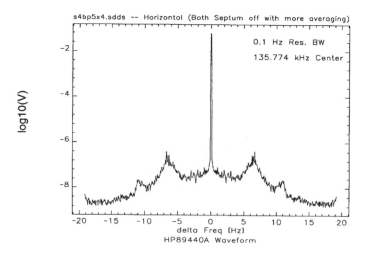

FIGURE 3. Frequency analysis of observed beam motion using the vector signal analyzer processing the rf BPM signals.

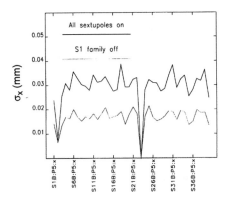

FIGURE 4. Comparison of rms beam motion with sextupole family S1 on and off. The lighter curve (S1 off) shows motion below 20 μm rms.

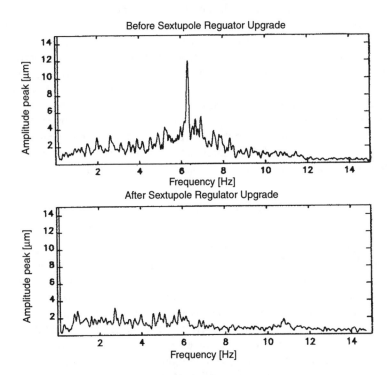

FIGURE 5. Observations of horizontal beam motion data before (upper) and after (lower) the sextupole power supply upgrade.

TABLE 2. Summary of Beam Motion Contributions Before and After the
Sextupole Power Supply Corrective Step

Bandwidth	μm rms	Condition
1 → 15	26.9	Before
	15.5	After
5 → 8	21.9	Before
	8.5	After
9 → 12	5.9	Before
	5.3	After

Transverse Size

As reported previously (8), the first qualitative transverse measurements were
in the single-turn, single-bunch beams of early operations. As the number of turns
increased, the neutral density filters in front of the imaging camera were adjusted.
The first stored beam at APS was actually done at ~ 4.2 GeV rather than 7 GeV
due to rf power limitations in the first few months of 1995. By the summer of 1995
routine single-bunch studies were underway. Initial data from both the S35 OSR
station and the sector-1 x-ray pinhole station were consistent with the 8.2-nm rad
natural emittance, about 0.1% energy spread, and a vertical coupling noticeably
lower than the 10% baseline design. As shown in Fig. 6, the horizontal size of σ_{xT}
~ 145 ± 25 μm includes contributions from the betatron emittance, the approximate 80 to 100 μm per 0.1% energy spread at this dispersive point in the lattice,
and the limiting resolution terms. Equation 2 represents this:

$$\sigma_{xT} = [\beta_x \varepsilon + (\eta_x \sigma_\varepsilon)^2 + (\sigma_{res})^2]^{1/2}. \tag{2}$$

For these data the vertical size of $\sigma_{yT} = 84 \pm 20$ μm has a significant contribution
from the $\sigma_{res} = 65 \pm 10$ μm term which is evaluated with a 550 × 40 nm bandpass
filter. The deconvolved vertical size is $\sigma_y \approx 50$ μm which implies a vertical coupling of about 3%. At the baseline 10% coupling, a vertical size of 117 μm would
be expected, and at 1% coupling a $\sigma_y = 39$ μm would be expected. These are beam
size projections only, and the β functions are assumed to be within 10% of their
design values. Subsequent tests included a parallel tracking of observed beam size
with coupling, with the vertical size ranging from 50 to 120 μm. Measurements
have also been performed as a function of single-bunch current and no significant
size increase was observed from 0.2 to 7.7 mA. These measurements were all
done with OSR imaging using a water-cooled Moly pick-off mirror whose properties were adequate up to 10 to 20 mA. The higher current runs will be addressed in
a later section.

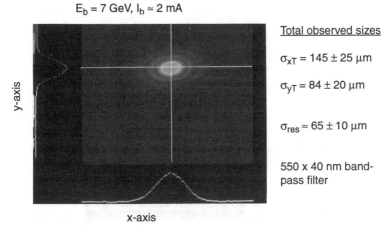

E_b = 7 GeV, I_b ≈ 2 mA

Total observed sizes

σ_{xT} = 145 ± 25 µm

σ_{yT} = 84 ± 20 µm

σ_{res} ≈ 65 ± 10 µm

550 x 40 nm band-pass filter

FIGURE 6. Transverse beam size measurements from OSR images. After removing the system resolution term contribution, the horizontal size and vertical size are consistent with baseline values.

Bunch Length

The early bunch length and longitudinal profile measurements were done initially in single-bunch mode. Initial data shown in Fig. 7 at 3-mA stored beam current are from a run on July 27, 1995. Here the dual-sweep streak camera covers about 1000 ps on the fast axis and 10 µs on the slow axis. Three successive turns in the SR are displayed with 3.68-µs spacing. The nominal bunch length is ~ 102 ps (FWHM) or 43 ps (σ). With this transverse image being rotated in the transport optics to the streak camera, the horizontal axis actually displays the y profile of the beam distribution during a single micropulse. The longitudinal profile is near Gaussian although the data appear to fall at a slightly faster rate from the peak on the later-time side.

Such measurements were performed versus single-bunch beam current during experiments in August and October 1995. Figure 8 shows a plot of bunch length in ps versus current on the horizontal axis. The extensive elongation of the bunch (tripling) from 0.2 mA to 7.7 mA is noted. Analysis of this dependency based on the Chao-Gareyte parameter led to an estimate of the ring impedance of about 0.5 Ω (10). The estimated rf gap voltage was about 6.5 MV compared to the design goal of 9.5 MV.

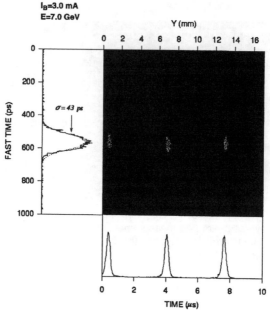

FIGURE 7. Dual-sweep streak camera images of single-bunch beam on three successive turns. The vertical axis is the fast time axis, and the horizontal axis has both 10-µs span and y-profile information.

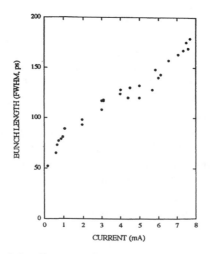

FIGURE 8. Plot of bunch length versus single-bunch current. Significant bunch lengthening with increased current was observed.

These effects are explored further in Fig. 9 where both the horizontal beam size and rms bunch length were tracked together in October 15, 1995 tests. Notable features include the relatively constant horizontal size of the beam in the dipole source point where the dispersion and energy spread product would produce the approximate horizontal sizes of 117, 147, and 215 μm for 0%, 0.1%, and 0.2% energy spread, respectively (see the dashed lines in the figure). The bunch again lengthens by almost a factor of two in comparison to the energy spread's feature. The low-current bunch length of 35 ps is consistent with even a lower rf gap voltage than the 6.5 MV in the August 19, 1995 run. Early assessments of these results indicate that the potential-well-distortion (PWD) model is more appropriate than the microwave instability model which would predict an energy spread larger than 0.2% at 7 mA stored beam current (10).

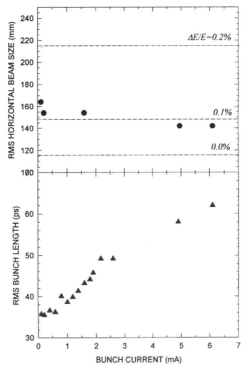

FIGURE 9. Measurements of rms horizontal beam size and rms bunch length versus single-bunch currents.

Vertical, Head-Tail Instability Observations

In an attempt to explore both measurement capability of the dual-sweep streak camera and the stable parameter space of the ring, the vertical, head-tail instability was purposely induced by decreasing the normal settings of the sextupole fields controlling vertical chromaticity. Although we initially injected and stored beam with all sextupoles off, they were needed for stored beam currents in a single bunch in excess of 1 mA to avoid the head-tail instability. The experiment was performed by first observing a dramatic increase in the transverse y-profile size in a standard OSR CCD-camera/image at about 3-mA stored beam current and decreased sextupole fields. Then the signature of the head-tail instability in a single micropulse was assessed with the dual-sweep streak camera. Figure 10(a) shows the streak images under stable conditions, and Fig. 10(b) shows the clear blurring of the y shape and the tilt in y-t space of the beam. Additionally the precession of the head-tail angle is tracked turn-by-turn in the 10-μs span or the 20-μs span. Figure 11 shows the y-t angle precesses one cycle in about four turns, and this may be connected with the fractional vertical betatron tune of 1/4. It is estimated that this instability will be controlled up to 10 to 15 mA in a single bunch with the installed sextupoles and power supplies.

HIGH CURRENT COMMISSIONING PHASE

The next thrust was to go for the baseline (100 mA) high stored beam currents. The practical aspect of machine protection against the high-powered x-ray beams from the dipole bending magnets and particularly the 10-kW insertion device (ID) beams immediately came to the forefront (11). In this section some aspects of the APS machine protection system (MPS) will be discussed as well as beam lifetime vs. vacuum chamber height, qualification of the DCCT, upgrades to the OSR photon monitor station, and actual ID closed-gap operations.

The APS storage ring is designed to be passively safe up to 100 mA from the bending magnet x-ray sources when all absorbers are properly in place and the water flow to the chambers and absorbers is as prescribed. In Fig. 12 a block diagram illustrates the architecture of the MPS which includes all the sensor inputs, the local MPS modules, the 1-MHz optical heartbeat, the main MPS module, and the rf abort features with an interruption of the low-level rf for 100 ms as the first action. The beam should coast in to the scraper in about 300 μs. Beam missteering in an ID straight is particularly threatening since the worst-case scenario allows about a 3-ms dwell time before the vacuum chamber integrity would be compromised.

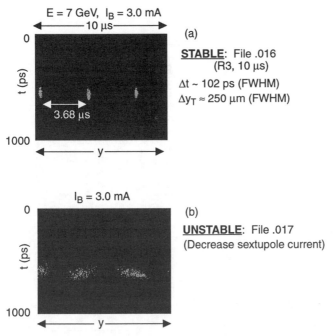

E = 7 GeV, I_B = 3.0 mA

10 μs

0

t (ps)

3.68 μs

1000

y

(a)

STABLE: File .016
(R3, 10 μs)

Δt ~ 102 ps (FWHM)

$\Delta y_T \approx 250$ μm (FWHM)

I_B = 3.0 mA

0

t (ps)

1000

y

(b)

UNSTABLE: File .017
(Decrease sextupole current)

FIGURE 10. Dual-sweep streak images of the beam (a) without and (b) with the vertical, head-tail instability.

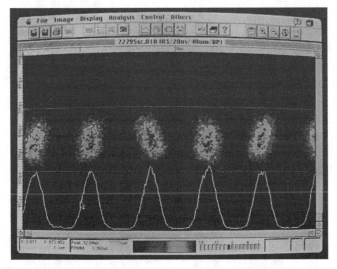

FIGURE 11. Dual-sweep streak image showing the precession of the y-t angle in about four turns.

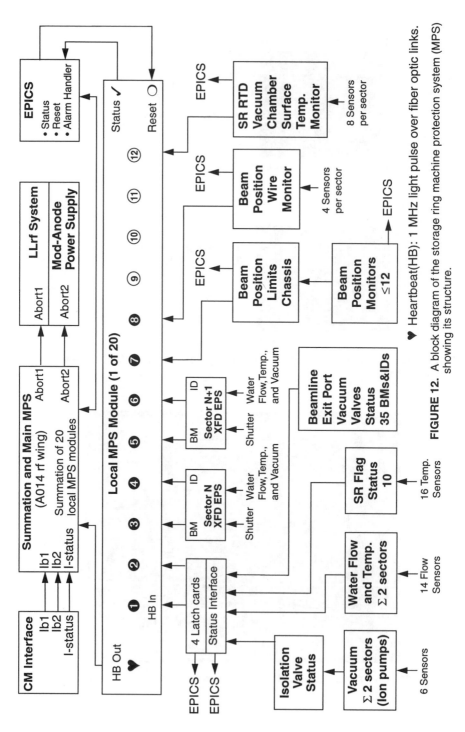

FIGURE 12. A block diagram of the storage ring machine protection system (MPS) showing its structure.

♥ Heartbeat(HB): 1 MHz light pulse over fiber optic links.

The beam position limits detector (BPLD) compares the rf BPM position readings to the preset limit of approximately ± 1.5 mm horizontally and ± 1 mm vertically. The first ID was operated with closed gap and at 100-mA stored beam current on January 26, 1996. At the time of this Workshop about six IDs and BPLDs had been commissioned. The phase I plan is 20 user IDs and one ID for particle beam characterizations.

One critical aspect of the BPLDs was the validation of the actual rf BPM offsets. These were determined by a beam-based technique using the variations of the nearby quadrupole strength and steering through the quad to determine the effective offset. Data are collected from 80 BPM locations in the ring. For the insertion device straight mini-BPMs, the standard rf BPM locations BP2 and AP2 on either side of the ID are evaluated, and a line is established to deduce the offsets for the mini-BPMs. Recent results are given in Table 3. The sector 1 ID is bounded by sector 1 and sector 2 rf BPMs. Offsets generally are in the 100- to 200-μm regime which are small compared to the trip limits at 1 to 1.5 mm. Cross-comparisons to x-BPMs in front ends were also done.

TABLE 3. Recent Results for Total Offsets of rf BPMs

ID#	BPM Location	X (mm)	Y (mm)
1	1BP1	0.132	−0.026
	2AP1	−0.267	−0.109
2	2BP1	0.046	0.137
	3AP1	0.020	−0.179
19	19BP1	−0.657	0.134
	20AP1	0.186	−0.043

As part of the evaluation of expected lifetime effects for the smaller gap chambers of 12-mm and 8-mm full gap, the vertical scrapers in the SR were used to simulate these apertures. Figure 13 shows the plot of the lifetime versus the half aperture. Lifetimes greater than five hours at low currents were achieved with the effective 8-mm full aperture. There was no measurable difference from the baseline 12-mm full aperture, so the project decided to proceed with the smaller-aperture chambers. This allows ID gaps down to about 10.5 mm and a much smoother coverage of x-ray energies from 5 to 70 keV. Lifetimes greater than five hours at low currents were achieved with the effective 8-mm full aperture.

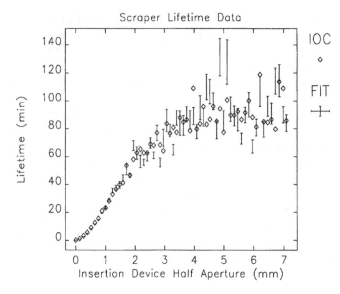

FIGURE 13. A plot of stored beam lifetime versus the half aperture. These data supported operations with the 8-mm full-aperture chamber.

Another commissioning point involved the measurement of the effects of the ID field error on the particle beam orbit. This is illustrated in Fig. 14 where the vertical orbit change around the circumference of the ring is observed between the open (45 mm) and closed (12.5 mm) gap condition. The data were measured and fit by M. Borland using the rf BPM system to detect the variation of orbit effects (1 to 10 μm) with the lattice positions and the vertical tune of about 14.3 (12).

Beam size measurements were taken with the x-ray pinhole camera at 70 to 100 mA. Transverse sizes of $\sigma_x = 158$ μm and $\sigma_y \sim 117$ μm now include the system resolution limit of about 80 μm in this first configuration as described elsewhere (9). The commissioning phase culminated with the successful storing of 100 mA in the ring as shown in the DCCT data of Fig. 15. The stored beam is electrons so the negative current is indicated in the figure around 22.6 and 23.1 to 23.4 hr on January 12, 1996. By May 1996 the baseline 10-hr lifetime at this current level also had been attained.

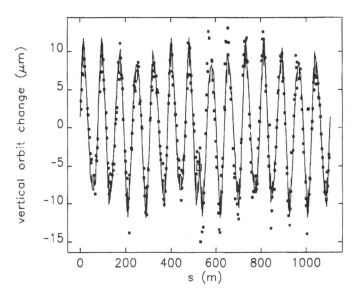

FIGURE 14. Comparison of measurement and simulation of the effects of the small ID field errors on the orbit.

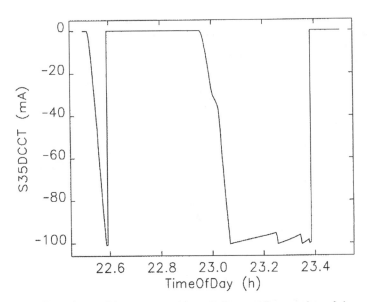

FIGURE 15. Plot of stored beam current versus time on the evening of January 12, 1996. First operations at 100-mA, the baseline objectives, are shown.

NEXT PHASE

Our next phase will include addressing the beam orbit stability with the use of a state-of-the-art digital feedback system. This system has been described previously (13), and its principles have been tested on other rings. Global tests were performed at SSRL, NSLS, and ESRF. A local feedback test was performed at ESRF. The hardware configuration is shown in Fig. 16 with additional discussion in another paper in these proceedings (14). The key features of on-board digital signal processors and the use of a reflective memory network to provide rf BPM information and corrector magnet power supply settings to each of 20 locations around the ring are important aspects of the system. As stated earlier, the orbit stability is near specifications for the baseline operations, but lower vertical coupling plans carry with them the need to stabilize the orbit even more in the future. The global feedback system is expected to be in commissioning in the Fall of 1996. A simulation of the global feedback system response is given in Fig. 17. The closed-loop noise rejection is –20 dB at 10 Hz.

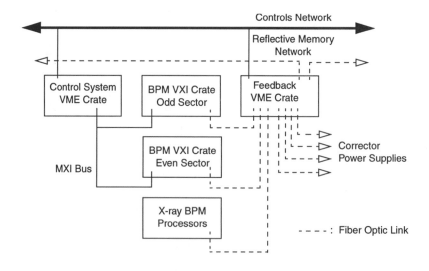

FIGURE 16. Hardware configuration of an orbit feedback station.

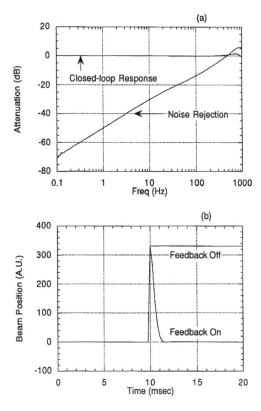

FIGURE 17. A simulation of the global feedback response to a step function displacement: (a) the attenuation vs. frequency and (b) the 1-ms response time.

SUMMARY

In summary, both the machine and the installed SR diagnostics systems have now been commissioned for high current (80 to 100 mA) runs. The rf BPMs have demonstrated the required resolution and are now being evaluated for offset effects with bunch current or bunch pattern and long-term drift. The MPS system is operational, the phase 1 x-ray pinhole imaging station is installed, and enhanced measurement capabilities with a diagnostics undulator are in the future. The investigations of bunch length with stored beam current and head-tail instabilities have already been successfully initiated. It is expected that with additional rf power installed, operations at greater than 100 mA will follow.

ACKNOWLEDGMENTS

The author acknowledges the contributions of the ASD Diagnostics staff and technicians, the many hours of overlap with Mike Borland, Louis Emery, Glenn Decker, Steve Milton, and Nick Sereno in the main control room, and the support of John Galayda of the Accelerator Systems Division.

REFERENCES

1. Galayda, J. N., "The Advanced Photon Source," Proceedings of the 1995 Particle Accelerator Conference and International Conference on High-Energy Accelerators, Dallas, TX, May 1-5, 1995, 4-8, (1996).

2. Lumpkin, A. H. et al., "Overview of Charged-Particle Beam Diagnostics for the Advanced Photon Source (APS)," Proceedings of the Fourth Accelerator Instrumentation Workshop, Berkeley, CA, Oct. 27-30, 1992, *AIP Conference Proceedings* **281**, 150-157 (1993).

3. Lumpkin, A. H. et al., "Initial Diagnostics Commissioning Results for the Advanced Photon Source (APS)," Proceedings of the 1995 Particle Accelerator Conference, Dallas, TX, May 1-5, 1995, 2473-2475 (1996) and references therein.

4. Kim, S. et al., "Investigation of Low-Frequency Beam Motion at the Advanced Photon Source," Proceedings of the Conference on Synchrotron Radiation Insrumentation, Oct. 17-20, 1995, *Rev. Sci. Instrum.* **67** (9), September 1996.

5. Kahana, E. and Chung, Y., "Commissioning Results of the APS Storage Ring RF Beam Position Monitors," these proceedings.

6. Wang, X, Lenkszus, F., and Rotela, E., "The Development of Beam Current Monitors in the APS," Proceedings of the 1995 Particle Accelerator Conference, Dallas, TX, May 1-5, 1995, 2464-2467 (1996).

7. Patterson, D., "Initial Commissioning Results from the APS Loss Monitor System," these proceedings.

8. Lumpkin, A. and Yang, B., "Status of the Synchrotron Radiation Monitors for the APS Facility Rings," Proceedings of the 1995 Particle Accelerator Conference, Dallas, TX, May 1-5, 1995, 2470-2472 (1996).

9. Yang, B. and Lumpkin, A., "Particle-Beam Profiling Techniques on the APS Storage Ring," these proceedings.

10. Chae, Y., (Argonne National Laboratory) private communication, October 1995.

11. Fuja, R. et. al. "The APS Machine Protection System (MPS)," these proceedings.

12. Borland, M., (Argonne National Laboratory) private communication, April 1996.

13. Chung, Y., "Beam Position Feedback System for the Advanced Photon Source," Proceedings of the 1993 Beam Instrumentation Workshop, *AIP Conference Proceedings* **319**, 1-20 (1994).

14. Barr, D. and Chung Y., "Hardware Design and Implementation of the Closed-Orbit Feedback System at APS," these proceedings.

Measurement of Subpicosecond Bunch Profiles Using Coherent Transition Radiation*

(1996 FARADAY CUP AWARD INVITED PAPER)

Walter Barry

Lawrence Berkeley National Laboratory, University of California, Berkeley, California 94720

Abstract. A technique for measuring the longitudinal profile of subpicosecond electron bunches based on autocorrelation of coherent transition radiation is reviewed. The technique uses sub-millimeter/far-infrared Michelson interferometry to obtain the autocorrelation of transition radiation emitted from a thin conducting foil placed in the beam path. The theory of coherent radiation from a charged particle beam passing through a thin conducting foil is presented for normal and oblique incidence. Michelson interferometric analysis of this radiation is shown to provide the autocorrelation of longitudinal bunch profile. The details of a noninvasive technique for measuring longitudinal bunch profile using coherent diffraction radiation are discussed.

INTRODUCTION

Transition radiation is a well-known phenomenon useful for measuring various charged-particle beam parameters including energy, emittance, and transverse profile (1). Typically, these measurements are made in the visible region where the radiation, due to individual particle effects, is incoherent.

In recent years, a technique which utilizes coherent transition radiation for measuring longitudinal profiles of short bunches has come into use (2,3). This technique uses Michelson interferometry to obtain the autocorrelation of coherent transition radiation emitted by short bunches passing through a thin foil. The autocorrelation of the radiation is a direct measure of the autocorrelation of the longitudinal bunch profile. For picosecond to femtosecond bunch lengths, the measurements are made in the submillimeter to far-infrared range where the bunches radiate collectively.

In this paper, a review based on reference (2) of the theoretical aspects of this longitudinal bunch profile measurement technique is presented. In addition, a noninterceptive technique for measuring bunch profile with coherent diffraction radiation is outlined (4). A companion paper (5) in these proceedings addresses experimental results of the coherent transition radiation bunch length measurement technique.

*Work supported by U.S. Department of Energy under Contract No. DE-AC03-76SF00098.

RADIATION FROM A CONDUCTING FOIL

In order to present a clear picture of the origin and properties of coherent transition radiation, the backward radiation from an arbitrary charge distribution striking a conducting foil is derived here in some detail.

Consider a beam of charged particles with velocity βc in the $\hat{z}$ direction striking a conducting foil located at the $z = 0$ plane in the cylindrical coordinate system (Fig. 1). The foil is assumed to be perfectly conducting and the beam filamentary along the z axis. For radiation wavelengths that are large compared to particle spacing, the beam is well approximated by a moving line charge distribution $q_\ell(t - z/\beta c)$, where $q_\ell(t)$ is the charge distribution of the beam measured as a function of time as the beam passes the $z = 0$ plane. Accordingly, the coherent beam current is defined as $I_0 (t - z/\beta c) = \beta c q_\ell(t - z/\beta c)$.

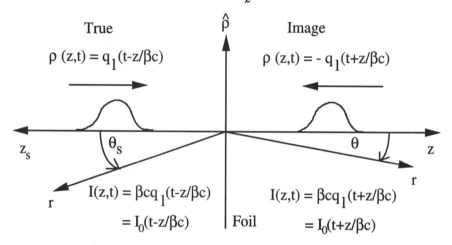

FIGURE 1. Image method for radiation from a thin foil.

In order to satisfy boundary conditions at the conducting plane, surface currents are induced in response to the fields of the incident beam. These currents in turn radiate electromagnetic energy. As indicated in Fig. 1, a convenient technique for solving the foil radiation problem makes use of image theory. Here, the conducting foil is removed and an image current $I_0(t + z/\beta c)$ is inserted for $z \geq 0$ so that

$$
I(z, t) = \begin{cases} I_0(t + z/\beta c) & z \geq 0 \\ I_0(t - z/\beta c) & z \leq 0 \end{cases} . \tag{1}
$$

With this current, it is clear that only the z component of vector potential exists and from symmetry, the vector potential and the fields derived from it are independent of ϕ.

The vector potential must satisfy the homogeneous wave equation for $\rho > 0$:

$$\nabla^2 A_z(\rho, z, t) - \frac{1}{c^2} \frac{\partial^2 A_z(\rho, z, t)}{\partial t^2} = 0 \quad \rho > 0 \tag{2}$$

subject to the excitation condition

$$\lim_{\rho \to 0} \int_0^{2\pi} H_\phi(\rho, z, t)\rho \, d\phi = I(z, t) . \tag{3}$$

The procedure for solving Eq. (2) is greatly simplified by introducing the two-dimensional Fourier transform

$$\tilde{f}(\rho, \eta, \omega) = \int_{-\infty}^{\infty} \int_{-\infty}^{\infty} f(\rho, z, t) \, e^{-j(\omega t + \eta z)} \, dt \, dz , \tag{4}$$

which has an inversion given by

$$f(\rho, z, t) = \frac{1}{4\pi^2} \int_{-\infty}^{\infty} \int_{-\infty}^{\infty} \tilde{f}(\rho, \eta, \omega) \, e^{j(\omega t + \eta z)} \, d\omega \, d\eta . \tag{5}$$

Applying Eq. (4) to Eqs. (1), (2), and (3) results in a statement of the problem in the transform domain:

$$\frac{\partial^2 \tilde{A}_z(\rho, \eta, \omega)}{\partial \rho^2} + \frac{1}{\rho} \frac{\partial \tilde{A}_z(\rho, \eta, \omega)}{\partial \rho} + (k_0^2 - \eta^2)\tilde{A}_z(\rho, \eta, \omega) = 0 \tag{6}$$

$$\lim_{\rho \to 0} \int_0^{2\pi} \tilde{H}_\phi(\rho, \eta, \omega) \, \rho d\phi = \frac{-j2k\tilde{I}_0(\omega)}{\eta^2 - k^2} , \tag{7}$$

where $k = k_0/\beta = \omega/\beta c$ and $\tilde{I}_0(\omega)$ is the frequency spectrum of the current defined by

$$\tilde{I}_0(\omega) = \int_{-\infty}^{\infty} I_0(t)e^{-j\omega t} \, dt \tag{8}$$

with inversion

$$I_0(t) = \frac{1}{2\pi} \int_{-\infty}^{\infty} \tilde{I}_0(\omega) e^{j\omega t} \, d\omega \ . \tag{9}$$

Equation (6) is Bessel's equation of order zero. Anticipating outward propagating wave characteristics, a solution to Eq. (6) is

$$\tilde{A}_z(\rho, \eta, \omega) = c(\eta, \omega) H_0^{(2)} \left(\rho \sqrt{k_0^2 - \eta^2} \right) \ . \tag{10}$$

The ϕ component of the $\vec{H}$ field can be derived from the vector potential using

$$\tilde{H}_\phi(\rho, \eta, \omega) = -\frac{1}{\mu_0} \frac{\partial \tilde{A}_z(\rho, \eta, \omega)}{\partial \rho} \ . \tag{11}$$

Using Eqs. (10), (11), and (7) and the small argument approximation for $H_1^{(2)}(x)$ yields the following expression for $c(\eta,\omega)$:

$$c(\eta, \omega) = \frac{-\mu_0 \tilde{I}_0(\omega)}{2} \left(\frac{k}{\eta^2 - k^2} \right) \ . \tag{12}$$

Therefore, from Eq. (10),

$$\tilde{A}_z(\rho, \eta, \omega) = \frac{-\mu_0 \tilde{I}_0(\omega)}{2} \left(\frac{k}{\eta^2 - k^2} \right) H_0^{(2)} \left(\rho \sqrt{k_0^2 - \eta^2} \right) \ . \tag{13}$$

Taking the $\eta \rightarrow z$ part of the inversion given in Eq. (5) yields

$$\bar{A}_z(\rho, z, \omega) = \frac{-\mu_0 \tilde{I}_0(\omega)}{4\pi} \int_{-\infty}^{\infty} \left(\frac{k}{\eta^2 - k^2} \right) H_0^{(2)} \left(\rho \sqrt{k_0^2 - \eta^2} \right) e^{j\eta z} \, d\eta \ . \tag{14}$$

Equation (14) gives, in integral form, the exact frequency domain expression for the vector potential. For the purpose of evaluating the fields in the radiation zone (ρ and z large), the integral in Eq. (14) may be approximated by the method of stationary phase giving:

$$\bar{A}_z(r, \theta_s, \omega) = \frac{-j\mu_0 \tilde{I}_0(\omega) e^{-jk_0 r}}{2\pi r} \left(\frac{k}{k_0^2 \cos^2\theta_s - k^2} \right) \quad k_0 r \gg 1 \ , \tag{15}$$

where $z = -r \cos \theta_s$, $\rho = r \sin \theta_s$. As indicated in Fig.1, r and θ_s are spherical coordinates with θ_s measured from the $-z$ axis (axis of specular reflection).

The electric and magnetic fields may be obtained in the typical manner using

$$\vec{H}(r, \theta_s, \omega) = \frac{1}{\mu_0} \nabla \times \vec{A}(r, \theta_s, \omega) \tag{16}$$

and

$$\vec{E}(r, \theta_s, \omega) = -j\omega\vec{A}(r, \theta_s, \omega) + \frac{\nabla\nabla \cdot \vec{A}(r, \theta_s, \omega)}{j\omega\mu_0\varepsilon_0}. \tag{17}$$

Resolving Eq. (15) into $\hat{r}$ and $\hat{\theta}_s$ components, substituting into Eqs. (16) and (17), and retaining only 1/r terms, expressions for the radiation fields are found:

$$\bar{E}_{\theta_s}(r, \theta_s, \omega) = \frac{Z_0\bar{I}_0(\omega)e^{-jk_0r}}{2\pi r}\left(\frac{\beta\sin\theta_s}{1 - \beta^2\cos^2\theta_s}\right) \tag{18}$$

and

$$\bar{H}_{\theta_s}(r, \theta_s, \omega) = \frac{\bar{E}_{\theta_s}(r, \theta_s, \omega)}{Z_0}, \tag{19}$$

where $Z_0 = \sqrt{\mu_0/\varepsilon_0} = 377\Omega$. Equations (18) and (19) are the frequency domain field expressions for backward transition radiation emitted from a beam striking a conducting foil.

Several important features of coherent transition radiation are evident from Eqs. (18) and (19). It is seen that the frequency spectrum of the radiation is identical to that of the beam current, at least for a foil of infinite extent. It is this property that makes the autocorrelation measurement, to be described, possible. Bandpass properties of finite foils will be addressed in a later section. Another important characteristic of transition radiation is its spatial distribution. From Eq. (18), it is clear that the angular distribution of energy or power density is given by the function

$$S^2(\theta_s) = \left(\frac{\beta\sin\theta_s}{1 - \beta^2\cos^2\theta_s}\right)^2. \tag{20}$$

This function has a single, very sharp maximum at $\theta_s = 1/\beta\gamma$. Therefore, for relativistic beams, virtually all of the radiation is in the vicinity of this extremely small angle. In this case, an excellent approximation for $S^2(\theta_s)$ is

$$S^2(\theta_s) \approx \left(\frac{\theta_s}{1/\gamma^2 + \theta_s^2}\right)^2 \qquad \gamma \text{ large } . \tag{21}$$

From Eq. (21) it is noted that the energy or power density is proportional to γ^2 at $\theta_s \approx 1/\gamma$.

For the autocorrelation technique to be described, we require $I_0(t)$ to be a simple periodic current corresponding to a train of bunches passing through the foil. By use of Eqs. (18) and (19) and the definition of the Poynting vector, the total average power per unit solid angle radiated by a periodic beam current $I_0(t)$ striking a foil can be found:

$$\frac{dP}{d\Omega} = \frac{P_{Z_0}}{4\pi^2} S^2(\theta_s) \quad \text{watts/steradian} , \tag{22}$$

where

$$P_{Z_0} = \frac{Z_0}{T} \int_T I_0^2(t) \, dt \quad T = \text{period of } I_0(t) . \tag{23}$$

The quantity P_{Z_0} is recognized as the rms power dissipated by $I_0(t)$ in a 377-Ω (free space) resistor. By integrating Eq. (22) over the backward radiation half space, the total radiated power for a relativistic beam is obtained:

$$P \approx \frac{P_{Z_0} \ln\gamma}{2\pi} \quad \text{watts} . \tag{24}$$

The frequency spectrum of the radiated power consists of discrete lines at integer multiples of $1/T$ with amplitudes proportional to the square of the Fourier transform of the bunch profile. In this case, it is clear that the critical frequency components for determining bunch profile and length are in the $1/\tau_b$ region. This region covers the far-infrared/submillimeter range for the approximate range of bunch lengths, $0.03 \text{ ps} \le \tau_b \le 3 \text{ ps}$.

AUTOCORRELATION OF BUNCH PROFILE THROUGH MICHELSON INTERFEROMETRY

A simple system for obtaining the autocorrelation of the beam current and therefore bunch profile is shown in Fig. 2. Here, the beam current $I_0(t)$ passes through a thin conducting foil at an incident angle of 45°. Backward transition radiation is then emitted about the axis of specular reflection and directed into an infrared/submillimeter wave Michelson interferometer. By measuring power at the output port of the interferometer as a function of Δ in the delay path, the autocorrelation, or equivalently, the power spectrum of $I_0(t)$ may be obtained. It is important to note that in general, the transition radiation field expressions for oblique incidence are considerably more complicated than those for normal incidence. However, as shown in the Appendix, for large γ, the field expressions about the axis of specular reflection are well approximated by the normal incidence expressions derived in the previous section.

The interferometer, illustrated in Fig. 2, basically consists of a fixed mirror, a movable mirror, and a splitter/combiner. These elements are arranged so that the incoming radiation is split into two beams. One of the beams is then delayed by a distance Δ before recombination takes place at the output port. As indicated in Fig. 2, the reflection and transmission coefficients for the splitter/combiner are designated as S_{11} and S_{21}, respectively. Both mirrors are assumed to have reflection coefficients of 1. If the divergence of the radiation is small (γ large), the total electric field at the power detector is given by

$$\vec{E}(r, \theta, \omega) = \vec{E}_1 + \vec{E}_2 = \frac{S_{11}S_{21}Z_0\bar{I}_0(\omega)}{2\pi r}e^{-jk_0r}(1 + e^{-jk_0\Delta})S(\theta)\hat{\theta} . \qquad (25)$$

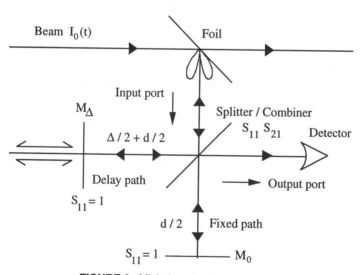

FIGURE 2. Michelson interferometer.

Dropping the overall phase factors in Eq. (25) and transforming to the time domain yields

$$E_\theta(r, \theta, t) = \frac{|S_{11}S_{21}|Z_0S(\theta)}{2\pi r}[I_0(t) + I_0(t - \tau)] , \qquad (26)$$

where $\tau = \Delta/c$. Because the radiation is confined to cone of half angle $1/\gamma$, it may be assumed that the detector measures the total available power. In this case, the total power detected as a function of τ becomes

$$P_d(\tau) = \frac{|S_{11}S_{21}|^2 ln\gamma}{\pi}\left[P_{Z_0} + \frac{Z_0}{T}\int_T I_0(t)I_0(t-\tau)\,dt\right], \qquad (27)$$

where P_{Z_0} is defined in Eq. (23).

Clearly, the second term in the brackets of Eq. (27) represents the autocorrelation of the beam current, which in turn is proportional to the autocorrelation of the bunch profile repeated periodically with period T. Thus by measuring power as the movable mirror is moved, over a distance equal to the bunch length, the longitudinal bunch profile is obtained. The resolution of this technique, for large foils, is limited only by the dispersive properties of the interferometer and the response of the power detector as discussed in ref. (5). The effect of finite foil dimensions on resolution is discussed in the next section.

DIFFRACTION RADIATION—A NONINVASIVE ALTERNATIVE

In some cases, perhaps in electron storage rings, it is desirable to have a noninvasive technique for measuring longitudinal bunch profile. This requirement can be easily accommodated by replacing the solid foil with a foil that contains a circular aperture centered on the beam axis as shown in Fig. 3. The most straightforward technique for finding the backward radiated fields in this case treats the conductor as a scatterer or backward diffractor of the incident fields of the beam.

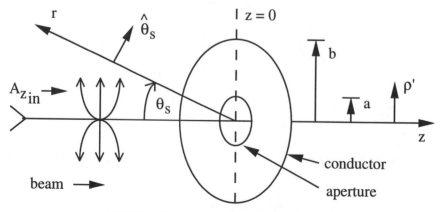

FIGURE 3. Geometry for diffraction radiation.

In the far field, i.e., $r \gg 2b^2/\lambda$, Fraunhofer diffraction theory can be used to find the fields radiated by the conductor. Here, b is the outer radius of the conductor which could conceivably be as large as the inner radius of a circular beam chamber. Based on the results of the previous analysis for a solid foil, the radiated fields are expected to be of most interest in the region about $\theta_s = 1/\gamma$. Thus, for relativistic beams, the far-field scattered vector potential can be found from the incident vector potential of the beam at the conductor using the paraxial Fraunhofer diffraction integral:

$$\bar{A}_{zs}(r, \theta_s, \omega) = \frac{-jk_0 e^{-jk_0 r}}{r} \int_a^b \rho' J_0(k_0 \rho' \sin\theta_s) \bar{A}_{zi}(\rho', \omega) \, d\rho' . \tag{28}$$

The incident vector potential of the beam has a $\hat{z}$ component only given by the well known expression

$$\bar{A}_{zi}(\rho', \omega) = \frac{\mu_0 \bar{I}_0(\omega)}{2\pi} K_0\left(\frac{k_0 \rho'}{\beta\gamma}\right) . \tag{29}$$

Substituting Eq. (29) into Eq. (28) and imposing the condition $a < \gamma\lambda/2\pi < b$, the scattered vector potential in the $\theta_s \approx 1/\gamma$ region is obtained:

$$\bar{A}_{zs}(r, \theta_s, \omega) \approx \frac{-j\mu_0 \bar{I}_0(\omega)}{2\pi r k_0}\left(\frac{1}{\theta_s^2 + 1/\gamma^2}\right) \times \tag{30}$$

$$\left[J_0(k_0 a \theta_s) - \sqrt{2}\cos(k_0 b \theta_s) e^{-\frac{k_0 b}{\gamma}} \right] \quad a < \gamma\lambda/2\pi < b .$$

The radiation fields are then found from Eqs. (16) and (17):

$$\bar{E}_{\theta_s}(r, \theta_s, \omega) \approx \frac{Z_0 \bar{I}_0(\omega)}{2\pi r} e^{-jk_0 r}\left(\frac{\theta_s}{\theta_s^2 + 1/\gamma^2}\right) \times \tag{31}$$

$$\left[J_0(k_0 a \theta_s) - \sqrt{2}\cos(k_0 b \theta_s) e^{\frac{-k_0 b}{\gamma}} \right] \quad a < \gamma\lambda/(2\pi) < b$$

and

$$\bar{H}_{\phi_s}(r, \theta_s, \omega) = \frac{\bar{E}_{\phi_s}(r, \theta_s, \omega)}{Z_0} . \tag{32}$$

For $b \to \infty$ and $a \to 0$, Eqs. (31) and (32) reduce to expressions (18) and (19) for the solid foil in the paraxial approximation.

In the ideal case, an exact measure of longitudinal bunch profile with the auto-correlation technique requires the spectrum of the radiated fields to be identical to that of the beam current $\bar{I}_0(\omega)$. Clearly, from the square-bracket term in Eq. (31), this is not the case for finite b, and $a \neq 0$. To see how this term comes about, consider the effective radius of the incident field of the beam $\rho_e \approx \gamma\lambda/2\pi$. For short wavelengths (high frequencies) the incident fields are concentrated around the beam axis. If a is larger than $\gamma\lambda/2\pi$, the incident fields pass through the aperture with very little energy scattered by the conductor. This is the origin of the low-pass term $J_0(k_0 a \theta_s)$ in Eq. (31). Similarly, if b is small compared to $\gamma\lambda/2\pi$ for long wavelengths, very little low frequency energy is scattered. This is the origin of the cosine/exponential high-pass term in Eq. (31). Together, these terms form the bandpass function in the bracket:

$$H(\omega) = J_0(k_0 a \theta_s) - \sqrt{2}\cos(k_0 b \theta_s)e^{-\frac{k_0 b}{\gamma}} . \tag{33}$$

An examination of Eq. (33) at $\theta_s = 1/\gamma$ indicates that the high- and low-frequency cutoff points occur approximately for $\lambda/2\pi = a/\gamma$ and $\lambda/2\pi = b/\gamma$, respectively. In addition, $H(\omega)$ rolls off at a rate of approximately 12 dB per octave at very low frequencies and exponentially at very high frequencies. The low-frequency rolloff will distort the measured bunch profile with tilt. Some admittedly subjective but conservative reasoning indicates that the tilt effect will be acceptable for a bunch of duration τ_b if the following condition on b is met:

$$b > 2\pi\gamma\ell_b \qquad \ell_b = c\tau_b . \tag{34}$$

The high-frequency rolloff will limit the resolution of the technique. Again, some conventional, albeit subjective, reasoning based on rise-time/bandwidth concepts indicates a minimum resolution of

$$x_\varepsilon = \frac{2\pi a}{\gamma} . \tag{35}$$

As pointed out at the end of the previous section, the bandpass characteristics of the interferometer and power detector must also be taken into account when estimating resolution. In any case, if conditions of Eqs. (34) and (35) are met, the diffraction technique should perform as well as the solid foil technique. Finally, it is mentioned that the comments in the Appendix regarding oblique incidence apply in the diffraction radiation case as well.

SUMMARY

A review of the theory of coherent transition radiation and its application in longitudinal bunch profile measurements has been presented. Through far-infra-red/submillimeter wave Michelson interferometry, the autocorrelation of longitudinal bunch profile may be obtained to a high degree of resolution using coherent transition radiation. In addition, it has been shown that the noninvasive technique, diffraction radiation, holds great promise as a standard longitudinal bunch diagnostic.

APPENDIX

Referring to Fig. A, consider a line charge distribution, $\rho_1(z_1,t) = q_\varrho(t + z_1/\beta c)$, traveling along the z_1 axis at an angle θ_i from the normal to the foil. Corresponding to ρ_1, there will be an image charge distribution, $\rho_2(z_2,t) = -q_\varrho(t + z_2/\beta c)$, traveling toward the foil along the z_2 axis. Following the conventions in the main text, the corresponding currents are

$$I_1(z_1 t) = -\beta c q_\varrho(t + z_1/\beta c) = -I_0(t + z_1/\beta c) \tag{A.1}$$

and

$$I_2(z_2 t) = \beta c q_\varrho(t + z_2/\beta c) = I_0(t + z_2/\beta c) \ . \tag{A.2}$$

Repeating the analysis in the main text for each current individually yields the following vector potentials:

$$\vec{A}_1(r_1, \theta_1, \omega) = \frac{-j\mu_0 \bar{I}_0(\omega) e^{-jk_0 r_1}}{4\pi r_1} \left(\frac{1}{k + k_0 \cos\theta_1} \right) \hat{z}_1 , \tag{A.3}$$

$$\vec{A}_2(r_2, \theta_2, \omega) = \frac{-j\mu_0 \bar{I}_0(\omega) e^{-jk_0 r_2}}{4\pi r_2} \left(\frac{1}{k + k_0 \cos\theta_2} \right) \hat{z}_2 . \tag{A.4}$$

The total vector potential is the sum of (A.3) and (A.4).

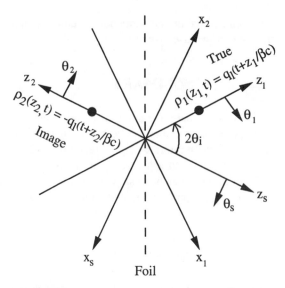

FIGURE A. Image geometry for oblique incidence.

The vector potential may be written in terms of the axis of specular reflection, z_s and its associated coordinate system by making the following coordinate and unit vector transformations:

$$\left\{ \begin{array}{c} r_1 = r_2 = r_s = r \\ \cos\theta_2 = -\cos\theta_s \\ \cos\theta_1 = \cos\theta_s\cos 2\theta_i - \sin\theta_s\cos\phi_s\sin 2\theta_i \\ \hat{z}_2 = -\hat{z}_s \\ \hat{z}_1 = -\hat{z}_s\cos 2\theta_i - \hat{x}_s\sin 2\theta_i \end{array} \right\} . \tag{A.5}$$

Substituting these transformations into (A.3) and (A.4) results in a total vector potential possessing $\hat{x}_s$ and $\hat{z}_s$ components given as follows:

$$\bar{A}_{x_s}(r, \theta_s, \phi_s, \omega) = \frac{j\mu_0\bar{I}_0(\omega)}{4\pi r}\left[\frac{\sin 2\theta_i}{k + k_0(\cos\theta_s 2\theta_i - \sin\theta_s\cos\phi_s\sin 2\theta_i)}\right]e^{-jk_0 r} \tag{A.6}$$

$$\bar{A}_{z_s}(r, \theta_s, \phi_s, \omega) = \frac{-j\mu_0\bar{I}_0(\omega)}{4\pi r}\left[\frac{1}{k - k_0\cos\theta_s} + \frac{\cos 2\theta_i}{k + k_0(\cos\theta_s\cos 2\theta_i - \sin\theta_s\cos\phi_s\sin 2\theta_i)}\right]e^{-jk_0r}. \tag{A.7}$$

For normal incidence ($\theta_i = 0$), $\bar{A}_{x_s}$ goes to zero and $\bar{A}_{z_s}$ reduces to Eq. (15).

It is clear from Eqs. (A.6) and (A.7) that the fields show some ϕ variation about the axis of specular reflection for the case of oblique incidence. However, for large γ, a careful study of the fields resulting from Eqs. (A.6) and (A.7) reveals that the first term in the brackets of Eq. (A.7) greatly dominates all other terms so that the vector potential is given approximately by

$$\bar{A}_{z_s}(r, \theta_s, \omega) \approx \frac{-j\mu_0\bar{I}_0(\omega)}{4\pi r}\left[\frac{1}{k - k_0\cos\theta_s}\right]e^{-jk_0r}. \tag{A.8}$$

Equation (A.8) in the small angle approximation becomes

$$\bar{A}_{z_s}(r, \theta_s, \omega) \approx \frac{-j\mu_0\bar{I}_0(\omega)}{2\pi r k_0}\left[\frac{1}{1/\gamma^2 + \theta_s^2}\right]e^{-jk_0r}. \tag{A.9}$$

This expression agrees with Eq. (15) in the main text for small θ_s. Therefore, for relativistic beams, the fields for oblique incidence are well approximated by normal incidence fields.

REFERENCES

1. R. B. Fiorito and D. W. Rule, "Optical Transition Radiation Beam Emittance Diagnostics," Beam Instrumentation Workshop, Santa Fe, NM, October 1993, *AIP Conference Proceedings* **319,** 21-37 (1994).
2. W. Barry, "An Autocorrelation Technique for Measuring Sub-Picosecond Bunch Length Using Coherent Transition Radiation," *Proceedings of the Workshop on Advanced Beam Instrumentation*, KEK, Tsukuba, Japan, April 22-24, 1991, KEK Proceedings 92-1, 224-235, June 1991.
3. H. Lihn, P. Kung, H. Wiedemann, and D. Bocek, "Measurement of 50-fs (rms) Electron Pulses," Beam Instrumentation Workshop, Vancouver, B. C. Canada, October 1994, *AIP Conference Proceedings* **333,** 231-237 (1995).
4. Y. Shibata et al., "Observation of Coherent Diffraction Radiation From Bunched Electrons Passing Through a Circular Aperture in the Millimeter- and Submillimeter-Wavelength Regions," *Physical Review E* **52** (6), 6787-6794, December 1995.
5. H. Lihn, "Measurement of Subpicosecond Electron Pulse Length," these proceedings.

Measurement of Subpicosecond Electron Pulse Length*

(1996 FARADAY CUP AWARD INVITED PAPER)

Hung-chi Lihn[†]

Applied Physics Department and Stanford Linear Accelerator Center
Stanford University, Stanford, California 94305

Abstract. A new frequency-resolved bunch-length measuring system has been developed at the Stanford SUNSHINE facility to characterize subpicosecond electron pulses. Using a far-infrared Michelson interferometer, this method measures the spectrum of coherent transition radiation emitted from electron bunches through optical autocorrelation. The electron bunch length is obtained from the measurement with a simple and systematic analysis that includes interference effects caused by the beam splitter. This method demonstrates subpicosecond resolving power that cannot be achieved by existing time-resolved methods. The principle of this method and experimental results are discussed.

INTRODUCTION

An interesting direction in the recent developments of particle accelerators is the production of very short electron bunches. Many future accelerator designs, such as next-generation synchrotron light sources, future linear colliders, free-electron lasers, and high-intensity coherent far-infrared light sources, demand electron bunches of subpicosecond duration. Hence, a bunch-length measuring system capable of characterizing subpicosecond pulses will provide a powerful diagnostic tool to assist this development.

Normally a time-resolved method, which resolves the beam-generated signal in the time domain, is applied to measure the bunch length. However, existing fast time-resolved methods such as the streak camera cannot provide enough resolution for subpicosecond electron pulses, and the hardware has become very complicated and expensive. In contrast, a frequency-resolved technique, which extracts the frequency content of a beam-generated signal, does not require fast processing speed and complex hardware. From the frequency information, the particle distribution can be deduced. Since the necessary broad bandwidth required for subpicosecond pulses can be achieved by optical methods, a good subpicosecond resolution can be easily obtained. This is a well-known technique used in the characterization

---

* Work supported by U. S. Department of Energy under Contract No. DE-AC03-76SF00515.

† Present address: Tencor Instruments, 2400 Charleston Road, Mountain View, California 94043.

of femtosecond laser pulses (1) and has been suggested for electron bunch-length measurements (2).

At the SUNSHINE facility, we have developed a new bunch-length measuring system based on this frequency-resolved technique using electron pulses generated at SUNSHINE (3-5). We have demonstrated that this simple, low-cost system is suitable for subpicosecond bunch-length measurements. In this paper, we will describe the principle of this technique, an analysis of bunch-length measurements, and experimental results.

PRINCIPLE OF AUTOCORRELATION METHOD

This frequency-resolved method uses a far-infrared Michelson interferometer to measure the spectrum of coherent transition radiation through optical autocorrelation. Coherent transition radiation emitted from electron pulses carries the information of bunch distribution in its frequency content. By analyzing the frequency information, the bunch length can be derived.

Coherent Transition Radiation

Transition radiation is generated when an electron passes the interface of two media of different dielectric constants (6). For a vacuum-metal interface, the spectrum of transition radiation is approximately flat in the far-infrared regime due to the almost perfect conductivity of the metal. The radiated intensity from a relativistic electron has a zero at $\theta = 0$ and reaches maximum at $\theta \sim 1/\gamma$, where θ is the angle between the radiation and the electron direction and γ is the Lorentz factor (5-7).

When a bunch of N electrons passes the interface, the resulting total intensity at angular frequency ω can be expressed as (8)

$$I_{\text{total}}(\omega) = N[1 + (N-1)f(\omega)] I_e(\omega) , \tag{1}$$

where $I_e(\omega)$ is the intensity of transition radiation emitted by an electron at frequency ω. In the far-infrared regime, $I_e(\omega)$ is constant. The form factor $f(\omega)$ is given by the three-dimensional Fourier transform of the electron bunch distribution (5,7,8). If the radiation can be observed in the forward direction ($\theta = 0$) from a transversely symmetric beam, the form factor $f(\omega)$ is only determined by the longitudinal bunch distribution.

However, transition radiation does not produce radiation in the forward direction. In order to use transition radiation to measure the bunch length, it is necessary to observe the radiation in an off-axis direction ($\theta \neq 0$). In this case, the transverse bunch distribution will contribute to the form factor even for a transversely symmetric beam. For large angles or large transverse beam sizes, the transverse contribution will result in an apparent bunch-length measurement that is longer than the actual one (5,7). This transverse contribution, however, can be ignored if the condition $\rho \tan \theta \ll l$ is satisfied, where ρ is the transverse beam size and l is the bunch length (5,7). This condition is assumed in the following analysis and is automatically obtained when the beam conditions are optimized in experiments. Hence, proper focusing to produce small transverse beam size and a reasonable angular acceptance for the detector is crucial for accurate subpicosecond bunch-length measurements.

Far-infrared Michelson Interferometer

Since the spectrum of coherent transition radiation emitted by subpicosecond electron bunches is in the far-infrared regime (4,7), a far-infrared Michelson interferometer (shown in Fig. 1) is used to measure the spectrum via optical autocorrelation. It consists of a beam splitter, a fixed and a movable mirror, and a detector. When light enters the Michelson interferometer, the beam splitter splits its amplitude into two mirror arms. As these two rays are reflected from the mirrors, they are recombined at the beam splitter and sent into the detector.

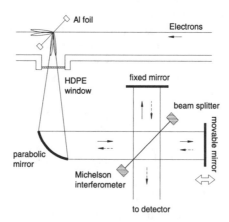

FIGURE 1. Schematic diagram of a Michelson interferometer for bunch-length measurements.

An ideal beam splitter has constant amplitude reflection (R) and transmission (T) coefficients over all frequencies, which satisfy $|R|^2 = |T|^2 = 1/2$. As shown in Fig. 1, for an incoming light pulse of electric field E with intensity proportional to $|E|^2$, the light pulse split by the beam splitter and reflected by the fixed mirror has a field amplitude of $T(RE)$ when it reaches the detector. The light pulse reflected by the movable mirror has an amplitude of $R(TE)$ at the detector. Perfect reflection on the mirrors is assumed. At zero optical path difference, the pulses completely overlap at the detector, and the total intensity reaches the maximum $|2RT E|^2 = |E|^2$. As the path difference increases but remains shorter than the bunch length, the two pulses overlap partially, and the total intensity decreases. When the path difference of two arms is longer than the bunch length, the two pulses are totally separated in time, and the resulting intensity at the detector is $2|RT E|^2 = |E|^2/2$. The intensity is constant over all path differences greater than the bunch length and is called the *baseline*. The variation of intensity about the baseline as a function of optical path difference is called the *interferogram*. Therefore, the width of the peak in the interferogram can be used to measure the bunch length.

The intensity of the recombined radiation at the detector can be expressed in the time domain with an additional time delay δ/c for the movable arm by

$$I(\delta) \propto \int_{-\infty}^{+\infty} \left| T\, RE(t) + RT\, E\!\left(t + \frac{\delta}{c}\right) \right|^2 dt$$

$$= \underbrace{2|RT|^2\ \text{Re}\int_{-\infty}^{+\infty} E(t)E^*\!\left(t + \frac{\delta}{c}\right)dt}_{S(\delta)} + \underbrace{2|RT|^2\int_{-\infty}^{+\infty} |E(t)|^2 dt}_{I_\infty}\ , \qquad (2)$$

or in the frequency domain with a phase factor $e^{-i\omega\delta/c}$ for the movable arm by

$$I(\delta) \propto \int_{-\infty}^{+\infty} \left| T\, R\tilde{E}(\omega) + RT\, \tilde{E}(\omega)e^{-i\omega\delta/c} \right|^2 d\omega$$

$$= \underbrace{2\ \text{Re}\int_{-\infty}^{+\infty} |RT|^2|\tilde{E}(\omega)|^2 e^{-i\omega\delta/c} d\omega}_{S(\delta)} + \underbrace{2\int_{-\infty}^{+\infty} |RT|\ |\tilde{E}(\omega)|^2 d\omega}_{I_\infty}\ , \qquad (3)$$

where δ is the optical path difference, c is the speed of light, $S(\delta)$ denotes the interferogram, and I_∞ is the baseline defined as the intensity at $\delta \to \pm\infty$. Equations (2) and (3) are related by the Fourier transform $\tilde{E}(\omega) = \frac{1}{\sqrt{2\pi}}\int_{-\infty}^{+\infty} E(t)e^{i\omega t}dt$. Therefore, the interferogram $S(\delta)$ is the autocorrelation of the incident light pulse (cf., Eq. (2)), and its Fourier transform is the power spectrum of the pulse (cf., Eq. (3)). Solving for $|\tilde{E}(\omega)|^2$ in Eq. (3) and using Eq. (1) with the relation $I_{\text{total}}(\omega) \propto |\tilde{E}(\omega)|^2$, the bunch form factor can be obtained by

$$f(\omega) \propto \frac{1}{N-1}\left[\frac{1}{4\pi c|RT|^2 NI_e(\omega)}\int_{-\infty}^{+\infty} S(\delta)e^{i\omega\delta/c}d\delta - 1\right] , \qquad (4)$$

where $|\tilde{E}(\omega)|^2 = |\tilde{E}(-\omega)|^2$ is used since $E(t)$ is a real function. Hence, the interferogram contains the frequency spectrum of coherent transition radiation and can be used to derive the bunch length.

Mylar Beam-splitter Interference Effects

Suitable beam splitters for the far-infrared regime (a Mylar foil in our design) do not provide constant and equal reflectance and transmittance for all frequencies. This is caused by the interference of light reflected from both surfaces of the beam splitter, which is equivalent to thin-film interference in optics (9). The total amplitude reflection and transmission coefficients for a Mylar foil of thickness t and refractive index n mounted at a 45° angle to the direction of incoming light are given respectively by (7,10)

$$R = -r\frac{1-e^{i\phi}}{1-r^2 e^{i\phi}} \quad \text{and} \quad T = (1-r^2)\frac{e^{i\phi/2}}{1-r^2 e^{i\phi}} , \qquad (5)$$

where r is the amplitude reflection coefficient of the air-to-Mylar interface at a 45° incident angle and ϕ is defined as $4\pi t\sigma\sqrt{n^2 - (1/2)}$ at wave number $\sigma = \omega/2\pi c$ (7,11). No absorption in the foil is assumed, and the refractive index is assumed to be constant ($n = 1.85$) over all frequencies (11).

The efficiency of the beam splitter defined as $|RT|^2$ is shown in Fig. 2. The efficiency becomes zero at frequencies where light reflected from both surfaces of the beam splitter interferes destructively. Derivations in the time domain, such as Eq. (2), are no longer valid for varying efficiency; however, those in the frequency domain, such as Eq. (3), still hold. The width of the interferogram cannot be directly used for bunch-length estimation unless interference effects on the interferogram are included.

Bunch-length Analysis

The interference effects are studied numerically for both Gaussian and rectangular bunch distributions, and the bunch length is then derived from this study.

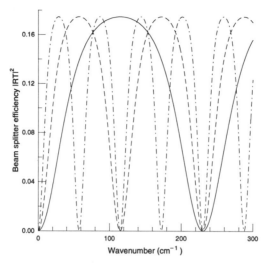

FIGURE 2. The efficiency of Mylar beam splitter as a function of frequency for different thicknesses: 12.7 (solid), 25.4 (dashed), and 50.8 μm (dash-dotted line).

Although real bunch distributions are neither Gaussian nor rectangular, the bunch lengths deduced from the two distributions will give reasonable bounds for the real one.

Examples of the beam-splitter interference effects on a rectangular bunch distribution are shown in Fig. 3. For an ideal beam splitter, the interferogram has the expected triangular peak with its FWHM equal to the bunch length (cf., Fig. 3(a)). For Mylar beam splitters, negative valleys appear in the interferograms, which are due to suppression of the low frequency area by the first zero of the beam-splitter efficiency. These valleys move closer to the main peak as the beam-splitter thickness (t) decreases (cf., Fig. 3(b)-(d)). For beam splitters thinner than about half the bunch length (l_b), they merge with and narrow the main peak (cf., Fig. 3(d)). The effects are similar for a Gaussian distribution. Detailed results on how the FWHM values of the interferogram change with the bunch length for both Gaussian and rectangular distributions are shown in Fig. 4. Once the beam splitter is chosen, the bunch length can be derived from the measured interferogram width through Fig. 4.

EXPERIMENTAL SETUP

To demonstrate this method, the SUNSHINE facility was operated to produce 1-μ pulse trains at 10 Hz containing about 3000 electron bunches at an energy of 30 MeV. Each bunch had about 3.5×10^7 electrons. As shown in Fig. 1, transition radiation is generated when the electrons pass through an Al foil. The divergent

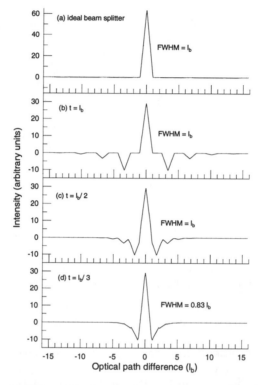

FIGURE 3. The simulation of the beam-splitter interference effects on a rectangular bunch distribution with different beam splitters: (a) an ideal beam splitter and   Mylar beam splitters of thicknesses (t), (b) equal to, (c) half, and (d)   one-third of the bunch length (l_b).

backward transition radiation is extracted from the evacuated beamline into air via a high-density polyethylene (HDPE) window and is converted into parallel light by an off-axis paraboloidal mirror (12). The parallel light then enters a far-infrared Michelson interferometer.

The interferometer consists of a Mylar beam splitter mounted at a 45° angle to the direction of incident light, a fixed and a movable first-surface mirror, and a room-temperature detector. The movable mirror is moved by a linear actuator via a PC. The detector consists of a Molectron P1-65 LiTaO$_3$ pyroelectric bolometer and a pre-amplifier (3,4,12). The detector signal is digitized and stored in the computer. The autocorrelation measurements are performed automatically via computer interfaces and the control program running in the LabVIEW environment.

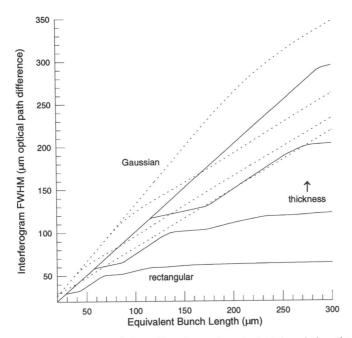

FIGURE 4. Interferogram FWHMs as functions of equivalent bunch lengths of both Gaussian (dotted lines) and rectangular (solid lines) bunch distributions for different Mylar beam-splitter thicknesses: 12.7, 25.4, 50.8, and 127 μm. Within the same distribution, the lines are shown from the bottom to the top in increasing order of thickness.

RESULTS AND DISCUSSION

By measuring the detector signal as a function of the position of the movable mirror via the computer program, the interferograms of 2.2 mm long with 5-μm mirror step size are measured for four different Mylar beam-splitter thicknesses and shown in Fig. 5. This 5-μm mirror step size corresponding to a 33-fs time resolution is good enough for the measurements; however, a subfemtosecond resolution can still be achieved by the actuator with a submicron step size. The beam parameters are kept the same when different beam splitters are used. The valleys around the main peak are separated farther apart as the beam-splitter thickness increases. This widens the main peak (cf., Fig. 5(a)-(c)) until the valleys are out of the peak (cf., Fig. 5(c),(d)). The base of the peak can even be seen in Fig. 5(d). The measured interferogram FWHMs and the estimated equivalent bunch lengths deduced from Fig. 4 for Gaussian and rectangular distributions are shown in Table 1. The estimated bunch lengths provide bounds for the real bunch length and are consistent over a 10-fold change in the beam-splitter thickness. The real bunch length is estimated about 100 μm (0.33 ps) long.

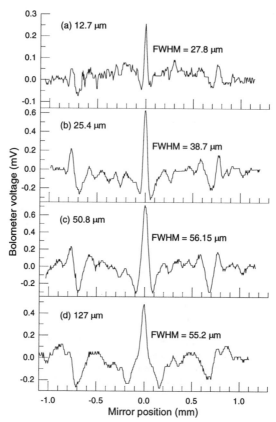

FIGURE 5. Interferograms measured for different Mylar beam-splitter thicknesses: (a) 12.7, (b) 25.4, (c) 50.8, and (d) 127 mm. The FWHMs of the main peaks are measured in mirror movement, which is only half of the actual optical path difference.

TABLE 1. Measured Interferogram FWHMs in Optical Path Difference (OPD) for Different Beam-splitter Thicknesses and the Corresponding Estimated Equivalent Bunch Lengths Deduced from Fig. 4 for Gaussian and Rectangular Distributions

Beam splitter thickness (μm)	Interferogram FWHM OPD (μm)	Estimated equivalent bunch length (μm)	
		Gaussian	Rectangular
12.7	55.6	60.8	100.6
25.4	77.4	72.8	103.2
50.8	112.3	86.9	111.0
127.0	110.4	83.1	109.1

Nondestructive methods such as using a bending magnet to generate coherent synchrotron radiation and using a metal foil with a hole to generate coherent diffraction radiation are suitable for applications where keeping good beam quality is essential. Unlike the case for transition radiation, the single-electron spectrum of these radiating processes is not constant in frequency. In order to extract bunch information $f(\omega)$, the measured spectrum has to be numerically corrected for the single-electron spectrum $I_e(\omega)$.

In principle, the measured spectral information can be used to reconstruct the bunch distribution and give a better bunch-length measurement. However, there are some practical difficulties in reconstructing the electron distribution for this experiment. The spectrum is contaminated by water absorption lines caused by humidity in ambient air (4), and the zeros of the beam-splitter efficiency produce artifacts when the spectrum is numerically corrected for the beam-splitter interference effects (4). Although one-dimensional phase-retrieval methods have been suggested for this reconstruction problem (13,14), they cannot guarantee the uniqueness of the solution (7). Alternatively, a new method described in ref. (7) using two-dimensional autocorrelation scans with a three-dimensional phase-retrieval method may be used to obtain the three-dimensional bunch distribution with the guaranteed uniqueness of the solution.

CONCLUSION

In conclusion, a new frequency-resolved bunch-length measuring method suitable for subpicosecond electron pulses has been developed at the SUNSHINE facility. This method measures the autocorrelation of coherent transition radiation emitted from electron bunches through a far-infrared Michelson interferometer. Measurements have verified this method by showing consistent results over a broad range of beam-splitter thicknesses. Based on a low-cost, easy-to-operate, compact, and transportable Michelson interferometer, this autocorrelation method demonstrates a subpicosecond resolving power far better than existing time-resolved methods.

ACKNOWLEDGEMENTS

The author would like to thank Prof. H. Wiedemann at Stanford University for his advice and support of this work. Thanks also go to D. Bocek, P. Kung, J. Sebek, C. Settakorn, and R. Theobald for their technical assistance.

REFERENCES

1. Fork, R. L., Greene, B. I., and Shank, C. V., *Appl. Phys. Lett.* **38**, 671 (1981).
2. Barry, W., in *Proceedings of the Workshop on Advanced Beam Instrumentation* **1**, KEK, Tsukuba, Japan, April 22-24, 1991, KEK Proceedings 91-2, 224-235, June 1991.
3. Kung, P., Lihn, H.-C., Bocek, D., and Wiedemann, H., *Proc. SPIE* **2118**, 191 (1994).
4. Kung, P., Lihn, H.-C., Bocek, D., and Wiedemann, H., *Phys. Rev. Lett.* **73**, 967 (1994).
5. Lihn, H.-C., Kung, P., Settakorn, C., Bocek, D., and Wiedemann, H., *Phys. Rev. E* **53**, (6), 6413-6418 (1996).
6. Ginsburg, V. L., and Frank, I. M., *Zh. Eksp. Teor. Fiz.* **16**, 15 (1946).
7. Lihn, H.-C., *Stimulated Coherent Transition Radiation*, Ph.D. thesis, Stanford University, Stanford, 1996.
8. Nodvic, J. S., and Saxon, D. S., *Phys. Rev.* **96**, 180 (1954).
9. Hecht, E., and Zajac, A., *Optics*, Addison-Wesley, Reading, Massachusetts, 1974, Sec. 9.7.
10. Chantry, G. W., *Submillimetre Spectroscopy*, Academic Press, London, 1971, App. A.
11. Bell, R. J., *Introductory Fourier Transform Spectroscopy*, Academic Press, London, 1972, Ch. 9.
12. Lihn, H.-C., Kung, P., Bocek, D., and Wiedemann, H., *AIP Conference Proceedings* **333**, 231-237 (1995).
13. Fienup, J. R., *Opt. Lett.* **3**, 27 (1978).
14. Lai, R., Happek, U., and Sievers, J., *Phys. Rev. E* **50**, R4294 (1994).

CONTRIBUTED PAPERS

Overview of Coupled-Bunch Active Damper Systems at FNAL*

James Steimel, Jim Crisp, Hengjie Ma, John Marriner,
and Dave McGinnis

Fermi National Accelerator Laboratory
P.O. Box 500 Batavia, IL 60510

Abstract. Beam intensities in all of the accelerators at Fermilab will increase significantly when the Main Injector becomes operational and will cause unstable oscillations in transverse position and energy. Places where the coupled bunch oscillations could dilute emittances include the Booster, Main Injector, and Tevatron. This paper provides an overview of the active feedback system upgrades which will be used to counteract the problem. It will explain the similarities between all the systems and will also explain design differences between longitudinal and transverse systems, fast sweeping systems, and systems for partially filled machines. Results from operational systems will also be shown.

INTRODUCTION

The accelerators at Fermilab provide a wide range of challenges for designing and constructing beam feedback systems. The basic concept of any beam feedback mechanism involves detecting an error in position, time, or energy of a bunch at the pickup; processing this signal; and holding it until the bunch arrives at the kicker for correction. Challenges in the design of the system include dealing with the large dynamic range of the pickup signal, maintaining the proper timing and phase advance from pickup to kicker, providing a large enough kick to the bunch, and providing a means to diagnose mismatches in delay and phase as accelerator parameters are tuned.

Table 1 illustrates the differences between the three accelerator rings at Fermilab. In particular, one should note the total change in acceleration voltage frequency, which determines the total change in delay from pickup to kicker, and also note the percentage of buckets filled, which determines the kind of front-end processing required. All of the rings have or will have coupled bunch dampers for every plane. The rest of this paper discusses the similarities and differences between damper systems for different planes and different machines.

---

* Operated by the University Research Association, Inc. under contract with the U.S. Department of Energy.

199

TABLE 1. Fermilab Accelerator Parameters

Ring	Energy	RF Frequency	RF Buckets	Bunches
Booster	0.4 - 8 GeV	37-52.8 MHz	84	84[a]
Main Ring[b]	8-150 GeV	52.8-53.1 MHz	1113	1,12,84
Main Ring[c]	8-150 GeV	52.8-53.1 MHz	1113	1092
Main Inj[b]	8-150 GeV	52.8-53.1 MHz	588	1,12,84
Main Inj[c]	8-150 GeV	52.8-53.1 MHz	588	504
Tevatron[b]	150-1000 GeV	53.1 MHz	1113	1 - 36
Tevatron[c]	150-1000 GeV	53.1 MHz	1113	1092

[a]Booster may have a one-bucket gap for firing the extraction kickers.
[b]Collider Mode: Main Ring/Inj is used for coalescing and stacking.
[c]Fixed Target Mode: Machines are full.

TRANSVERSE DAMPERS

Construction of transverse dampers is not as costly as longitudinal dampers because the gain of the system does not need to be as high (1). Consequently, the power amplifiers used to drive the kickers have lower power requirements, and the kickers can have a much wider bandwidth. Because the bandwidth of the system can be so great without extraordinary cost in power amplifiers, most transverse systems are designed to handle all possible coupled-bunch modes. Figure 1 shows a block diagram of the transverse damper system that is used or will be used in all of the accelerators. All of the systems include a stripline pickup and kicker, an analog front end/auto-zero, a digital notch filter, and a wideband power amplifier. The only differences between the systems for the different accelerators are the bandwidth of the stripline components, the bandwidth and processing of the analog front end, the trigger generation for the digitizers, and the power capabilities of the power amplifiers.

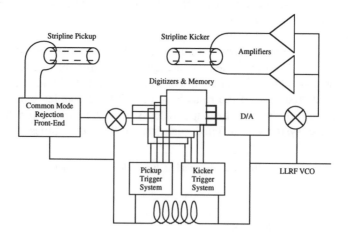

FIGURE 1. Transverse damper block diagram.

Stripline Components

The specifications of the stripline length are not critical for the damper system. The pickup stripline needs to be short enough to have a passband as wide as the value of the rf frequency, but it must be long enough not to have too much attenuation at the operating frequency of the damper. All of the current transverse dampers operate at the rf frequency. The quarter-wave frequencies of the Booster, Main Ring, and Tevatron pickups are 120MHz, 212MHz, and 53MHz, respectively. The pickups can be short in this case because the signal is so large. Even at the rf frequency, the signal from the pickup is attenuated to accommodate the maximum input level of the analog front end. The quarter-wave frequencies of the stripline kickers for the Booster, Main Ring, and Tevatron are 75MHz, 75MHz, and 53MHz, respectively.

Analog Front End / Auto-Zero

Signals due to betatron motion of the beam are at least 60dB down from the fundamental signal in the Fermilab accelerators. No digitizer has the combination of speed and dynamic range to process the signal directly, so an analog circuit was designed to enhance the betatron signal (2). The circuit enhances the betatron signal by creating a virtual electrical center in the pickup which tracks slow variations in beam position while passing variations in beam position associated with betatron motion. This technique will stop signal due to closed orbit error, bunch-to-bunch intensity variations, and mode-0 synchrotron motion at high dispersion.

Figure 2 shows a block diagram of the system. The signals from the two sides of the stripline pickup are input into a 0°/180° hybrid. From the hybrid, the sum and difference signal are filtered with matched filters, attenuated, and input

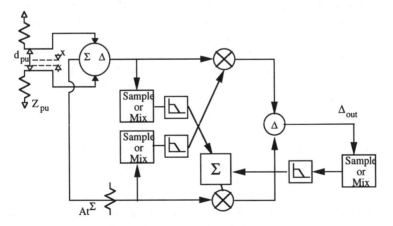

FIGURE 2. Auto-zero block diagram.

into a pair of linear analog multipliers. The difference between the two products gives the difference signal with the electrical center displaced. Tracking is provided by the other factors on the multipliers which are derived from the input signals in the feedforward case or the final difference signal in the feedback case. The method used to derive the tracking signal varies with accelerator parameters.

Tracking with Every Bucket Filled

Every bucket in the Fermilab Booster ring is full during acceleration. The dominant signals, from an uncompensated pickup, are due to the fundamental rf and harmonics of the rf. The tracking for the multipliers can be derived by mixing the error signals with the fundamental rf frequency and low-pass filtering the signals just above the maximum synchrotron frequency. If phase errors between the sum and difference legs of the system cause insufficient rejection of the higher harmonics of the rf, extra feedback multipliers can be added which cancel out only the higher harmonics.

Tracking in a Partially Filled Machine

Reference (2) explains the details of the auto-zero circuit for a partially filled machine. It is important to realize that the mixers used to derive the tracking signal for the full machine will not work as well for the partially filled machine because the beam spectrum will be more evenly distributed among the revolution harmonics. Thus, it is better to use sample and hold modules, which can track a single bunch, in place of the mixers.

Digital Processing

The purpose of the digital processing is to provide the proper delay and phase shift at all of the betatron coupled-bunch frequencies to damp all of the modes. It also notches out any remaining common mode signal from the auto-zero circuit, as well as notching out any signal due to coupled-bunch synchrotron oscillations which the auto-zero circuit cannot detect. The digital processing circuit also acts as an advanced damper diagnostic tool capable of measuring beam spectrums and beam transfer functions.

After the auto-zero circuit, the difference signal is mixed down to baseband, anti-alias filtered, amplified, and sent to the digitizers. The digital processing consists of from three to seven interleaved, 8-bit digitizers with dual port memories and two trigger systems to control the digitizers. One trigger system controls the sampling time of all the digitizers, while the other trigger system controls the output data latch of the digitizers. The trigger systems are referenced to the accelerator voltage controlled oscillator with different delays, and each of the trigger systems can operate asynchronously with respect to one another. With the delays set properly, the system will maintain the proper timing from pickup to kicker, even if the revolution frequency of the beam varies rapidly (3).

The digitizers also contain an adjustable notch filter which controls the phase shift on the betatron lines to compensate for the phase advance from pickup to kicker. This filter's properties can be changed as a function of time in the acceleration cycle in order to track changes in phase advance due to changes in tune.

Output data from the digitizer cards are bussed to a digital-to-analog (D/A) card which contains a dual port look-up table for adjusting digital gain and a 10-bit digital-to-analog converter (DAC). The output of the DAC is then sent to the power amplifiers.

Timing Considerations for Large Radius Accelerators

The digitizers, trigger systems, and DAC rely on the accelerator reference oscillator to remain synchronous with the beam. The reference oscillator must adjust its frequency prior to a change in beam velocity in order for the triggers to remain synchronous. The lead time of the oscillator is determined by the electrical delay from the oscillator to the accelerator cavity. If the electrical delay from the oscillator to the digitizers exceeds this lead time by an amount on the order of the total change in oscillator frequency, synchronization will be lost.

In the case of the Tevatron, the total change in oscillator frequency is so small that the oscillator signal can be distributed anywhere in the ring without significant phase slippage. The Booster, on the other hand, has an extremely large frequency sweep, but the lead delay of the oscillator is about one revolution period. This means the oscillator signal can be distributed anywhere in the Booster <u>ahead</u> of the lead time, making it possible to synchronize the digitizers through the cycle.

The Main Ring and Main Injector have significant frequency sweeps, short lead delays, and large radii. When the oscillator signal is cabled to the other side of the ring, the phase slips more than five full buckets relative to beam through the acceleration cycle. This causes a delay error on the damper system of more than 10 buckets which is unacceptable. Therefore, an rf phase unwinder system was designed and built for use by the Main Ring and Main Injector dampers (4).

Results and Measurements

The Booster digital transverse dampers have been in operation for over a year. Detailed results and analysis of the Booster dampers can be found in (5). The Booster system has also been programmed to function as a tracking network analyzer for measuring beam transfer functions. The methods used for the measurement and the results are also shown in the reference.

TRACE A: D1 Spectrum

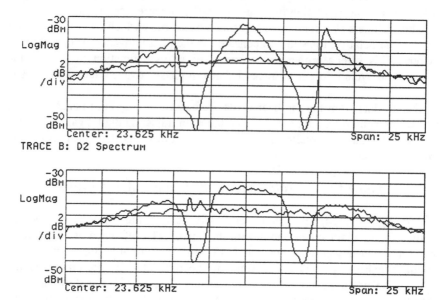

FIGURE 3. Comparisons of noise spectrums from Tevatron damper input. Top graph is horizontal and bottom graph is vertical. Smooth traces are with dampers off, and traces with valleys are with dampers on.

Using the damper system is an ideal way to measure transfer functions in a fast frequency sweeping machine, because commercial analyzers cannot lock to the reference frequency. However, the resolution of the system is limited by the amount of memory available. For relatively fixed frequency machines, such as the Tevatron, it is more advantageous to use commercial analyzers and take advantage of the finer resolution. Figure 3 shows the spectrum of the Tevatron dampers using a vector signal analyzer with the feedback loop open and closed. The noise spectral density of the damper at the betatron frequency is reduced by the squared ratio of the damping time over the decoherence time when the loop is closed (6). The plot shows that the horizontal Tevatron dampers have a damping rate of about three times the decoherence rate.

LONGITUDINAL DAMPERS

Longitudinal impedances are usually much greater than transverse impedances which cause an equal growth rate. Therefore, the gain of longitudinal systems will be greater and require more voltage on the kicker than a transverse system. For example, the electronic gain of the longitudinal system needs to be better than

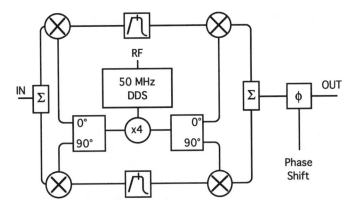

FIGURE 4. Block diagram of narrowband longitudinal damper low level.

70dB in the Booster just to counteract instability growth rates (7). Calculations show that it would require 10kV on the kicker to have dampers which damp injection oscillations in the Tevatron on the order of 10 synchrotron periods. This corresponds to 1MW of power into a wideband 50Ω cavity, and it would be extremely expensive to build. Therefore, the longitudinal dampers are primarily narrowband, high impedance systems.

Narrowband Systems

Narrowband damper systems concentrate on one or a small number of coupled-bunch modes. Figure 4 shows a block diagram of a single-mode, longitudinal, coupled-bunch damper (3). The direct digital synthesizer (DDS) keeps the reference frequency of the circuit locked to the frequency of the coupled-bunch mode. The bandpass filter is configured so that the dipole motion synchrotron frequency passes with a 90° phase shift while DC and frequencies well above the synchrotron frequency are filtered out. The bandwidth of the signal after the filter is on the order of a few kHz or smaller, so errors in delay from pickup to kicker look like a phase shift at the coupled-bunch mode frequency. A phase shifter is added just before the power amplifiers to compensate for delay errors which may change as a function of time in the cycle.

The Booster currently has narrowband dampers working on coupled-bunch modes 1, 49, 50, and 51 after transition. Because these dampers operate after transition, the maximum synchrotron frequency is limited to 2kHz. This makes the design of the analog bandpass filter easier, reducing the chances of driving higher-order modes with the system. The damper uses the power amplifiers and cavities of the acceleration system as its final amplifier and kicker. This saves a great amount of time and money, because there is no need to design and purchase a specialized, high voltage amplifier and cavity. Unfortunately, this technique only works where the amplifiers have a good response and the cavities have a high impedance. In the case of the Booster, the first mode is close enough to resonance to have a high impedance, and the other modes correspond to a parasitic resonance in the cavity.

Wideband Systems

It is quite easy to convert a digital transverse system into a longitudinal system by changing the analog front end to look at phase deviations instead of amplitude variations and by running the stripline kicker in common mode instead of differential mode. Again, the amount of power required to damp injection oscillations would be prohibitive. If, however, there are some or many slow growing instabilities that need to be controlled, the digital system could be a viable solution depending on the beam intensities and the instability growth rates.

Another application for the longitudinal wideband system is in a partially filled accelerator with bunches evenly spaced. In the Tevatron during colliding mode, there will be 36 proton bunches, spaced almost evenly around the ring. In this case, the bandwidth of the system is greatly reduced, and it is again possible to use the accelerating amplifiers and cavities to damp the beam. A digital, wideband, longitudinal damper system which uses the accelerating cavities is currently being designed for the Tevatron.

REFERENCES

1. Chao, A. W., *Physics of Collective Beam Instabilities in High Energy Accelerators* (John Wiley & Sons, Inc., New York, 1993), Ch. 4, pp.162-175.
2. Ma, H., and Steimel, J., "A Closed-Orbit Suppression Circuit for Main Ring Transversal Damper," in these proceedings.
3. Steimel, J., and McGinnis, D. P., "Damping in the Fermilab Booster," in *Proceedings of the 1993 Particle Accelerator Conference* (IEEE, Piscataway, NJ, 1993) pp. 2100-2102.
4. Steimel, J., "Trigger Delay Compensation for Beam Synchronous Sampling," in these proceedings.
5. Steimel, J., "Fast Digital Dampers for the Fermilab Booster," in *Proceedings of the 1995 Particle Accelerator Conference* (IEEE, Piscataway, NJ, 1996), pp. 2384-2388.
6. McGinnis, D. P., "A Transverse Damper System for the SPS in the Era of the LHC," CERN, July 13, 1995.
7. McGinnis, D. P., "Coupled Bunch Mode Instabilities Measurement and Control," *Conference Proceedings of the 1991 AIP Accelerator Instrumentation Workshop*, (AIP, New York, 1992) pp. 78-81.

Observation, Control, and Modal Analysis of Longitudinal Coupled-Bunch Instabilities in the ALS via a Digital Feedback System

J. D. Fox, R. Claus, H. Hindi, I. Linscott,
S. Prabhakar, W. Ross, D. Teytelman*

Stanford Linear Accelerator Center, P.O. Box 4349, Stanford, CA 94309

A. Drago, M. Serio

INFN - Laboratori Nazionali di Frascati, P.O. Box 13, I-00044 Frascati (Roma), Italy

J. Byrd, J. Corlett, G. Stover

Lawrence Berkeley Laboratory, 1 Cyclotron Road, Berkeley, CA 94563

Abstract. The operation of a longitudinal multibunch damping system using digital signal processing (DSP) techniques is shown via measurements from the Lawrence Berkeley Laboratory (LBL) Advanced Light Source (ALS). The feedback system (developed for use by PEP-II, ALS, and DAΦNE) uses a parallel array of signal processors to implement a bunch-by-bunch feedback system for sampling rates up to 500 MHz. The programmable DSP system allows feedback control as well as accelerator diagnostics. A diagnostic technique is illustrated which uses the DSP system to excite and then damp the beam. The resulting 12-ms time domain transient is Fourier analyzed to provide the simultaneous measurement of growth rates and damping rates of all unstable coupled-bunch beam modes.

SYSTEM DESCRIPTION

An electronic system to add damping to the longitudinal bunch motion of a storage ring requires components to measure the energy of particle bunches, a processing system to compute a correction signal from the energy errors, and a power stage to correct the energy of the particles. We have developed a damping system that uses a digital signal processing (DSP) formalism to compute correction signals. Such a technique is extremely flexible, allowing many operating modes and diagnostic measurements, and general purpose in that a single hardware design can be used by many accelerator facilities (the specific operating requirements of

* Work supported by U.S. Department of Energy, contract DE-AC03-76SF00515.

each system are implemented via software parameters). Figure 1 presents a block diagram of the processing system, which has been developed for use at the Stanford Linear Accelerator Center (SLAC) PEP-II accelerator. The heart of the signal processing is a scalable array processor (composed of commercial digital signal processing microprocessors) which provides a flexible signal processing block (1,2,3). The digital processing array is composed of 40 processors which provide an aggregate multiply-accumulate rate of 1.6×10^9 operations/s.

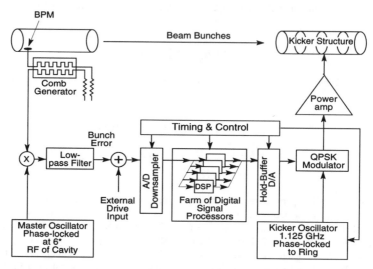

FIGURE 1. Block diagram of the longitudinal feedback system. The array of digital signal processors operate in parallel to compute correction signals on a bunch by bunch basis.

EXPERIMENTAL RESULTS

The prototype longitudinal feedback system has been installed at the Advanced Light Source (ALS) at Lawrence Berkeley Laboratory. The ALS machine is a third-generation storage ring, with an rf frequency of 500 MHz and a harmonic number of 328. The ALS implementation requires a 500-MHz bunch crossing and error correction rate. The ALS machine has displayed evidence of strong longitudinal instabilities since commissioning in April 1993 (4). Table 1 lists relevant parameters for the accelerator and the feedback system. The system was commissioned for routine operation in September 1995 and has been used for production running of the light source at both 1.5- and 1.9-GeV energies. Operated in conjuction with an all-mode transverse feedback system (5), the feedback systems have demonstrated increased intensity and reduced linewidths of the emitted higher-order undulator radiation. The feedback systems suppress coupled-bunch instabilities up to the full (400 mA) operating current of the storage ring.

TABLE 1. Parameters of the Accelerator and the Feedback System

Parameter	Description	Value
E	Beam energy	1.5 GeV
f_{rf}	rf frequency	499.65 MHz
h	Harmonic number	328
f_{rev}	Revolution frequency	1.5233 MHz
α	Momentum compaction factor	1.594×10^{-3}
I_0	Operating current	400 mA
f_s	Synchrotron frequency	11 kHz
-	Bunch sampling rate	499.65 MHz
-	Downsampling factor	21-31
N	Digital filter length	6
-	Feedback loop gain	6-28 dB
-	Feedback output power	200 W
-	Output amplifier bandwidth	1-2 GHz

The feedback error processing is implemented in a bunch-by-bunch discrete time system, in which the error signal for a particular bunch is computed using several past measurements of only that bunch. The algorithm used is a finite impulse response (FIR) discrete time filter (6),

$$u_i(n) = \sum_{k=1}^{N} h(k)\phi_i(n-k) , \tag{1}$$

with coefficients selected to provide maximum gain at the synchrotron frequency and with the filter phase shift adjusted to provide net negative feedback at the synchrotron frequency. The action of the parallel processing array and the parallel computation of all the populated bunches creates an all-mode global transfer function which folds the single-bunch transfer function around all the revolution harmonics in the spectrum of the bunch motion.

The digital processing scheme allows the computation of linear or nonlinear (saturated) correction signals. One interesting feature of the programmable system is the capability to record the bunch motion as the feedback system operates (7). Such digital data records provide a very powerful diagnostic capability to observe and quantify the motion of the particles in the storage ring. In particular, observing the motion as the feedback system is turned off displays the growth rates of the unstable modes, while recording the motion when the feedback signals are turned back on reveals the net damping in the system. Frequency domain information can be computed from these time-domain data sets via use of Fourier transform techniques.

An example measurement using this grow-damp technique is illustrated in Fig. 2. In this experiment the beam is initially stable under the action of the feedback system. Under software control the feedback gain is set to zero and, after a holdoff interval, the DSP processors start recording the bunch motion. For $t_{holdoff} < t < t_{on}$ the growing bunch oscillations are recorded, and at t_{on} the feedback gain is restored to the operating value. The motion (now a damping transient) continues to be recorded until t_{max}, at which time the DSPs stop recording but continue computing the feedback signal (controlling the beam). The data records are stored in a dual-port memory which is accessible to an external processor (the total record length is 960 samples of each of 324 bunches). The recorded data is processed off-line. After the data array is read, the DSP processors can be triggered again to record another transient.

Figure 2(a) shows the envelopes of the synchrotron oscillations of each of 328 bunches for such a transient. We see a complicated growth of motion up to 6 ms, followed by damping of the motion. In this representation the phase relationship of the individual bunch oscillations is not obvious. If the data is arranged in a two-dimensional array of bunch number vs. sample number, each row (containing the oscillation coordinate of each of 328 bunches sampled on a single turn)[*] can be Fourier transformed to reveal modes of oscillation of the bunches. This Fourier transform is computed for each sampling time (turn number) in the array. The resulting data is displayed in Fig. 2(b), which shows the presence of two modes of oscillation in the transient.[†] The growth rates and damping rates for each mode can be found from this data set via the numeric fitting of exponential curves to the data, as shown in Figs. 2(c) and (e). Figures 2(d) and (f) present the effective growth rate and damping rate vs. mode number. We see here the action of the feedback system in turning the net growth rate from positive to negative.

This figure also reveals the gain margin in the damping system. The instability growth rate per mode is due to a net effective impedance (found by a summation over revolution harmonics) (10)

$$\frac{1}{\tau_l} = \frac{I_0 f_{rf} \alpha}{2(E/e)Q_s} \ \text{Real}(Z_l^{eff}) \tag{2}$$

$$Z_l^{eff} = \sum_{p = -\infty}^{+\infty} \frac{\omega_p}{\omega_{rf}} \ \exp(-\omega_p^2 \sigma_\tau^2) \ Z(\omega_p) \tag{3}$$

$$\omega_p = (pN + l + Q_s) \ \omega_{rev} \ , \tag{4}$$

---

[*]   Due to the action of the downsampled processing, the data for the various bunches are actually sampled on different turns corresponding to the sampling pattern and downsampling factor. In the postprocessing an interpolating filter is used on each bunch's column in the raw data array to compute the oscillation coordinates over the non-sampled turns, effectively time-aligning all the bunch coordinate data before the Fourier transforms are taken.

[†]   The Fourier transform of a spatial record of length N is decomposed into N/2 spatial frequencies. Thus in this representation the 328 modes of the bunch system are folded into the 164 modes of the Fourier transform of the turn oscillation coordinates. Each computed "mode" actually contains two true beam modes corresponding to the upper and lower synchrotron sidebands around a particular revolution harmonic (8,9).

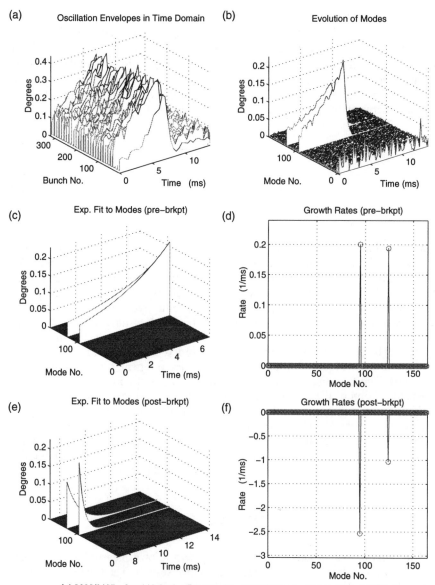

feb0696/2105: Io= 118.4mA, Dsamp= 22, Shift Gain= 4, Nbun= 320, Gain1= 0,
Gain2= 1, Phase1= 30, Phase2= −140, Brkpt= 496, Calib= 24 cnts/mA−deg.

FIGURE 2. (a) Plot of the oscillation envelopes of 328 bunches in the ALS over a 12-ms time interval. The feedback system is on for t < 0 and t > 6 ms, and off for 0 < t < 6 ms. Unstable bunch motion grows during the feedback-off interval. (b) is a turn-by-turn FFT of the bunch data, revealing two unstable modes (mode -95 and -124). (c) is an exponential fit to first 6 ms of (b). (d) is the fitted growth rate vs. mode #. (e) shows the fit damping exponentials (last 6 ms of (b)). (f) is the net damping rate vs. mode # (showing the action of the feedback system).

where τ_l is the growth time of mode l due to the ring impedance, I_0 is the beam current, $Z(\omega)$ is the total ring impedance, e is the charge of an electron, Q_s is the synchrotron tune (f_s/f_{rev}), σ_τ is the rms time-of-arrival variation of particles within a bunch, and N is the number of bunches (see Table 1 for more symbol definitions).

The action of the feedback system provides a damping rate which is proportional to the gain at that mode; the net response of the system is determined by the difference between the two (natural growth and feedback damping) rates

$$1/\tau^{fb} = \frac{f_{rf}\alpha}{2(E/e)Q_s} G_{fb} \tag{5}$$

$$1/\tau_l^{net} = 1/\tau_l - 1/\tau^{fb} \tag{6}$$

which is plotted per excited mode in Fig. 2(f).

This approach directly measures the unstable modes of the ring—with a variant of this technique it is possible to measure damping/growth rates for naturally stable modes. For this measurement a narrowband excitation at the desired mode frequency is injected into the feedback system at the error summing node (see Fig. 1). The feedback system passes this excitation signal onto the beam via the power stage and kicker, and the bunch motion at that specific frequency is excited (a single mode of the beam). If the excitation and feedback is turned off, the excited mode decays (or grows!), and as the feedback is turned on the net damping rate increases. The grow-damp transient can be recorded and processed just as for the self-excited case. Figures 3(a-f) show such a measurement for an excitation of mode # 61. The figure reveals a natural damping rate of the mode (–0.01 1/ms) as well as the action of the feedback system to provide a net damping rate of –0.9 1/ms. If this measurement is repeated for several modes (or an excitation is applied which excites several modes), the effective gain of the feedback system vs. mode can be determined. Measurement of such a frequency-dependent effective gain is a useful system check and is particularly helpful in the determination of the frequency responses of the wideband power amplifier and kicker stages.

SUMMARY

The programmable nature of the DSP system allows measurement of beam motion as well as computation of feedback control signals. The time-domain transient techniques illustrated require only a few ms of beam motion and are essentially invisible to users of the storage ring. The information obtained from the analysis of the time-domain data reveals growth rates and damping rates of beam modes and is very useful in adjusting the feedback system or as a quick check on its operation.

The time-domain techniques, in conjunction with off-line FFT analysis, are complementary to narrowband detection in the frequency domain in that they allow the quantification of many unstable modes in a single non-destructive tran-

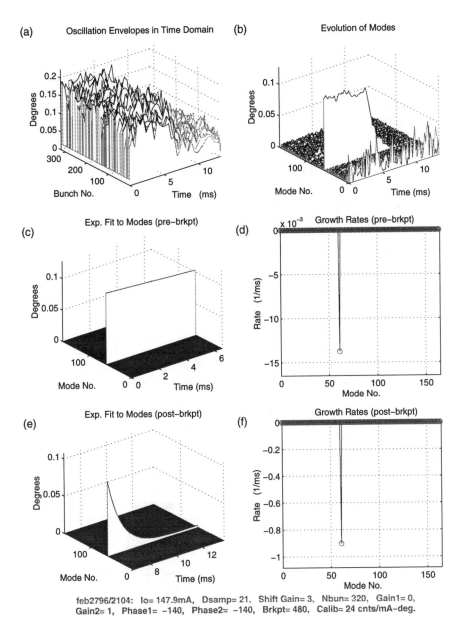

(a) Oscillation Envelopes in Time Domain

(b) Evolution of Modes

(c) Exp. Fit to Modes (pre–brkpt)

(d) Growth Rates (pre–brkpt)

(e) Exp. Fit to Modes (post–brkpt)

(f) Growth Rates (post–brkpt)

feb2796/2104: Io= 147.9mA, Dsamp= 21, Shift Gain= 3, Nbun= 320, Gain1= 0,
Gain2= 1, Phase1= −140, Phase2= −140, Brkpt= 480, Calib= 24 cnts/mA−deg.

FIGURE 3. Plot in the same manner as Figure 2, but with a narrowband excitation of mode #61 applied to the summing node of the feedback system. The natural damping rate is seen in (c) and (d), the damping with feedback in (e) and (f).

sient and do not require repetitive narrowband steady-state swept frequency domain measurements of each potentially unstable mode. The essential advantage of the time-domain technique is speed—successive narrowband sweeps of each possibly unstable mode can reveal the same information. However, for an accelerator with potentially hundreds or thousands of unstable modes and constantly drifting parameters (such as cavity temperature or cavity tuning) the speed advantage makes the time-domain techniques very useful as accelerator diagnostics.

ACKNOWLEDGMENTS

The authors thank J. Hoeflich, J. Olsen, G. Oxoby of SLAC, and G. Lambertson of LBL for numerous thoughtful discussions and direct contributions of technical expertise. We also thank Boni Cordova-Gramaldi of SLAC for her patient fabrication of electronic components.

REFERENCES

1. J. Fox et al., "Feedback Control of Coupled-Bunch Instabilities," Proc. of the 1993 Particle Accelerator Conference, Washington, DC, May 17-20, 1993, 2076 (1993).
2. R. Claus et al., "Software Architecture of the Longitudinal Feedback System for PEP-II, ALS and DAΦNE," Proc. of the 1995 Particle Accelerator Conference, Dallas, TX, May 1-5, 1995, 2660 (1996).
3. D. Teytelman et al., "Operation and Performance of the PEP-II Prototype Longitudinal Damping System at the ALS," Proc.of the 1995 Particle Accelerator Conference, Dallas, TX, May 1-5, 1995, 2420 (1996)
4. A. Jackson et al., "Commissioning and Performance of the Advanced Light Source," Proc. of the 1993 Particle Accelerator Conference, Washington, DC, May 17-20, 1993, 1432 (1993).
5. W. Barry et al., "Commissioning of the ALS Transverse Coupled-Bunch Feedback System," Proc. of the 1995 Particle Accelerator Conference, Dallas, TX, May 1-5, 1995, 2423 (1996)
6. A. Oppenheim, R. Schafer, *Discrete-Time Signal Processing*, Prentice Hall, 1989.
7. D. Teytelman et al., "Feedback Control and Beam Diagnostic Algorithms for a Multiprocessor DSP System," (these proceedings).
8. F. Sacherer, "A Longitudinal Stability Criterion for Bunched Beams," *IEEE Trans. Nucl. Sci.*, **NS-20-3**, 825 (1973).
9. J.M. Wang, "Modes of Storage Ring Coherent Instabilities," AIP Conf. Proc. 153 (1985).
10. M. Zisman et al., Lawrence Berkeley Lab. Report No. LBL-21270.

Towards the Limits of Frame Transfer CCDs in Beam Instrumentation

R.J. Colchester, R. Jung, P. Valentin

CERN, SL Division, CH1211 Geneva 23, Switzerland

Abstract. Charge-coupled devices (CCDs) are used for beam position and beam size measurements with screens and synchrotron radiation monitors. They have a large spatial resolution and a high dynamic range which make them interesting for high resolution two-dimensional measurements. In most applications, the video signal is digitised with a frame grabber and then the beam parameters are computed. The results are limited in resolution, usually to 8 bits, and in rate. These restraints were seen as major limitations for fast and precise beam observations with synchrotron light monitors in LEP, where the revolution frequency is 11 kHz. They have been overcome by using the characteristics of frame transfer CCDs. Advantage is taken of the separate image and memory areas on the CCD chip to control independently the integration and digitising periods to achieve a 12-bit amplitude resolution at the individual pixel level. Using the CCD as a buffer memory, together with a pulsed intensifier, it is possible to make 9 to 18 two-dimensional beam spot acquisitions up to a rate of 11 kHz. The most recent development uses the CCD as a summing device to perform on-chip horizontal or vertical beam projections; several hundred successive profiles can be acquired in this mode.

INTRODUCTION

Synchrotron light telescopes are the basic operational instruments for transverse emittance measurements in the Large Electron Positron (LEP) Storage Ring. The instruments, called BEUV for Beam Emittance in the UV, are planned not only for TV observation by the Control Room operators, but also for precision emittance measurements. When looking more precisely into the CCD architecture, it was realised that a CCD is not only good for TV observation, but can also be considered for powerful and economic transient beam behaviour observations down to the LEP turn-by-turn rate of 11 kHz, by combining a frame transfer CCD with a fast intensifier (1). Various CCDs were investigated and a modular system was built over the years as development time became available and needs were expressed. For the LEP start-up in July 1989, four BEUV telescopes and fifteen luminescent screens (LSs) were operational for TV observation (2,3). Soon after, the BEUV images were digitised and profiles calculated. In 1990 the fast shutters and the 11 kHz "burst" acquisition for up to 9 turns were put into operation. The following years were mainly devoted to increasing the telescope's precision from the original aim of 1 nm to the 0.1-nm range (4). Finally, in 1996, the fast turn-by-turn profile acquisition is being put into operation. No major change of the system was necessary over this period, and only the original proximity-focused diode intensifier was replaced by a more modern microchannel plate (MCP) intensifier.

UNCONVENTIONAL USES OF FRAME TRANSFER CCDs

Several types of solid-state detectors have been developed for the TV market. The main types are the frame transfer (FT) and the interline (IL) CCDs. Another type, the charge injection device (CID) X-Y matrix is used for high radiation applications. The frame transfer CCD was chosen because of its versatile signal processing and operational possibilities. These detectors are used primarily in scientific applications as they are more expensive, they use larger silicon areas than the IL detectors (which are the main technology used in the consumer market), and they are produced in large quantities. Cameras with X-Y CID matrices are the most expensive of the three types. Different types of detectors are used in the Super Proton Synchrotron (SPS) and the LEP and have been compared for resolution as shown in Fig. 1.

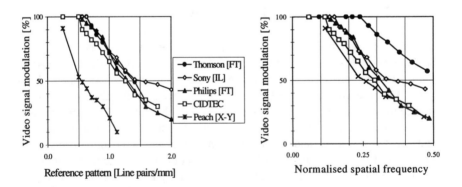

FIGURE 1: Resolution test: the video signal modulation is plotted as a function of the spatial frequency of a reference plate used in a standard test set-up.   Left: direct measurement result; right: result normalised by the pixel size.

Most cameras, except the low-cost metal oxide semiconductor (MOS) X-Y matrix surveillance camera, have comparable resolutions which justifies the choice of the more economical IL CCD cameras for LS observation and standard video acquisitions (5,6). In these applications, the CCIR standard signals are digitised by frame grabber-type converters with an 8-bit resolution, i.e., 254 grey levels. For the more demanding application of the LEP Synchrotron Light Telescope's BEUV (3), an FT CCD chip, Thomson type TH7863, was selected. This chip was well documented, had peripheral chips available, and had a better figure of merit in normalised resolution (see Figure 1), which is important for fast profile measurements with a minimum number of pixels. The amplitude resolution of this chip is close to 12 bits, i.e., 4095 grey levels, and has improved over the years well beyond the 8 bits available with the usual video frame grabbers. Frame grabbers normally operate on the composite video signal which results in a loss of horizontal resolution. It was therefore decided to digitise the pixels individually over at least 12 bits. A compromise had to be found between digitising speed and resolution taking into account the long distances (up to 500 m) between the camera and the ADC, the loss of charges as a function of shift rate, and the thermal noise integration (7). A 12-bit analog-to-digital converter (ADC) was chosen for precision digitisation and is operated at 450 kHz; however, when speed is the major requirement, as it is for the fast projection scheme, a 1-MHz rate is used. The thermal noise is minimised by cooling the CCD chip with Peltier cells; the whole telescope has cooled air circulation. The 12 bits also relax the constraints on the light level control for precision profile calculations and ease operation. In certain applications, like beam tail measurements, the 12-bit resolution is not sufficient. The range is extended by attenuating the dense part of the beam core by a factor of twenty with a so-called "corona filter," while acquiring directly the low-density beam tails (4). The dynamic range is then equivalent to 10^5, with the normal signal-to-noise ratio. Tails up to 12 sigmas from the central core could be measured in this way (8).

The FT CCD technology has many possibilities which are of special interest for the BEUV project. It uses two identical areas: an image area on which the scene is imaged during the integration time (typically 20 ms) and a memory area to which the image area is shifted 200 µs after integration. The image information is transferred from the memory area into the output register line by line and read out at the CCIR line frequency at a frame rate of 50 Hz. In normal video applications, these operations are controlled by custom chips locked to a well-defined, crystal-controlled frequency and a whole cycle takes two mains periods. But one of the aims of BEUV is to measure turn-by-turn profiles in LEP for beam instability studies. A fast intensifier/shutter can select one particular bunch and turn, but the acquisition rate will still be 50 Hz and the correlation of successive pictures is not guaranteed at the LEP turn resolution. As the two areas of the TH7863 chip are made of 288 (V) times 384 (H) square pixels of 23 µm in size and as far less pixels are needed for extracting the beam size information in good conditions, the image-to-memory transfer is subdivided into blocks of 30 lines shifted successively in 18 µs, each shift locked to the LEP turn clock. In this way, it is possible to accumulate nine profiles, taken at well-defined turn intervals, in the CCD memory if the beam is imaged just above the image/memory separation, as shown in Fig. 2. This mode is called the "burst mode" for obvious reasons.

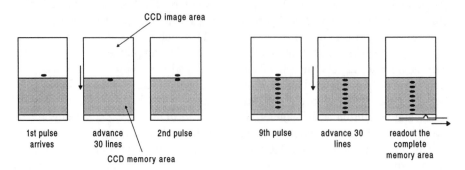

FIGURE 2: Principle of the burst mode.

As several bunches—up to 16 of the same type—can circulate in LEP, it is necessary to gate out the unwanted ones. This is done with an electro-optical fast shutter. The number of memorised images can be doubled by imaging the beam at the top of the CCD image area. As still only a fraction of the CCD is used in the horizontal direction, several beam images could be memorised on one chip (9).

It was felt since the start that 9, and even 18, images were somewhat restricting and that in certain cases it was more interesting to acquire many more profiles, i.e., projections, over many turns. It is possible to do so by using the FT CCD to perform on-chip projections. The "fast projection mode" is the logical extension of the burst mode. As in the latter mode, the beam is imaged onto the image area of the CCD and the beam image is shifted down towards the memory area between two successive beam passages, in blocks of 30 lines, corresponding to a beam height of 3.3 mm in the BEUV case. The difference with the burst mode is at the memory-to-output register shift. Instead of shifting the memory into the output register line by line and reading out each line separately, the block of 30 lines is shifted all at once into the output register, which then contains the sum over 30 lines, i.e., the beam projection onto this axis. A block of 40 columns containing the whole beam projection is then read out in 40 µs, and so on, as shown in Fig. 3. In this way, fast projections at the LEP revolution frequency of 11 kHz can be taken continuously with the 1-MHz conversion rate of the 12-bit ADC, as shown in Fig. 4. The projection on the other axis is obtained by rotating the beam image by 90°. If simultaneous H and V projections are required, they could be obtained at once by splitting the light beam in two, rotating one beam by 90° and projecting it at the side of the direct beam, and reading out 80 columns instead of 40 per projection . This doubles the read-out time and limits the fastest acquisition rate to one out of two LEP turns. The fast projection mode naturally has a limitation along with its speed advantage. As the projection is done on the CCD itself, the dynamic range is limited to the full well capacity of the output register, whereas previously this limit applied only to the pixel with maximum amplitude. For the vertical profile in LEP, which is the most demanding one, this ratio is of the order of 20. On the other hand, the thermal noise integrated is approximately 30 times smaller than in the other modes, and hence the signal-to-noise ratio is not degraded. The profile resolution is now comparable to the one obtained by calculating projections with an 8-bit frame grabber, but with a rate of 11 kHz instead of 50 Hz!

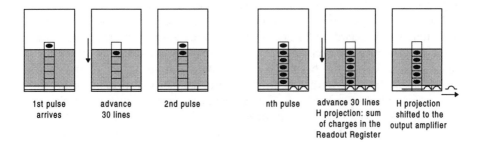

| 1st pulse arrives | advance 30 lines | 2nd pulse | | nth pulse | advance 30 lines H projection: sum of charges in the Readout Register | H projection shifted to the output amplifier |

FIGURE 3: Principle of the fast projection mode.

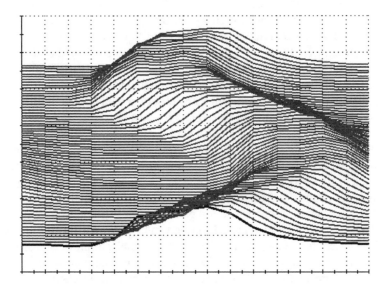

FIGURE 4: Fast projection display. Seventy profiles (out of 3000) of the horizontal projection of the image of a moving light source (pulsed LED) are displayed.

THE CCD CAMERA SYSTEM

The fast shutter gates out unwanted bunches, selects the time interval between measurements, and provides some gain for single-shot measurements. It is also used as a wavelength shifter to convert low-wavelength UV light into the CCD bandwidth which starts at 450 nm for a 20% quantum efficiency. This is interesting to decrease the image broadening by diffraction for the small LEP emittances (4). But the finite resolution of the MCP becomes a limiting factor for the whole system. A single-stage MCP with an S20 photocathode on quartz is presently used. The use of short wavelengths down to 313 nm compensates the resolution degradation introduced. Going to a shorter wavelength of 254 nm was not beneficial; this may be due to the higher energy of the incoming photons. The major limitation with standard MCPs is a saturation effect in the electron multiplication at the 11-kHz rate. The consequence is an increasing amplitude limitation of the dense central part of the beam which deteriorates the precision of the profile measurement. It can nevertheless provide useful contour information for instability studies with the Burst Mode. The saturation disappears when decreasing the observation rate to 2 kHz, or every fifth LEP turn. A more interesting solution is to use a high-current MCP. This was tried and gave better results in laboratory tests. Beam tests will be carried out as soon as LEP starts up in 1996.

A system has been built progressively around the TH7863 chip with the aim to make optimum use of its capabilities for beam instrumentation. The first and still major application of the system is in the LEP synchrotron light telescopes, for precision emittance measurements. However, the system is now finding other applications such as the individual profile and position measurement of the four lepton bunches injected from SPS into LEP at 156-μs intervals and observed with OTR screens, or the monitoring of the matching of the proton beams injected into SPS, and in future into LHC, based on the measurement of the beam size oscillations during the first turns after injection and observed with fast screens (10).

The system has been split into three parts for best reliability and performance. The camera head is comprised of a small card for the CCD and the phase drivers, with Peltier cell cooling of the CCD so as to reduce the thermal dark current to a minimum. This head is the only part mounted inside the BEUV telescopes. It is linked via a cable up to 2 m in length to a small crate which contains the specific CCD phase generators, a Sample&Hold for the pixel signal, line drivers to the control and read-out electronics, and a regulator set to generate locally the necessary voltages. These two elements, head and mini-crate, are located inside the normally inaccessible radiation area. The rest of the electronics is located in an accessible underground area, up to 500 m away from the camera head. These electronics (3) are in the VME standard and are comprised essentially of two CPUs, an 8-bit video frame grabber for permanent beam size monitoring (4, 5), the slower 1-MHz/12-bit ADC with full frame memory, a control card for choosing the various camera modes, timing units for controlling the MCP triggering in relation with the beam, video generators for mixing digital information, and beam TV images for fast and direct Control Room information. The various modes are set by the VME control card which controls the phase generator card in the mini-crate. This card includes a sequencer chip which was, at the start, the standard Thomson chip for controlling the TH7863. With this chip it is possible to implement the following modes: TV, full frame 1-MHz or 450-kHz digital acquisitions, and burst. It is not possible to implement the fast projection mode with this chip. A field-programmable gate array (FPGA) is now used in place of the original sequencer chip; the rest of the system remains unchanged. The acquisition rate is under the control of the ADC. Its buffer memory can contain the image of a full frame of 110×10^3 pixels with one or nine two-dimensional beam images, or a string of 3000 fast 12-bit projections. The software resident in the CPU calculates the projections and extracts the beam sizes with a Gaussian fit, the results of which are superimposed on the beam image on the TV screens or transmitted to workstations for display and logging. Another CPU works with the VME frame grabber and projection calculator for generating a permanent real-time beam size video display.

RESULTS WITH BEAM

The real time beam observation in TV mode and the video display of the beam dimensions in the Control Room provide the operators with valuable qualitative information on the LEP beam behaviour. The precision beam emittance measurement is used for machine optimisation and studies. The major difficulty in this application comes from the poor knowledge of the real Twiss parameters of LEP during the various phases of operation. These parameters were not available up to now with sufficient precision and were the major source of error of the "measured" LEP emittances and calculated luminosities. Once it was possible to measure the Twiss parameters with sufficient precision (11) and to take into account various perturbations like beam-beam beta-beating, it was possible to measure vertical emittances down to 0.2 nm with a precision of ± 0.1 nm when compared to the wire scanner measurements or the experimental luminosities (12). Beam tails have been measured with the "Corona filters" up to 12 sigmas away from the core, demonstrating exponential tails beyond 3 sigmas (8). Beam instabilities could easily be observed and quantified with the burst mode. Non-Gaussian beam cores have been observed in single-shot mode, averaging out to Gaussian shapes when integrated over the standard 20 ms. Different behaviours of the various bunches in the bunch train operation mode were easily observed with the MCP shutters, both in TV and digital mode.

The fast projection will be put in operation in 1996 and should provide useful, coherent, turn-by-turn information over hundreds of turns.

CONCLUSION

A flexible system has been built around a frame transfer CCD chip. This system is modular and has been completed over the years to fulfill evolving needs. It goes far beyond the possibilities of a standard TV system and frame grabber digitiser. We consider it at the limit of what can be achieved with available CCDs in beam instrumentation applications. The system has a spatial resolution at the pixel level, i.e., 23 μm, and an amplitude resolution of 12 bits matching the chip's capability. It can acquire from 9 to 18 time-coherent, two-dimensional LEP beam images and up to 3000 beam projections at a rate up to 11 kHz. It uses the CCD in these last two modes as a buffer memory and an image processor. High-precision results have been obtained in profile measurements and only the LEP wire scanners have achieved comparable or better results (12).

ACKNOWLEDGEMENTS

B. Halvarsson and J. M. Vouillot have actively participated in the design of the system and J. J. Gras coordinated the software production at the various stages. Their contributions are acknowledged with gratitude.

REFERENCES

1. Minutes of the 112th Meeting of the LEP Commissioning Com., October 1986.
2. G. Burtin et al., "The LEP injection monitors: design and first results with beam," Proc. of the 1989 IEEE PAC, Chicago, 1492-1494 (1989); and CERN/SL/89-06(BI), March 1989.
3. C. Bovet et al., "The LEP synchrotron light monitors," CERN SL/91-25 (BI), April 1991; ext. vers. of Proc. of the 1991 IEEE PAC, San Francisco, 1160-1162 (1991).
4. R. Jung, "Precision emittance measurement in LEP with imaging telescopes," CERN SL/95-63 (BI), June 1995, and KEK Proc. 95-7, A, Sept. 1995.
5. VME Frame Grabber designed by G. Ferioli and J. Provost, CERN, SL Division.
6. J. Camas et al., "Screens versus SEM Grids for single pass measurements in SPS, LEP and LHC," Proc. of DIPAC 95, Travemünde, DESY M 95 07, June 1995.
7. R. Jung, "Image acquisition and processing for beam observation," Proc. of DIPAC 93, Montreux, CERN PS/93-35, CERN SL/93-35 (BI), August 1993.
8. G. Ferioli et al., "Non-intercepting beam tail measurements in LEP with BEUV," SL-MD Note 209, March 1996.
9. G. Baribaud et al., "Three dimensional bunch observation in LEP," Proc. of 1992 Int. Acc. Conf., Hamburg, and in CERN SL 92-33, July 1992.
10. C. Bovet, R. Jung, "A new diagnostic for betatron matching at injection into a circular accelerator," LHC Project Report 3, March 1996.
11. P. Castro et al., "Betatron function measurement at LEP using the BOM 1000 turns facility," Proc. of the 1993 IEEE Particle Acc. Conf., Washington D.C., 2013-2015 (1993).
12. P. Castro et al., "Cross-calibration of emittance measuring instruments in LEP," SL-MD Note 202, February 1996.

Beam Monitors Based on
Residual Gas Ionization

Jean-Luc P. Vignet, Rémy M. Anne, Yvon R. Georget, Robert E. Hue,
Christian H. Tribouillard

Grand Accélérateur National d'Ions Lourds (GANIL), BP 5027, 14021 Caen-cedex, France

Abstract. At GANIL we have developed two beam monitors based on ionization of the residual gas of the beam transport lines under the impact of the high-energy heavy ion beams. One provides the beam profile, i.e., its spatial distribution, and the second one measures its time structure, i.e., the length of the beam bunches delivered by the 10-MHz cyclotron. They are operated for beam intensities between a few nanoamperes and some microamperes and in a beamline vacuum better than 10^{-6} mbar. The charged ions or the electrons produced in the residual gas drift by means of an electrostatic field onto a microchannel plate (MCP) which amplifies the primary current. The MCP output signal is collected on a multistrip anode to get the beam profile, whereas the fast time structure of the beam is measured by means of a 50-Ω anode. Calculations and results are presented as well as discussions about the different parameters of both devices.

A NONINTERCEPTIVE BEAM PROFILE MONITOR

Introduction

The profiles of the heavy ion beams at GANIL are usually measured by means of secondary emission multiwire monitors or by gas-filled proportional chambers, according to the intensity of the beams. These beams range from carbon to uranium, and their intensities range from a few hundred ions per second to some μA. They provide the two horizontal and vertical profiles but, as they are interceptive, they deteriorate the beam properties or become themselves destroyed under high intensity beams. Thus, we have been developing and testing a noninterceptive profile monitor at GANIL, using the signal provided by the residual gas ionized under the passage of heavy ions (1).

Monitor Principle

All along the beam transfer lines, the pressure of which is around 10^{-6} mbar, ions are produced by ionization of the residual gas under the impact of the beam. If collected, these ions—mainly hydrogen and water—provide information about the position and the size of the beam. Positive ions are guided by a transverse electro-static field onto a microchannel plate (MCP) amplifier, the output of which is con-nected to a multistrip anode, thus providing signals proportional to the spatial beam density (Fig. 1). The number of ions produced in collisions between beam and residual gas depends on energy loss of the beam, i.e., its nature and energy, but also on the pressure in the beam transfer lines. For example, in the case of a 200-nA argon beam at 44 MeV/u circulating in a beam transfer line where the pressure is around 10^{-6} mbar, the positive ion current collected along 5 mm of the beamline is around 2 pA.

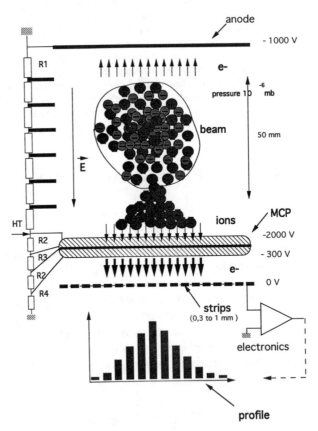

FIGURE 1. General layout of the monitor.

Mechanical Features

The monitor is mainly composed of two parts: the drift space and the amplifier stage. The 50-mm-long electrostatic drift space for the residual gas ions is obtained by means of a series of copper electrodes, the electrostatic potential of which is regularly distributed in order to have a homogeneous field in the central area. A stack of two microchannel plates is used as the amplifier; in front of this a mask limits a useful area, a few-mm slit, perpendicular to the beam direction. The multi-strip anode is made by gold deposit on a Teflon-glass support. The space s between the centres of gravity of two contiguous strips can be as low as 0.2 mm, which con-tributes to a good resolution of the beam size measurement.

Electronics

The charges collected on each anode strip at the output of the MCP are inte-grated by means of a capacitance C, each one of which is connected to an analog multiplexer with a high input impedance R (10^9 Ω). A 50-kHz clock sequentially reads all the strips with a recurrence time t_i, which ranges between 10 ms and 5 s, according to the beam intensity (2). This variable integration time, together with the adjustable gain of the MCP, enables one to detect beams with intensities that can vary within a large range. Analog signals are converted in a microprocessor unit and sent to a computer, thus enabling one to do calculations on the profiles such as the centre of gravity and the width of the beam (Fig. 2).

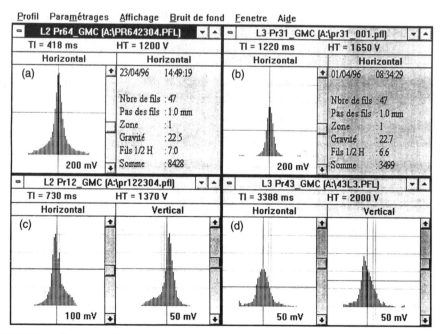

FIGURE 2. Beam profiles: (a) Ar 70 MeV/u, I = 2.6 µA, s = 1 mm, Ti = 418 ms, HV = 1200 V; (b) Sn 75 MeV/u, I = 300 nA, s = 1 mm, Ti = 1220 ms, HV = 1650 V; (c) Sn 75 MeV/u, I = 300 nA, s = 0.5 mm, Ti = 730 ms, HV = 1370 V; (d) Te 40 MeV/u, I = 1 nA, s = 0.5 mm, Ti = 3388 ms, HV = 2000 V.

Tests and Results

We have performed a series of measurements using different beams of various intensities with pressure better than 10^{-6} mbar. Some examples of profiles are displayed in Figure 2(a-d) for the following beams: Ar, Sn, Te, and with different parameters for the monitor and its associated electronics.

Discussion

Simulations have been carried out assuming that recoil ions are mainly produced in a $q = 1$ charge state, that their kinetic energy is around 0.03 eV, and that they are isotropically emitted in a plane perpendicular to the beam direction (3). If E is the electrostatic drift field, m is the mass of a recoil ion, q is its charge, θ and E_0 are its angle and kinetic energy of emission, respectively, t is the time, and x and y are the axes parallel and perpendicular to E, respectively, the motion equations of a recoil ion are

$$x = (qE/\ 4E_0)\ y^2/\sin^2\theta + y\ /tg\ \theta + x_0 , \tag{1}$$

$$y\ = (2E_0/m)^{1/2}\ t\ \sin\theta. \tag{2}$$

The trajectories of recoil ions emitted at different interaction points within the residual gas volume have been calculated according to Eqs. (1) and (2) with $E_0 =$ 0.03 eV, $x_0 = 0$, and an angular spectrum $\theta = 0$ to $360°$. A standard deviation of $\sigma_y = 0.2$ mm has been estimated in these conditions (Fig. 3).

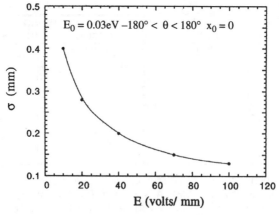

FIGURE 3. Theoretical standard deviation versus electrostatic drift field.

BEAM TIME-STRUCTURE MONITOR

Introduction

The GANIL cyclotrons deliver heavy ion bunched beams, the frequence of which are between 8 and 14 MHz with bunch lengths less than 2 or 3 nanoseconds, depending on the energy and nature of the accelerated ions. The high-frequency (HF) signal taken from the resonant cyclotron cavities (Fig. 4(a)) is usually used for nuclei velocity measurements based on "time of flight" or "start-stop" techniques involved in reaction products identification processes. As a consequence, beam bunches as short as possible are needed. The aim of the monitor is to measure and check on line the time distribution of the accelerated bunched beams.

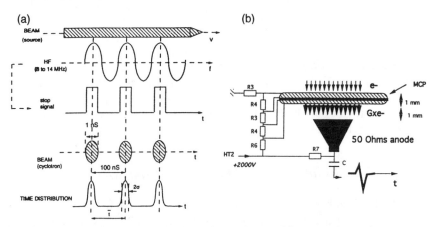

FIGURE 4. (a) Timing of the GANIL beam, (b) output electronics.

Principle

The signal is obtained from residual gas ionization in the same conditions as previously described for the profile monitor, but the localization strips are replaced by a 50-Ω anode (Fig. 4(b)).

According to the direction of the transverse electrostatic field E, electrons or ions produced by ionization of the residual gas under the impact of ion beams are guided onto a microchannel plate amplifier which provides a signal characteristic of the timing of the beam.

Expected Resolution

With the above-mentioned parameters, the general motion equation of recoil ions or electrons produced by ionization of the residual gas in the transverse electrostatic drift field of the detector is

$$t = (-(E_0)^{1/2}\cos\theta + (E_0\cos^2\theta + qE(d - x_0))^{1/2})(2m)^{1/2} / qE\,c \qquad (3)$$

with m and E_0 in eV, q in charge units, and E in V/m. Ions or electrons can be collected, but which of them provides the best resolution? According to vacuum gas analysis, the greater part of residual gas is hydrogen, and the smaller part is water. We will compare, for example, the motion of an atom of ionized hydrogen with that of an electron and calculate for both of them the time dispersion motion Δt according to different emission parameters, especially their x_0 coordinate. We will take, for example, a beam of size $W = 4$ mm with distance to the microchannel plate of $d = 30$ mm and mean kinetic energies of 0.03 eV and 2.0 eV for the emitted hydrogen ion and electron, respectively. Suppose in all cases that $\theta = 0°$, $E = 40$ V/mm, and $x_0 = \pm\, W/2 = \pm\, 2$ mm. The derivative of t versus $x = (d - x_0)$ gives

$$dt/dx_0 = (1/2\,c)\,(2\,m/(E_0 + q\,E(d - x_0))^{1/2} \qquad (4)$$

which results in Δt (hydrogen) $= 3.12 \times 10^{-9}$ s and Δt (electron) $= 74 \times 10^{-12}$ s. This succinct calculation indicates that it is much better to take the time signal from electrons than from hydrogen ions, especially if we notice that the ions from the residual gas are not only hydrogen but heavier atoms like water, carbon dioxide, nitrogen, argon, and so on.

Resolution as a Function of Beam Size

Equation (4) with the above parameters values shows that for electrons, $\Delta t = 18$ 10^{-12} s for a 1-mm beam size. The variation of Δt for a 0- to 6-mm beam size is represented in Fig. 5(a).

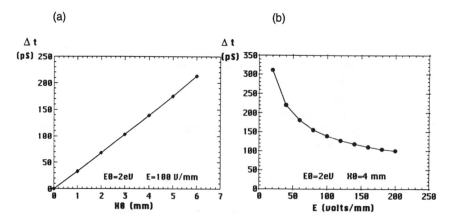

FIGURE 5. (a) Δt versus beam size, (b) Δt versus electrostatic field.

Resolution as a Function of Beam Velocity

In order to reduce the time jitter, we delimit the active area of the monitor by a 3- to 5-mm width slit *l*, perpendicular to the beam direction. For example, with $\beta = 0.2$ dt/dl $= 16.7 \times 10^{-12}$ s/mm.

Resolution as a Function of Electrostatic Field

$$dt/dE = (2mE_0)^{1/2} \cdot E^{-2}/qc - ((2m)^{1/2} / 2qc)(2E_0E^{-3} + q\,(d - x_0)\,E^{-2}) \cdot$$
$$(E_0E^{-2} + q\,(d - x_0)E^{-1})^{-1/2}$$

with $d = 30$ mm, $E = 10^5$ V/m, $E_0 = 2$ eV, $dt/dE = -0.5\ 10^{-12}$ V/m shows that as high a field E as possible is required, as can be seen in Fig. 5(b). For a Gaussian beam, $E = 100$ V/mm, $x_0 = 4$ mm, and the standard deviation is around 25×10^{-12} s.

Distance d Between the Beam Axis and the Microchannel Plate

The distance *d* is, in fact, limited by a compromise between the maximum high voltage *V* that can be applied to the device, the minimum distance *d* between the beam axis and the MCP, and the needed high value of $E = V/d$.

We have chosen $d = 30$ mm even though, in the case of a larger beam size, *d* should be higher in order to increase the resolution.

Electronics

Each electron produced in the vacuum is sent onto the MCP stack, giving rise to a fast signal with charge equal to G e⁻. This signal, collected on a 50-Ω anode (Fig. 6), is amplified, sent to a constant fraction discriminator, and finally sent to a time-to-amplitude converter (TAC) as a start signal. The stop signal is taken as the cyclotron HF signal. The TAC output signal, sent to a multichannel analyser, provides the time distribution of the cyclotron beam bunches.

Results

On-line measurements have been performed during physics experiments in order to check the stability of the beam (Fig. 7(a-c)) or during cyclotron beam tuning (Fig. 7(d)).

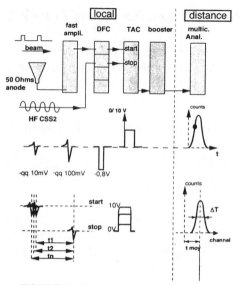

FIGURE 6. Signal processing synoptic.

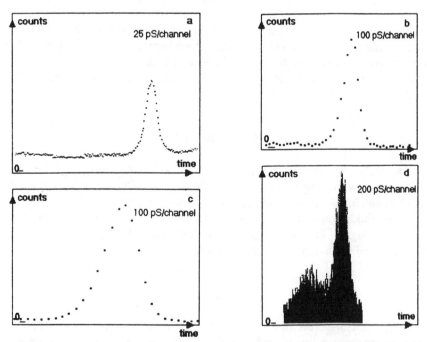

FIGURE 7. On-line beam measurements: (a) Sn 63 MeV/u, I = 350 nA, Ti = 300 s, Δt (FWMH) = 1.5 ns; (b) O 63 MeV/u, I = 1.7 μA, Ti = 560 s, Δt = 550 ps; (c) O 63 MeV/u, I = 200 nA, Ti = 7600 s, Δt = 600 ps; (d) Ar 30 MeV/u, I = 90 nA, Δt = 900 ps.

REFERENCES

1. Anne, R.M., Georget, Y.R., Hue, R.E., Tribouillard, C.H., Vignet, J.L.P., *Rapport GANIL*, R 87-10, 1987.
2. Anne, R.M., Georget, Y.R., Hue, R.E., Tribouillard, C.H., Vignet, J.L.P., "A noninterceptive heavy ion beam profile monitor based on residual gas ionization," *Nucl. Instrum. Methods* **A329**, 21-28, (1993).
3. Grandin J.P., Hennecard, D., Husson, X., Leclerc, D., Lesteven-Vaisse, and Lisfi D., *EURO-PHYS.LETT.* **6** (8), 683 (1988).

High-Brightness Electron Beam Diagnostics at the ATF[*]

X.J. Wang and I. Ben-Zvi

Accelerator Test Facility, NSLS
Brookhaven National Laboratory, Upton, NY 11973

Abstract. The Brookhaven Accelerator Test Facility (ATF) is a dedicated user facility for accelerator physicists. Its design is optimized to explore laser acceleration and coherent radiation production. To characterize the low-emittance, picoseconds-long electron beam produced by the ATF's photocathode rf gun, we have installed electron beam monitors for transverse emittance measurement and developed a new technique to measure electron beam pulse length by chirping the electron beam energy. We have also developed a new technique to measure the ps slice emittance of a 10-ps-long electron beam. Stripline beam position monitors were installed along the beam to monitor the electron beam position and intensity. A stripline beam position monitor was also used to monitor the timing jitter between the rf system and laser pulses. Transition radiation was used to measure electron beam energy, beam profile, and electron beam bunch length.

INTRODUCTION

The Brookhaven Accelerator Test Facility (ATF) has been in operation for several years as a user facility for accelerator physicists, equipped to study the interaction between high power electromagnetic fields and bright electron beams. Most of the ATF's experiments (1) fall into one of two categories. The first type of experiments explore new particle acceleration techniques using the ATF's powerful CO_2 laser, including the Inverse Cerenkov Acceleration (ICA) and Inverse Free-Electron Laser accelerator (IFEL) experiments. The second class of experiments use the ATF's bright electron beam to generate intense electromagnetic radiation, as in the Free-Electron Laser (FEL) and Smith-Purcell radiation experiments.

The ATF's experimental program requires electron beam parameters that range widely in charge, pulse length, and pulse train length. The single-bunch charge can be varied from pC to nC, with a normalized rms emittance from 10 mm-mrad to 0.02 mm-mrad. The electron beam bunch length can be varied (by adjusting the relative phase between the laser and rf gun) from 10 ps to 500 fs. The number of bunches in a macropulse can be varied continuously from one to a few hundred micropulses. The electron beam diagnostics system at ATF was designed in such a way that both transverse and longitudinal emittance of electron beams can be characterized. The main components of the ATF diagnostics system are beam

[*] Work is supported by the U.S. Department of Energy, contract no. DE-AC02-76CH00016.

profile monitors and stripline beam position monitors. Both a dipole magnet spectrometer and transition radiation were used to measure the electron beam energy. Transition radiation was also used for electron beam profile and bunch length measurements.

We have developed a new technique to measure the electron bunch longitudinal charge distribution. The method uses a slit in a dispersive line to pass part of an energy-chirped electron beam. The resolution of the measurement is half a picosecond. A somewhat similar diagnostic (with a quadrupole-scan emittance setup past the slit) was developed to measure the 1-ps slice emittance of a 10-ps-long electron beam. Using this slice emittance diagnostic, we have made the first direct demonstration of emittance compensation.

In the following sections we will introduce the ATF's photoinjector and accelerator, its main electron beam diagnostic elements, our picosecond-resolved charge distribution and slice emittance diagnostics, transition radiation measurements, and the use of a stripline beam position monitor to measure the timing stability of a single bunch with subpicosecond resolution.

THE ATF ACCELERATOR

The ATF's S-band 1.5-cell photocathode rf gun (or photoinjector) (2) is followed by a solenoid magnet. Although its quantum efficiency is low, copper cathode has been chosen as the photoemitter due to its unlimited lifetime at a reasonable vacuum. A second solenoid magnet behind the rf gun bucks the field of the first solenoid magnet. Magnetic measurements were performed to insure that the longitudinal magnetic field on the cathode surface is nearly zero. Following the solenoid magnet is a 45° aluminum mirror for monitoring the laser beam profile on the cathode and carrying out optical transition radiation (OTR) measurement. A phosphor-coated copper block on an actuator is next in line. It is used for charge measurements and can also be used for beam profile and position monitoring. The beam is then injected into the linac. The distance between the cathode and the entrance of the linac is 81 cm, optimal for emittance compensation. The linac consists of two sections of SLAC-type traveling wave linac. The rf gun and linac are powered separately by two XK-5 klystrons.

The ATF's photoinjector and two linac sections are laid out in a straight line with nine quadrupole lenses, numerous steering magnets, and many beam position and profile monitors (Fig. 1). The quadrupoles are used for beam matching or emittance measurements, using two beam profile monitors with a five-meter separation. A 20° dipole magnet bends the beam into a dispersive section with more magnets and diagnostics, including a momentum analysis slit for energy spread measurement or energy selection. Additional 20° dipoles in this line can bend the beam into three dispersion-free beamlines in a shielded experiment hall.

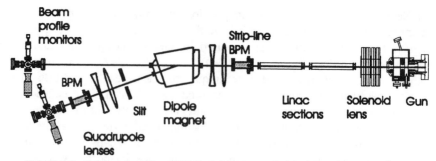

FIGURE 1. Schematic of the ATF photoinjector and elements relevant to slice measurements.

The photocathode is irradiated at a wavelength of 266 nm with a FWHM pulse length of 9 ps and energy up to 0.2 mJ. This light is derived from a laser system comprising a diode-pumped Nd:YAG oscillator followed by two multiple-pass amplifiers and a pair of frequency-doubling crystals. By Fourier relaying an aperture at the beginning of the laser system onto the cathode, a position stability of 0.5% has been achieved, and the peak to peak energy stability is 4%. The laser spot size on the cathode can be changed by varying the focal length and location of one of the imaging lenses. The laser beam incidents at an angle of 72° to normal of the cathode. The beam ellipticity caused by this oblique incidence is corrected by a pair of cylindrical lenses as the final imaging optics.

THE ATF ELECTRON BEAM DIAGNOSTICS

The main electron beam diagnostic devices at the ATF are Faraday cups, phosphor screen beam profile monitors (BPMs), and stripline beam position monitors. For electron beam energy measurement, several spectrometers consisting of dipole magnets, quadrupole lenses, and beam profile monitors are in use. Transition radiation is used extensively at the ATF for measuring beam energy, transverse beam profile, and bunch length.

Figure 2 is the schematic of an ATF beam profile monitor (3). It consists of an imaging system, motorized control system, and image analysis. The imaging system is comprised of a Gd_2O_2S:Tb phosphor deposited on a 1-mil-thick aluminum foil, an aluminum mirror mounted at 45°, and a pair of lenses. The phosphor screen is perpendicular to the electron beam, improving the resolution of the BPM. Two lenses image the electron beam image with a magnification of f1/f2. The focal length f2 is short, resulting in a large light collection angle. The intensity of the image is controlled by a motorized aperture on the f2 lens. Our measurement showed that this BPM has a resolution of 50 μm and a sensitivity of 5 pC with a 0.5-mm-diameter spot.

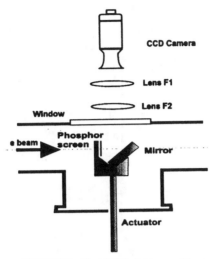

FIGURE 2. The ATF's beam profile monitor.

The ATF's BPM can be easily converted into an optical transition radiation beam profile monitor by simply removing the phosphor screen. The CCD cameras currently in use are Pulnix 745E or WAT902A; the latter has been used for OTR measurements. A good OTR image is obtained at a beam energy of 40 MeV for a charge of 0.25 nC. The BPM was used for transverse emittance measurement, employing both the quadrupole magnet scan and the multiple screen methods.

The ATF beam position monitor consists of a beam pick-up element with four stripline electrodes in a 1.5"-diameter stainless steel pipe mounted with 2.75"-diameter Con-flat flanges and a local receiver. This device is shown in Fig. 2.

The electrodes are shorted at the downstream end, resulting in a bipolar output signal with a separation of 550 ps, twice the length of the electrode. The signals of an opposing pair of electrodes are combined in a hybrid to produce sum and difference signals.

The sum is used as a measure of the beam charge (or phase, see below) and the difference over sum provides a normalized beam position. The bipolar sum and difference signals are converted to unipolar video signals in a pair of balanced mixers driven by a reference. Passband filters (not shown) are used to select the desired Fourier component of the bipolar pulses. The current reference frequency is 2856 MHz, which is about the third harmonic in the bipolar signal's spectrum. Currently we are in the process of changing over to a ~ 1-GHz reference, where better performing components are available. The video output is amplified and sent to charge-sensitive analog-to-digital converters (ADCs).

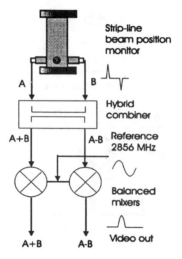

FIGURE 3. The stripline beam position monitor system.

PICOSECOND RESOLUTION SLICE EMITTANCE AND CHARGE DISTRIBUTION MEASUREMENTS

We have developed a new technique for making subpicosecond time-resolved measurements of local (or 'slice') properties within the bunch. This technique was used to make slice emittance measurements and longitudinal charge distribution measurements. The layout of the slice measurement system is shown in Fig. 3.

For a given rf gun phase (referenced to the laser timing), the linac rf power and phase are adjusted to produce a 52-MeV electron beam on crest. Then the second linac section is dephased to 30° off crest. This produces an energy chirp of 0.44% per ps within the electron beam. The electron beam is then transported to the momentum slit. The horizontal beam size observed on the momentum slit is given by

$$x = \sqrt{\beta\varepsilon} + D\frac{\Delta p}{p}, \tag{1}$$

where β is the Courant-Snyder β function, ε is the horizontal emittance, D = 5.4 mm/% is the dispersion function at the momentum slit, and $\Delta p/p$ is the relative energy spread.

To perform the electron beam energy spread measurement and selection, the dispersive term must be larger than the emittance term. This is accomplished by tuning the quadrupole magnets upstream of the dipole magnet to produce a small value of the β function. The opening of the slit is set to 0.5%. At this setting the horizontal beam size is dominated by the energy spread. The dipole magnet is adjusted to a current corresponding to 48.8 MeV (corresponding to a phase of 30°). The electron beam bunch length and the charge distribution within the bunch are measured by varying the second section linac rf phase and measuring the charge past the momentum slit using the stripline beam position monitor behind the momentum slit for measuring the slice charge. Each degree of rf phase corresponds to about 0.97 ps. The timing jitter between the rf system and linac limits the bunch-length measurement accuracy to ± 0.5 ps. The electron beam bunch length was measured for three values of the cathode electric fields: 90, 100, and 110 MV/m. The results are shown in Fig. 4.

The slice emittance measurement (4) is done in a similar manner to the bunch distribution measurement. The difference is that at each slice (selected as above by the phase of the second linac section) we do a quadrupole scan (on a lens downstream of the slit), measuring the vertical beam size as a function of quadrupole current. The measurement can be done for various settings of the emittance compensation solenoid, yielding significant insight on the process of emittance correction.

We have measured the relative rotation of the beam slices in transverse phase space with the change of the solenoid current. This relative rotation makes it possible to counter the effect of the integrated emittance growth in the rf gun due to space-charge and rf effects. The integrated emittance growth is due to the lack of overlap of misaligned slices.

This diagnostic should make it possible to do nonlinear emittance corrections, since the beam matrix of arbitrary slices can be measured and corrected individually.

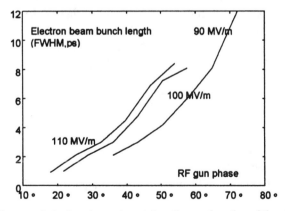

FIGURE 4. Measured electron beam bunch length as a function of the rf gun phase for three values of the gun cathode fields.

TIMING JITTER AND TRANSITION RADIATION
MEASUREMENTS

Maintaining the timing stability between the rf system of the photocathode rf gun and its driving laser system is crucial for the performance of the rf gun. Since the pulse length of the laser is on the order of a few picoseconds, the timing jitter is required to be on the order of subpicoseconds. We have developed a new technique (5) of measuring the subpicosecond timing jitter between the rf and laser using a stripline beam position monitor. As we discussed earlier, one of the resonant frequencies of the BPM is the ATF rf system frequency of 2856 MHz. Using the ATF low-level rf signal as a local oscillator, the output from the mixer with a stripline sum signal contained both the intensity and relative phase information of the photoelectron beam. The output of the mixer is plotted in Fig. 5 as a function of local oscillator phase; near the zero crossing the output is almost linearly proportional to the relative phase between the rf system and electron beam. The sensitivity of our system was measured to be 6.5 mV/ps. Using this technique, we have measured the rms timing jitter between the laser and the rf system to be 0.5 ± 0.25 ps.

Transition radiation is extensively used at ATF for electron beam diagnostics. The coherent transition radiation was observed using a pyroelectric detector (6); the observed square dependency of radiation power on the electron beam intensity (BPM sum signal) confirmed the ATF electron beam bunch length is short. Transition radiation was used to measure the energy of the electron beam from the rf gun. Since the space between the rf gun and the linac is very limited, transition radiation provided an easy solution. An aluminum mirror was installed inside a 4-inch, six-way cross to produce transition radiation. A mirror on the rotational stage outside the vacuum window reflected the radiation to a photomultplier. Figure 6 shows the angular distribution of the transition radiation for two different input rf powers into the rf gun. The 6-MeV electron beam corresponds to a peak acceleration field of 130 MV/m on the rf gun cathode.

ACKNOWLEDGEMENTS

Many people have made significant contributions to the ATF electron beam diagnostics, especially K. Batchelor, R. Malone,   J. Rogers, J. Sheehan, P. Russell, and Z. Segalov. The technical support of W. Cahill, M. Montemagno, and R. Harrington made the ATF electron beam diagnostics work.

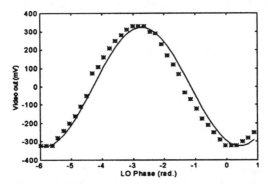

FIGURE 5. Stripline BPM sum video out vs. local oscillator phase.

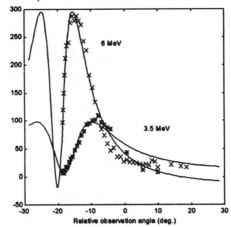

FIGURE 6. OTR angular distributions and their best fit.

REFERENCES

1. I. Ben-Zvi, "The BNL Accelerator Test Facility and Experimental Program," Proc. of the 1991 Particle Accelerator Conference, San Francisco, CA, May 6-9, 1991, 550 - 554 (1991).
2. X.J. Wang et al., Proc. of 1993 Particle Accelerator Conference, Washington, DC, May 17-20, 1993, 3000 - 3002 (1993).
3. D.P. Russell and K.T. McDonald, Proc. of the 1989 Particle Accelerator Conference, 1510 - 1512 (1989).
4. X. Qiu, K. Batchelor, I. Ben-Zvi and X.J. Wang, accepted for publication in Phys. Rev. Lett.
5. X.J. Wang, Z. Segalov and I. Ben-Zvi, Proc. of the 1996 European Particle Accelerator Conference, to be published.
6. E.B. Blum, "Observation of Coherent Transition Radiation at the ATF," BNL-62738 (1996).

Electron Beam Diagnostics Using Synchrotron Radiation at the Advanced Light Source*

Roderich Keller, Tim Renner, and Dexter J. Massoletti

Advanced Light Source
Ernest Orlando Lawrence Berkeley National Laboratory
University of California
Berkeley, California 94720, U.S.A.

Abstract. Synchrotron light emitted from a bend magnet is being used to diagnose the electron beam stored in the main accelerator of the Advanced Light Source (ALS) at Berkeley Lab. The radiation has maximum intensity in the soft x-ray region and is imaged by a Kirkpatrick-Baez mirror pair from the source point inside the ring onto a Bismuth/Germanium-Oxide (BGO) crystal, converted into visible light, and magnified by an attached microscope. The final image is captured by a TV camera-tube and digitized by a frame-grabber device to obtain records of parameters such as beam size, center location, and profile. Data obtained from this diagnostic beamline have been very useful in day-to-day operation of the ALS storage ring to assess the quality and repeatability of the stored beam. The line has further been utilized in several dedicated research activities to measure bunch lengths under various conditions and observe transverse beam instabilities. A summary of obtained results is given in this paper, together with a description of the technical features of the diagnostic beamline.

INTRODUCTION

The main accelerator of the Advanced Light Source (ALS) is a third-generation electron storage ring that produces synchrotron radiation in bend magnets and insertion devices over the spectral range from ultraviolet to the soft x-ray region. One of the bend-magnet photon beamlines is exclusively used to obtain on-line information about the electron beam to provide quality assurance in day-to-day operations as well as support for beam development activities and troubleshooting in case of a malfunction. The synchrotron-light image of the beam cross section carries information on horizontal and vertical beam sizes, positions, and density profiles. With time-resolved measurements, dynamic effects such as bunch lengthening and instabilities can be observed.

The design features of the diagnostic beamline have been described in detail elsewhere (1). In this paper, after giving a brief overview of its characteristics we

---

* This work was supported by the Director, Office of Energy Research, Office of Basic Energy Sciences, Material Sciences Division, U.S. Department of Energy, under Contract No. DE-AC03-76SF00098.

concentrate on discussing applications of the beamline to accelerator diagnostics and the main results so far obtained. Currently some minor modifications are being implemented that will allow faster and safer switching between various observation modes and reduce the setup time for infrequently used special detectors such as streak camera and single-photon counting system.

BEAMLINE LAYOUT AND ELEMENTS

A schematic of the diagnostic beamline layout is given in Figure 1. The source point, representing a cross section through the electron beam, is located inside a storage ring bend magnet. A Kirkpatrick-Baez (K-B) mirror pair (2) images this object plane into the image plane where a Bismuth/Germanium-Oxide (BGO) single-crystal scintillator converts the soft x-ray radiation to visible light. This visible-light image is magnified and recorded by a charge-coupled device (CCD) camera. A set of carbon foils of different densities is inserted in front of the BGO crystal to absorb low-energy photons and to attenuate the transmitted soft x-ray radiation to a level suitable for the CCD camera. In this way the full resolution of the CCD camera can be utilized without being subject to the larger diffraction effects of the visible or ultraviolet fractions. A pair of silicon mirrors, retractably mounted in front of the carbon-foil attenuator, can deflect the visible light into a second optical axis to perform time-resolved measurements with either a streak camera, a fast photodiode, or a single-photon counter.

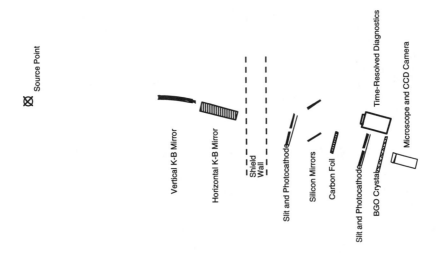

FIGURE 1. Schematic elevation view of the diagnostic beamline.

The actual object-mirror and mirror-image distances are 6.1 and 6.1 m, horizontally, and 5.7 and 6.5 m, vertically, resulting in magnification factors of 1.00 and 1.14, respectively. The horizontal grazing angle is 1.5° and the vertical one 2.0°. Both mirrors are mounted inside a common vacuum tank that can be pivoted around the transverse symmetry axis of the vertical mirror. The first of the two silicon mirrors can be retracted manually to allow passage of the radiation to the end station for the standard operating condition when no time-dependent measurements are being performed.

The two slit/photocathode assemblies in front of the silicon mirrors and in front of the BGO crystal consist of two orthogonal, 200-μ-wide slits and AL_2O_3 scintillator screens. The slits can be moved across the light beam in horizontal or vertical direction to generate intensity profiles, and both assemblies can be completely retracted when they are not in use. The first slit assembly is used to verify the alignment of the K-B mirrors with respect to the incoming photon beam, and the second one serves to monitor the electron-beam density-profiles.

The carbon foils are mounted on a manually operated wheel and range in thickness from 0.1 μ through 20 μ in steps of about a factor of two, corresponding to about a factor of 10 in transmission reduction for every step. Foils thicker than 5 μ are made from pyrolytic graphite; the thinner ones consist of evaporated carbon. The filter system allows one to work at optimum signal-to-noise ratio for a wide range of electron beam currents and energies without saturating the CCD camera. The limiting aperture of the beamline admits photons within an angle of 1.0 mrad, and under diffraction-limited conditions the desired spatial resolution of 10 μ implies that the wavelength of the utilized radiation be shorter than 12.5 nm, corresponding to photon energies above 99 eV. The carbon foils actually transmit at energies above 280 eV, and thus the resolution requirement is fulfilled.

The end station elements, consisting of the second slit assembly, the BGO crystal, and the CCD camera with its imaging optics, can be moved longitudinally on a common stage. In this way the beamline can be made to provide a stigmatic image of the source point by slightly rotating the vertical K-B mirror. Adjusting a grazing-incidence mirror angle mostly changes its effective focal length; the slight vertical shift in image position can be accommodated by mechanically adjusting the end station.

Time-dependent measurements can be performed along the secondary photon-beam axis, with the first silicon mirror inserted into the main axis. We have performed measurements using a fast photodiode (18-GHz frequency limit) and a streak camera.

OBSERVATIONS AND MEASUREMENTS

For most of the time, when the ALS is running for synchrotron-light users, the CCD camera is providing an on-line image of the electron-beam cross section on a TV monitor in the ALS control room. Operators can observe the stability of beam shape and position and take the necessary actions if the storage ring performance is not adequate. Quite often, sudden changes in the beam conditions are noticed first on this beam image and can then be verified with other diagnostic elements such as capacitive beam position monitors. Based on its usefulness for operations, the on-line beam image is now being patched to the various experimental stations in the storage ring building to help the users distinguish between photon-beamline and electron-beam motion.

The monitor image can also be captured by frame grabbing utilities and digitized for post-processing. Correlating beam-size or center-location variations off-line with archived data, such as power supply currents, helps identify defective accelerator components. The processed frame grabber data are also very useful for a multitude of beam studies performed in dedicated accelerator development shifts. Some examples of observed beam shapes are given in Figure 2.

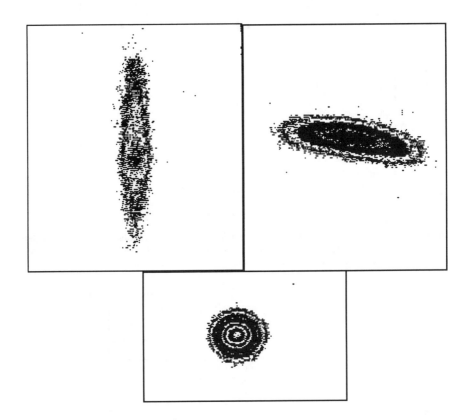

FIGURE 2. Digitized electron beam images as recorded by the CCD camera for various conditons of the fast feedback systems (3). Top left, vertical feedback OFF, horizontal and longitudinal feedback ON. Top right, vertical and horizontal feedback ON, longitudinal feedback OFF. Bottom, all feedback systems ON; in this case the horizontal beam size is $\sigma_x = 51\ \mu$. All three shapes are reproduced in the same scale. The original monitor images are displayed in false colors representing zones of different intensities; this explains the rings seen on the black-and-white reproductions of this paper.

Figures 3, 4, and 5 show measured horizontal and vertical beam sizes as functions of beam current for different feedback conditions and bunch filling patterns.

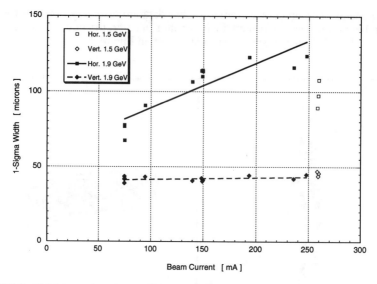

FIGURE 3. ALS beam sizes for multibunch fills (320 out of 328 buckets filled) and energies of 1.522 and 1.9 GeV. The feedback systems were switched OFF for this measurement series. Linear fits are applied to the 1.9-GeV data.

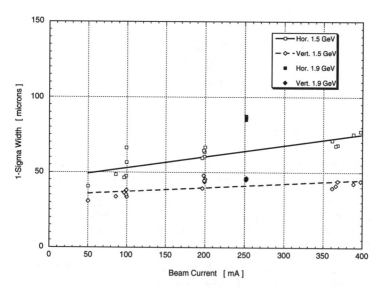

FIGURE 4. ALS beam sizes for multibunch fills (320 out of 328 buckets filled) and energies of 1.522 and 1.9 GeV. The feedback systems were switched ON for this measurement series. Linear fits are applied to the 1.522-GeV data.

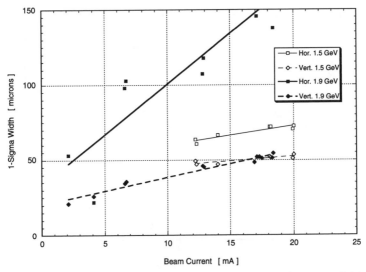

FIGURE 5. ALS beam sizes for single-bunch fills and energies of 1.522 and 1.9 GeV. The feedback systems were switched OFF for this measurement series. Linear fits are applied to all data.

Examples of time-dependent measurements are given in Figures 6 and 7.

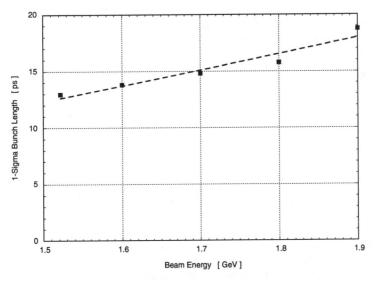

Figure 6. Bunch lengths as a function of electron beam energy as measured by a streak camera for a 200-mA multibunch fill, 0.625 mA per bunch (4).

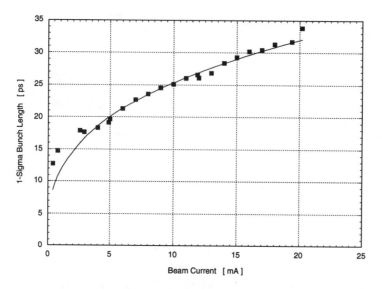

Figure 7. Bunch lengths as a function of electron beam current, I, as measured by a fast photodiode for a single-bunch fill at 1.522-GeV energy. 20 ps are subtracted from all measured lengths to account for the cable delay in the measurement circuitry. A curve proportional to $I^{1/3}$ is fitted to the data with $I \geq 2.9$ mA.

The bunch-length data σ_l in Figure 7 are well fitted by a curve $\sigma_l \propto I^{1/3}$, as is expected when the threshold for turbulent bunch lengthening is exceeded (5). The deviation of the data from the fit at very low currents is due to increasingly poor signal-to-noise ratio, whereas the deviation at 20 mA marks the onset of another effect, the transverse mode-coupling instability (6) that represents an absolute current limit for the ALS storage ring with single-bunch fills.

IMMINENT UPGRADES

The diagnostic beamline already has proven its usefulness for day-to-day operations as well as for accelerator development and troubleshooting activities. Some of its present features, however, are either somewhat uncomfortable to work with or even pose a risk for the machine safety. Therefore, three upgrade projects have been devised and are currently being implemented. The first one addresses the danger that a carbon filter might break when high photon intensity is transmitted through the beamline. In that case, our calculations show that the BGO crystal will be overheated and might break as well, leaving the rather thin exit window to absorb most of the radiation power. A failure of this window will expose not only the beamline itself but also parts of the ALS storage ring to humid air at atmospheric pressure, a clearly undesirable prospect. The remedy consists of mounting a mirror behind the BGO crystal which will absorb or pass the x-ray component of the incident radiation straight through while deflecting the visible light 90° upwards into the CCD camera.

Secondly, rotating the filter wheel to accommodate different electron beam intensities presently requires insertion of a photon shutter because the wheel has empty spaces between the filters. A protective mask will circumvent this problem and ultimately allow remote control of the wheel orientation.

A third modification affects the mounting of the first silicon mirror that can deflect visible light into the secondary beamline axis reserved for time-resolved measurements. With the new mount, the mirror can be precisely inserted into its previous operating position, without the need for realignment as is presently required. This upgrade will also allow remote control of the insertion and retraction tasks.

ACKNOWLEDGMENTS

We would like to thank Howard Padmore and Alan Jackson for their leadership shown in the effort to bring the diagnostic beamline into its present configuration. It is also a pleasure to express our gratitude towards the ALS operators for their help in carrying out all beam measurements and towards Pat McKean for technical assistance. Jim Hinkson and Mike Chin gave valuable support with the time-dependent measurements, and John Byrd contributed some material and advice regarding the bunch-length measurements.

REFERENCES

1. T. Renner, H.A. Padmore, and R. Keller, "Design and Performance of the ALS Diagnostic Beamline," 9th Nat. Conf. on Synchrotron Radiation Instrumentation, Argonne, IL, Oct. 17-20, 1995, *Rev. Sci. Instru.* **67** (9), September 1996.
2. P. Kirkpatrick and A. V. Baez, *J. Opt. Soc. Am.* **38,** 766 (1948).
3. W. Barry, J. Byrd, and J. Corlett, LBNL Berkeley, and J. Fox, H. Hindi, I. Linscott, and D. Teytelman, SLAC Stanford, unpublished work (1995).
4. R. Holtzapple, SLAC Stanford, unpublished work (1995).
5. M. Furman, J. Byrd, and S. Chattopadhyay in H. Winick, ed., *Synchrotron Radiation Sources*, (Singapore: World Scientific, 1994), Ch. 12, p. 328.
6. M. Furman, J. Byrd, and S. Chattopadhyay, l.c. , ref. (5), Ch. 12, p. 329.

Diagnostics Development for the PEP-II *B* Factory*

A.S. Fisher, D. Alzofon, D. Arnett, E. Bong, E. Daly,
A. Gioumousis, A. Kulikov, N. Kurita, J. Langton, E. Reuter,
J.T. Seeman, H.U. Wienands, and D. Wright

Stanford Linear Accelerator Center, Stanford University
Post Office Box 4349, Stanford, California 94309

M. Chin, J. Hinkson, D. Hunt, and K. Kennedy

Lawrence Berkeley National Laboratory, University of California
1 Cyclotron Road, Berkeley, California 94720

Abstract. PEP-II is a 2.2-km collider with a 2.1-A, 3.1-GeV positron ring 1 m above a 1-A, 9-GeV electron ring; both are designed for a maximum of 3 A. Several diagnostics are now in preparation for commissioning the rings. The beam size and pulse duration are measured using visible synchrotron radiation from arc dipoles. Grazing-incidence, water-cooled mirrors that must withstand up to 200 W/cm extract the light. The sum signal from a set of four pickup buttons, normalized to a DC current transformer's measurement of the ring current, is processed to measure the charge in each bunch. This enables us to fill 1658 of the 3492 buckets per ring to a charge that must be equal within ±2%. For diagnostics and machine protection, 100 photomultiplier-based Cherenkov detectors measure the beam-loss distribution.

INTRODUCTION

The PEP-II *B* Factory (1) is a 2.2-km-circumference, two-ring, e+e- collider under construction at the Stanford Linear Accelerator Center (SLAC) in the tunnel of the original PEP single-ring collider. The project is a collaboration with the Lawrence Berkeley and Lawrence Livermore National Laboratories (LBNL and LLNL). To produce *B* mesons with nonzero momentum in the lab frame, the design involves two rings at different energies; both rings require large currents for high luminosity. The 2.1-A, 3.1-GeV positron ring (the low-energy ring, or LER) runs 1 m above the 9-GeV, 1-A electron ring (the high-energy ring, or HER). At one collision point, the LER comes down to the height of the HER, and the two beams collide with zero crossing angle. Table 1 lists several PEP-II parameters.

* Supported by the U.S. Department of Energy under contracts DE-AC03-76SF00515 for SLAC and DE-AC03-76SF00098 for LBNL.

© 1997 American Institute of Physics

TABLE 1. PEP-II Parameters.

Parameter	HER	LER	Unit
Circumference	2199.318		m
Revolution frequency	136.312		kHz
Revolution time	7.336		μs
RF frequency	476		MHz
Harmonic number	3492		
Number of full buckets	1658		
Bunch separation	4.20		ns
Luminosity	3×10^{33}		cm^{-2}·s^{-1}
Center-of-mass energy	10.58		GeV
Current	0.99 (3 max)	2.16 (3 max)	A
Energy	9.01 (12 max at 1 A)	3.10 (3.5 max)	GeV
Betatron tunes (x,y)	24.617, 23.635	38.570, 36.642	
Synchrotron tune	0.0449	0.0334	
Emittances (x,y)	49.18, 1.48	65.58, 1.97	nm·rad
Bend radius in arc dipoles	165	13.75	m
Critical energy in arc dipoles	9.80	4.83	keV

Because the HER reuses the PEP-I magnets (although with a new, low-impedance vacuum chamber), it will be ready for commissioning in 1997. LER commissioning will follow one year later. The BABAR detector will be installed in 1999. Since both rings will use similar diagnostics, their development is proceeding on the HER schedule. Several of these diagnostics are discussed below. A separate paper (2) in these Proceedings describes the beam position monitors (BPMs).

SYNCHROTRON-LIGHT PROFILE MONITOR

Synchrotron radiation (SR) in the visible and near ultraviolet regime (600-200 nm) will be used to measure beam profiles in both transverse dimensions and, with a streak camera, in the longitudinal direction. A measurement of each ring will be made in the middle of Arc7; this is a high-dispersion point for the HER and a low-dispersion point for the LER. A second HER measurement will be made near the start of the arc, where the dispersion is low. No suitable high-dispersion bend is available in the LER.

The high SR power on the first mirror, due to the high current (3A maximum) in each ring, presents a major challenge. PEP-II is not a synchrotron light source: the design offers limited access to SR, and ports must present a low impedance to the beam. Space in the narrow tunnel does not permit backing the mirror away to reduce the heat load. The beam will be incident on the mirror at 4° to grazing, reducing the maximum power along the SR stripe to 200 W/cm in the HER and 50 W/cm in the LER. We first describe the arrangement for the HER.

HER arcs are almost entirely occupied by magnets (see Fig. 1(a)). A 5.4-m dipole fills most of each 7.6-m half cell, with a quadrupole, corrector, and sextupole taking up much of the rest. The intense SR fan is dumped on the water-cooled outer wall of the chamber. The mirror (in Fig.2), mounted in the vacuum chamber on the arc's outer wall, reflects the light horizontally across the chamber to the downstream inner corner. The mirror is rotated so that its upstream edge is slightly recessed behind the opening in the chamber wall, which then shades the mirror edge. The downstream end sticks slightly into the chamber, shading the leading edge of the chamber as it resumes downstream of the mirror. Thus, by reflecting back into the chamber rather than outward, we avoid a much higher heat load at these edges, which would otherwise be normal to the beam.

At 200 W/cm, the mirror cannot be cooled sufficiently to obtain adequate flatness for good imaging. Instead, we make use of the fact that the SR fan at the 9.8-keV critical energy is 15 times narrower than the visible fan we want to image. A 4-mm-high slot will be cut in the mid-plane of the mirror to a maximum depth of 5mm. The x-ray fan fits into the slot when the electrons are traveling on axis, while the visible beam reflects from the mirror surfaces above and below. Because of grazing incidence, the xrays never reach the bottom of the slot as they traverse the full 7-cm length of the mirror. Instead, they continue past the mirror to dump their heat into a thermally separate absorber just downstream (Figure 2). Since the x-ray heat load is not deposited onto the mirror, it remains flat for good imaging. The residual heat load of $1W/cm^2$, due largely to scattered SR, causes a temperature variation across the surface of less than 1°C.

However, the electron beam will not always be correctly positioned. Then we do not demand that the mirror be suitable for imaging, but only that it not exceed its yield strength. We can then steer the electrons back to their proper orbit and

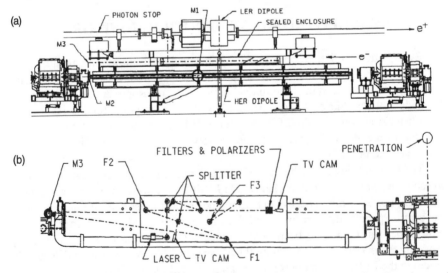

FIGURE 1. HER and LER beamlines in the middle of Arc 7 in (a) an elevation view and (b) a plan view, showing the path of the synchrotron light. For clarity, only the HER optics are shown in the sealed optics chamber on top of the HER dipole.

wait for the mirror to cool. Both the mirror and the dump are made of Glidcop® (copper strengthened with a dispersion of fine aluminum-oxide particles) with water-cooling channels, following techniques (3) developed for the Advanced Light Source. An ANSYS thermal analysis of a beam hitting the mirror 2.5mm above the top of the slot shows that the temperature for a 3-A HER beam rises from 35°C to 160°C, and the stress rises to 90% of yield.

After this first mirror, the light reflects upward from a 45° second mirror and passes through a fused-silica window to exit the beamline vacuum into an argon-filled optics chamber on the HER dipole. Figure 1(b) shows the imaging scheme, which is designed to compensate for the effect of the slot. In geometric optics, a slot or aperture placed in the plane of a lens (like a camera iris) serves only to restrict uniformly the amount of light reaching the image plane without otherwise affecting the image. Here, the first focusing mirror F1 images the slot onto the second focusing mirror F2, which then images the beam onto a charge-coupled device (CCD) camera.

We also consider two effects of diffraction. Table 2 shows the loss in resolution due to the small vertical dimension of the beam. To reduce this effect, the light at all three emission points is taken near horizontally defocusing quadrupoles, where the beams are large vertically. For a point source, diffraction from the slot narrows the image of the beam somewhat at the half maximum point and creates tails, but the effect is small for our 4-mm slot. However, when this pattern is convolved with a narrow ($\sigma_y/\sigma_d = 2$) Gaussian electron beam, the tails disappear and the distribution broadens by 6% over the full width at half maximum without a slot. The increase is less with PEP's broader beams.

In the LER, the SR diverging from the beam downstream of each dipole enters an antechamber; 2/3 of these photons strike a water-cooled photon stop 6m beyond the bend (see Fig. 1(a)). The remainder, emitted in the downstream part of the bend, do not deviate enough from the positron orbit to hit this photon stop and continue to the next one. Their intensity is then lower, since they have diverged further, but their broad visible fan is clipped passing through the narrow antechamber of the intervening magnets. Consequently, we use the light from the closer dipole. When incident on a mirror at a 4° grazing angle, the maximum SR power density is 50 W/cm—well below that of the HER. To get the photons out, we take a 15-mm-wide vertical slice out of the photon stop, and let this light pass through to a mirror similar to the HER

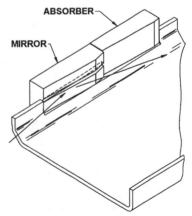

FIGURE 2. The slotted first mirror and the x-ray absorber, both mounted in the wall of the HER chamber.

TABLE 2. Resolution of the SR Profile Monitor for Measurements at 300 nm. Here, the diffraction spot size, $0.26(\rho\lambda^2)^{1/3}$, uses a larger experimental coefficient rather than the calculated value of 0.21. The image size is given by the quadrature addition of the source size and the diffraction size.

	HER Mid-Arc	HER Start of Arc	LER Mid-Arc
Radius of curvature in dipole [m]	165	165	13.75
Diffraction spot size σ_d [μm]	64	64	28
Electron/positron beam size σ_y [μm]	179	199	217
σ_y / σ_d	2.8	3.1	7.8
Image size σ_{image} [μm]	190	209	219
$\sigma_{image} / \sigma_y$	1.06	1.05	1.01

mirror, inclined at 4° and with a central slot. The light is deflected horizontally, but this time away from the positron orbit, since the photon stop shades the leading edge of the mirror. A 45° mirror then sends the beam down to the common optics box on the HER dipole.

The beams are imaged onto CCD video cameras, and a computer with a frame grabber determines the transverse beam size. We put the optics and cameras (with shielding) inside the PEP tunnel in order to get good resolution from a short, stable optical path with few surfaces. In the middle of Arc7, the collected HER and LER light beams are split, with only half used for this local imaging. The other half of the light will be sent upwards through a 10-m penetration to an optics room at ground level, where we can make more elaborate measurements with expensive equipment, such as the streak camera, away from radiation in the tunnel. Figure 1(b) shows both optical paths for the HER beam.

TUNE MONITOR

The tune of each ring will be monitored with a spectrum analyzer processing the signal from dedicated BPM-type pickup buttons. The analyzer includes a tracking generator to excite the beam with a swept sine or broadband noise. No separate excitation structures are needed. Instead, for transverse excitation, this signal will be summed with the input to the power amplifiers for the striplines of the transverse feedback system (4). For longitudinal excitation, the signal will be added to a signal from the longitudinal feedback system (5) that modulates the ring rf to control low-frequency modes.

PEP's high currents require a broad dynamic range, from a ring with one bunch of 5×10^8 electrons or positrons to 1658 bunches of 8×10^{10}, 4.2 ns apart. We must also measure the tune while the feedback systems damp oscillations. The bunch-by-bunch feedback systems for PEP have been prototyped at LBNL's Advanced Light Source. The sensitive front end of the transverse feedback system can measure transverse and longitudinal tune lines with the feedback on, even without exciting the beam. The tune monitor design, shown in Fig. 3, derives from this front end. The button signals are combined, then attenuated or amplified as needed, all with broadband components to keep the pulses narrow, so that a fast GaAs switch can gate the signal for measurements of specific bunches or pass the signal from the entire ring. In the normal fill pattern, 5% of the circumference is left empty as an ion-clearing gap. Bunches near the gap will experience different beam-beam kicks and so have a tune shift. Also, the gate allows us to measure the tune while turning off feedback for a specific bunch.

The two-channel spectrum analyzer uses digital signal processors (DSPs) and a fast Fourier transform (FFT) to compute spectra from 0 to 10 MHz. To bring the signals from the pickups into range, the front end includes mixers at $2f_{RF}$. The analyzer incorporates an ethernet interface for control by an Xterminal, using a functional image of the front panel. It has peak tracking to automatically follow the tune and runs user programs, allowing control, for example, of the beam-excitation signal to measure the peak with the minimum drive.

Other systems will be available to follow the beam's response. Each ring has a second dedicated set of buttons reserved for special measurements, such as high-frequency spectra to examine bunch dynamics. Also, the DSP on each BPM processor card can record signals from 1024 turns and calculate an FFT. Using all the BPMs, we can follow oscillations of a bunch around 1024 turns.

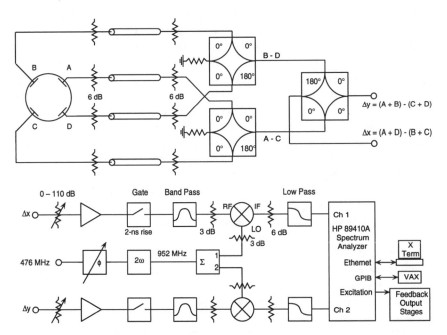

FIGURE 3. The rf front end of the tune monitor.

We can excite the beam at the tune frequency and have the DSP for the kth BPM fit the 1024-turn data for x_k to

$$x_k = A_{x,k} \cos(v_x \omega_{rev} t + \phi_{x,k}) , \tag{1}$$

giving the phase advances $\phi_{x,k}$ and beta functions $\beta_{x,k} = A_{x,k}^2 / \varepsilon_x$ around the ring. A better approach (6), independent of BPM calibrations, solves for $\beta_{x,k}$ using

$$\phi_{x,k} - \phi_{x,k-1} = \int_{s_{k-1}}^{s_k} \frac{ds}{\beta_x(s)} \tag{2}$$

and the measured phase advances from BPM k to BPMs k-1 and k+1.

CURRENT MONITORS

The DC current in each ring is measured by a commercial (7) DC current transformer (DCCT). The device has a 5-μA resolution over a 1-s integration time and a full-scale value of 5A. For two comparisons, a 1-A current with a 3-hour lifetime drops by 93μA/s and injecting 5×10^8 e$^\pm$ adds 11μA.

We have built the necessary housing for the DCCT, which places it outside the vacuum envelope, provides an electrical gap directing the DC wall currents around the transformer core, and capacitively bypasses the gap for higher frequencies to provide a low impedance to the beam. Impedance measurements are underway.

A second system measures the charge in each of the 3492 rf buckets. In a normal fill (see Table 1), 1658 buckets, 4.2ns apart (two rf periods), are filled equally within ±2% to as much as 8×10^{10} for a 3-A beam. With a relative accuracy of 0.5%, the bunch-current monitor must update measurements of each ring at its 40-Hz injection rate (into both rings; 60Hz for e$^+$ only, and 120 Hz for e$^-$ only) to control the fill. For lifetime measurements of individual bunches, we need an accuracy of 0.05% in 1s, allowing the detection of a lossy bunch (≤10 minute lifetime) at a rate useful for operator adjustments.

In each ring we sum and filter the signals from a set of four BPM-type buttons, using a combiner with a 2-period comb filter at $3f_{RF}$. The filter is designed to avoid crosstalk from adjacent bunches. A 1428-MHz mixer brings this signal down to baseband. It is then sent into a VXI crate, where an 8-bit track & hold analog-to-digital converter (ADC) clocked at f_{RF} digitizes the signal from every bucket. This data stream is divided among 12 Xilinx logic arrays. Even after this "decimation," the rate remains too high for processing. Each Xilinx downsamples by processing only one bucket out of eight in each turn, so that it takes eight turns to sample the entire ring. There is enough time in a 60-Hz interval to make 256 samples of the each bucket, for increased resolution. The averages are written into a table in a reflected (dual-port) memory. The VXI processor maintains a second table with sums over 1-s intervals.

Another VXI-based system, the bunch-injection controller, reads this memory whenever needed. It also reads the DCCT to normalize the individual bunch currents, and performs lifetime calculations. These results are reported to the control system. The bunch-injection controller determines the injection sequence for a third system, the master pattern generator, which controls the timing of the injector linac to fill the appropriate buckets in the rings.

BEAM LOSS MONITORS

A network of 100 beam loss monitors (BLMs) localizes and records beam losses at selected points (collimators, septa, and selected quadrupoles) around the rings. The BLMs must operate over a wide dynamic range, covering well stored beams, lossy beams, and injection. The output will be used for machine tuning, for loss histories, and for the rapid detection of high losses requiring a beam abort. They must discriminate against synchrotron radiation and be positioned or shielded for preferential sensitivity to HER or LER.

We have chosen a Cherenkov detector, using a miniature (16-mm diameter) photomultiplier, with an 8-mm-diameter, 10-mm-long fused-silica cylinder placed over the fused-silica photomultiplier tube (PMT) window for a Cherenkov radiator. The small tubes have 2-ns output pulses, comparable to the bucket spacing, that will stretch to about 5ns after the output cable. The assembly will be enclosed in 6mm of lead to reduce synchrotron background. By taking advantage of the large HER dipoles for shielding, we can provide HER/LER discrimination in many places; additional shielding will be needed at other points. The detectors are small enough to be moved around a ring and between HER and LER, for commissioning each ring and for troubleshooting.

Ten-channel CAMAC modules in crates around the rings (shared with the BPMs) process each BLM signal through two input circuits that together provide a wide dynamic range. For low loss rates, the PMT pulses are sent through a discriminator and counted over 1-s or 8-ms intervals, for data logging and plotting. At higher loss rates, the pulses can overlap and are not suitable for counting. Instead, the signal goes through an *RC* filter with a 10-μs time constant; this integrated signal is used twice. First, it passes through a peak detector and is digitized every 8ms for output to the control system. Second, if the 10-μs signal exceeds a programmable threshold, then one or both rings may be aborted. The processor records the channel that triggered the abort, and all BLMs around the rings freeze their most recent readings.

During a brief interval ($\approx 100\mu s$) after injection into a ring, the BLM network is inhibited from aborting the stored beam, since faulty injection is a more likely source of high loss rates. The output of the peak detector remains available, and since it is digitized at the end of this inhibit interval, it measures injection loss.

REFERENCES

1. *PEP-II: An Asymmetric* B *Factory*, Conceptual Design Report, LBL-PUB-5379, SLAC-418, CALT-68-1869, UCRL-ID-114055, UC-IIRPA-93-01, June 1993.
2. Aiello, R., et al., "Beam Position Monitor System for PEP-II," in these proceedings.
3. DiGennaro, R., and Swain, T., *Nucl. Instrum. Methods* **A291**, 313–318 (1990).
4. Barry, W., Byrd, J., Corlett, J., Fahmie, M., Johnson, J., Lambertson, G., Nyman, M., Fox, J., and Teytelman, D., "Design of the PEP-II Transverse Coupled-Bunch Feedback System," in *Proc. IEEE Particle Accelerator Conf.,* Dallas, TX, May 1995, 2681-2683 (1996).
5. Oxoby, G. et al., "Bunch-by-Bunch Longitudinal Feedback System for PEP-II," in *Proc. European Particle Accelerator Conf.,* London, 27 June–1 July 1994 (World Scientific, Singapore, 1994) 1616–1618; Teytelman, D., et al., "Feedback Control and Beam Diagnostic Algorithms for a Multiprocessor DSP System," in these proceedings.
6. Castro, P. et al., "Betatron Function Measurement at LEP Using the BOM 1000-Turns Facility," in *Proc. IEEE Particle Accelerator Conf.,* Washington, DC, May 1993, 2103–2105 (1993).
7. Parametric Current Transformer, Bergoz Precision Beam Instrumentation, Crozet, France.

Design, Commissioning and Operational Results of Wide Dynamic Range BPM Switched Electrode Electronics[*]

Tom Powers, Lawrence Doolittle, Rok Ursic and Jeffrey Wagner

Continuous Electron Beam Accelerator Facility,
12000 Jefferson Ave., Newport News, VA 23606

Abstract. The Continuous Electron Beam Accelerator Facility (CEBAF) is a high-intensity, continuous-wave electron accelerator for nuclear physics. Total acceleration of 4 GeV is achieved by recirculating the beam through two 400-MeV linacs. The operating currents over which the linac beam position monitoring system must meet specifications are 1 µA to 1000 µA. A system was developed in 1994 and installed in the spring of 1995 that switches four electrode signals at 120 kHz through two signal-conditioning chains that use computer-controlled variable gain amplifiers with a dynamic range greater than 80 dB. The system timing was tuned to the machine recirculation period of 4.2 µs so that components of the multipass beam could be resolved in the linacs. Other features of this VME-based system include long-term stability and high-speed data acquisition, which make it suitable for use as both a time-domain diagnostic tool and as part of a variety of beam feedback systems. The computer interface has enough control over the hardware to make a thorough self-calibration and verification-of-operation routine possible.

INTRODUCTION

The CEBAF accelerator linacs are designed to operate with linac currents ranging from 1 µA to 1 mA. The beam position monitoring (BPM) system is required to detect the beam with a relative accuracy of 100 µm for a position range of ± 5 mm. It must maintain this accuracy during both pulsed and continuous wave (CW) operation. It was also required that the system have the capability of distinguishing the position of each of the five beamlets during pulsed operation for machine tune-up. Additionally, a technique to determine the position of the five beamlets during CW operation was desired.

The original BPM electronics (1) that were used for commissioning the machine did not have sufficient dynamic range to operate at total currents outside of a range between 10 µA and 100 µA. This four-channel system suffered from differing drifts in the gain between the plus and minus channels for each beam axis. Additionally, there were no provisions for detecting multipass beam. The switched electrode electronics beam position monitoring (SEE BPM) system described in this paper overcame problems encountered with the previous BPM system by employing a single amplifier-detector chain for each of the two X and Y channels. Many of the features of the system described in this paper have been

---

[*] Supported by U.S. Department of Energy, Contract #DE-AC05-84ER40150.

257

used in similar system developed at other laboratories (2-5). The significant feature of this system is its pulsed multipass detection scheme which is achieved by tuning the SEE timing to match the machine recirculation period of 4.237 μs. This, along with operating the machine with 4.2-μs pulsed beam, allowed us to resolve the components of a multipass beam. This same type of operation can be used to detect the difference in the beam centroid when an inverse snake (a reduction in beam current for 4.2 μs) is applied to the CW beam.

SYSTEM REQUIREMENTS AND PERFORMANCE

The design team started with the requirements summarized in Table 1. The dynamic range is defined as the range of currents for which the system operates within all other specified limits. At the end of the design phase the performance of the hardware was measured on the bench and to a lesser extent in the machine. The high current end of the dynamic range will be tested on the machine when it is operated at rated currents. Laboratory testing indicates that the lower end of the dynamic range is limited by thermal noise, and the high end is limited by signal compression in the electronics. The current dependence, beam position range, and rms fluctuation measurements are also based on laboratory measurements which have not been refuted by operating experience.

SYSTEM DESCRIPTION

The installed linac SEE BPM system consists of six VME crates each controlling seven to ten BPM channels (56 channels total). These crates are located in the service buildings approximately 10 meters above the beamline. Each channel consists of a BPM detector, a radio frequency (rf) module located in the tunnel and an intermediate frequency (IF) module located in the VME crate. Each crate has a timing module which is used to synchronize the system to the accelerator and generate specific timing signals required by the IF, rf, and data acquisition modules. The VME crate also contains three commercial data acquisition modules and a single-board microcomputer which are used to acquire and process the position signals prior to transmitting them to the machine control system.

TABLE 1. Requirement and performance specifications

	Requirement specification	Performance specification
Dynamic range	1 - 1000 μA	0.4 - 2000 μA
Nominal measuring rate out of control system	1 meas./s	1 meas./s
Beam position range	lxl, lyl ≤ 5 mm	lxl, lyl ≤ 5 mm
Resolution (rms. fluct. at nominal meas. rate)	≤ 0.1 mm	≤ 0.1 mm
Current dependence	≤ 0.1 mm	≤ 0.1 mm
Multipass capability	some kind	"snake" pulse
Measuring bandwidth	10 – 100 kHz	120 kS/s

The rf Module

The detector in the system is a four-wire stripline antenna system previously described (6). It is connected to a four-input (X+, Y+, X–, and Y–) two-output (X and Y) rf module, which is located about 1 m off of the beamline axis. The rf module switches between the plus and minus channels, amplifies the signal by 23 dB if necessary, down converts the 1497 MHz rf signals to 45 MHz, and transmits them to the IF module via coaxial cables. The range of input signals for the rf module is –77 dBm to –11 dBm (1 μA to 1 mA, off centered 5 mm). The conversion gain in the high-gain mode is 25 ± 0.2 dB and 1.5 ± 1 dB in the low-gain mode. The rf module is also capable of injecting a calibration signal of variable amplitude into any of the four electrodes for calibration purposes. Communication with and control of the rf module is done via the IF module using a two-way, RS-485 serial link. The TTL level control signals for the attenuator rf switch and the rf boards are produced on the rf logic board which is located in the rf module. To improve radiation hardness of the system, the rf logic board design was implemented using bipolar logic.

The rf switching, amplification, down conversion, and IF amplification were implemented on the rf board. A simplified schematic is shown in Figure 1. In this diagram the X+ channel is being switched through a pair of GaAsFET switches (Mini Circuits, YSWA-2-50DR) to the input of an isolator. At the same time, the Calibration switch and the X– switch are connected to double the isolation between the channels. From the output of the isolator, that was used to keep the input impedance constant throughout the dynamic range, the signal passes through a 75-MHz bandwidth crystal filter which rejects unwanted beam-induced harmonics. The filter is followed by a pair of switches which allow one to switch in a pair of 11.5-dB amplifiers (Mini Circuits, MAR-6) that provide gain during low-current (<100 μA) operations. The signal is then down-converted to 45 MHz, low-pass filtered, and amplified. The variable gain IF amplifier (Analog Devices, AD603) was adjusted during the module calibration procedure such that the output gain of all of the rf modules provided 25 dB of conversion gain when operated in the high-gain mode. The most difficult part of the design was to reduce the VSWR of the input in all modes of operation to a value below 1.2. This was accomplished by using a line transformer which was further tuned during production by adding capacitive stubs to the input traces located between the rf inputs and the first rf switches. This resulted in VSWR values between 1.03 and 1.19 at all rf inputs.

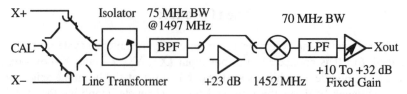

FIGURE 1. Switched electrode electronics rf chain schematic diagram.

The IF Module

The function of the IF module is to amplify the IF signal generated by the rf module and down convert it to a baseband signal that can be digitized with a commercial data acquisition module.   A simple schematic drawing is shown in Figure 2. The input range of the IF module is –59 dBm to –27 dBm.   The first bandpass filter is an LC filter that reduces the broad-band noise. It is followed by a pair of variable-gain IF amplifiers (AD603) which have 84 dB of dynamic gain control. To further reduce the noise level, the IF bandwidth is limited to 1 MHz using a commercial LC filter. This is followed by a single-stage amplifier whose gain is adjusted at the time of calibration. The 45-MHz signal is down-converted to baseband using a low-level video detector (Motorola, MC1330). The baseband signal is buffered and filtered by using an 860-kHz low-pass filter followed by a gated integrating filter which integrates the signal for 2.9 μs of the 4.2-μs cycle time. A sample and hold amplifier maintains the signal level at the end of the integration period while the signal is multiplexed with the signal from the Y channel. This produces a series of X+, Y+, X–, Y– signals which are applied to the input of a commercial, 12-bit, high-speed data acquisition module (VMIC-3115).

In addition to attention to cross talk and low-level noise issues associated with the design and layout of the IF module printed circuit board, there are two subtle details associated with the design of the IF module. The signal levels at each stage in the IF chain were calculated and the fixed gain IF amplifier was

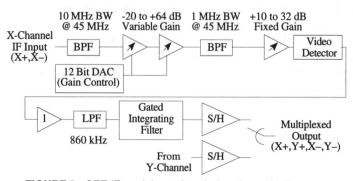

FIGURE 2. SEE IF module analog chain schematic diagram.

adjusted such that none of the earlier stages saturate prior to reaching the maximum gain. The final critical item was the timing of the gated integrating filter, switch clock, data clock, and the multiplexer. Cable transient times and beam delay along the linac (because several channels share the same clock) had to be calculated so that the gated integrating filter was charging only when there was valid data present at the IF module. Additionally, the falling edge of the data clock signal was timed to occur when there were no other digital transitions occurring on the IF card. For this reason, communication between the local VME computer, the timing module, and the IF modules is done using the two bi-directional data ports on one of the VMIC-3115 modules. The advantage of this is that there was control of all of the logical signals on the IF module which insured that the transitions generated by the bussed control, data, and address signals did not interfere with the low-level 45-MHz signals.

The Timing Module

System timing and synchronization is controlled by the timing module which produces the timing signals for all of the IF modules in a given VME crate, the acquisition clock for the data acquisition modules, the delays required for multipass operation, and the pulse count per acquisition trigger. The clock on the timing module was selected such that the system operates at the revolution time of the machine (4.237 μs) to within 0.02%. The delay control, which was required for multipass operation, allowed a variable control between 109 ns and 27 μs in 109-ns steps. The timing module is a straightforward digital design implemented with a combination of PLDs and discrete logic.

Operational Software

During normal operations the data is synchronously acquired at a 248-kHz rate and processed locally by a Motorola MV162 single-board VME computer (68LC040 based). This computer runs a custom-written data acquisition task under VxWorks, which processes the raw digital data into beam position. This task also implements a digital gain control loop so that the video detectors operate in their linear range. The X and Y beam positions that this task computes are passed to an EPICS (7) network database, which also runs on the local computer. From there, the beam position is made available over ethernet to high-level applications and displays that run on Unix host computers.

The low-level data acquisition task is also set up to acquire calibration data on demand: without beam, the calibration oscillator is turned on and routed to one of the electrodes, and the output signals from the other axis are recorded as a function of IF amplifier setting. This data is used to determine offset between electrodes due to VSWR mismatch, cable differences, and detector imbalance. Finally, the system can be used to acquire a trace of beam motion for as much as 68 ms at the full 128-kS/s data rate. In one case, five BPMs in one crate have been simultaneously sampled in this mode, providing complete characterization of the beam properties (current, x, x', y, y', E) with a measurement bandwidth of 60 kHz.

LABORATORY RESULTS

Figure 3a shows the beam position as a function of beam current, measured under laboratory conditions. The beam was simulated by using a power splitter with a 4.5-dB attenuator inserted in the X+ line. The input power was varied from −100 dBm to −10 dBm; the AGC circuit in the IF module was controlled and the data was acquired using a VMIC-3115 module controlled by a Macintosh computer running LabVIEW®. Approximately 2000 position readings were acquired for each data point on the graphs. The thermal noise and bandwidth of the system limit the low current operation within the specified ± 100 μm to 400 nA in high-gain mode and 5 μA in low-gain mode where the high and low gain refer to the gain setting in the rf module. Below these values the video detector in the IF module is not capable of reliably locking into the 45-MHz IF signal because of the level of the broad-band noise. Figure 3b shows the standard deviation of the position as a function of beam current. This indicates that signal averaging must be employed at currents below 10 μA (high gain) and 100 μA (low gain) to insure reasonable data. At the high current end of the operations the IF module gains saturate at 125 μA and 2.5 mA depending on the gain setting of the rf module.

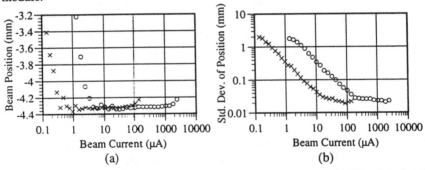

FIGURE 3. (a) Beam position and (b) standard deviation of beam position as a function of beam current in a laboratory setup which includes a SEE rf module, a SEE IF module and a SEE timing module. An "X" indicates rf module high-gain operation and an "O" indicates rf module low-gain operation. The data on (b) is position data taken at 124 kS/s.

OPERATIONAL EXPERIENCE

Figure 4 shows the measured beam position as a function of location for seven different currents ranging from 500 nA to 100 μA. These data were taken by varying the current of a single-pass beam in the CEBAF accelerator while keeping all other machine parameters fixed. One hundred forty-four readings were averaged for each point shown. Figure 5 is another representation of the same data. In this figure each data point represents the difference between the recorded values and the average of all the points for which the current exceeded 10 μA at each machine position. With the exception of 1.7% of the data, the drift of the position as function of beam current is < 100 μm when the beam current is > 1 μA. Additionally, only one data point is > 100 μm from the mean value when the current is > 10 μA.

Two factors contribute to the uncertainty of the measurements with beam. The first is the number of readings taken. For the laboratory measurements, 2000 readings were averaged for each data point. This becomes more important as the beam current decreases below 1 μA where the standard deviation of the position readings is greater than 400 μm. The second factor is the uncertainty of the actual beam position stability during the 1-1/2 hours that it took to ramp the current and take the measurements. Unfortunately this factor cannot be quantified. However, further examination of the data shown in Figures 4 and 5 indicates that positions 10, 11, 32, 40, and 42 are the only positions in the machine for which the current dependence exceeds ±100 μm when the current is greater than 1 μA. The fact that the worst offenders are grouped geographically tends to indicate beam motion.

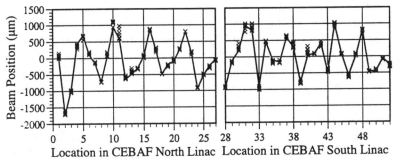

FIGURE 4. Measured beam position as a function of location in the CEBAF linacs for seven different values of beam current between 500 nA and 100 μA.

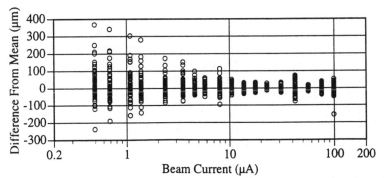

FIGURE 5. Difference from average of the beam position at each location when the current is > 10 μA as a function of beam current as measured in the CEBAF linacs.

During normal operations five passes of continuous beam are present in the linacs. One machine parameter that has to be controlled is the re-injection properties of the second and subsequent passes (position and angle). The technique that was developed is known as sending a "snake" around the machine. A beam pulse just shorter than the single-pass time for the machine, 4.2 μs, is injected into the machine. A beam synchronization pretrigger is applied to the timing module, and the delays on the SEE BPM system are set such that the beam pulse is captured on each of the five passes without overlap from the previous ones. Each beam pulse is acquired five times as it repeatedly passes through the BPM. Normally the sequence is X1+, X2−, X3+, X4−, X5+ on the first beam pulse followed by X1−, X2+, X3−, X4+, X5− on the subsequent beam pulse, where the number indicates the pass through the accelerator. Thus, data from two successive pulses must be combined before full multipass beam position can be computed. For the SEE system to be operated simultaneously with the 4-channel electronics, installed on the arcs, the actual beam pulse structure must consist of a 250-μs pulse followed by a 100-μs pause, followed by a 4.2-μs snake pulse. This is now the normal tune-up mode of operation at CEBAF.

To adjust the delays required for multipass operations, a single-pass beam pulse of 4.2 μs is injected into the machine. The delay of each timing module is varied while the outputs of all of the channels in the same VME crate are recorded with the gain control held constant. Figure 6 shows the results of this operation for three of the eight channels in the first VME crate in the machine. Effectively, the charge control signal on the gated integrating filter (2.9 μs) is being convoluted with the beam pulse signal as it arrives at the IF module. The flat top of each of the curves represents the delay time when the integration time occurs coincident with the beam pulse. The difference in the relative position of the flat top is related to the length of the IF module to rf module cables (30 m to 150 m) and the relative position of the BPM on the beamline. The solid vertical line indicates the optimum delay for that group of channels. The dashed vertical lines indicate the time that the next pass beam pulse would be present. Since both the beam pulse and the gated integrator have sharp time-domain cutoffs, cross talk between the passes is kept to a minimum.

Time-domain beam position data has been taken using an EPICS-based interface. The data shown in Figure 7 was taken in a dispersive region of the machine in order to evaluate the need for a beam-based, fast-feedback system. The raw position data, which was acquired at a 124-kHz rate, was filtered using a 20-point running average. The position data represents an energy variation ($\Delta E/E$) of 2×10^{-4} with major harmonic content at 60 Hz, 120 Hz, and 180 Hz (8). To reduce this energy fluctuation, a beam-based, fast-feedback system is being implemented with the SEE BPM electronics.

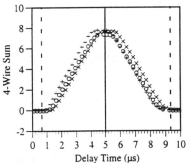

FIGURE 6. Four-wire sum as a function of delay time for three channels in the same VME crate. The solid line indicates the optimum delay time.

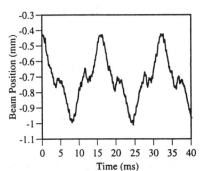

FIGURE 7. Beam position as a function of time in a dispersive region of the accelerator. The data was filtered using a 20-point running average (8).

CONCLUSION

The SEE BPM system installed in the linacs at CEBAF has been presented. The system provides ± 100 μm accuracy with a position range of ± 5 mm for currents between 400 nA and 2 mA. Beam tests with currents between 500 nA and 100 μA have been presented; they are consistent with laboratory results. The time-domain data which have been presented demonstrate the usefulness of the system for capturing the beam motion with an acquisition rate of 124 kHz. Future enhancements to the system include using it in a beam-based, fast-feedback system; increasing the dynamic range; and lowering the minimum specified current to levels below 200 nA.

ACKNOWLEDGMENTS

The authors would like to express their appreciation to the machine installation staff who were instrumental in making it possible to bring this system on line 14 months after the beginning of the design process.

REFERENCES

1. Hofler, A. S. et al., "Performance of the CEBAF Arc Beam Position Monitors," Proceedings of the 1993 Particle Accelerator Conference, Washington DC, 2298-2300 (1993).
2. Cardenas, J., Denard, J-C., "Mesure de Position et d'Intensity," LURE internal report NI/9-75, Orsay, 1975.
3. Biscardi, R., Bittner, J. W., "Switched Detector for Beam Position Monitor," Proceedings of the 1989 Particle Accelerator Conference, Chicago, Illinois, 1516-1518 (1989).
4. Kleman, K. J. "High Precision Real Time Beam Position Measurement System," Proceedings of the 1989 Particle Accelerator Conference, Chicago, Illinois, 1465-1467 (1989).
5. Ursic, R. et al., "High Stability Beam Position Monitoring of ELETTRA," 1st DIPAC, Montreux, Switzerland (1993).
6. Barry, W., *Nucl Instrum Methods A,* **301**, 407-416 (1991).
7. Dalesio, L. R. et al., "The EPICS Architecture," ICALEPCS Proceedings, 278 (1991).
8. Chowdhary, M., CEBAF, personal communication.

Commissioning Results of the APS Storage Ring RF Beam Position Monitors

Emmanuel Kahana and Youngjoo Chung†

Accelerator Systems Division, Advanced Photon Source,
Argonne National Laboratory, Argonne, IL 60439

Abstract. The commissioning of the 360 rf beam position monitors (BPMs) in the Advanced Photon Source (APS) storage ring (SR) is nearing completion. After using the single-turn capability of the BPM electronics in the early ring commissioning phase, resolution measurements versus current and bandwidth were successfully performed. In the standard SR vacuum chamber geometry, the resolution was measured with beam as 0.16 μm·mA/$\sqrt{Hz}$. For the insertion device vacuum chamber geometry, the resolution was measured to be 0.1 μm·mA/$\sqrt{Hz}$.

Since the photon beam stability requirement for the users is only 4.5 microns rms in the vertical direction, investigations of rf BPM offset versus current and bunch pattern have also been initiated. Both single bunch and multibunch beam patterns with varying intensity were used to determine offset stability for both the global and the local orbit feedback applications.

INTRODUCTION

The stability of the x-ray beam is an essential requirement of the APS. The current requirement is 4.5 microns rms. In order to achieve this stability, we need a BPM system with a very good resolution that serves as a sensor for local and global feedback systems. The BPM resolution is inversely proportional to the beam current and directly proportional to the square root of the processing bandwidth (1, 2). Another consideration is the beam stability dependence upon bunch pattern and current. This paper presents the results of BPM resolution measurements, their relation to beam current and bandwidth, and beam stability as a function of current and bunch pattern.

---

* Work supported by the U.S. Department of Energy, Office of Basic Energy Sciences, under Contract No. W-31-109-ENG-38.

† Present address: Kwangju Institute of Science and Technology, South Korea.

RESOLUTION MEASUREMENTS

The BPM samples the beam position N consecutive times, with the time interval Δt, using a 12-bit A/D converter. The revolution time is 3.68 microseconds, and, since in the normal mode we alternate between x and y, the sampling is at half this rate (3). The rms beam position error Δx has two components: one due to real beam motion and one due to electronic noise. Since the two components are assumed to be uncorrelated, the rms beam position error is given by:

$$\Delta x^2 \approx \Delta x_b^2 + \Delta x_e^2 , \tag{1}$$

where Δx_b is the rms error due to real beam motion and Δx_e is the rms error due to BPM electronics noise.

It was determined experimentally that Δx_e is inversely proportional to the single bunch current and inversely proportional to the square root of N. Therefore,

$$\Delta x_e = \frac{k}{I_b \sqrt{N}} , \tag{2}$$

where k is a constant, I_b is the single bunch current, and N is the number of samples averaged.

Since we only want to measure Δx_e, we need to cancel the effect of the beam motion. The beam position is measured by comparing the voltages induced by the beam on four symmetrically placed capacitive "buttons" inserted into the vacuum tube. Therefore, we can cross-connect the cables attached to the BPM buttons and measure x and y separately. For example, if the top or bottom two cables are crossed, the beam motion effect on the x measurement will be negligible.

For this work, a BPM in sector 6 had the top buttons crossed, a BPM in sector 9 had the inboard buttons crossed, and a BPM in sector 7 was normally connected. Figures 1 through 3 show the results of the measurements of the BPM position error, as a function of N, for the single-bunch current $I_b = 0.22$, 0.91, and 7.1 mA. The dependence of the beam position measurement error on the beam current is shown in Figures 4 and 5 for N=2 and N=1024.

For the BPMs with crossed connections, the dependence of the beam position measurement error decreases according to $1/I_b$. The error decreases with $1/\sqrt{N}$ as shown in Eq. (2). For the normal BPMs in sector 7, the error decreases initially, but approaches asymptotically about 23 microns rms for x and 5 microns rms for y, due to actual beam motion. Since the specification for x is 17 microns rms, there was a problem which was later determined to be a result of power supply ripple (4). From the plots we can determine the constant k=40 for Eq. (2). In the case of insertion device (ID) BPMs, the calculation gives k=25 due to the smaller chamber size and smaller buttons (1).

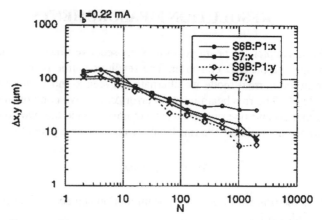

FIGURE 1. Beam position measurement error Δx and Δy, as a function of N, for a single-bunch current $I_b = 0.22$ mA.

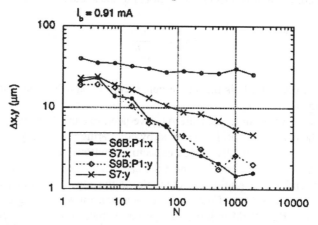

FIGURE 2. Beam position measurement error Δx and Δy, as a function of N, for a single-bunch current $I_b = 0.91$ mA .

Since the average number N is related to the processing bandwidth, it is convenient to calculate Δx_e in terms of $[\mu m \cdot mA / \sqrt{Hz}]$. From the frequency response of the averager and the turn rate, we obtain (1):

$$\Delta x_e [\mu m] = \frac{\sqrt{BW[Hz]}}{I_b[mA]} \times \left\{ \begin{array}{ll} 0.16 & \text{regular BPM} \\ 0.1 & \text{ID BPM} \end{array} \right\} , \qquad (3)$$

where BW is the processing bandwidth.

In the APS storage ring, the rf BPMs will be used for global and local beam position feedback to stabilize the particle and x-ray beams. The correction bandwidth is expected to be approximately 100 Hz, with a sampling frequency of

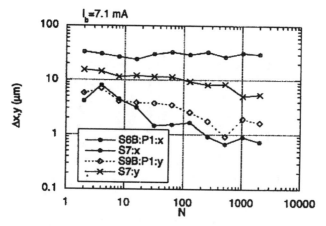

FIGURE 3. Beam position measurement error Δx and Δy, as a function of N, for a single-bunch current I_b = 7.1 mA.

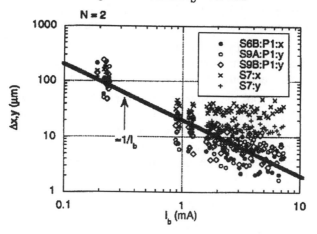

FIGURE 4. Beam position measurement error Δx and Δy, as a function of the single-bunch current, for N =2.

4 kHz. With this bandwidth and with the single-bunch current of 1 mA, the resolution will be better than 1.6 microns for regular BPMs and 1 micron for ID BPMs. Beam position perturbations larger than this resolution will be corrected by feedback.

ORBIT STABILITY TEST

An important characteristic of the BPM system is the offset stability as a function of bunch pattern and beam current. In order to measure this characteristic, we measured the beam orbit, using all 360 BPMs in the storage ring. We eliminated

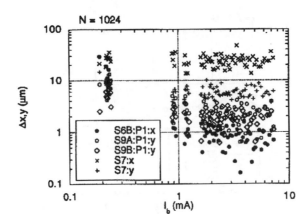

FIGURE 5. Beam position measurement error Δx and Δy, as a function of the single-bunch current, for N =1024.

from the data some "bad" BPMs which showed unusually large offsets or no response. The orbit was measured for the following conditions:

a. 5 bunches, every third bucket, current 2.5 to 14.5 mA
b. 15 bunches, every third bucket, current 2.8 to 17.5 mA
c. 25 bunches, every third bucket, current 3.5 to 18.5 mA

As a reference, we took the 25-bunch pattern at 16.3 mA. In each case we averaged 2048 samples of each BPM measurement and averaged 30 orbits to reduce the effect of beam motion. We calculated the difference between each averaged orbit and the reference, and then calculated the average and standard deviation of the differences for each bunch pattern and each current. Figures 6 through 9 show the results.

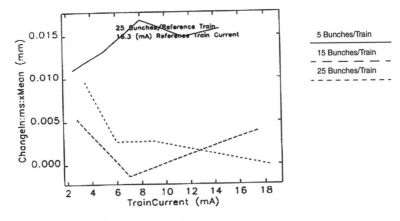

FIGURE 6. Average horizontal difference vs. bunch pattern and current.

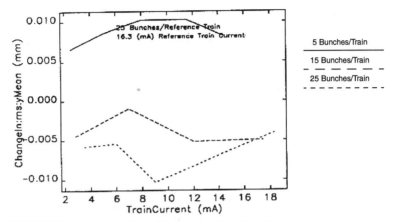

FIGURE 7. Average vertical difference vs. bunch pattern and current.

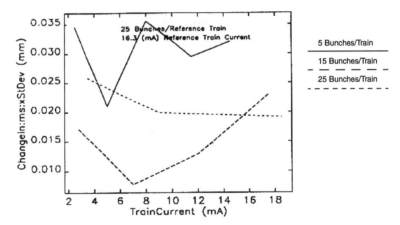

FIGURE 8. Horizontal difference standard deviation vs. bunch pattern and current.

The figures show that the short bunch train case differs from the long bunch train cases. This is expected, since the location of the sampling gate is more critical for short bunch trains. Since the bunch is stretched by the input filter to about 100 ns, for bunch trains longer than 100 ns, the change in average offset is less than 5 microns. Including bunch trains shorter than 100 ns, we have a change in average offset of less than 20 microns. As a function of current, the change in average offset is about 5 microns.

SUMMARY

Both the resolution and the offset stability are compatible with the requirement of 4.5 microns rms photon beam stability for the ID. More measurement will be

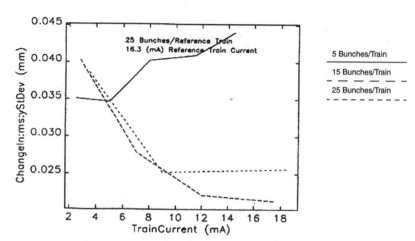

FIGURE 9. Vertical difference standard deviation vs. bunch pattern and current.

taken for long-term drift, orbit stability at high current, and closed-loop feedback performance.

ACKNOWLEDGMENT

Thanks to Nick Sereno, Mike Borland and others in the Operations Analysis Group for their continuous support in taking and analyzing the data.

REFERENCES

1. Y. Chung and E. Kahana, "Performance of the Beam Position Monitor for the Advanced Photon Source," Proc. of the SRI'95, Argonne, IL, October 16-20, 1995, to be published.
2. Y. Chung and E. Kahana, "Resolution and Drift Measurements on the Advanced Photon Source Beam Position Monitor," Proc. of the 6th Accelerator Instrumentation Workshop, Vancouver, Canada, October 2-6, 1994, AIP Conf. Proc. No. 333, 328-334 (1995).
3. E. Kahana, "Design of the Beam Position Monitor for the APS Storage Ring," Proc. of the 3rd Accelerator Instrumentation Workshop, Newport News, VA, October 28-31, 1991, AIP Conf. Proc. No. 252, 235-240, (1992).
4. A. Lumpkin et al., "Commissioning Results of APS Storage Ring Diagnostics Systems," Proc. of the 7th Accelerator Instrumentation Workshop, Argonne, IL, May 5-9, 1996, these proceedings.

A Submicronic Beam Size Monitor for the Final Focus Test Beam

P. Puzo, J. Buon, J. Jeanjean, F. LeDiberder[*], V. Lepeltier

*Laboratoire de l'Accélérateur Linéaire, IN2P3-CNRS et
Université de Paris-Sud, F91405 Orsay Cedex, France*

Abstract. A gas-ionization beam size monitor has been used to optimize and measure the transverse electron beam dimensions of the Final Focus Test Beam at SLAC. The ultimate values measured with the monitor are 1.5 μm × 70 nm. Moreover, the high sensitivity of the monitor has been used to detect and cancel detrimental 'banana' shapes of the beam.

INTRODUCTION

A prototype focusing system for a future linear collider, the Final Focus Test Beam (FFTB), has been built at SLAC and operated by a collaboration between DESY (Hamburg), FNAL (Batavia), KEK (Tsukuba), INP (Novosibirsk), LAL (Orsay), MPI (Munich), and SLAC (Stanford) (1). The SLC 50-GeV electron beam is injected in the 200-m-long FFTB line where it is demagnified by a factor of 380 down to a vertical size smaller than 100 nm. One of the most challenging problems was to measure transverse beam sizes smaller than 100 nm. For that purpose, two new detectors have been installed at the final focus of the FFTB: a laser-Compton monitor (2) and a gas-ionization monitor (3). The former (built at KEK) is based on Compton scattering of the electron beam against a fringe pattern of two laser beams, while the latter (built at LAL) is based on scattering of ions by the space-charge field of the electron beam.

PRINCIPLE

The principle of the gas-ionization monitor has already been described in detail (3). A pulse of helium gas is injected into the beam pipe, near the focal point of the FFTB line, at the passage of each electron bunch. The electrons create He^+ ions by ionization of the gas atoms. These ions are kicked by the space-charge electric field of the bunch during the time of the bunch passage that follows the ion creation. As the beam electrons are ultrarelativistic, the electric field is transverse.

---

[*] Present address : LPNHE, 4 Place Jussieu, Tour 33, 75252 Paris Cedex 05, France.

The ion velocity being much lower than the light velocity ($\beta_{ion} \approx 10^{-3}$–$10^{-4}$), the magnetic force of the space-charge field can be neglected. Finally, the kick on the ions is essentially transverse. After the passage of the beam, the ions drift towards a detector surrounding the beam pipe.

For the typical values of the FFTB parameters ($\sigma_x = 1.8$ μm, $\sigma_y \approx 100$ nm, $\sigma_z = 650$ μm, and $N_{e^-} = 7.0 \times 10^9$), the maximum space-charge field reaches 0.8 V/Å, leading to a maximum ion energy of 10 keV.

The information on the transverse beam sizes is carried out by the properties of the ion momentum (direction and velocity), as it reflects the space-charge field inside the bunch where the ion has been kicked. This is essentially a probe of the space-charge field of the electron bunch. From now on, we will assume the beam charge N_{e^-} and the bunch length σ_z to be known, which is the case for the FFTB.

The maximum velocity of the ions (or their minimum time of flight T_{min} towards the detector) is obtained for ions created by the head of the electron bunch and at the transverse edge of the bunch where the field is maximum. That field is inversely proportional to the largest transverse beam size, the horizontal dimension σ_x for the FFTB. Its dependence on the other transverse dimension σ_y becomes smaller as σ_y decreases (4). The measurement of the minimum time of flight allows determination of the horizontal dimension σ_x of a flat beam with a small correction for the vertical dimension (see Fig. 1).

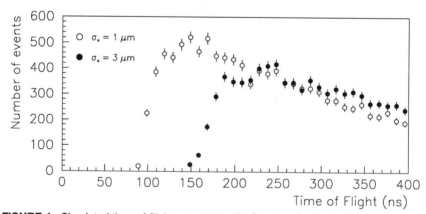

FIGURE 1. Simulated time-of-flight spectrum of He$^+$ ions for two horizontal beam sizes $\sigma_x = 1$ μm (o) and $\sigma_x = 3$ μm (•). The time of flight is computed for a detector located at 6.5 cm from the beam. Each bin is 10 ns ($\sigma_y = 200$ nm, $\sigma_z = 650$ μm, and $N_{e^-} = 7.0 \times 10^9$).

The space-charge field is null on the bunch axis and maximum at the edge. There is then a potential well in which the He$^+$ ions oscillate for high values of the field. Inside a flat bunch, the ions oscillate with horizontal and vertical amplitudes proportional to the respective coordinates of their creation point with respect to the bunch axes. On average, horizontal amplitudes are larger than the vertical ones for a horizontally flat beam. The ions are then preferentially kicked towards the two horizontal directions (left and right): the azimuthal ion distribution shows twin peaks in the left and right directions. This leads to an *even* anisotropy in the azimuthal distribution (see Fig. 2) that can be measured by the amplitude of the second-order Fourier coefficient A_2 of this distribution. When the vertical dimension σ_y decreases, the Fourier coefficient A_2 increases, allowing a measurement of σ_y.

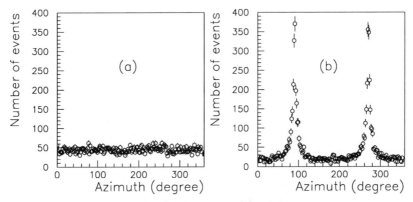

FIGURE 2. Simulated azimuthal distribution of He^+ ions for two vertical beam sizes $\sigma_y = 1\ \mu m$ (a) and $\sigma_y = 60\ nm$ (b). The origin of the azimuth is in the upward direction ($\sigma_x = 1\ \mu m$, $\sigma_z = 650\ \mu m$, and $N_{e^-} = 7.0 \times 10^9$).

Moreover, the azimuthal distribution of He^+ ions is also sensitive to any asymmetry in the electron distribution inside the bunch. In particular, a frequent asymmetry in high-energy linac beams is a transverse tail of the longitudinal distribution, i.e., a transverse displacement of the back with respect to the front of the bunch, the so-called 'banana' shape. That asymmetry leads to a sizeable effect in the ion azimuthal distribution: all the ions created by the head of the bunch are kicked in the direction of the transverse tail at the passage of the back. The azimuthal distribution shows a single peak in that direction. This leads to an *odd* anisotropy in the azimuthal distribution, that can be measured by the amplitude of the first-order Fourier coefficient A_1.

As an example, Fig. 3 shows the effect of a horizontal tail at 90° with the same conditions as Fig. 2.

EXPERIMENTAL SET-UP

The monitor is made of a gas injector and an ion detector surrounding the beam pipe (see Fig. 4), both being linked to a control and data acquisition system.

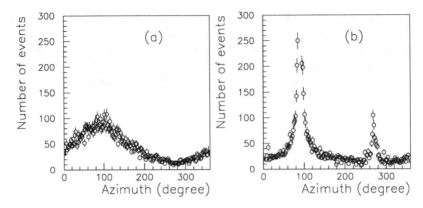

FIGURE 3. Simulated azimuthal distribution of He$^+$ ions with a tail at 90° for two vertical beam sizes $\sigma_y = 1$ µm (a) and $\sigma_y = 60$ nm (b). The origin of the azimuth is in the upward direction ($\sigma_x = 1$ µm, $\sigma_z = 650$ µm, and $N_{e^-} = 7.0 \times 10^9$).

Gas Injection System

To inject He gas pulses into the beam pipe, a fast electromagnetic valve, triggered at the passage of the beam, is mounted on the pipe near the focal point (see Fig. 4). The small amount of injected gas (about 60 cm^3 at 10^{-4} Torr) is taken out of the beam pipe by two 150 l/s turbomolecular pumps located 30 cm away on both sides of the monitor.

The width and the delay of the pulse triggering the electrovalve allow adjustment of the number of detected ions per electron bunch by varying the maximum pressure in the pulse up to 10^{-3} Torr.

The rise time of the gas pulses is about 150 µs, depending on mechanical adjustments, and the static pressure is recovered in a few ms. The amplitude stability is about 15% FWHM. Due to gas injection, the vacuum increases locally from 5×10^{-9} to 10^{-7} Torr, which is acceptable for the FFTB line.

Ion Detector

The ion detector is an array of eight pairs of microchannel plates (MCPs) surrounding the beamline at a distance of 6.5 cm (see Fig. 4). It covers nearly 360° in the transverse direction, the fraction of the geometrical dead angle being less than 20%. Each MCP pair covers an area of 40×50 mm^2. The MCPs are located in a vessel having its own turbo pump to avoid pollution from the injected gas. They are protected from the beam-induced background by a 2-cm-thick tungsten shielding.

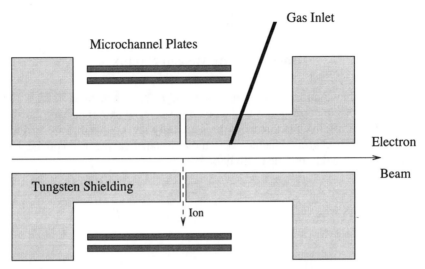

FIGURE 4. Longitudinal schematic view of the monitor (not to scale).

A narrow slit in front of the detector allows selection of only those ions emitted along a longitudinal length comparable with the β_y^* value of the beam. The width of this slit is adjustable from 0 to 1.2 mm, in steps of 10 μm, in order to adapt its spacing to the actual β_y^* value.

The charges delivered by the microchannels are collected by a set of 80 strip-lines parallel to the beam direction. There are eight strips behind each of six MCP pairs to measure the vertical profile and sixteen strips behind each of two MCP pairs to measure the horizontal profile. The latter two MCP pairs receive a higher flux of ions because of the strong azimuthal variation expected for a flat beam.

The signals collected on the 80 strips are locally amplified, then transmitted through 35-m-long cables to shapers. The 30-ns-wide shaper pulses are sampled at 200 MHz and stored in analog memories (5) before being digitized.

It is worthwhile noting that the flight distance is much larger than any possible beam jitter, making the monitor insensitive to it.

Data Analysis

A microcomputer is used to control the monitor and to read, store, and treat the data. The individual ion signals are recognized and the time of flight and the azimuthal angle of each ion are determined.

An on-line treatment allows determination of the parameters (T_{min}, A_1, and A_2) that are used to extract the beam dimensions.

RESULTS

Time-of-Flight Measurements

Figure 5 shows an experimental time-of-flight spectrum of He^+ ions. The time-of-flight origin corresponds to the beam passage. It shows a sharp threshold at about 100 ns. Below the threshold there are only few counts that are attributed to H^+ ions. This background being 1%, it allows a statistical precision on the determination of the minimum time of flight of 2%.

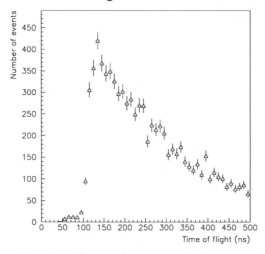

FIGURE 5. Time-of-flight spectrum of He^+ ions for 400 pulses, with a lower cut at 50 ns. The time-of-flight origin corresponds to the electron bunch passage.

Not shown on Fig. 5 are the signals in coincidence with the beam passage, due to some background of fast particles (x rays or electrons). After minimization, these background signals usually have an amplitude on each strip that is comparable or smaller than the mean amplitude delivered by an individual He^+ ion. As the ions drift for at least 100 ns towards the detector, this background, localized at $t = 0$, can easily be removed from the analysis.

Azimuthal Distribution Measurement

Figure 6 shows an experimental azimuthal distribution of He^+ ions. The origin of the azimuth is in the upward direction. Each entry corresponds to a stripline. It shows two peaks in the horizontal direction (90° and 270°), as expected for a flat beam.

Due to the fact that 50% of the ions arrive on the detector in 150 ns and that the shaper pulse is 30 ns wide, one has to decrease the gas pressure to limit the counting rate and the pile up of the signals. At about 30 ions per pulse, the maximum pile up correction on the peaks is 40%.

The high voltages of the MCPs are adjusted to equilibrate their relative efficiencies. Their discrepancies are about 2%.

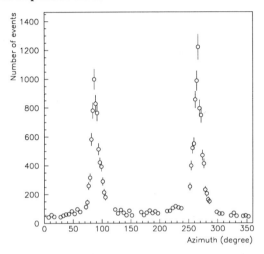

FIGURE 6. Azimuthal distribution of He$^+$ ions (with time of flight less than 500 ns) for 400 pulses, corrected for the pile up. The azimuth origin is the upward direction.

Tail Cancellation

Figure 7 shows the result of a controlled experiment with the banana tails. The origin of the azimuth is in the upward direction. Each entry corresponds to a stripline. Two peaks are observed in the horizontal direction but, opposite to Fig. 6, both peaks do not have the same height. This leads to a non-negligible first-order Fourier coefficient A_1. From Fig. 7(a) to Fig. 7(b), the tail was reversed from 90° to 270° using a linac bump. By using an intermediate value for the bump, we would have equilibrated both peaks and removed the banana tail.

Beam Size Measurement

From the knowledge of T_{min} and A_2 (computed for time of flight less than 500 ns), one deduces σ_x and σ_y from comparison with values obtained by Monte Carlo. Knowing the bunch length and the bunch charge, one can deduce that both Fig. 5 and Fig. 6 correspond to transverse beam sizes of $\sigma_x = 1.5 \pm 0.2$ μm and $\sigma_y = 73 \pm 10$ (stat) ± 10 (syst) nm.

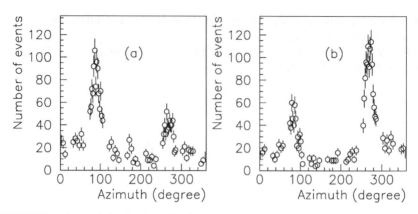

FIGURE 7. Azimuthal distribution of He$^+$ ions with a horizontal tail at 90° (a) and at 270° (b).

CONCLUSION

We have built and succesfully operated a nondestructive submicronic beam size monitor which ultimately measured a vertical dimension of 70 nm. It has also been useful to correct banana tails.

ACKNOWLEDGMENTS

We would like to thank the FFTB collaboration for its support, and in particular D. Burke, C. Field, and P. Tenenbaum. Many thanks are also due to F.J. Decker for his help in the understanding of the banana shape. We are grateful to the LAL staff, led by M. Davier, for its continuous support and help.

REFERENCES

1. V. Balakin et al., *Phys. Rev. Lett.* **74**, 2479-2482 (1995).
2. T. Shintake, *Nucl. Instrum. Meth.* **A 311**, 453-464 (1992) and these proceedings.
3. J. Buon et al., *Nucl. Instrum. Meth.* **A 306**, 93-111 (1991) and P. Puzo, PhD thesis, Orsay University (1994).
4. P. Chen, SLAC-PUB 5186 (1990).
5. G.M. Haller et al., SLAC-PUB 5317 (1990).

A Laser-Based Beam Profile Monitor for the SLC/SLD Interaction Region*

M. C. Ross, R. Alley, D. Arnett, E. Bong, W. Colocho, J. Frisch,
S. Horton-Smith, W. Inman, K. Jobe, T. Kotseroglou,
D. McCormick, J. Nelson, M. Scheeff, S. Wagner

Stanford Linear Accelerator Center, Stanford, California, 94309

Abstract. Beam size estimates made using beam-beam deflections are used for optimization of the Stanford Linear Collider (SLC) electron-positron beam sizes. Typical beam sizes and intensities expected for 1996 operations are 2.1×0.6 μm (x,y) at 4.0×10^{10} particles per pulse. Conventional profile monitors, such as scanning wires, fail at charge densities well below this. Since the beam-beam deflection does not provide single beam size information, another method is needed for interaction point (IP) beam size optimization. The laser-based profile monitor uses a finely focused, 350-nm, wavelength-tripled yttrium-lithium-fluoride (YLF) laser pulse that traverses the particle beam path about 29 cm away from the e+/e- IP. Compton scattered photons and degraded e+/e- are detected as the beam is steered across the laser pulse. The laser pulse has a transverse size of 380 nm and a Rayleigh range of about 5 μm. This is adequate for present or planned SLC beams. Design and preliminary results will be presented.

INTRODUCTION

The Stanford Linear Collider (SLC) is the first of a new generation of colliding beam machines that rely on micron-sized beams colliding at a relatively low repetition rate (1). The ultimate performance of the collider is limited by the control of emittance dilution effects that increase the transverse beam size ($\sigma_{x,y}$) at the interaction point (IP), where the two beams meet each other. A useful estimate of $\sigma_{x,y}$ is obtained from the deflection seen when sweeping one beam across the other (2). The two principle drawbacks of the deflection technique are its sensitivity to shifts in beam centroid and a lack of indication of which beam is changing.

The latter has the most significant impact on emittance dilution and optics-related optimization. A single beam (e+ or e-) diagnostic that can operate over the full range of SLC beam intensities from 0.3×10^{10} to 4×10^{10} particles per bunch is needed. Unfortunately, wire scanners cannot be used for beam sizes smaller than 1.4 μm or for intensities greater than 0.6×10^{10}. A wire scanner equipped with 4-μm-diameter carbon wire is used to measure $\sigma_{x,y}$ for beams safely away from these thresholds. If either threshold is exceeded, the energy deposited in the wire from a single pulse severs it. The wire scanner is installed deep inside the so-

*Work supported by U.S. Department of Energy, Contract DE-AC03-76SF00515.

lenoidal SLC Large Detector (SLD) system and is virtually inaccessible, so routine replacement of the carbon wires is not practical.

The SLC IP laser-based beam profile monitor will be used to measure $\sigma_{x,y}$ of individual beams inside the SLD over the full range of operating intensities with about 10% accuracy. Some key features of the device are similar to the laser-based beam size monitor developed for the FFTB at SLAC (3,4). A finely focused laser pulse is brought into a 90° crossing angle collision with the electron and positron beam and the Compton-scattered photons and degraded beam particles are detected. As the e+/e- beams are steered across the laser pulse on a succession of pulses, the amplitude of the scattered radiation is recorded and used to estimate the beam sizes, in a manner similar to that used with SLC wire scanners (5).

Future linear colliders (NLC) will employ beams of higher charge density than those of SLC (6). In most designs, conventional wire scanner limits will be exceeded for all damped beam regions. Sets of laser-based monitors will serve as emittance monitors. Lessons learned from the operation of the device described in this paper will be useful for the development of NLC beam profile monitors.

PRINCIPLE OF OPERATION

We considered three basic 'optical scattering structures': 1) a diffraction-limited, finely focused waist (TEM'00' mode), 2) an interference fringe pattern similar to that used at FFTB, and 3) the finely focused waist of a first-order, transverse mode laser beam (TEM'01' mode).

The minimum transverse size, σ_0, of a diffraction limited laser beam is (7)

$$\sigma_0 = \frac{\lambda f}{4\pi\sigma_{in}},$$

where λ is the wavelength of the light, f is the focal length of the lens, and σ_{in} is the Gaussian beam sigma as defined by the photon density. The effective length of the focused section is twice the Rayleigh range (z_R)

$$2z_R = \frac{8\pi\sigma_0^2}{\lambda}, \qquad \text{where } \sigma(z_R) = \sqrt{2}\sigma_0.$$

The incoming beam size, σ_{in}, and f combine to give an $f\#$, with the aperture roughly $\pm 3 \sigma_{in}$ giving $\sigma_0 \sim 1/2 \; f\# \; \lambda$. For typical SLC parameters, with the required beam-stay-clear distance of 25mm and an $f\#$ of 2, λ must be shorter than 500 nm in order to have $\sigma_0 < \sigma_y$ and $z_R \sim 5\mu m$. This is the option selected.

The second optical scattering structure, the interference fringe pattern, is useful for much smaller beam sizes than those expected for SLC and, unless the pattern pitch is controllable, can be used to measure only a small range of beam sizes. Scanning with this system involves measuring the modulation depth of the scattered radiation as the beam is moved across the pattern. We could not find room in the confines of SLD for fringe pitch control.

The third option mentioned above, the TEM '01'-mode laser beam, with a field null at the center of the spot, may prove useful. It is easy to implement since the mode would be generated at the laser and the modest increase in σ_{in} is accommodated in the transport and IP optics. A double-lobed result is generated from a TEM'01' mode scan, and the spacing between the lobes can be used as a laser beam diagnostic.

As the particle beam is swept over the laser beam with a varying impact parameter y_0, the number of scattered photons, $N_\gamma(y_0)$, is given by

$$N_\gamma(y_0) = \frac{PN_e\sigma_c}{chv\sqrt{2\pi}\sigma_s} \exp(\frac{-y_0^2}{2\sigma_s^2}),$$

where P is the power of the laser beam intercepted by the particle beam ($\sigma_z = 750$ μm), v is the frequency of the laser light, N_e is the number of electrons in the beam, σ_c is the Compton scattering cross section, and σ_s is the overlap size. For peak laser power of 10MW and with $\sigma_y = 1$ μm and $N_e = 10^{10}$, N_γ is ~ 5000. A correction, nominally about 12%, is required since the particle beam has an aspect ratio (σ_y/σ_x) ~ 5 and the laser spot does not have an ellipsoidal cross section.

The energy distribution of the scattered photons and degraded beam particles is relatively flat and, for $\lambda = 350$ nm (3rd harmonic YLF), has a peak gamma radiation energy of 29 GeV for SLC. Detectors for monitoring the gamma rays and the degraded e+/e- particles are located along the beamline that extends to the beam dump area (8,9). Backgrounds in the degraded particle detectors are typically about 50 particles per beam crossing.

DESIGN

Optics

Optics design goals were to achieve the minimum σ_0 with the highest transmitted power and the lowest possible aberrations. The most important mechanical constraints were the minimum beam transport diameter of 25 mm, the available beamline length of 52 mm, and mass and density restrictions. Figure 1 shows an elevation view of the IP layout. Within the cone subtended by active SLD segments, the mass of the optics and related supports must cause minimal scattering of the decay particles the SLD is intending to analyze.

The IP $f2$ optics (Fig. 2) are catadioptric, with minimal geometric aberrations, in which a diverging meniscus lens is coupled with a spherical reflector (10). Two such systems are required, for both σ_x and σ_y. For laser pulse lengths longer than about 150 ps, the particle beam will also scatter some of the incoming photons before they reflect from the spherical mirror.

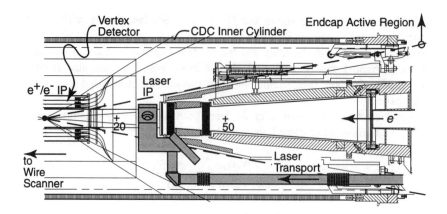

FIGURE 1. Cutaway elevation view of the inside of the SLD. The new vertex detector (11), surrounding the e+/e- IP with the wire scanner on one side (not shown) and the laser profile monitor on the other, is itself surrounded by the 13-inch inner cylinder of the SLD central tracking drift chamber (CDC). Access is not available beyond a few inches from the end of the cylinder. The laser transport line (shaded) enters from the bottom right of the figure and passes through two bellows before terminating in the IP optics bench. The numbers along the central axis indicate the distance in cm from the e+/e- IP.

The incoming laser beam has σ_{in} of about 1.3 mm. As σ_{in} is increased, diffraction scattering from the edges of the input optic produces non-Gaussian tails effectively increasing σ_0. At the end of the transport, in the IP 'optics bench' (Fig. 3), a compact switch system is used to select which of the two possible paths through the IP the laser will follow. A brewster polarizer is used in conjunction with a linear polarizer at the laser to do the selection. A compact construction was required in order to minimize the mass obscuring the SLD end cap detector segments and in order to fit around SLD internal masking and supports.

An estimate of the deviation from diffraction limit caused by surface figure distortions yielded a per element mirror and lens surface figure tolerance of $\lambda/40$.

Laser-Induced Optical Component Damage

Since we require reliable operation for several years (about 100 million pulses), and since much of the system is sealed and inaccessible, considerable design effort was concentrated on damage prevention. In order to prevent inadvertent high power density related damage, no lenses or other focusing elements are present in the transport line that carries the light from the laser room to the IP.

The most threatening source of damage is from chemical contaminants. Studies of damage in sealed laser systems have shown that organic chemical contaminants are a leading cause (12). Studies of infrared lasers pinpoint certain trace contaminants, such as silicone sealers, as root causes. There is little information

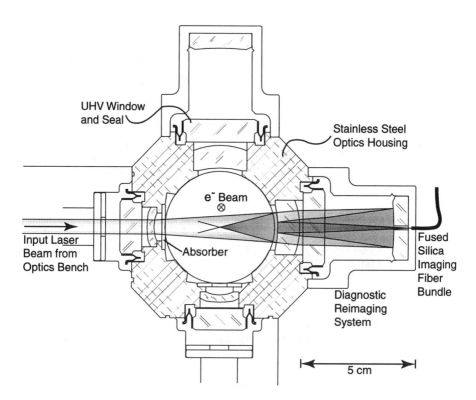

FIGURE 2. Central cross section of IP optics. The parallel laser beam (z_R>> 20 m) enters at right or from underneath with a σ_{in} of about 1.3 mm. The four 8-mm-thick UHV seal windows with their weld rings are placed on each side of the IP, with the precision optics installed inside the vacuum chamber. 1% of the light is transmitted and re-imaged on the far side of the IP for diagnostic purposes. The spent light is absorbed by glass absorbers located inside and outside of the vacuum system.

available concerning the long-term operation of UV lasers where it is likely that the sensitivity to trace organic chemicals will be greater. For this reason, our design transport σ is as large as possible and all IP, IP bench, and transport assembly took place in Class 100 clean rooms. Non-ultra-high vacuum (UHV) volumes are purged with an Ar/O$_2$ (90/10) mixture as suggested in the report noted above.

Bulk multi-photon damage is another serious concern for long-term operation of the UV system with several transmissive optics. Darkening in fused silica has been seen following long-term exposure to high-fluence, short-wavelength light (13). We interpolated between results at 248 nm and at 550 nm and set our tolerance a factorfr the threshold wherthese effects are first seen.

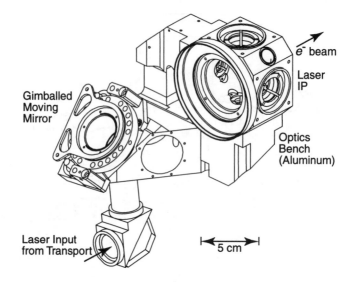

FIGURE 3. IP optics bench. The bench, a fairly complex machining, has 2.2-cm-diameter machined internal passages for light transport. The IP is in the top right of the figure, shown without the diagnostic re-imaging optics installed. Eight mirror retaining cap mounting holes are clearly visible in the front of the figure.

Mechanical

Three subsystems—the IP and its mirror bench, the 30-m transport line, and the laser itself—comprise the profile monitor. The laser is housed in an external clean room. The transport line passes through the beamline radiation shielding and then along the superconducting quadrupole triplet before entering the SLD vertex area.

The beam enters the IP (Fig. 2) from the right or from the bottom, for y or x scans, respectively, and passes through the UHV fused silica window that separates the gas-filled optics bench from the beamline vacuum. A seal design was developed that allowed the window surfaces to be ground to $\lambda/40$ surface quality after attaching the weld ring. The window is coated with a graded index coating.

After passing through the focal point the spent laser light is absorbed using glass absorbers. This prevents possible reflection from metal surfaces that might make secondary ghost images or sputter nearby metal surfaces and thereby cause damage to optical components. The reflector coating allows 1% of the incident light to pass through. Its rear surface, together with a second spherical reflector outside of the vacuum chamber, generate an image of the IP spot that is used as a diagnostic.

The IP optics bench directs the light from the transport to the IP using one controllable and four fixed mirrors. It terminates the 1-inch-diameter laser transport line. The transport line is evacuated for 90% of its length (10^{-6} Torr) and slightly pressurized, along with the optics bench, for the remaining 10%.

The IP housing has a required vacuum performance of 5 nTorr. The catadioptric optics are most sensitive to lens and mirror centering and coplanarity errors which lead to machining tolerances of 5 μm.

A gimbaled moving mirror mount was developed for the transport line that has angular stability of 30 μrad and compensated bellows vacuum forces. Fine alignment adjustments are made remotely using compact piezoelectric motors.

Diagnostics

Steering and alignment diagnostics are provided by four miniature CCD profile monitors located behind transport line and IP bench mirrors. They view the laser light directly through the partial transmission of the mirror. A frame-grabber-based video analysis system is used to monitor the position and size of the spot at each of the monitors.

At the monitor beyond the IP, where reliability and camera radiation damage are concerns, a 50-μm spot is generated and a 10,000-fiber fused silica bundle is used to transmit its image from the high radiation area to a monitor table. The fiber fluoresces and shifts the UV into visible. The fiber bundle has a diameter of 0.5mm and a single fiber diameter of 10 μm. The IP assembly also has two fast diodes and a beam pickup electrode that are used for finding the relative timing of the laser and particle beams.

The transport line is protected from stray laser light using a CW He-Cad 350-nm λ alignment laser that is mode matched to the high power laser on the laser table. If the signal from the CW laser at the monitor beyond the IP is lost, the high power laser shutter will be closed.

Laser

The laser system consists of a mode-locked, 119-MHz, 100 ps-pulse-length, yttrium-lithium-fluoride (YLF) seed laser and a YLF regenerative amplifier laser. We chose YLF as the lasing material because of its reduced thermal lensing compared to yttrium aluminum garnet (YAG). The amplifier has a gain of 10^6 in 20 passes and produces a 10-mJ pulse of 1047-nm λ light with a repetition rate of 60 Hz. One mJ of UV light at the IP gives a peak power of 10 MW.

PERFORMANCE

Scanning is done using controls similar to those used for the beam-beam deflection. The waist of the particle beam is first moved 29 cm from the e+/e- IP to the laser IP. Initial timing setup first requires the use of the beam pickup electrode and the fast diode. Following a scan, the beam size is deconvoluted from the laser size in order to provide a measurement. Unfortunately, it is not possible to measure directly the high power laser beam size using independent techniques. At low e+/e- beam intensities, where the carbon wire scanner can be used, a calibration is done using the TEM'01' mode.

Low power test results from the IP are shown in Figure 4.

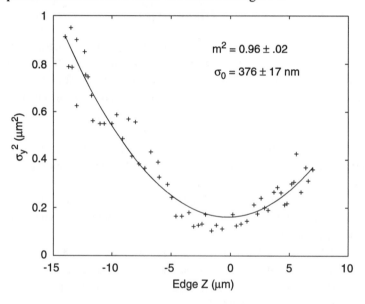

FIGURE 4. Laser IP waist test data. A 350-nm, 1-mW CW HeCad laser was used for this test. At each point a laser beam size measurement was made using a resolution target edge. The position of the edge is then moved along the laser beam direction in order to obtain the waist scan. The parameter m^2 indicates the deviation from ideal, diffraction-limited conditions; σ_0 is the minimum spot size.

ACKNOWLEDGEMENTS

We would like to acknowledge support for this project from the SLD collaboration, D. Burke and J. Sheppard. Technical support from K. Ratcliffe, C. Rago, O. Millican, A. Farvid, and H. Baruz is also greatly appreciated. D. Meyerhofer of the University of Rochester gave much appreciated review and technical assistance during the course of the project.

REFERENCES

1. Emma, P., "The Stanford Linear Collider," Proc. of the 1995 Particle Accelerator Conference, Dallas, Texas, 1-5 May 1995, 606-610 (1996).
2. Koska, W. et al., *Nucl. Instrum. Methods A* **286**, 32 (1990).
3. Shintake, T. et al., "Design of Laser Compton Spot Size Monitor," Proc. of the Int. Conf. on High-Energy Accelerators, Hamburg, Germany, July 20-24, 1992, 215-218 (1992).
4. Burke, D.L., "The Final Focus Test Beam Project," Proc. of the 1991 Particle Accelerator Conf., San Francisco, CA, May 6-9, 1991, 2055-2057 (1991).
5. Ross, M. C. et al., "Wire Scanners for Beam Size and Emittance Measurements at the SLC," Proc. of the 1991 Particle Accelerator Conf., San Francisco, CA, May 6-9, 1991, 1201-1203 (1991).
6. Urakawa, J., Ed., "Sixth International Workshop on Linear Colliders, LC95," KEK 95-5, 1995.
7. Siegman, A., *Lasers*, University Science Books, 1986.
8. Field, R. C., *Nucl. Instrum. Methods A* **265**, 167-169 (1988).
9. Shapiro, G. et al., "The Compton Polarimeter at the SLC," Proc. of the 1993 Particle Accelerator Conference , Washington, DC, May 17-20 1993, 2172-2174 (1993).
10. Bouwers, A., *Achievements in Optics*, New York: Elsevier, 1946; and Matsukov, D.D. *J. Opt. Soc. Am.* **34**, 270-284 (1944).
11. SLD Collaboration (S. Hedges et al.), "VXD3: The SLD Vertex Detector Upgrade Based on a 307 MPixel CCD System," Proc. of the International Europhysics Conference on High Energy Physics, Brussels, Belgium, 27 Jul - 2 Aug, 1995, World Scientific Press, 1996; also SLAC-PUB-95-6950, July 1995.
12. Sharps, J., "Package Induced Failure of Semiconductor Laser and Optical Telecommunication Components," Annual Symposium on Optical Materials for High Power Lasers, 1995; and Guch, Steven Jr. et al. "Beyond perfection: the need for understanding contamination effects on real-world optics," Proceedings of SPIE, **2114**, 505-511 (1994).
13. Schermerhorn, Paul, "Excimer laser damage testing of optical materials," Proceedings of SPIE, **1835**, 70-79 (1993).

Quartz Wires versus Carbon Fibres for Improved Beam Handling Capacity of the LEP Wire Scanners

C. Fischer, R. Jung, J. Koopman

CERN, SL Division, CH1211 Geneva 23, Switzerland

Abstract. After the first investigations were performed in 1994, the study of thermal effects on carbon and quartz wires was pursued in 1995. Carbon wires of 8 µm have been studied. Light emission resulting from the two heating mechanisms, electromagnetic fields and collision losses with the beam, were observed. Quartz wires of 10 and 30 µm were investigated and light emission due to the heating by collision with the beam was observed. The heat pattern differs completely from that of carbon fibres. The quartz wires withstood circulating currents of at least 8 mA at 20 GeV, the 1995 operational level in LEP. Quantitative evaluations and the influence of various dissipation processes are presented with the aim of evaluating a beam current limit.

INTRODUCTION

Investigations made in 1994 with the LEP wire scanners have shown that the major heating mechanism in a conducting wire comes from the electromagnetic fields (1). The effects are felt well before the lepton beam traverses the wire and outside the portion of the wire crossed by the beam. These observations led to the use of quartz, an isolating material, which is only sensitive to energy deposition by collision with the beam and hence withstands higher currents despite a lower fusion point. The comparison between carbon and quartz wires was pursued in 1995 with new CCD cameras.

TESTS DESCRIPTION

Two multi-fork tanks were installed in 1995 and were equipped with three horizontal and two vertical mechanisms. The forks were made of ceramics and could be observed with CCD cameras through viewing ports (1). These multi-fork tanks included the operational vertical profile wire which was made of 30-µm quartz. The three horizontal mechanisms were redundant and were used for material studies. One ceramic fork was fitted with two wires, and during the first part of the year two 8-µm carbon wires with 12-mm spacing were used. This set-up was intended to evaluate very low diameter fibres, so as to limit the effect of beam blow-up. One of the two wires broke at the top early in the period under unrecorded conditions, hence only one wire remained operational. Scans were performed during LEP fills at 20 GeV with beam currents going from 0.5 to 7.5 mA in 1-mA increments. During a scan, the light emitted from the wire was memorised on a video recorder, i.e., at regular intervals of 20 ms. The second wire broke during a scan with a total beam current of 6 mA soon after a scan performed at 5 mA. Both wires broke at the top as seen on the recorded light patterns.

The carbon wires were replaced by two quartz wires of 10 and 30 μm with the same 12-mm spacing, and the same tests were resumed. In this case, scans were made up to the operational beam current limit of LEP which was around 8 mA. From light emission it could be observed that the 10-μm wire broke probably from fatigue or wearing out after having undergone about 10 scans with a circulating current of 7.3 mA. The thicker 30-μm wire survived and was used for further tests. At the end of the run, the wires were inspected and it appeared that the 10-μm half wires had been bent towards the extremities of the 30-μm wire. This wire was still intact, but the centre was thinned down to a few microns over the beam height, and the melted material had collected in a quartz pearl just below.

This confirms clearly the very different behaviour of carbon and quartz wires.

CARBON LIGHT EMISSION PATTERN

The digitised video recording of the light emitted by the 8-μm carbon wire when scanning horizontally a 0.8-mA beam is shown in Figure 1. The wire lights up to the points where it is fixed to the ceramic fork. Hence the illuminated length is a precise indication of the total wire length (29 mm). The time dependence of the radiation pattern is given by the time structure of the video signal, i.e., 20 ms per profile. This time corresponds to a fork displacement of 5 mm at a wire speed of 0.25 m/s. The two large hot areas at the right and left sides of each profile are characteristic of the heating of the wire under the influence of electromagnetic

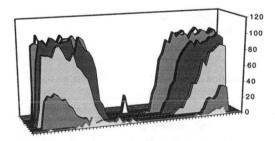

FIGURE 1. Digitised video recording of an 8-μm carbon wire scanning a 0.8-mA beam. The wire is parallel to the horizontal axis, and the light intensity is plotted along the vertical axis (arbitrary units). Successive profiles are separated by 20 ms. The central spot corresponds to the passage of the wire through the beam.

fields. They correspond to portions of the wire which are not traversed by the beam and appear well before the wire crosses the beam. This heating is induced by the vertical component of the electric field, direct and wakefield, generated by the passage of the 50 ps (rms) short bunches. The effect of collision losses is only observed on one curve which exhibits a reduced intensity central spot with a width of about 2 mm. Its amplitude is lower by more than a factor of 5 with respect to the two radiating ends. The two adjacent curves do not have this central spot, indicating that within 20 ms the temperature has decreased below the sensitivity threshold of the camera.

The temperature of this spot cannot be scaled with that of the extremities which are well above the saturation level of the camera. The light emission is stronger on the second half of the scan, after beam traversal, as the wire is heating up regularly. The level of radiation of the wire extremities and their length increase with beam current. The structure of the wakefields is clearly visible at high currents; see Figure 2, made during a sweep at a current of 5 mA.

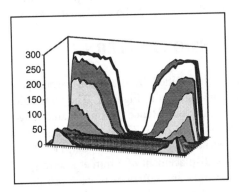 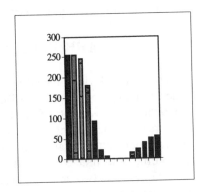

FIGURE 2. Light emission of an 8-μm carbon wire scanning a 5-mA beam. Left: Light intensity pattern along the carbon wire as a function of its position. Right: Maximum light emission evolution. The wire/beam encounter takes place at the left. The light emission hole indicates an area with vanishing wakefields.

The central hot spot generated by the beam interaction is barely visible on this scan as the gain was reduced by a factor of 100 to limit saturation effects. Wakefield calculations have been made with the computer program MAFIA to evaluate this type of pattern. The calculations show that the wakefields last for approximately 250 ps. The computed voltage differences, shown in Figure 3, have been used to estimate the temperature pattern along the wire. It is close to the measured one and a maximum temperature of 3300 K has been found for a beam of 5 mA (2).

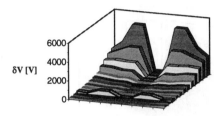

FIGURE 3. Voltage differences calculated along the wire (5-mm steps) as a function of wire position during a scan of 32 bunches with a total intensity of 5 mA (bunch trains).

The previous observations show clearly that the main heating effect in conducting wires is by the electric component of the wakefields. It explains also why the wires always break towards an extremity. Since the vacuum tank was as well-made as possible from a wakefield point of view (it has a loss factor of 0.1 V/pC), there is little to be gained by minimizing the wakefields and increasing the beam handling capacity beyond 6 mA. The only possible way to increase this current limit would be to install the scanner at the crossing point (IP1) where the electric fields of the electron and positron beams would cancel under ideal conditions.

QUARTZ WIRE LIGHT EMISSION

Quartz wires have been used to replace the carbon fibres, because they are the most easily available nonconducting wires with diameters from 5 to 30 µm. They have a melting point around 1700 °C and are acceptable from beam blow-up and energy deposition points of view. A typical radiation pattern is given in Figure 4 taken at a circulating beam of 7.3 mA. Only the central region where beam and wire interact is now radiating, and nothing is observed before crossing the beam. The first curve corresponds to the spot emitted by the 30-µm wire when crossing the beam. The amplitude of the next curves are slightly reduced as the wire starts to cool down. However, 60 ms later (fourth curve), the amplitude is again increased as the second wire (10 µm) starts to cross the beam, and its signal adds to the remnant radiation of the first wire. The subsequent curves give an indication of the thermal conduction time constant of the quartz wires. Within 60 to 80 ms the amplitude decreases by a factor of 10. The thermal conduction is a slow process in quartz. The curves have a width of approximately 2 mm which fits well with the rms beam size of 800 µm.

The same pattern is obtained when scanning in the other direction where the 10-µm wire is the first to encounter the beam. Its emission characteristics, in width and amplitude, are very similar, which corroborates the calculations performed in the next sections.

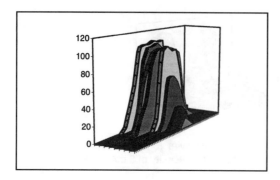 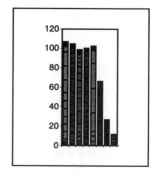

FIGURE 4. Radiation patterns from the two quartz wires (enlarged scale) versus wire position for a 7.3-mA beam. The first beam encounter is at the left in both graphs. Right: Maximum light emission evolution, with the two beam encounters clearly visible.

HEATING FROM COLLISION LOSSES

Detailed calculations can be found in (2). The basic features are recalled here. The ionisation losses for high-energy electrons are given by (3)

$$\frac{1}{\rho}\frac{dE}{dx} = \frac{0.153}{\beta^2}\frac{Z}{A}\left(F(\beta) - 2\ln I - \delta(X)\right) \quad [\text{MeV g}^{-1}\text{ cm}^2],$$

where $\delta(X)$ describes the density effect of the medium and is given by $\delta(X) = 4.605\,X + C$, with $X = \log(\beta\gamma)$ and C to be found in Table 1. The density effect decreases the ionisation losses by some 36% in both materials. The energy lost by a beam of N particles during a scan is

$$\Delta E(\text{scan}) = N <x> dE/dx,$$

with $<x>$ the average wire thickness seen by each particle during a scan (2)

$$<x> = \frac{\pi R^2}{vT},$$

where R is the wire radius, v the wire speed (0.25 m/s), and T the beam revolution period (88.9 μs in LEP).

The fraction of the energy actually deposited in the wire has to be calculated. The energy deposited between E and E + dE at a depth x by an incident particle is:

$$dW = \frac{d^2N}{dEdx} \; E \; x \; dE,$$

TABLE 1: Comparative Data for Carbon and Quartz Wires for a 7.5-mA Total Intensity

	Carbon	Quartz	
Z/A	0.4995	0.4993	
I [eV]	78	139	
ρ [g cm^{-3}]	2.26	2.2	
c_p [J g^{-1} K^{-1}]	1.647*	1.124*	
k [W m^{-1} K^{-1}]	120	2.7	
Sublimation/Melting point [°C]	3 700	1 700	
C	2.868	4.003	
δ(X)	18.28	17.15	
1/ρ dE/dx [MeV g^{-1} cm^2]	2.335	2.333	
D [μm]	**8**	**10**	**30**
<x> [μm]	2.3	3.5	31.8
ΔE (scan) [MeV]	46 10^8	70 10^8	630 10^8
W [MeV]	28.3 10^8	42 10^8	374 10^8
η_1	61%	59%	59%
η_2	75%	76%	68%
η	54%	55%	59%
ΔT [K]	**1070**	**1600**	**1700**

* averaged from 300 to 1800 K

and between I, the binding electron energy, and E_{max}, each particle will deposit

$$W = \int_I^{E\,max} dW \, .$$

The distribution of secondary electrons with energy E is given by (4)

$$\frac{d^2N}{dEdx} = 0.153 \frac{Z}{A} \frac{z^2}{\beta^2} \rho \frac{F}{E^2} \, .$$

For 20-GeV leptons, E_{max} = 10 GeV and F = 1 (4). For N particles with average path <x>, one gets for W the values of Table 1 and hence the fraction η_1 of energy transmitted to knock-on electrons. Some knock-on electrons escape from the wire and will not contribute to the wire heating. Their fraction η_2 is

$$\eta_2 = \frac{\ln(E_{max} / < Ei >)}{\ln(E_{max} / I)} \, ,$$

where <Ei> is the average ionisation threshold

$$<Ei> = \frac{1}{1.296} (\rho <x>)^{0.58} \, .$$

Finally, the energy actually deposited inside the wire is: $E_d = \Delta E(scan)\ \eta$, with η the overall wire heating efficiency given by

$$\eta = (1 - \eta_1) + (1 - \eta_2)\ \eta_1$$

varying from 54 to 59% for the different wires (see Table 1). If σ_v is the rms vertical beam dimension (800 μm at 20 GeV), the wire volume V heated during a horizontal scan is

$$V = \pi R^2 \sqrt{2\pi}\ \sigma_v,$$

and if c_p is the heat capacity averaged over 1500°, then the temperature rise is

$$\Delta T\ =\ E_d\ /\ (V \rho c_p).$$

The final results are given in Table 1 and show temperature increases of 1100° for carbon and 1600 to 1700° for quartz. They confirm that losses by collision are large enough to bring the quartz wires close to their fusion point, but will not destroy the carbon fibres.

EFFECT OF THERMAL CONDUCTION

The thermal time constant in the wire has been estimated with a simple model. The wire section length which is heated by the beam is

$$L = \sqrt{2\pi}\ \sigma_v.$$

Considering one-half of the wire starting at its centre (hot point), the average path along this section is L/4 and hence its thermal time constant τ is

$$\tau = R_{th}\ C_{th} = c_p\ \rho\ (\ L/4)^2\ /\ k,$$

where k is the thermal conductivity (Table 1). The results are $\tau = 15$ ms for carbon and 500 ms for quartz. In carbon, $\tau = 15$ ms explains why the central spot disappears within consecutive video acquisitions. For quartz, the value of τ corroborates also the observations. It leads to a temperature decrease of 15% within 80 ms which in turn corresponds to a reduction of the emitted power by a factor of 5 in the CCD spectrum according to Planck's law. During a complete wire sweep through the beam (typically 7 ms) the loss by thermal conduction in quartz is about 1% of the deposited energy and will not induce any beneficial cooling.

EFFECT OF RADIATION LOSSES IN QUARTZ

From Stephan-Boltzmann's law, the total emitted power is $P = \varepsilon\sigma AT^4$ with $\sigma = 5.672\ 10^{-8}$ Wm^{-2}K^{-4}, the radiating area is $A = \pi DL$, and the quartz emissivity is $\varepsilon =$

0.9 at room temperature. At 1700 K one finds P = 0.03 W and P = 0.08 W for the 10- and 30-µm quartz wires, respectively. Assuming the emission to take place mainly in the second half of the beam traversal (high temperature), the energy absorbed during a scan would be reduced by 17% and 4.5%, respectively. Thermoionic emission can be neglected below 2000 K, i.e., for quartz.

CONCLUSION

Thermal effects generated in the LEP wire scanners by electromagnetic fields have been observed directly and quantified for conducting materials such as carbon. They lead to temperature increases four times higher than losses by collisions for a beam of 5 mA and are the major source of wire heating. Nonconducting wires are not affected by them, and quartz wires withstood beam currents up to 8 mA. Both the calculations of energy deposition by collision with the beam and the inspection of the wires confirm that this current level is close to the possible operational threshold at 20 GeV. This threshold has to be reduced for horizontal scans at higher energies because the vertical beam emittances will be much smaller. The experience gained with quartz wires indicates a safe operational beam current level close to 5 mA at the LEP 2 energy. Tests will be pursued in 1996 with quartz to confirm this limit and also with 15-µm SiC wires. The use of diamond wires is also considered and the installation of a horizontal wire scanner at IP1 will be evaluated.

ACKNOWLEDGEMENTS

The ingenuity of J. Camas in handling and installing all kinds of fibres is much appreciated. P. Valentin performed together with A. Wagner the calculations with MAFIA and their help was precious. We also appreciate the collaboration of our colleagues of the Vacuum Group for the recurrent installation work. It is a pleasure to acknowledge their contribution in the completion of this study.

REFERENCES

1. J. Camas et al., "Observation of Thermal Effects on the LEP Wire Scanners," CERN SL/95-20 (BI), May 1995, and Proc. of the 1995 IEEE PAC, Dallas, TX, 2649-2651 (1996).
2. C. Fischer, P. Valentin, "An Evaluation of Various Heating Mechanisms in the LEP Wire Scanners," note to be published.
3. R. M. Sternheimer, M. J. Berger, S. M. Seltzer, Atomic Data and Nuclear Data Tables 30, 261-271 (1984).
4. K. Hikasa et al., "Review of Particle Properties," *Phys. Rev. D* **45** (11), June 1992.

High-Current CW Beam Profile Monitors Using Transition Radiation at CEBAF*

P. Piot, J.-C. Denard, P. Adderley, K. Capek, E. Feldl

Continuous Electron Beam Accelerator Facility,
12000 Jefferson Avenue, Newport News, VA 23606

Abstract. One way of measuring the profile of CEBAF's low-emittance, high-power beam is to use the optical transition radiation (OTR) emitted from a thin foil surface when the electron beam passes through it. We present the design of a monitor using the forward OTR emitted from a 0.25-μm carbon foil. We believe that the monitor will resolve three main issues: i) whether the maximum temperature of the foil stays below the melting point, ii) whether the beam loss remains below 0.5%, in order not to trigger the machine protection system, and iii) whether the monitor resolution (unlike that of synchrotron radiation monitors) is better than the product $\lambda\gamma$. It seems that the most serious limitation for CEBAF is the beam loss due to beam scattering. We present results from Keil's theory and simulations from the computer code GEANT as well as measurements with aluminum foils with a 45-MeV electron beam. We also present a measurement of a 3.2-GeV beam profile that is much smaller than $\lambda\gamma$, supporting Rule and Fiorito's calculations of the OTR resolution limit due to diffraction.

INTRODUCTION

The optical transition radiation (OTR) phenomenon is a very convenient and reliable means for beam profile and position measurement in a quasi-nondestructive way. However, there are limitations in the use of such monitors. First, depending on the beam current of the considered beam, the screen can melt. Second, at high temperature the screen becomes a source of black body radiation whose power can be greater than that of the OTR. Finally, the plural scattering that electrons experience while passing through the foil is a source of emittance degradation and beam loss. For the CEBAF electron beam, this last issue is of importance because of the machine protection system (MPS) that shuts the injector down when beam losses exceed 1 μA out of a total beam current that can reach 200 μA. In this paper we describe a monitor which uses the forward OTR produced at the surface of a very thin (0.25 μm) carbon foil. This device has the potential to monitor and study the continous electron beam delivered to the experimental halls.

---

* Work supported by U.S. Department of Energy, Contract DE-AC05-84ER40150.

PLURAL SCATTERING THROUGH FOILS

Experimental Results Using Aluminum Foils

In this section, we present the results of the measurements of the scattering angle distribution of an electron beam through thin aluminum foils of various thicknesses, (0.25, 0.8, and 1.5 μm). The foils are mounted on a support that is inserted into the beam by means of a stepping motor. We perform this test in the 45-MeV region of the CEBAF injector. We measure the beam size using a wire scanner installed 7.4 m downstream from the foils. There is no powered quadrupole magnet between the foils and the wire scanner. Therefore, by knowing the distance between the two instruments, we can easily calculate the distribution angle of the particles at the exit of the foil from the beam profile distribution measured by the wire scanner. The results (Fig. 1(B)) show that, within the range of the thicknesses tested, the scattering angle containing 70% of the particles decreases in proportion with the foil thickness.

However, the angle that contains 90% of the particles does not decrease as much. Measuring angles that contain more than 90% of the beam would require scanning larger angles. The curves in Fig. 1 are accurate up to 90%; beyond this value we have to use theoretical formulae or Monte Carlo simulations.

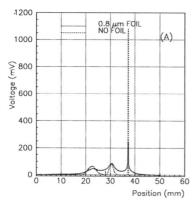
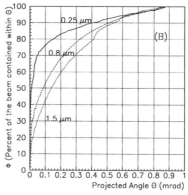

FIGURE 1. Beam scattering angle distribution through thin aluminum foils, measured at 45 MeV; before inserting the foil, we set the optics for a minimum vertical beam size at the wire scanner location. (A) Wire scanner vertical profile taken 7.4 m after the foil set up, with and without foil; the wire scanner has three 50-μm-diameter wires (one vertical, one horizontal, and one at 45°). (B) Fraction of the beam contained within the vertical beam scattering angle. In order to deduce these curves from the profiles: (i) we assumed that 100% of the beam has been scanned; (ii) we considered only the right side of the third peak corresponding to the horizontal profile in order to reduce the error due to the signal received by the other wires; and (iii) we deconvoluted the relevant profiles with respect to the beam.

Comparative Results with Keil's Model and
Monte Carlo Simulation

In materials so thin that the mean number of collisions is less than 20 (domain of plural scattering), Molière's model of multiple scattering is not valid anymore. We used two techniques to calculate the scattering angle distribution. One is a calculation based on Keil's model (1), which is the only analytical model available. The other is a Monte Carlo technique, the GEANT (2) computer code. The angular distribution generated by the Monte Carlo code is narrower than Keil's model by a factor of 2. The measured distribution is narrower than the Monte Carlo distribution by approximately another factor of 2 (Fig. 2(A)).

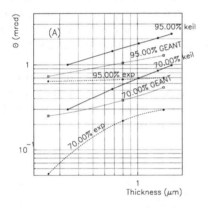
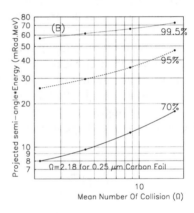

FIGURE 2. (A) Comparative results of experiment with Keil's model and GEANT computation for thin aluminum foils. The curves show the effects of the foil thickness on the semi-angle containing 70% and 90% of the beam. (B) Projected scattering angle Θ of highly relativistic electrons in thin foils according to GEANT simulation code and normalized at 1GeV. Θ is the half angle at which the distribution contains the indicated percentage $\mathcal{P}$ of the beam. The projected angle of a scattered particle is defined on any plane that contains the incoming trajectory. It is the angle between the incoming and outgoing trajectory projected on that plane. The percentage of the beam loss (1- $\mathcal{P}$) can be computed at any energy for any screen material and thickness, by knowing the machine acceptance angle and the beam energy.

The beam energy in the transport line ranges from 0.8 GeV to 4 GeV. The beam scattering angle is inversely proportional to beam momentum. The worst-case beam loss occurs at the lowest energy. At 0.8 GeV, CEBAF's beam angle acceptance is around 100 µrad at the location we wish to install a viewer. Table 1 lists the smallest commercially available thickness of various foil materials that can be mounted on an 8-mm-diameter hole and the corresponding mean number of collisions Ω per µm. We used the formula

$$\Omega = 0.78 \times \rho \times \frac{(Z+1)Z^{1/3}}{A} ,$$

where ρ is the density in g/cm^3, Z is the atomic number, and A is the atomic weight.

TABLE 1. Survey of the materials commercially available for monitoring intense beams with OTR screens. We excluded the materials with low thermal conductivity and mean number of collisions greater than 30 in their smallest thickness.

Material	Be	C	Mg	Al	Ti	Fe	Cu	Ag	Au
Number of collisions per µm thickness	7.5	8.7	8.3	13.6	28	51	62	90	127
Thinnest foil available (µm) for 8-mm-dia.	0.25	0.25	0.5	0.5	1	0.5	0.25	0.25	0.1
Corresponding Ω	1.9	2.2	4.2	6.8	28	26	15	22.6	22

The best candidates are beryllium and carbon; they are equivalent but we prefer the latter because of the toxicity of beryllium. From results of the GEANT simulation (Fig. 2(B)), we can estimate the percentage of beam lost at any energy for any screen material and thickness by knowing the machine acceptance angle. Previous calculations (3) showed that aluminum foils could stand a 200 µA continous beam of 50 µm rms transverse dimensions. Carbon, with its equivalent thermal conductivity and higher melting point, will be even superior. We present the results of a thermal analysis in the last section of this paper.

OTR PROFILE MONITOR WITH A 0.25-µm CARBON FOIL

Description of the Monitor

The backward OTR depends on the reflection characteristics of the surface. There are problems with very thin foils: surface inhomogeneity makes their coefficient of reflection nonuniform, and it is also difficult to stretch them enough to obtain a flat surface. Furthermore, materials such as aluminum and gold present wrinkles that affect locally both reflection coefficient and angle on the scale of the beam size. These problems disappear with forward OTR since it is emitted around the beam direction regardless of the angle or coefficient of reflection of the foil. The forward OTR has two additional advantages over the backward OTR with a foil at 45°: the depth of field error becomes negligible, and the incoming beam at normal incidence has a shorter path through the material which reduces the beam loss due to scattering.

We built a profile monitor, shown in Fig. 3, which looks at the forward OTR emitted from a 0.25-μm carbon foil mounted on a support that can be inserted into the beam. To prevent the beam from damaging the support, the support is U-shaped and open on the side crossing the beam path. A mirror sends part of the forward OTR to the optical system through a vacuum window. With the OTR that is strongly directional in a 1/γ angular cone, we need to collect the light emitted at small angles from the beam trajectory. We did this by locating the mirror on the same insertion mechanism as the foil. The mirror is 17.5 cm downstream from the foil; this insertion mechanism brings its edge close to the beam trajectory (4 mm).

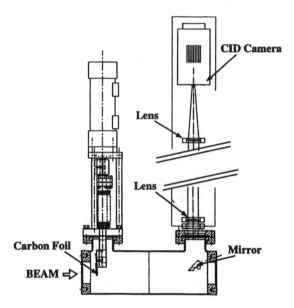

FIGURE 3. Simplified view of the forward OTR profile monitor.

The mirror collects the light and sends it to a charge-injected device (CID) camera via two achromatic doublets. Despite its lower sensitivity, we preferred a CID camera to the more commonly used charge-coupled device (CCD) cameras because of its good radiation resistance (specified to 100 krad). The CID array has 768×575 square pixels of 11.5 μm. The camera is sensitive to the whole visible spectrum. By choosing achromatic lenses, we can improve the monitor's sensitivity without any loss in resolution. We illuminated the support from outside the vacuum chamber so that its 10-mm-square shape can be used to precisely calibrate the CID image. The magnification of the optics is such that the 10-mm-square support encompasses about 500 pixels which yields a pixel size of approximately 20 μm in the object plane.

Optical Resolution

The beam profile has a Gaussian shape with an rms value $\sigma \cong 100$ μm . We measured the resolution of the instrument in the laboratory, replacing the foil with a resolution target. We found the minimum resolvable distance between two black lines to be about 80 μm. We believe the camera is the limiting element since this number is within a factor of 2 of the manufacturer's specifications. The resolution measured with such a target corresponds approximately to the full width of a Gaussian beam (4σ). Hence, the resolution expressed in terms of "one sigma" is about $80/4 = 20$ μm, which is reasonably small compared to the expected beam size. The resolution measurement in the lab does not take into account the pupil function of the transition radiation which has a strong angular dependence. The resolution due to the OTR pupil function is a controversial issue. It has been suggested that it was limited to $1.22 \times \lambda\gamma$. However, Rule and Fiorito calculated (4) an OTR diffraction limit very close to that of a standard source which has a uniform intensity with angle. Although the mirror center is not on the OTR revolution axis, we believe that the results reported in the following section support their calculations.

Experimental Results

We tested the monitor in the transport line of an experimental hall with a 3.2-GeV electron beam. Figure 4 shows the horizontal beam profile measured with a 3-μA continuous beam, whose dimension is $\sigma_x \cong 100$ μm. The camera has its maximum sensitivity at 500 nm (1.22 $\lambda\gamma = 3.9$ mm). The 1.22 $\lambda\gamma$ product represents the distance between the two minima of the diffraction pattern. A fit to the Airy diffraction function with a Gaussian distribution yields a σ that is 5.8 times smaller than the distance between the two minima of the Airy function. The hypothetical "OTR diffraction limit" at that beam energy would yield $\sigma = 3.9/5.8 = 0.7$ mm which is much greater than the 0.1 mm actually measured.

A test done at 25 μA shows that the beam scattering through the foil does not introduce losses greater than 0.5% of the beam. The electrode-sum signals of beam position monitors located upstream and downstream of the foil provided the intensity information for beam loss calculation. With fluctuations of the intensity measurements at the 0.1% level, we were not able to detect any beam loss. In the near future we will test the monitor in a 180-μA continuous beam to confirm that the beam loss is below the MPS threshold of 1 μA.

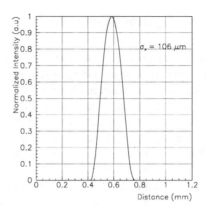

FIGURE 4. Horizontal profile of a 3.2-GeV, 3-μA beam measured with the OTR monitor.

THERMAL CONDUCTION

An electron beam going through matter deposits ionization energy that heats the foil and knocks out secondary particles. Calculation of the energy deposition (5) or its computation using the Monte Carlo computer code EGS4 (6) yields similar results. The heat is lost through conduction (3) and radiation. Figure 5 shows the maximum current required to melt various materials. For CEBAF, the maximum tolerable beam loss due to beam scattering limits actual application of OTR to lower currents.

CONCLUSION

We described an OTR profile monitor using a very thin carbon foil that can operate in a 200-μA continuous electron beam. The measurement of a 100-μm (rms radius) beam profile at 3.2 GeV supports Rule and Fiorito's calculations and confirms the possibility of imaging micron-size beams of high energy. It also supports the concept of using the OTR monitors at CEBAF where the combination of high average beam power, small emittance, and high energy makes other profile diagnostics unsuitable. The present construction of a free-electron laser should be an opportunity to test this kind of monitor at higher continuous currents.

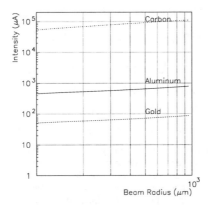

FIGURE 5. Beam intensity and radius that bring various foil materials to their melting point, taking conduction and radiation phenomena into account. The calculation considers 25-mm-diameter foils whose outside edge is at ambient temperature, a reasonable assumption for small energy deposition.

REFERENCES

1. Scott, W.T., "The Theory of Small-Angle Multiple Scattering of Fast Charged Particles," *Rev. Mod. Phys.* **35** (2), 288-290, April 1963.
2. Groosens et al., "GEANT, Detector Description and Simulation Tool, User Guide," CERN Application Software Group, CERN Geneva, Switzerland.
3. Denard, J-C. et al., "Experimental Diagnostics Using Optical Transition Radiation at CEBAF," Beam Instrumentation Workshop, Vancouver, B.C., Canada, October 1994, *AIP Conference Proceedings 333*, 224-230 (1995).
4. Rule, D.W. and Fiorito, R.B., "Imaging Micron-Sized Beams with Optical Transition Radiation," 2nd Annual Workshop, Batavia, IL, *Particles and Fields Series 44, Accelerator Instrumentation*, p. 315 (1990).
5. Fisher, C., private communication, November 1995.
6. Nelson, W.R., Hirayama, H., Rogers, D.W.O., "The EGS4 Code System," Stanford Linear Accelerator Center report SLAC-210, 1978.

Squids, Snakes, and Polarimeters: A New Technique for Measuring the Magnetic Moments of Polarized Beams

P. R. Cameron, A. U. Luccio, T. J. Shea, and N. Tsoupas

Brookhaven National Laboratory, Upton, NY 11973 USA

D. A. Goldberg

Lawrence Berkeley Laboratory, Berkeley, CA USA

Abstract. Effective polarimetry at high energies in hadron and lepton synchrotrons has been a long-standing and difficult problem. In synchrotrons with polarized beams it is possible to cause the direction of the polarization vector of a given bunch to alternate at a frequency which is some subharmonic of the rotation frequency. This can result in the presence of lines in the beam spectrum which are due only to the magnetic moment of the beam and which are well removed from the various lines due to the charge of the beam. The magnitude of these lines can be calculated from first principles. They are many orders of magnitude weaker than the Schottky signals. Measurement of the magnitude of one of these lines would be an absolute measurement of beam polarization. For measuring magnetic field, the Superconducting Quantum Interference Device, or squid, is about five orders of magnitude more sensitive than any other transducer. Using a squid, such a measurement might be accomplished with the proper combination of shielding, pickup loop design, and filtering. The resulting instrument would be fast, non-destructive, and comparatively cheap. In addition, techniques developed in the creation of such an instrument could be used to measure the Schottky spectrum in unprecedented detail. We present specifics of a polarimeter design for the Relativistic Heavy Ion Collider (RHIC) and briefly discuss the possibility of using this technique to measure polarization at high-energy electron machines like LEP and HERA.

INTRODUCTION

The idea of making a direct magnetic measurement of beam polarization has been under consideration for many years (1,2,3). Given the sensitivity of the squid, the amount of magnetic flux present due to the aligned dipole moments in the polarized beam in the Relativistic Heavy Ion Collider (RHIC) is sufficient to permit a quick and accurate measurement of beam polarization. In the case of charged particle beams, these magnetic dipoles are inextricably bound to electric monopoles, whose current generates magnetic fields which, depending upon the specifics of a given measurement configuration, are ten to fifteen orders of magnitude larger than those due to the magnetic dipoles. It is possible to take advantage of the difference in direction between the beam current field and the field due to beam polarization to design a pickup loop which couples primarily to the beam polarization. Elimination of the beam current field via loop orientation is limited by position stability of the beam and also by the effect of statistical fluctuations (Schottky noise) within the beam. As will be explicitly shown in the pickup loop section, the best one might reasonably hope for is to improve the ratio of the signal to this background to about −160 dB. Historically, this overwhelming background has rendered impossible the task of measuring the beam magnetic moment. However, with the increasing sophistication of hardware and techniques used to accelerate and manipulate polarized beam, it now appears possible to separate the signal due to the magnetic moment from the signal due to the charge. This separation must be accomplished in the frequency domain and is possible because the charge is a monopole whose field has the same sign from turn to turn in the accelerator, whereas the moment is a dipole whose field direction might alternate from turn to turn, thus generating lines in the beam spectrum where none exist due to charge.

THE BEAM SPECTRUM

Consider RHIC to be uniformly filled with sixty bunches of equal intensity. The spectrum would then be a comb of lines spaced by the bunching frequency which, in this case, is about 4.7 MHz. The envelope of the lines is determined by the Fourier transform of the longitudinal bunch profile. To accomplish our measurement we must utilize the dipole nature of the magnetic moment to create lines in the beam spectrum where none exist due to the beam current. We might begin by alternating the direction of the beam polarization from bunch to bunch, creating lines in the spectrum at half the bunching frequency, every 2.35 MHz. With a uniformly filled RHIC, there would be no lines present due to beam current at odd multiples of 2.35 MHz. Unfortunately, it is not possible to have a uniformly filled RHIC. There is not only intensity variation from bunch to bunch, but also the 4-bunch abort gap required by kicker rise time. As a result, signals will appear at all harmonics of the 78-kHz revolution frequency and not just those corresponding to multiples of the 4.7-MHz bunching frequency. One cannot reasonably expect these lines to be down in magnitude from the lines at the bunching frequency by more than about 40 dB. Our lines at odd multiples of 2.35 MHz will lie directly under these rotation harmonics. Alternating polarization direction from bunch to bunch results in a modest improvement in signal to background, from about −160 dB to perhaps −130 dB. To gain any further improvement requires that we examine the effect of devices like snakes and spin flippers.

THE SPECTRUM WITH SNAKES AND SPIN FLIPPERS

In order that the polarization signal be observable, it must be made to appear at a frequency whose fundamental periodicity is different from that of the revolution frequency. One way to accomplish this is by alternating the polarization direction from turn to turn. In a machine with one or more full snakes, there is a stable spin direction which is independent of energy (4). When the polarization points in the stable spin direction, its direction does not change from turn to turn, and the polarization signal has the periodicity of the revolution frequency and cannot produce lines in the beam spectrum which can be used to measure beam polarization. In RHIC the stable spin direction is the y (vertical) direction. Horizontal magnetic field in the x (radial) direction will rotate the polarization vector toward the horizontal plane, producing a component which is initially in the z (beam) direction, and which will then precess about the vertical. Because RHIC has a fractional spin tune of one half, the direction of this component will reverse from turn to turn, and the effect of the horizontal field will cancel from turn to turn. However, if the direction of the magnetic field is reversed from turn to turn, the effect of these kicks on the polarization direction is then cumulative. In this manner it is possible to flip the spin (5). To measure polarization it is not necessary to fully flip the spin. We need only perturb the beam slightly to produce a horizontal polarization component, measure the polarization, then rotate it back to the vertical.

An ingenious method has been proposed for such a spin flipper in RHIC (6). It is based upon the fact that the spin tune and the betatron tune are not harmonically related, so that it is possible to kick the beam to flip the spin, and simultaneously kick the beam to undo the betatron oscillations resulting from previous spin flip kicks. It requires only a single fast kicker, able to deflect a single beam bunch vertically by an angle δ, that can be placed at any available section in the accelerator. This deflection results in rotation of the polarization direction about the x-axis by an angle

$$\psi = G\gamma\delta \tag{1}$$

where $G = 1.793$ is the magnetic anomaly of the proton. If the kicker operates at a frequency $f_{flip} = \nu_s f_r$ where ν_s is the fractional spin tune and f_r is the revolution frequency, then the resonance condition is satisfied and the spin direction of each particle in the beam bunch will be rotated about the x-axis by the angle ψ every time the beam bunch passes the kicker. In RHIC we have $\nu_s = 1/2$ and $f_r = 78$ kHz, so that $f_{flip} = 39$ kHz. Bunch deflection with such a frequency will result in large betatron oscillations. In order to minimize these oscillations and the resulting emittance growth, another deflection pulse with a 'proper' frequency, amplitude, and phase can be superposed on the spin flip pulse. As an example, let the spin kick deflection be $\delta = 1$ µrad, the fractional betatron tune of the accelerator be $v_b = 0.6$, the betatron function at the location of the deflector be $\beta = 25$ m, and the frequency and amplitude of the superposed pulse be $v_k = 2v_b - v_s$ and 2δ. Figure 1 shows the result of simultaneous spin flipping and correction.

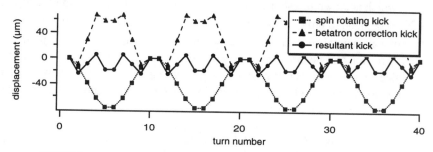

FIGURE 1. Beam displacement due to spin flipping and correction.

Only that component of the polarization which is rotated away from the stable spin direction is subject to the depolarization which results from the effect of spin tune spread over many turns. In a typical polarization measurement the beam might be kicked until the component normal to the stable spin direction has a value of 10% of the total polarization. Because the components add vectorially, the component remaining in the stable spin direction would have a value of 99.5% of the total polarization, so that even in the unlikely worst case of complete decoherence of the perturbed component of the polarization, the total beam polarization would be only minimally affected.

Suppose that the beam is kicked as outlined above to produce a horizontal component. This component will initially be parallel to the beam, whereas polarization in the transverse x direction is required to make the measurement. As the beam proceeds through the main dipoles, the longitudinal component precesses about the y axis. If we locate the kickers at 12 o'clock in the RHIC ring, and the pickup loops at 10 o'clock and 2 o'clock, the amount of precession between kickers and pickups will be given by Eq. (1), with $\delta = \pi/3$. The condition for which this component lies precisely in the x direction is $G\gamma = 3(2n - 1)/2$. For $\gamma = 266$ this gives $n = 160$. While it is not necessary to have the normal component precisely in the x direction to measure polarization, it is clear that this condition is satisfied at frequent intervals in the RHIC energy range.

THE PICKUP LOOP

The field due to the dipole falls off with distance more quickly than the field due to the charge. To maximize both the magnitude of the signal and the signal-to-background ratio, it is desirable to bring the loop as close as possible to the beam. A possible loop configuration is shown in Fig. 2. The beam direction is along the z axis, which is drawn foreshortened in this figure. The ends of the loop are bent out of the y-z plane to provide 10 σ of clearance for the beam. The loop is configured to provide maximal coupling to magnetic field due to magnetic moment pointing in the x direction and minimal coupling to field due to beam current. The amount of flux due to magnetic moment which might be captured by the loop if all the polarization were kicked into the x direction can be found by integrating the vector potential over the perimeter (7)

$$\Phi = \frac{\mu_0}{4\pi} \int_0^{z/2} \frac{4 \cdot \mu_p \cdot n_p \cdot n_b \cdot n_l \cdot P \cdot \gamma \cdot 10 \cdot \sigma_y}{64 \left(x^2 + \left(10 \cdot \sigma_y \right)^2 + \gamma^2 z^2 \right)^{3/2}} dz . \tag{2}$$

For $n_p = 10^{11}$, $n_b = 60$, $P = 0.7$, $\gamma = 266$, $\sigma_y = 0.35$ mm, $x = 0$, $z = 0.5$ m, a bunch spacing of 64 m, $\Phi_0 = 2 \times 10^{-15}$ T-m^2, and the number of loops $n_l = 3$ to impedance match to the 2-μH input inductance of the squid, the flux captured per turn is about $\Phi = 0.01 \, \Phi_0$. As will be discussed in more detail in the squid section, typical squid flux noise levels are between 10^{-5} and $10^{-6} \, \Phi_0/\text{Hz}^{1/2}$.

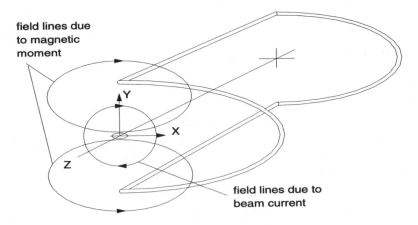

field lines due
to magnetic
moment

field lines due to
beam current

FIGURE 2. The pickup loop.

The background flux due to the beam current is zero when the loop is perfectly aligned. The flux captured by the loop as the result of a small vertical misalignment δ is

$$\Phi_b = \frac{\mu_0}{4\pi} \int_{-z/2}^{z/2} \int_{10\cdot\sigma_y-\delta}^{10\cdot\sigma_y+\delta} \frac{n_l \cdot n_p \cdot n_b \cdot \beta \cdot \gamma \cdot q}{64 \cdot (x^2 + (10\cdot\sigma_y)^2)} dydz, \tag{3}$$

where q is the charge of the proton. For a displacement $\delta = 1$ micron, the signal-to-background ratio is a few 10^{-9}, and the signal is about 170 dB down from the background. If we choose a signal line between the weakest revolution lines, this background might be reduced to about +140 dB. It can appear either at the revolution lines as a result of closed orbit distortions or at the betatron lines as a result of coherent motion.

Another source of background is Schottky noise within the beam. The flux captured by the loop as a result of statistical fluctuations in position is

$$\Phi_b = \frac{\mu_0}{4\pi} \int_{-z/2}^{z/2} \int_{10\cdot\sigma_y-\sigma_y}^{10\cdot\sigma_y+\sigma_y} \frac{n_l \cdot n_b \cdot \sqrt{n_p} \cdot \beta \cdot \gamma \cdot q}{64 \cdot (x^2 + (10\cdot\sigma_y)^2)} dydz \tag{4}$$

Here the signal-to-background ratio is about 10^{-8}, and to first order this is independent of beam position. The signal is about 160 dB down from the Schottky noise in the beam. Again by choosing the location of the signal line this background might be reduced to about +130 dB. It will appear as satellites of the revolution lines and the betatron lines.

SQUIDS

The squid measures magnetic flux; in this it is unlike most beam instrumentation, which measures the *time derivative* of electric or magnetic flux (8). When flux is applied to the squid loop, voltage proportional to the flux appears across the loop at the same frequency as the applied flux. The squid comprises a superconducting loop which is interrupted in two places by weak links, or Josephson junctions. When magnetic flux is applied normal to the plane of the loop, a shielding current flows in the loop to resist penetration of the flux into the superconductor and thence into the loop. Because the shielding current must tunnel to flow through the junctions the phase of the Cooper pair electron wave function changes across the junctions and a voltage tries to appear across each junction. This voltage is of opposite sign at the ends of the superconductors which join the junctions into a loop, resulting in no measurable effect. However, if a bias current is applied across the junctions, it will add to the current through one of the junctions and subtract from the current through the other. The magnitude of the voltage across the junctions then no longer cancels and can be measured. The DC response of the squid is important in our application; it permits efficient measurement at low frequency without excessively large pickups. The upper limit of frequency response is set by the self-resonant frequency of the squid loop, and is typically in the range of 10 to 100 GHz. In most squid applications the squid is operated in a flux locked loop configuration. The bandwidth of flux locked loops is typically a few hundred kHz, although recent development has raised the upper limit to about 5 MHz, and further extension to 20 MHz is predicted (9). Finally, the squid has good dynamic range, is extremely sensitive, and has, most importantly, an extremely low noise floor. When operated in the flux locked loop mode, the upper limit of the approximately 100 dB of squid dynamic range is the flux quantum, and this upper limit corresponds to the noise floor of any other available magnetic field transducer. The noise floor of the squid sits better than 100 dB below that level, at about 10^{-5} to 10^{-6} $\Phi_0/\text{Hz}^{1/2}$. Typical transfer functions for squids are in the range of 1 to 100 mV/Φ_0.

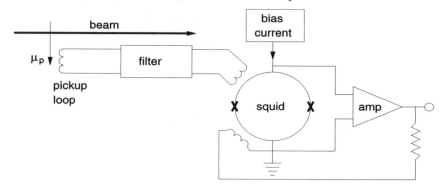

FIGURE 3. Block diagram.

Figure 3 is a block diagram showing the squid as it might be used in our application. The signal from the magnetic moment of the beam is coupled through the pickup loop into a filter. The filter prevents the betatron signals and revolution harmonics from being coupled into the squid loop. The amplifier output is coupled back into the squid, closing the flux locked loop. As mentioned above, there is recent improvement in flux locked loop bandwidth. It is also possible to lock the loop at a lower bandwidth and pick off high-frequency signals from the squid before the flux locked loop.

IMPLEMENTATION

It is desirable to have the pickup loop shown in Fig. 2 at room temperature. This reduces cost and simplifies installation, maintenance, and modifications. With reasonable conductor size, the loop resistance might be about 1 milliohm. At room temperature the Johnson noise current due to this resistance would correspond to about 10^{-5} $\Phi_0/\text{Hz}^{1/2}$, which suggests that it is feasible to consider a room temperature pickup. In the simplest implementation this pickup would be surrounded by room temperature shielding. At the closest practical distance a cryostat would be positioned immediately adjacent to the beamline. The filter and squid would be surrounded by superconducting shielding within the cryostat. It is essential to minimize the number and maximize the quality of connections between the pickup and the cryostat. The pickup would be free to move in the x direction, towards and away from the beam. Fine tuning of the relative position of beam and pickup could be accomplished with steering magnets, by translating the pickup vacuum chamber horizontally and vertically, or by their combination.

As previously outlined, our signal shows up as AM sidebands of bunching subharmonics. Suppose that we look at the AM sideband which is 39 kHz away from the 2.35-MHz bunching subharmonic. If we build a fifth-order Butterworth superconducting bandpass filter with a Q of 10^4 and a center frequency of 2.389 MHz, the 3-dB passband will be about 240 Hz wide, and signals at the betatron lines 25 kHz away will be attenuated by about 200 dB (10). Such a filter would have a characteristic time of about 4 msec, corresponding to about 300 revolutions in RHIC. During this time the envelope of the output of the filter would rise as $e^{-t/4ms}$, so that after 300 revolutions the flux applied to the squid would be about 0.63 of the flux at the filter input from a single revolution. If we have 10% of the total polarization kicked into the horizontal plane, this will be about 10^{-3} flux quanta. A measurement might consist of the following sequence:

- move the loop in while servoing beam and loop position to null output;
- measure background;
- kick and damp, and measure signal plus background;
- kick and damp back, and measure background;
- move the loop out.

The measurement would be the difference between two curves, the background and the inverse exponential of signal plus background. Lineshape analysis, digital filtering, and further signal processing can then be applied.

OTHER POSSIBILITIES

In a machine with zero chromaticity, the Schottky spectrum at low frequencies yields less information than at high frequencies. Because of the large coherent signals at multiples of the revolution frequency, the Schottky signals at the betatron lines are observed more easily than those at the rotation harmonics. The envelopes of their synchrotron satellites are determined by Bessel functions whose argument contains the bunch length, which for nanosecond bunches means that the higher terms differ significantly from zero only at frequencies above several hundred MHz. As a result, the only information yielded by the low-frequency Schottky spectrum would be the (fractional) incoherent tune and the amplitude-dependent tune spread, which is manifested in the width of the central betatron satellite. Non-zero chromaticity both shifts and broadens the envelopes, so that many satellites are present at low frequencies (11, 12). This contrasts with the high-frequency case, where the width of the betatron side bands is determined primarily by the variation in revolution frequency. Hence the low-frequency Schottky signals complement the high-frequency ones, permitting direct measurement of the chromaticity. Moreover, given the extremely low noise levels obtained using the squid, with proper attenuation and perhaps a different filter we might hope to probe this Schottky spectrum in unprecedented detail.

At a large electron machine, like LEP or HERA, there are no snakes. However, it is possible to apply the technique presented here by adjusting the machine energy such that the natural fractional spin tune is 1/2. The additional required hardware is then only the fast vertical kicker and the squid polarimeter. At such a machine one gains the relativistic gamma factor, the ratio of electron to proton magnetic moment, and reduced transverse beam size, so that both the signal and the signal to background are about three orders of magnitude larger than at RHIC.

Finally, there is an additional interesting possibility which might be opened at RHIC by fast, high-resolution polarization measurement. In the discussion about spin flippers it was assumed that the spin tune was precisely known and that the spin flip kick was properly phased to the spin tune. The spin flip resonance is narrow, and kicking off resonance can produce depolarization. While this is not a problem for the polarization measurement—where only a small component of the polarization is rotated into the horizontal plane, and then only for a time short relative to the decoherence time—it could be a problem for spin flipping. One possible solution is to sweep the kick over the resonance. This requires approximate knowledge of the spin tune, so that the sweep frequency spans the resonance, and more detailed knowledge of the resonance width and shape, so that the effective integrated kick is correct. Another proposed solution is coherent spin flip, the 'beam maser effect' (3). Coherent spin flip would be accomplished by using the polarimeter output and the spin flip kicker to close a feedback loop around the polarized beam. This would ensure that the spin flip kick is properly phased to the spin tune and would result in a distinctive signature for the polarization signal, further helping to separate it from the background. It has been suggested that this might also remove the requirement that spin flip be accomplished in a short time relative to the decoherence time to avoid depolarization (3,13). The spin might then be flipped much more slowly, say in 10^5 turns, which would reduce kicker strength requirements and could open the door to a series of more detailed accelerator spin studies.

CONCLUSION

There is a good possibility that the technique described here could be used to measure polarization at RHIC. There is precedent for using squids in the magnetically noisy accelerator environment (14). However, the beam spectrum has never been explored at this level of sensitivity, and there is concern that unexpected noise or signals might exist. Because of the large dynamic range of the squid, the narrow filter bandwidth, and the possibility of background subtraction, a good measurement can be accomplished in the presence of significant noise and background. A next step would be to explore the spectrum in a machine similar to RHIC, perhaps with a squid-based Schottky detector at the Alternating Gradient Synchrotron (AGS).

ACKNOWLEDGEMENTS

The authors acknowledge helpful discussions with many people, including but not limited to the following individuals: M.M. Blascewicz, R.C. Connolly, Ya. S. Derbenev, M.A. Goldman, M.A. Harrison, H. Huang, A.D. Krisch, G.R. Lambertson, R. Lin, V.R. Mane, F.G. Mariam, W.W. MacKay, R.A. Phelps, A.G. Ratti, T. Roser, W.A. Ryan, R.L. Siemann, R.E. Sikora, G.A. Smith, A.N. Stillman, M.J. Syphers, S. Tepikian, and J. Wei.

REFERENCES

1. Cameron, P.R., unpublished notes, 1983.
2. Barber, D.P., Cabrera, B., and Montague, B.W., private communications, 1983.
3. Derbenev, Ya. S., "RF-resonance Beam Polarimeter," in *Proceedings of the Eleventh International High Energy Spin Physics Symposium,* AIP Conference Proceedings **343**, 264-272 (1994).
4. Derbenev, Ya. S. and Kondratenko, A.M., "Possibilities to Obtain High Energy Polarized Particles in Accelerators and Storage Rings," Symposium on High Energy Physics, Argonne, IL, October 26, 1978, *AIP Conference Proceedings* **51**, 292-306 (1979).
5. Caussyn, B. D. et al, "Spin Flipping a Stored Polarized Proton Beam," *Phys. Rev. Lett.* **73** (21), 2857-2859 (1994).
6. Roser, T., private communication, February, 1996.
7. VanBladel, J., *Relativity and Engineering,* New York: Springer-Verlag, 1984.
8. Clarke, J., "Squids: Theory and Practice," in *The New Superconducting Electronics* (The Netherlands: Kluwer, 1993) pp. 123-180.
9. Drung, D., Matz, H., and Koch, H., "A 5 MHz bandwidth SQUID magnetometer with additional positive feedback," *Rev. Sci. Instrum.* **66** (4), 3008-3015, April 1995.
10. Williams, A.B. and Taylor, F.J., *Electronic Filter Design Handbook,* New York: McGraw-Hill, 1988.
11. Siemann, R. L., "Spectral Analysis of Relativistic Bunched Beams," these proceedings.
12. Chattopadhyay, S., "Some Fundamentals of Fluctuations and Coherence in Charged Particle Beams in Storage Rings," CERN, 84-11.
13. Derbenev, Ya. S., private communication, May 1996.
14. Peters, A. et al., "A Cryogenic Current Comparator for Nondestructive Beam Intensity Measurements," Proc. of the Fourth European Particle Accelerator Conference, EPAC, 27 June to 1 July, 1994, 290-292 (1994).

Positron Beam Position Measurement for a Beam Containing Both Positrons and Electrons*

N. S. Sereno and R. Fuja

Advanced Photon Source, Argonne National Laboratory,
9700 South Cass Avenue, Argonne, IL 60439

Abstract. Positron beam position measurement for the Advanced Photon Source (APS) linac beam is affected by the presence of electrons that are also captured and accelerated along with the positrons. This paper presents a method of measuring positron position in a beam consisting of alternating bunches of positrons and electrons. The method is based on Fourier analysis of a stripline signal at the bunching and first harmonic frequencies. In the presence of a mixed species beam, a certain linear combination of bunching and first harmonic signals depends only on the position and charge of one specie of particle. A formula is derived for the stripline signal at all harmonics of the bunching frequency and is used to compute expected signal power at the bunching and first harmonic frequencies for typical electron and positron bunch charges. The stripline is calibrated by measuring the signal power content at the bunching and first harmonic frequencies for a single species beam. A circuit is presented that will be used with an APS positron linac stripline beam position monitor to detect the bunching and first harmonic signals for a beam of positrons and electrons.

INTRODUCTION

The 7-GeV Advanced Photon Source (APS) is a third-generation light source optimized to produce insertion-device x-ray radiation for materials science research. The APS linear accelerator is used to accelerate a positron beam to 450 MeV and deliver it to a positron accumulator ring (PAR). The beam is accumulated and damped in the PAR, transported to the booster synchrotron and accelerated to 7 GeV, and finally injected into a storage ring containing insertion device undulators. The large emittance of the positron beam in the linac requires accurate positron beam position measurement to minimize beam loss during transport to the PAR. In addition, positron position measurement is essential so that response matrix measurements can be performed to verify linac optics for various configurations used to produce and transport positrons.

* Work supported by U. S. Department of Energy, Office of Basic Energy Sciences under Contract No. W-31-109-ENG-38.

Figure 1 shows the layout of the APS linac. The top part of the figure (the electron linac) shows the 100-keV gun electron source, accelerating waveguides, and magnets up to the positron production target. The bottom part of the figure (the positron linac) shows the linac S-band (2856 MHz) accelerating waveguides and magnets after the target. The electron beam on the target is 200 MeV and produces both positrons and electrons through the bremsstrahlung process. Stripline beam position monitors (BPMs) are located throughout the linac between waveguides to monitor beam position. The difficulty of determining positron beam position downstream from the target is that both species of particle are captured and accelerated by the rf.

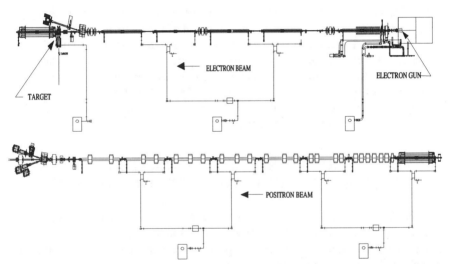

FIGURE 1. Layout of the APS linear accelerator showing the gun, accelerating waveguides, magnets, and target. The top part of the figure shows the linac layout from the 100-kV gun to the target. The bottom part of the figure shows the linac layout from the target to the end. Stripline BPMs are located between the accelerating waveguide sections.

STRUCTURE OF THE POSITRON LINAC BEAM IN THE TIME DOMAIN

The parameters that define the linac beam (1) are driven by the requirement that the PAR be completely filled with 6.6 nC of positrons in 0.5 seconds. The electron beam striking the target is a macropulse 30 ns long repeated at a 60-Hz rate with a peak current of 1.7 amperes. Each macropulse is made up of 86 micropulses repeated at a 2856-MHz rate. After the target, the beam consists of both positrons and electrons that are captured by the rf fields of the first accelerating waveguide. The positron and electron micropulses are separated by half an rf wavelength within each macropulse, forming an alternating microbunch charge pattern. The design peak current for the positron part of the macropulse is 8 mA. The electron current for each macropulse is typically around 6 mA but can be larger or smaller for different linac configurations.

Based on the comments in the previous paragraph, the current in each macropulse as a function of time can be written as

$$I(t) = \sum_{n=-N}^{N} Q_p f_p(t - n\tau) - Q_e f_e\left(t - \left(n + \frac{1}{2}\right)\tau\right) , \tag{1}$$

where Q_p and Q_e are the absolute values of the positron and electron charges in each micropulse (assumed constant as a function of the index n), $f_p(t)$ and $f_e(t)$ are the positron and electron normalized distributions, and τ is the temporal bunch spacing. The micropulse distribution for each particle is assumed different to take into account the fact that the bunching process is different for electrons and positrons. For the APS linac beam, there are $(2N + 1) = 86$ electron and positron micropulses per 30-ns macropulse and the temporal bunch spacing is determined by the S-band rf frequency (2856 MHz).

BPM STRIPLINE SIGNALS GENERATED BY THE POSITRON LINAC BEAM IN THE TIME DOMAIN

The APS linac BPMs consist of striplines designed to be a quarter wavelength long at the rf frequency of 2856 MHz. The stripline output is a bipolar voltage pulse for each beam micropulse that passes along its length (2). In the time domain, the stripline output voltage is written as

$$V(t) = Z\{I_w(t) - I_w(t - 2\alpha\tau)\} , \tag{2}$$

where α is the stripline length in units of wavelength at 2856 MHz, Z is the stripline impedance (50 Ω), and $I_w(t)$ is the wall current induced in the stripline electrode. The linac stripline electrical length is actually measured to be 0.21λ. The wall current is proportional to the beam current given by Eq. (1) where the proportionality constant is a function of transverse position and stripline geometry (3).

In the analysis that follows, the position dependence in Eq. (2) is suppressed and the wall current is simply written as the beam current in addition to some geometry factors. This is justified because the analysis presented here is with respect to the time and frequency domains, and the position of the particles in the beam is considered constant over the 30 ns it takes to measure the beam pulses. In the rest of the analysis presented here, one should keep in mind that the beam current terms in the equations contain implicitly the transverse position information being sought.

BPM STRIPLINE SIGNALS GENERATED BY THE POSITRON LINAC BEAM IN THE FREQUENCY DOMAIN

The bipolar stripline voltage pulses given by Eq. (2) are now analyzed in the frequency domain. Taking the Fourier transform of Eq. (2) yields

$$V(\omega) = \frac{Z\phi}{4\pi}(1 - e^{-2i\alpha\omega\tau})\left(Q_p F_p(\omega) - e^{-\frac{i\omega\tau}{2}} Q_e F_e(\omega)\right)k_N(\omega) , \tag{3}$$

$$k_N(\omega) = \frac{\sin\left\{(2N + 1)\frac{\omega\tau}{2}\right\}}{\sin\left(\frac{\omega\tau}{2}\right)} , \tag{4}$$

where $F_p(\omega)$ and $F_e(\omega)$ are the positron and electron bunch spectra and ϕ (typically on the order of 1 radian) is the angle subtended by the stripline electrode. The bunch spectra roll off at frequencies on the order of the reciprocal positron and electron rms microbunch length. For the APS linac, the rms bunch length is on the order of 5 ps resulting in rolloff of the electron and positron bunch spectra at frequencies on the order of 200 GHz. The quantity $k_N(\omega)$ is a strongly peaked function of frequency at the bunching frequency harmonics. A good measure of the width of each peak is given by the difference in frequency between the first zeros of the numerator of Eq. (4) on either side of the peak. This bandwidth is given by

$$\delta\omega = \frac{2\omega_0}{2N + 1} , \tag{5}$$

where $\omega_0 = 1/\tau$ is the rf bunching frequency (2856 MHz). Using Eq. (5), the APS linac macropulse length of 30 ns (86 micropulses of electrons and positrons) means that the stripline spectrum has appreciable power content over a bandwidth of 66.4 MHz around each harmonic of the rf frequency.

The stripline power spectrum is now computed for each bunching frequency harmonic for the bandwidth given by Eq. (4). The power spectrum is defined by (4)

$$P(n\omega_0) = \frac{1}{\Delta}\int_{n\omega_0 - \frac{\delta\omega}{2}}^{n\omega_0 + \frac{\delta\omega}{2}} \frac{|V(\omega)|^2}{Z}d\omega , \tag{6}$$

where $\Delta = 30$ ns is the width of the macropulse. Evaluation of this integral is greatly simplified by the fact that the amplitude spectrum is strongly peaked so that over the bandwidth given by Eq. (5) the bunch spectra are constant. The only factor that remains under the integral after this simplification is the function $k_N(\omega)^2$. Performing the integration results in

$$P(n\omega_o) = \left(\frac{Z\phi^2\Delta}{8\pi^2\tau}\right)\left(\frac{I_N}{(2N+1)^2}\right)\sin^2(2\pi n\alpha)F(n\omega_o)^2\left(I_p - (-1)^n I_e\right)^2 , \quad (7)$$

$$I_N = \frac{\tau}{2}\int_{n\omega_o - \frac{\delta\omega}{2}}^{n\omega_o + \frac{\delta\omega}{2}} k_N^2(\omega)d\omega = 243.9 , \quad (8)$$

where

$$Q_p = \frac{I_p\Delta}{2N+1} \quad (9)$$

and

$$Q_e = \frac{I_e\Delta}{2N+1} \quad (10)$$

are the average positron and electron current in the 30-ns macropulse. The electron and positron bunch spectrum $F(n\omega_o)$ is taken to be the same for both positrons and electrons because for short bunch lengths far from rolloff, these factors are equal to unity for any arbitrary bunch distribution. For the APS linac BPM application considered in this paper the fundamental and first harmonic stripline signals need to be detected. Using Eq. (7), the fundamental ($n = 1$) and first harmonic ($n = 2$) have a power content of 1.2 dBm (257 mV into 50 Ω) and −22 dBm (18 mV into 50 Ω) for typical positron and electron average currents of 8 mA and 6 mA, respectively.

For the APS linac striplines, the small first harmonic signal is unavoidable since the stripline length is so close to a quarter wavelength. For position detection of a mixed species beam, an improved stripline configuration would be to construct each stripline to be equal to one-eighth wavelength. This particular stripline length would make the geometry factor in Eq. (7) unity for the first harmonic signal and reduce the fundamental by 3 dB. This is not a problem because the fundamental is proportional to the algebraic sum of the average currents for each particle species in the beam and hence is always much stronger than the first harmonic signal.

The voltage amplitude as a function of harmonic number is related to the power according to Eq. (10) by

$$V(n\omega_0) \propto \sqrt{P(n\omega_0)} \propto \sin(2\pi n\alpha)(I_p - (-1)^n I_e) \,, \tag{11}$$

where we consider $F(n\omega_0) = 1$ for short microbunches and frequencies far from rolloff. This equation shows that for each stripline, if a linear combination of fundamental and first harmonic signals is made, the current and implicit position dependence of one or the other species of particle can be eliminated. Using Eq. (11), the linear combination of fundamental and first harmonic voltages becomes

$$V_p \propto \sin(2\pi\alpha)V(2\omega_0) + \sin(4\pi\alpha)V(\omega_0) \,, \tag{12}$$

$$V_e \propto \sin(2\pi\alpha)V(2\omega_0) - \sin(4\pi\alpha)V(\omega_0) \,, \tag{13}$$

where V_p and V_e are the effective stripline positron and electron signals that are proportional to the position of only a single species of particle. The stripline voltages defined in Eqs. (12) and (13) are used to perform a difference over sum calculation for opposite BPM striplines to obtain the transverse position of the positrons and electrons that make up the beam.

PROPOSED ELECTRONICS DESIGN TO DETECT THE FUNDAMENTAL AND FIRST HARMONIC STRIPLINE SIGNALS

Figure 2 shows a block diagram of a prototype electronics system including component specifications that will be used to detect the fundamental (2856 MHz) and first harmonic (5712 MHz) BPM stripline signals. The system must extract these components in addition to detecting the phase shift of the first harmonic signal. The front-end electronics separates and filters the fundamental and first harmonic signals using standard components. The 4:1 rf switch selects BPM striplines for signal analysis at the 60-Hz macropulse repetition rate of the linac beam.

The I/Q detectors are the basic component used to detect both amplitude and phase of the fundamental and first harmonic signals. The I and Q channels of the detector produce baseband signals proportional to the product of the amplitude of the input signal and the sine and cosine, respectively, of the phase difference between the reference and the input signal. For the first harmonic signal, if the last term on the right-hand side of Eq. (11) changes sign from one linac configuration to the next (either due to current changes, actual beam position changes, or some combination of both), both outputs of the I and Q detector will also change sign. The sum of the squares of the I and Q channels yields the harmonic signal ampli-

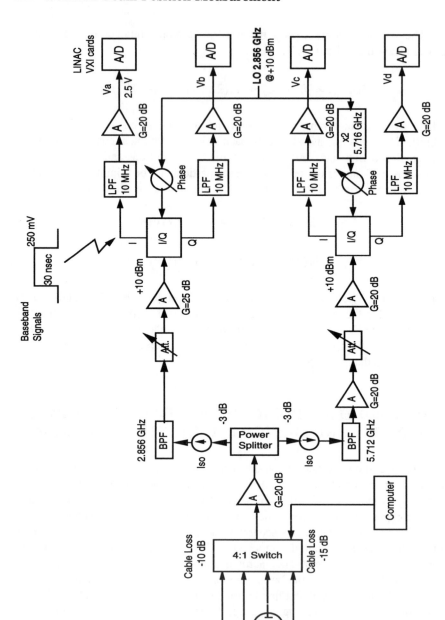

FIGURE 2. Layout of the prototype circuit used to detect fundamental and first harmonic BPM stripline signals.

tude (squared) and their ratio yields the phase angle between reference and input signal. The I/Q detector thus provides enough information to get each harmonic amplitude and their relative phase with respect to each other. The fundamental and first harmonic signals will be determined in software using the outputs of the I/Q detectors.

Calibration of the system is performed using a pure electron beam which is obtained by simply removing the target. The beam is first centered in the BPM so that the stripline voltages are all equal. The gain of each signal channel is adjusted using variable attenuators so that both fundamental and first harmonic signals have the same amplitude at the entrance to the A/D converters. This adjustment effectively takes into account the geometry factor present in the formula for the fundamental and first harmonic signals. The linear combinations given by Eqs. (12) and (13) as well as the final difference over sum position calculation will be performed by software.

CONCLUSION

A method of determining the position of a beam containing both positrons and electrons has been presented. It is based on the detection of both fundamental and first harmonic beam signals. A prototype detection circuit was presented based on I/Q detectors that will be tested with the standard quarter-wavelength stripline BPMs installed in the APS linac. Based on these tests, further tests would be made for a one-eighth-wavelength stripline optimized to maximize the first harmonic beam signal since this signal is attenuated by 6 dB due to the present quarter-wavelength stripline geometry.

ACKNOWLEGEMENTS

The authors would like to acknowledge W. Sellyey who proposed a similar idea to detect beam position of a mixed species beam. In addition, M. Borland, G. Decker, E. Kahana, J. Galayda, A. Lumpkin, and M. White provided valuable insight and comments regarding the ideas and analysis presented in this work.

REFERENCES

1. M. White, et al., "Performance of the Advanced Photon Source (APS) Linear Accelerator," in *Proceedings of the 1995 Particle Accelerator Conference*, Dallas, TX, 1073-1075, (1996).
2. R. E. Shafer, *AIP Conference Proceedings* **249**, 618 (1992).
3. R. E. Shafer, *AIP Conference Proceedings* **249**, 608 (1992).
4. R. J. Mayhan, *Discrete-Time and Continuous-Time Linear Systems*, Addison-Wesley Publishing Company, 1984, p. 433.

Development of Beam Position Detection Electronics for the KEKB Injector Linac

Tsuyoshi Suwada and Hitoshi Kobayashi

National Laboratory for High Energy Physics (KEK)
1-1 Oho, Tsukuba, Ibaraki 305, Japan

Abstract. A stripline-type beam position monitor is under development for the KEK B-Factory (KEKB) injector linac. This monitor reinforces the easy handling of the orbit of high-current, single-bunched electron beams (~10 nC/bunch) generating positron beams for the KEKB. About ninety monitors will be installed in the linac and the transfer line. The beam position detection electronics associated with wide-dynamic range analog and digital processing networks has been developed in order to detect the beam orbit with a position resolution of 0.1mm or better, which is required to suppress any beam blowup generated by a transverse wakefield. In this report, the design and preliminary bench-test results of the electronics are presented.

INTRODUCTION

The KEK B-Factory (KEKB) project (1) is in progress in order to test CP invariance violation in the decay of B mesons. The KEKB is an asymmetric electron-positron collider comprising 3.5-GeV positron and 8-GeV electron rings. The PF Injector Linac (2) is also being upgraded in order to inject single-bunched positron and electron beams directly into the KEKB rings. The beam currents are required to be 0.64 nC/bunch and 1.3 nC/bunch for the positron and electron beams, respectively. High-current primary electron beams (~10 nC/bunch) are required in order to generate positron beams. It is therefore important to easily handle the orbits of the primary electron beams so as to suppress any beam blowup generated by large transverse wakefields. A beam position monitor (BPM) system has been developed to perform this function since 1992. The goal of the beam position measurement is to detect the charge center of gravity within a resolution of 0.1 mm or better. Tests of a prototype BPM (3) using conventional stripline-type pickups were completed in the summer of 1995. The mass production of BPMs has been progressing since. The first 25% of the BPMs is in the process of being installed along the upgraded linac. A prototype detection electronics and a data-taking system have also been developed in parallel. The detection electronics was designed to improve the dynamic range, linearity, and resolution for the pickup signals. The dynamic range is required to be 40 dB, taking into account the detection of a low beam current for orbit tuning at the commissioning of the KEKB injector linac. The position resolution is required to be less than 0.1 mm along the entire linac. Table 1 shows the design parameters used for the detection electronics.

TABLE 1. BPM Design Parameters

e+ Beam Current	0.64 nC/bunch
e- Beam Current	1.28 nC/bunch
Primary e- Beam Current	~ 10 nC/bunch
Number of BPMs	~ 90
Position Resolution	0.1 mm
Dynamic Range	> 40 dB
Cross-Talk	< −40 dB
Detector SNR	< −40 dB
Linearity	< 1%

GEOMETRICAL STRUCTURE OF THE BPM

A drawing of the monitor geometry and a photograph are shown in Figs. 1(a) and (b), respectively. It is a conventional stripline-type BPM made of stainless steel (SUS304) with a $\pi/2$ rotational symmetry. The total length (195 mm) was chosen to make the stripline length (132.5 mm) as long as possible so that it can be installed into limited spaces in the new beamline of the linac. A 10-mm-long bellows is attached to one side of the monitor. Two types of quadrupole magnets (44-mm and 23-mm bore diameters) are used in the new beamline. The outer diameter of the monitor was fixed to be 39.9 mm; this allows them to be inserted into the quadrupole magnets. The monitors are fixed at the end of the pole piece for the 23-mm-type quadrupoles. The inner diameter (27.1 mm) of the electrode was chosen so as to comprise a 50-Ω transmission line. The angular width of the electrode was set to be 60 degrees in order to avoid a strong electromagnetic coupling between the electrodes (4). A 50-Ω SMA vacuum feedthrough is connected to the upstream side of each electrode, while the downstream ends are short-circuited to a pipe in order to simplify the mechanical manufacturing. Quick-release flange couplings (manufacturer's standard KF flange) are used at one end of the monitor for easy installation into the beamline. The other end was attached to a small flange (44 mm in diameter) connected with a vacuum pipe 22 mm in diameter.

BPM PICKUP VOLTAGE

The four pickup signals are sent directly to the detection electronics on the klystron gallery through 35-m-long coaxial cables (KEYCOM (5), CH5055) which are low-cost, double-shielded cables and have good frequency characteristics. The outer conductors are comprised of a thin aluminum foil and a braided wire wound

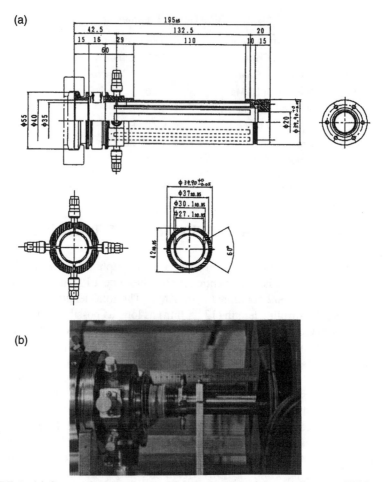

FIGURE 1. (a) Geometrical drawing and (b) photograph of the stripline-type BPM.

on a polyethylene dielectric. The induced noise level, which was mainly generated by a high-power klystron modulator, was measured in order to test the noise-shielding characteristics under typical linac operation by using several types of coaxial cables. The coaxial cable (CH5055) had better characteristics compared with a double-braided-wire shielded cable. The outer diameter of the CH5055 cable is almost equivalent to that of the RG-223 cable, and, therefore, the characteristic of the cable attenuation is almost equivalent to it in the low-frequency region. However, the CH5055 cable is better in the high-frequency region (≥ 20 MHz) because of the use of a highly-foamed low-loss dielectric. The frequency characteristics of these coaxial cables are shown in Fig. 2.

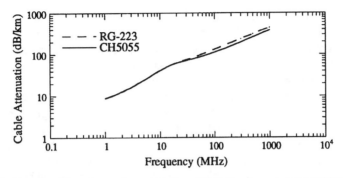

FIGURE 2. Measured frequency characteristics of the 35-m-long coaxial cables. The data of RG-223 cable is shown as a reference.

If a relativistic charged beam passes through in the center of a cylindrical BPM, the pickup voltage V(t) is induced on the inner surface of the electrode at the upstream port. This voltage can be given by the following formula (6) in the time domain:

$$V(t) = \frac{\phi Z}{4\pi}\left[I_b(t) - I_b\left(t - \frac{2l}{c}\right)\right],$$ (1)

where ϕ is the opening angle of the electrode, Z is the characteristic impedance of the transmission line, $I_b(t)$ is the beam current, l is the stripline length, and c is the speed of light. This formula shows a well-known bipolar-doublet of the form peculiar to stripline-type pickups. On the other hand, the impulse response v(t) of a coaxial cable can be approximately estimated, taking into account the skin effect and neglecting dielectric losses, by using the following formula (7):

$$v(t) = \frac{4}{k^2\sqrt{\pi}}\, t'^{-3/2} e^{-1/t'},$$ (2)

with

$$k = 6.6 \times 10^{-8}\frac{A \times l}{\sqrt{f}}$$ (3)

and

$$t' = \frac{4}{k^2} \times t,$$ (4)

where t is the real time in seconds, t' is the normalized time, A is the cable attenuation in dB/km, l is the cable length in meters, and f is the frequency (in MHz) at
which the cable attenuation is defined. Here, approximating a waveform of a single-bunched beam as impulse-like and taking into account the cable attenuation,
the impulse response can be calculated using Eqs. (1) through (4). Figures 3(a)
and (b) show the calculated impulse response normalized with the beam current of
1 nC/bunch and the pickup signal waveform measured for a single-bunched beam
(8.8 nC/bunch) through a coaxial 20-dB attenuator, respectively. The calculated
waveform well approximates that of the pickup signal induced by the real beams.
Thus, the fast bipolar-doublet signals (pulse width ~ 0.5ns at FWHM) are transmitted into the detection electronics. The range of the signal pulse height transmitted to the detection electronics is between 0.88 and 88V, which corresponds to the
range of the beam current from 0.2 to 20 nC/bunch through the 35-m-long coaxial
cable.

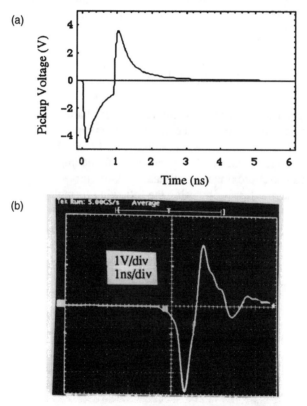

FIGURE 3. (a) Calculated impulse response of the BPM taking into account the cable
attenuation (CH5055, 35m). The pulse waveform is normalized with a beam current of
1 nC/bunch. (b) Pickup signal waveform measured for the single-bunched electron beam
(8.8 nC/bunch) by using a fast digital sampling oscilloscope (Tektronics, TDS684A).

DESIGN OF THE DETECTION ELECTRONICS

The detection electronics comprise a conventional difference-over-sum processing technique by using the peak detection of pickup signals, which can be
made to be low cost and simple. A block diagram of the prototype detection electronics is shown in Fig. 4.

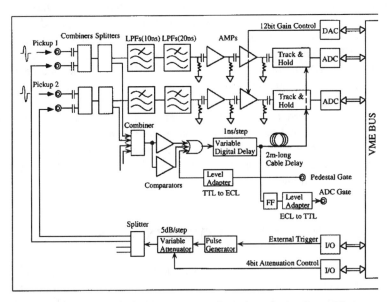

FIGURE 4. Block diagram of the prototype detection electronics. The component
identified as LPFs, AMPs, and FF are low-pass filters, two-cascaded amplifiers, and D-
type flip-flop, respectively. The four channel circuits are stored in a double-spanned NIM
box and several VME-based modules (ADCs, DACs, and digital I/Os) are in a VME crate.

The pickup signal is, first of all, divided into two signals by a signal splitter
(Mini-Circuit (8), PSC-2-5). One pickup signal comes into a rise-time filter (Picosecond Pulse Labs (9), model 5920). The other is delivered into a signal combiner
(Mini-Circuit, PSC-4-5) in which the four pickup signals are added in order to
generate a gate signal for a peak detection circuit. The filter comprises a 50-Ω-
matched low-pass filter (LPF) which has a quasi-Gaussian response with the constant pulse width for an impulse signal. Two cascaded LPFs were used to suppress
any signal overshoot and ringing and also to reject any high-frequency leakage of
the LPF. The low-passed signals are then fed to four two-cascaded, low-noise
amplifier chains (AMPs) (Analog Devices (10), AD603AQ); the first amplifier is
used as a variable-gain device controlled by the DC voltage, and the second is used
as a fixed-gain device. The DC voltage control is performed with a 12-bit VME-
based DAC (Internix (11), Profort PVME-323). The gain differences for each

channel, including the filters, combiners, and splitters, need to be matched by adjusting the gain of the second amplifier to within 1%. The total gain for each channel, which can change over 0 to 40dB, needs to be adjusted by the first variable-gain amplifier depending on the linac operation modes. The signal peak detection is performed with a fast track-and-hold circuit (Comlinear (12), CLC942) with a 12-bit resolution, which comprises a fast diode-bridge switch and driver circuit. The fast combined signal is fed into an ultrafast leading-edge-triggered comparator (Analog Devices, AD96685) which generates the gate signal for the track-and-hold circuit. The timing of the gate signal is adjusted in order to hold the signal peak by using a 2-m-long cable delay and variable digital delay line (JPC (13), EPD7NH) with a 1-ns step. The peak-detected signals are sent to a 12-bit VME-based ADC module (Internix, Profort PVME-303). The module includes eight individual ADCs, and thus, AD conversion for the eight signals can be simultaneously performed with a conversion time of 10 μs. The signal pedestal measurement is done with the beam trigger gate signal. The calibration system includes a bipolar signal generator, a variable-controlled attenuator, and a signal splitter. The calibration system monitors the gain variation and drift of the detection circuit. If the gain variation is measured, a gain correction is made by the on-line VME-based computer.

EXPERIMENTAL RESULTS IN A TEST BENCH

The detection electronics has been tested in a test bench by using a calibration pulse of the bipolar-doublet form. The dynamic range, linearity, circuit signal-to-noise ratio (SNR), and crosstalk between the channels of the detection electronics were precisely measured as a function of the test charges. The pulse height (V) of the calibration pulse was transformed to the test charge (nC/bunch), which corresponds to the beam intensity, by using the measured normalization factor (4.4V/(nC/bunch)) described in the previous section. Figure 5 shows the linearity measurement of the detection electronics as a function of the test charge. The overall dynamic range was restricted to about 35 dB for each channel, which corresponded to the range of the beam intensity (0.2 to 10 nC/bunch). The linearity was better than 1% in the dynamic range. Figure 6 plots the circuit SNR measurement. The circuit SNR (dB) was obtained by measuring the standard deviations applying 100 test pulses. The circuit SNR was about –36 dB at the test charge of 1 nC/bunch and was not very good in the region of less than 1 nC/bunch. We need to tune the fine gain control of the amplifiers and to switch to high-gain operation of the amplifiers in the low intensity region (< 1 nC/bunch). Figure 7 indicates the crosstalk measurement between the channels. The crosstalk was obtained by measuring the pulse heights of a nearest-neighbor channel induced by the signal-input channel. The crosstalk was about –27 dB for a test charge of 1 nC/bunch. This is attributed to complicated mixing and poor isolation on a printed circuit board for analog and digital circuits. The new version of the printed circuit board, taking into account the multi-layered printed circuit, is now in development.

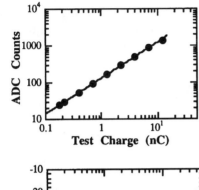

FIGURE 5. Linearity measurement of the detection electronics as a function of the test charge. This graph shows about a 35 dB linear range according to the range of the test charge of 0.2 to 10 nC/bunch.

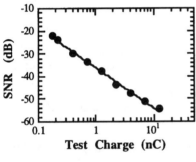

FIGURE 6. SNR measurement of the detection electronics as a function of the test charge. The solid curve shows an eye's guide line.

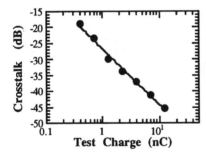

FIGURE 7. Crosstalk measurement of the detection circuit as a function of the test charge. The solid curve shows an eye's guide line. The data shows the ratio of the induced pickup of a nearest-neighbor channel to that of the input channel.

CONCLUSIONS

A prototype detection electronics using a conventional peak-detection technique has been developed in order to reinforce the beam position monitoring system for the KEKB injector linac. Several circuit characteristics were measured in the test bench. The overall dynamic range of the detection electronics was found to be about 35 dB, which corresponded to a range of the beam current of 0.2 to 10 nC/bunch. Good linearity of less than 1% was obtained in the dynamic range. The circuit SNR was about –36 dB for a test charge of 1 nC/bunch and was not very good in the region of less than 1 nC/bunch. The crosstalk was about −27dB for a test charge of 1 nC/bunch. The main reasons for this were attributed to the complicated mixing of analog and digital circuits due to poor isolation on a printed circuit board. An upgraded version of the printed circuit board, taking into account the multi-layered printed circuit, is now being developed. Beam tests of the new detection electronics are now being planned.

REFERENCES

1. S. Kurokawa et al., "Accelerator Design of the KEK B-Factory," KEK Report 90-24 (1991); "KEKB B-Factory Design Report" KEK Report 95-7 (1995).
2. A. Enomoto et al., "Linac Upgrade Plan for the KEK B-Factory," Proc. of the 1993 Particle Accelerator Conf., Washington, D.C., 590-592 (1993).
3. T. Suwada et al., "Development of a Stripline-Type Position Monitor," AIP Conf. Proc. **319**, 334-342 (1993).
4. T. Suwada, "Numerical Calculation of the Eelectromagnetic Coupling Strength," Proc. of the 10th Symposium on Accerelator Science and Technology, Hitachinaka, Japan, 269-271 (1995).
5. KEYCOM Corp., 4-21-8 Kotesashi, Tokorozawa, Saitama 359, Japan.
6. Robert. E. Shafer, "Beam Position Monitoring," AIP Conf. Proc. **212**, 26-58 (1989).
7. R. L. Wigington and N. S. Nahman, "Transient Analysis of Coaxial Cables," Proc. IRE 45, 166-174 (1957).
   T. Shintake, "Pulse Response of Coaxial Cables," Tristan Design Note, TN-86-0017 (1986).
8. Mini-Circuits, P.O. Box 350166, Brooklyn, New York, 11235-0003 USA.
9. Picosecond Pulse Labs., Inc., P.O. Box 44, Boulder, Colorado 80306 USA.
10. Analog Devices, Inc., P.O. Box 9106, Norwood, MA 02062-9106, USA.
11. Internix Inc., 7-4-7 Nishi-Shinjyuku, Shinjyuku-ku, Tokyo 160, Japan.
12. Comlinear Corp., 4800 Wheaton Drive, Fort Collins, CO 80525 USA.
13. JPC CO., LTD., 2-5-5 Saginomiya, Nakano-ku, Tokyo 165, Japan.

The DAΦNE Beam Position Monitors

A. Ghigo, F. Sannibale, M. Serio, C. Vaccarezza

INFN Laboratori Nazionali di Frascati - 00044 Frascati (Roma) - Italy.

Abstract. The beam diagnostics network of DAΦNE, the Frascati Φ-factory, includes more than 110 beam position monitors divided between button monitors and striplines. The shape of the vacuum chamber changes along the accelerator implying several different geometries for these monitors. Moreover, in the two interaction regions of the collider where the electron and positron beams pass into the same chamber, a six-button configuration has been used. A bench calibration of each family of BPMs and striplines is being performed. A polynomial correction function has been derived by fitting the calibration results. An analytical-numerical analysis of the buttons' geometry has been done in order to compare the experimental with the theoretical results.

INTRODUCTION

DAΦNE, the Frascati Φ-factory (1) being built at Laboratori Nazionali di Frascati (LNF), consists of two storage rings and a full energy injector, composed of a linac and a damping ring, for topping-up at 510 MeV. The stored positron and electron beams circulate in opposite directions, intersecting at a horizontal angle of 20 mrad in two interaction points. The first interaction region is dedicated to CP invariance violation experiments while the other is dedicated to hypernuclei experiments.

We present the DAΦNE beam position monitor (BPM) system, the bench calibration set-up, and an analytical analysis method for BPM geometries.

DAΦNE BPM GEOMETRIES

Injector BPM System

The 60-m-long linac has 14 capacitive BPMs. The approximately 140 meters of transfer lines connecting the different parts of the complex are equipped with 23 striplines (50 Ohms, 15 cm length). The 30-m-long damping ring (DAΦNE Accumulator) has four stripline BPMs for first-turn and low-current position measurements and eight button monitors for stored beam position measurements.

BPM System for the Main Rings

Each of the two 100-m-long Main Rings (MRs) is equipped with 32 BPMs (interaction regions excluded). Because of the strong requirement to keep the vacuum chamber impedance as low as possible, the Main Ring BPMs are all of the button type. Their geometries vary according to the vacuum chamber shape. Figure 1 shows some examples.

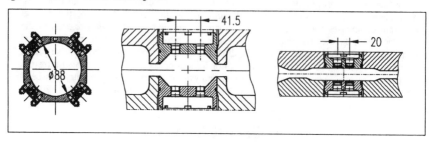

FIGURE 1. From left to right: straight section BPM, dipole BPM, wiggler BPM.

BPM System for the Interaction Regions

In each of the 'day-one' interaction regions (IRs), when the experiment detectors will not be present, five directional striplines and ten button monitors will be installed. A special configuration of monitors with six buttons will allow selective measurement of the position of the electron and positron beams where they are separated in the horizontal plane. Figure 2 shows some of the day-one interaction region BPMs. The final IR configurations will each include six four-button BPMs.

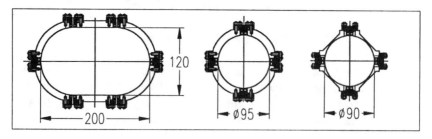

FIGURE 2. Examples of day-one interaction region BPMs.

BPM BENCH CALIBRATION

A measurement bench has been built for calibrating the different families of DAΦNE BPMs. In such a device the beam passage is simulated by a wire stretched inside the BPM. The system is shown schematically in Fig. 3. The rf output of a network analyzer (NA) is sent to a 0.3-mm-diameter silver-plated steel wire placed inside the BPM under measurement. This wire can be moved transversely inside the BPM by two perpendicular stages (Micro Controle UT 100). A resistive matching at the wire output end ensures that no rf power is reflected back from that point, so that it is possible to measure at port B of the NA the rf power that flows into the monitor under measurement. The BPM electrodes are connected to a multiplexer whose output goes into the NA port A. The value of A/B at 736 MHz (twice the Main Rings rf) is read and stored. A Macintosh computer controls all the system parts via GPIB, so that the measurement is completely automatic. The rms value of both the reading error and the repeatability error is less than 2 μm, giving an overall error better than 3 μm.

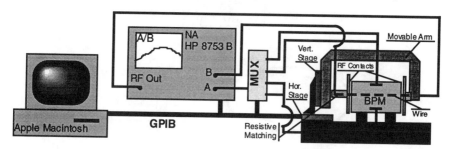

FIGURE 3. Calibration bench schematics.

Figure 4 shows, as an example, the measurements performed on one of the BPMs placed on the wiggler vacuum chamber. The geometry of this monitor is shown in Fig. 1. (For the meaning of the quantities U and V, see next paragraph.)

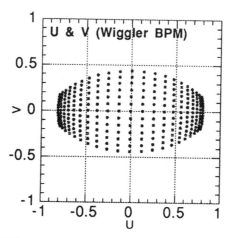

FIGURE 4. U and V measured values for the wiggler BPM.

Beam Position Reconstruction

The voltages of the BPM electrodes can be combined in various ways in order to give a couple of dimensionless quantities from which the position of the beam can be derived. For many of the BPM geometries with four electrodes, a convenient arrangement, shown in Fig. 5, is (2):

$$U = \frac{1}{2}\left(\frac{V_2 - V_3}{V_2 + V_3} + \frac{V_4 - V_1}{V_4 + V_1}\right), \tag{1}$$

$$V = \frac{1}{2}\left(\frac{V_2 - V_3}{V_2 + V_3} - \frac{V_4 - V_1}{V_4 + V_1}\right). \tag{2}$$

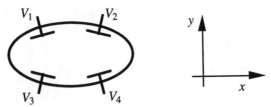

FIGURE 5. Four-electrode BPM nomenclature.

Because of the nonlinear response of such monitors, the experimental data need to be corrected by using a nonlinear fitting function in order to reconstruct the beam position. In the great majority of cases, ours included, a polynomial fitting function is used. Implicit and explicit polynomial fitting functions can be used:

$$x = x_0 + k_x(x,y)U \qquad\qquad y = y_0 + k_y(x,y)V, \tag{3}$$

$$x = x_0 + P_x(U,V) \qquad\qquad y = y_0 + P_y(U,V). \tag{4}$$

where k_x, k_y, P_x, and P_y are polynomial functions while x_0 and y_0 are the BPM offsets.

Because of symmetry considerations that can be applied when the mechanical tolerances of the BPM are small, not all of the polynomial terms are to be used. In the implicit case:

$$k_x(x, y) = \sum_{n=0}^{N/2} \sum_{m=0}^{N/2-n} a_{nm} x^{2n} y^{2m}, \tag{5}$$

$$k_y(x, y) = \sum_{n=0}^{N/2} \sum_{m=0}^{N/2-n} b_{nm} x^{2n} y^{2m}, \tag{6}$$

where N is an even number. In the explicit case:

$$P_x(U,V) = \sum_{n=0}^{(N-1)/2} \sum_{m=0}^{(N-1)/2-n} a_{nm} U^{2n+1} V^{2m}, \tag{7}$$

$$P_y(U,V) = \sum_{n=0}^{(N-1)/2} \sum_{m=0}^{(N-1)/2-n} b_{nm} U^{2m} V^{2n+1}, \tag{8}$$

where N is an odd number.

In our case the implicit functions in Eqs. (5) and (6) give more accurate results. The use of this kind of function implies that a recurring algorithm must be used to derive the beam position from the electrode voltages. Fortunately just a small number of iterations are necessary to achieve the necessary precision, and the consequent delay in obtaining the beam position is quite negligible.

Before applying Eqs. (5) and (6), the data obtained by wire measurements must be 'cleaned' of three different kinds of systematic errors:

- *Tilt error.* Angle θ between the BPM transverse reference frame (defined by the mechanical symmetries) and the wire reference frame.
- *Offset error.* Displacements x_0 and y_0 between the BPM longitudinal center line and the origin of the wire reference frame.
- *Perpendicularity error.* Non-perpendicularity by an angle ψ between the x and y axes of the wire reference frame.

The above errors can be derived by fitting the data in a small area around the BPM center (where the monitor is reasonably linear) by means of the linear functions

$$x = x_0 + AU + BV \qquad\qquad y = y_0 + CU + DV; \tag{9}$$

for small values of θ and ψ, it can be seen that

$$\theta = -C/A \qquad\qquad \psi = -B/D - C/A. \tag{10}$$

These quantities, together with the offset values, can be used to make the reference frame transformation necessary to apply Eqs. (5) and (6).

A FORTRAN routine has been developed in order to make the linear fit, the reference frame transformation, and the implicit polynomial fit. In the case of the wiggler BPM we used $N = 4$. For this monitor Fig. 6 shows the wire position after the systematic errors have been cleaned up and the reconstructed position. The rms fit error is 16 microns for the horizontal plane and 32 microns for the vertical one. The measured U and V values for this BPM are shown in Fig. 4.

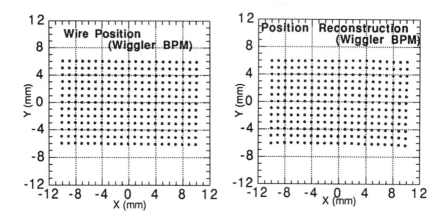

FIGURE 6. Wiggler BPM wire position and reconstructed position.

THEORETICAL ANALYSIS

With the Green function method we can write the general solution for the Dirichlet problem of a potential $\Phi(\mathbf{x},t)$ for an ideal conductive box of rectangular section and periodic boundary conditions along the longitudinal coordinate, where some portions of the surface are at a potential $V_i \neq 0$ while the rest is kept at $\Phi = 0$, and where an internal charge distribution $\rho(\mathbf{x},t)$ is present (3,4):

$$\Phi_{in}(\mathbf{x},t) = \iint \rho(\mathbf{x}',t')G_{in}(\mathbf{x},t;\mathbf{x}',t')d^3x'\,dt'$$

$$-\frac{1}{4\pi}\sum_i \int V_i(t')\oint_S \frac{\partial G_{in}(\mathbf{x},t;\mathbf{x}',t')}{\partial n'_{in}}da'\,dt'. \tag{11}$$

The above expression holds inside the box, and the following holds outside:

$$\Phi_{out}(\mathbf{x},t) = -\frac{1}{4\pi}\sum_i \int V_i(t')\oint_S \Phi(\mathbf{x}',t')\frac{\partial G_{out}(\mathbf{x},t;\mathbf{x}',t')}{\partial n'_{out}}da'\,dt'. \tag{12}$$

Here the $G_{in,out}(\mathbf{x}',t';\mathbf{x},t)$ are the solutions of

$$\left(\nabla_x^2 - \frac{1}{c^2}\frac{\partial^2}{\partial t^2}\right)G = -4\pi\delta(\mathbf{x}-\mathbf{x}')\delta(t-t') \tag{13}$$

with periodic boundary conditions in the longitudinal coordinate z and Dirichlet boundary conditions on the box surface in the (x,y) plane.

Considering the Fourier transform of the potential and the charge distribution

$$\Phi(\mathbf{x},\omega) = \frac{1}{2\pi}\int_{-\infty}^{\infty}\Phi(\mathbf{x},t)e^{i\omega t}\,dt,$$

$$\rho(x,\omega) = \frac{1}{2\pi}\int_{-\infty}^{\infty}\rho(\mathbf{x},t)\,e^{i\omega t}\,dt, \tag{14}$$

and knowing that, for each isolated portion of the box surface, the induced charge q_i satisfies

$$q_i^{out} = -\frac{1}{8\pi}\int_{S_i}\frac{\partial\Phi^{out}}{\partial n_{out}}da = -q_i^{in} = \frac{1}{8\pi}\int_{S_i}\frac{\partial\Phi^{in}}{\partial n_{in}}da, \tag{15}$$

in the case of a single button we have

$$V(\omega) = \frac{\displaystyle\int_S da\left\{\frac{\partial}{\partial n}\left[\int_V d^3x'\,\rho_w(\mathbf{x}')G_\omega^{in}(x,x',\omega)\right]\right\}}{\displaystyle\frac{1}{2\pi}\int_S da\int_S da'\frac{\partial^2 G_\omega^{in}(x,x',\omega)}{\partial n\,\partial n'}}, \tag{16}$$

where $\omega = \dfrac{k2\pi c}{L}$, c = light speed, L = machine length, k is an integer, and $\partial G_\omega^{in}/\partial n'_{in} = -\partial G_\omega^{out}/\partial n'_{out}$.

In general, with N electrodes we have

$$q_i = \sum_{j=1}^{N}C_{ij}V_j = f_{i0} + \sum_{j=1}^{N}f_{ij}V_j,\ (i=1,\ldots N) \tag{17}$$

where, from Eq. (15)

$$f_{i0} = f(\mathbf{x}_0^{beam}, \mathbf{x}_i^{BPM}),$$

$$f_{ij} = f(\mathbf{x}_i^{BPM}),$$

(18)

and C is the capacity matrix. Note that, for the symmetry of the Green function described above, one can prove that $C_{ij} = -f_{ij}$.

Equation (17) has been solved explicitly in the case of a four-electrode BPM in the DAΦNE wiggler configuration. The results obtained are reported in Fig. 7, where they are also compared with the results of the wire calibration, indicating fair agreement.

The case of the six-button geometry is presently under study with the aim to provide the most convenient linear combination of the button potentials which gives a representation of one beam position reasonably decoupled from the other in a region where the two beams are separated horizontally.

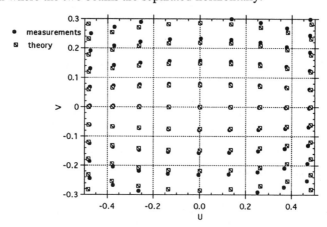

FIGURE 7. Wiggler BPM: Comparison between analytical and experimental results.

ACKNOWLEDGMENTS

The authors want to express thanks for C. Marchetti's efforts in optimizing the mechanical performances of the calibration bench.

REFERENCES

1. The DAΦNE Project Team, "Overview of DAΦNE, the Frascati Φ-factory," Proceedings of the VI International Conference on High Energy Accelerators, Hamburg, Germany, 20-24 July 1992, 115-117 (1992).
2. Borer J., Bovet C., "Computer response of four pick-up buttons in an elliptical vacuum chamber," LEP Note 461, July 28, 1983.
3. Jackson J. D., *Classical Electrodynamics*, New York: Wiley & Sons, 1975, Ch. 1, pp. 40-41; Ch. 6, pp. 226-235.
4. Cuperus J. H., "Monitoring of particles beams at high frequencies," *Nucl. Instrum. Methods* **145**, 219-231 (1977).

Beam Position Monitor System for PEP-II[*]

G. Roberto Aiello, Ronald G. Johnson, Donald J. Martin[+],
Mark R. Mills[#], Jeff J. Olsen, and Stephen R. Smith

SLAC, P.O. Box 4349, Stanford, CA 94309-4349
[+]*Now at Texas Instruments, Dallas, Texas.*
[#]*Now at Unisite, Richardson, Texas.*

Abstract. The beam position monitor (BPM) system for PEP-II, the B-Factory under construction at SLAC, is described in this paper. The system must measure closed orbit for a 3-A multibunch beam and turn-by-turn position for a low-current single bunch injected in a 200-ns gap in the multibunch beam. A system that combines broadband and narrowband capabilities and provides data at high bandwidth was designed. It includes a filter-isolator box (FIB) that selects a harmonic of the bunch spacing (952 MHz) and absorbs the other frequency components; a CAMAC-based wideband I&Q demodulator, ADC, and signal processor that provides beam position information to the control system; and a calibrator that must work even in presence of beam, correcting for electronic measurement errors. This paper describes the system requirements, the electronics design, and the laboratory tests.

INTRODUCTION

The beam position monitor (BPM) system for the B-Factory (1) under construction at Stanford Linear Accelerator Center (SLAC) is required to measure both multibunch beam on a turn-by-turn basis at 134 kHz and single-bunch beam injected in a 200-ns gap. The gap is present for ion clearing and the single bunch injected in it is used for injection tuning. The electronics must work both in narrowband mode, for multibunch, and in wideband mode, for single bunch. The single-bunch requirement makes the use of AM/PM technique (2) or synchronous detection (3) impractical and difficult, and the required accuracy makes the use of log-ratio technique (4) not feasible. The electronics must also have a very low reflection coefficient because the BPM button is a highly reflective signal source, and the measurement accuracy of the pilot bunch would be impaired by the multiple reflections through the cable. The good accuracy and wide dynamic range require that the system calibrate for electronic measurement errors such as channel offset and gain and error vector. The calibration must be performed in the presence of beam, given the 'factory' characteristics at which the accelerator is designed to perform. Both the High Energy Ring (HER) and Low Energy Ring (LER) are located in the same tunnel. The circumference is 2200 m and their energies are,

---

*Work supported by U. S. Department of Energy contract DE-AC03-76F00515.

respectively, 9 GeV and 3.1 GeV. Relevant machine parameters are given in Table 1.

TABLE 1. PEP-II Relevant Parameters

Parameter	Value
rf frequency	476 MHz
Number of bunches	1658
Total beam current (HER)	1 A
Total beam current (LER)	2.1 A
Revolution frequency	134 kHz
Bunch spacing	238 MHz

SYSTEM DESCRIPTION

The main issues that drive the design and that we present in this paper are:

1. maintain good accuracy over a wide dynamic range
2. measure a low-current single bunch injected in a gap in the circulating beam
3. combine the signals from the electrodes to realize x-only and y-only BPMs
4. multiplex HER and LER BPMs in the same electronic module.

These problems were addressed in several ways, in order to design a system based on commercial components. The system's block diagram is shown in Fig. 1. In most cases the signals from the four electrodes are filtered and combined by the filter-isolator box (FIB), located within a few feet from the buttons, which creates the x-only and y-only BPMs. Signals from a small number (25) of BPMs are not combined by the FIB, but only filtered, in order to keep both x and y information at the same location for machine study. The signals from the FIB are processed by the Ring I&Q Module (RInQ) and made available through the CAMAC interface.

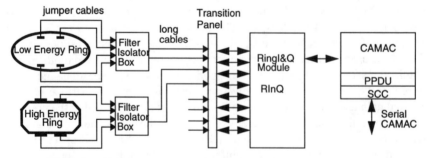

FIGURE 1. BPM system block diagram (only half the channels are shown).

The overall requirements are shown in Table 2. The resolution is the pulse-to-pulse reproducibility, assuming no beam motion. This parameter is not limited by the thermal noise, but by the ADC quantization error. The required accuracy (i.e., the offset between the measured position and the actual center) means that the maximum systematic error must be smaller than 0.25 mm rms at full current. This corresponds to 0.2 dB in the worst case, when the sensitivity is 0.9 dB/mm, and necessitates the use of a calibrator to compensate for electronics errors and drifts.

TABLE 2. Requirements and Expected Resolution

Beam Current	V (rms)	Power	Required resolution	Expected resolution	Accuracy
10^{11}, multibunch	3.5 V	24 dBm	15 µm	0.2 µm	250 µm
5×10^8, single bunch	1.3 mV	−45 dBm	1 mm	0.5 mm	1 mm

The measurement accuracy for a single bunch is limited by the reflection of the multibunch signal through the cable. The RInQ was designed to provide a good load, VSWR better than 1.2. An isolator was included in the FIB to reduce the reflection from the electronics to the button. The reflection contributions add up to different values for LER and HER, because of different cable length leading to different cable insertion loss with round trip time equal to the gap length. The total accuracy is therefore a function of the ratio between injected and circulating beam. The worst case is for LER, where the accuracy is 220 µm if single bunch and multibunch have the same current, or 2.2 mm if the single-bunch current is 10% of the circulating beam.

The same electronics processes both circulating beam and single-bunch beam. A bandpass filter at the beam fourth bunch frequency (952 MHz) with Q = 50 suppresses the out-of-band components for the circulating beam and produces a ringing when hit by the single-bunch impulse. This process reduces the power available to the electronics in single bunch by an extra 21 dB and limits the resolution due to the signal-to-noise ratio (5) as shown in Table 2, given an electronics noise figure of 16 dB and maximum cable attenuation of 7.2 dB.

BUTTON CHARACTERISTICS

There are three types of vacuum chambers for the HER and LER and consequently three types of BPM buttons that differ somewhat mechanically but have nearly identical electrical performance. The buttons have a 1.5-cm diameter, a matched impedance of 50 Ω and a measured capacitance of 2.6 pF. The three types of vacuum chambers and the effective radii for the buttons are: an octagonal copper chamber for the HER arcs (30.7 mm), a round stainless-steel chamber for the HER straight sections (44.5 mm), and an oval aluminum chamber for the LER (31.0 mm).

The buttons were designed to produce acceptable signals and to avoid significant contribution to long-range wake fields. Extensive simulation using MAFIA (6) to predict signal levels and the contribution to ring impedance were performed in this design effort.

The accuracy requirement on the BPM system makes it necessary to calibrate each BPM. Calibration test stands have been designed and built to measure the electrical center of each BPM to an accuracy of 100 μm. The response of the BPMs in the calibration test stands has been calculated by conformal mapping. The calculated and measured response differ by less than 2%. At this time approximately 15% of the BPMs for the HER have been calibrated. Production of LER chambers has not started.

FILTER-ISOLATOR BOX

The BPM filter-isolator box is a passive microwave instrumentation network which performs a filtering, combining, and impedance matching function to broadband, high-frequency signal energy. The input energy sources are 30-ps impulses with peak amplitudes exceeding 500 V.

TABLE 3. Filter-Isolator Box Requirements

Parameter	Requirement	Performance (typical)
Center frequency	952 MHz (nominal)	942.6 MHz
Bandwidth @ 3 dB	50 MHz < BW < 238 MHz	150 MHz
Insertion loss @ 952 MHz	< 2 dB	1.62 dB
Power handling	> 50 W	ok
Output return loss	< −19 dB	−21.5 dB
Input return loss	< −6 dB	−17.9 dB
Bias voltage	350 V	ok

The FIB consists of constant-resistance bandpass filters, low-pass filters, hybrid 2-way combiners, ferrite isolators, and DC bias circuitry. A photograph of a FIB assembly is shown in Fig. 2, and a functional diagram is shown in Fig. 3.

FIGURE 2. FIB assembly.

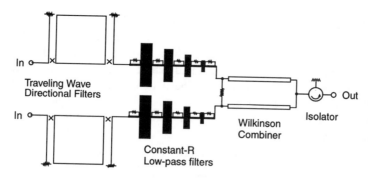

FIGURE 3. FIB block diagram.

I&Q PROCESSOR

The signals from the FIB, filtered and combined, are processed by the Ring I&Q Module (RInQ), that is based on baseband conversion using I&Q demodulators, as shown in Fig. 4. The RInQ also includes the calibrator, described in the following paragraph. The in-phase and quadrature (I&Q) components are demodulated, digitized, and the position is calculated (7).

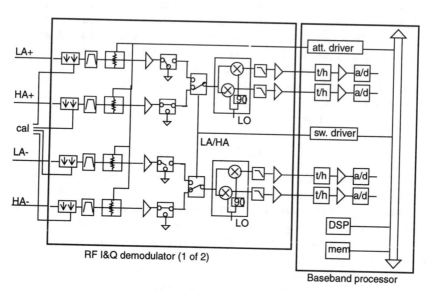

FIGURE 4. I&Q processor block diagram.

The 10-dB directional coupler accepts a signal from the FIB and from the calibrator through the coupling port. The 952-MHz bandpass filter selects the processing frequency in multibunch and generates an rf burst in single bunch. The programmable attenuator extends the dynamic range and the amplifier optimizes the signal to the I&Q demodulator. The switches multiplex HER and LER signals to the same detector. Track-and-hold and 14-bit ADCs acquire the signals. The TI320C31 DSP and dual-port memory preprocess the information and make it available through CAMAC.

TABLE 4. RF I&Q Demodulator Requirements

Item	Value
rf carrier	952 MHz
Maximum input level	+28 dBm
Minimum input level	−51 dBm
System bandwidth	10 MHz
Sampling rate	134 ksps
Linearity	0.1 dB
Channel matching	0.1 dB
Channel-to-channel isolation	86 dB

The multiplexer's 86-dB isolation requirement is achieved by two SPST switches followed by a high-isolation 2PST switch. This requirement comes from the fact that HER and LER BPMs are multiplexed to the same channel, and it may be necessary to measure a low-current single bunch from one ring when a high-current multibunch is present in the other ring.

Both digital and rf sections are laid out on an eight-layer CAMAC board. Most of the board is FR4, but the rf section is built on a separate teflon layer that partially covers the top layer.

CALIBRATOR

In order to achieve the required accuracy as shown in Table 1, several sources of errors must be considered: channel gain mismatch, channel offset, and amplitude and quadrature phase unbalance. Channel offset mismatch generates an apparent amplitude, while channel gain mismatch produces an error for beam off center or for different beam currents. These two errors are common to all systems that use linear processing, while phase and amplitude unbalance are typical of the data acquisition processing method chosen for PEP-II. The system must be able to compensate for these errors as well. The requirements are shown in Table 5.

TABLE 5. Calibrator Requirements

Item	Value
Channel offset mismatch	0.03 dB
Amplitude unbalance	0.08 dB
Quadrature phase unbalance	1 deg
Channel gain mismatch	0.08 dB
Total error	0.12 dB

Offset mismatch is measured by turning the calibrator off and measuring the apparent beam position. Gain mismatch is measured by turning on the calibrator, acquiring the signal, and fitting the channel transfer function to the best line. Phase and amplitude unbalance are measured by setting the calibrator frequency to a few kHz (i.e., 15 kHz) off the local oscillator frequency, digitizing at the maximum sampling rate (134 kHz), and then running the amplitude and phase calculation algorithm. This is a fixed-frequency curve fitting algorithm (8) that solves a system of linear equations for amplitude, phase, and offset. The algorithm is expected to provide accuracy better than 0.5 deg and 0.05 dB (9).

Since the calibrator must be able to perform when the beam is present, it works at a frequency different from the beam fundamental. When it is turned on, the local oscillator that provides the reference frequency to the rf processor for the baseband conversion is also detuned from the beam frequency. The specifications are shown in Table 6. The calibration is always performed during the gap in the multibunch beam to achieve more isolation. The residual beam frequency component is expected to be at –25 dBm in the worst case (30 dBm maximum power, 55 dB rf processor return loss), while the calibration power is expected to be 21 dBm.

TABLE 6. Calibrator Specifications

Item	Value
Frequency sweep	932 - 972 MHz in 300-Hz steps
Maximum phase jitter	1 deg at 100-kHz samples for 100 samples
Phase noise	–61 dBc/Hz at 1 kHz
Spurious	–61 dBc at 1-kHz to 100-kHz offset
Amplitude step	0 - 60 dB in 4-dB steps
Maximum power at the channel inputs	+21 dBm

The wide-band mode is calibrated by taking a series of measurements over the working bandwidth. Both the calibrator and the local oscillator are based on the DDS-driven PLL technique (10), as shown in Fig. 5. The reasons to choose this solution are: broad output bandwidth, small step sizes, low phase noise and spurious performance, and minimal complexity and cost. The DDS clock is derived by a prescaler working with a 119-MHz signal locked to machine reference as an input. It is programmed to provide an output frequency from 9.32 MHz to 9.72 MHz in

5-Hz steps. The PLL final output frequency is 932-972 MHz, which is given by the feedback loop divider, set to 100. The digital programmable attenuator allows the power to be swept in 4-dB steps over the 60-dB dynamic range for testing linearity. The medium power linear amplifier with turn off capability provides the required power to the rf processor inputs and guarantees the isolation during normal mode of operation. Switches and power splitter distribute the signal to the eight rf processor inputs.

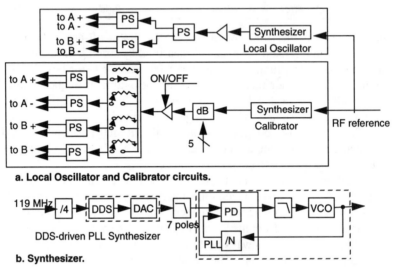

a. Local Oscillator and Calibrator circuits.

b. Synthesizer.

FIGURE 5. Calibrator circuit block diagram.

CONCLUSIONS

We have described the Beam Position Monitor System for PEP-II, the SLAC B-Factory. The system must be able to measure multibunch beam up to 10^{11} particles per bunch on a turn-by-turn basis and single-bunch beam as low as 5×10^8 for injection tuning. The single bunch is injected in a gap of the circulating beam. The system is designed to measure both multibunch beam and single-bunch beam injected in a gap.

The FIB is in production, the RInQ prototype is in house, and RInQ fabrication is scheduled to begin in September 1996, with installation in December 1996.

ACKNOWLEDGEMENTS

The authors wish to thank Linda Hendrickson, Alan Cheilek, Mike Zelazny, and Tony Gromme for their help with the software, Bob Noriega and Brooks Collins for their help in the laboratory, Vern Smith for FIB installation and cabling, and Tom Himel, Ray Larsen, and Alan Fisher for useful discussions.

REFERENCES

1. PEP-II, An Asymmetric B Factory, Conceptual Design Report, SLAC-418, June 1993.
2. Jachim, S.P., Webber, R.C., Shafer, R.E., "RF Beam Position Measurement for Fermilab Tevatron," *IEEE Trans. Nucl. Sci.* **28**, 2323 (1981).
3. Hinkson, J., "Advanced Light Source Beam Position Monitor," AIP Conf. Proc. **252**, 21-42 (1991).
4. Aiello, G.R., Mills, M.R., "Log-ratio Technique for Beam Position Monitor Systems," *Nucl. Instrum. Methods A* **346**, 426-432 (1994).
5. Shafer, R.E., Beam Position Monitoring, *AIP Conf. Proc.* **212**, 26-58 (1989).
6. The MAFIA Collaboration, *AIP Conf. Proc.* **297**, 291, (1993).
7. Smith, S.R., private communication.
8. IEEE Std. 1057, Digitizing Waveform Recorders (1989).
9. Aiello, G.R., "A Digital Approach for Phase Measurement Applied to Delta-t Tuneup Procedure," Proc. of the 1993 Particle Accelerator Conference, 2367-2369 (1993).
10. Application Note AN2334-4, QUALCOMM Inc., San Diego, California 92121-2779.

Quadrupole Shunt Experiments at SPEAR*

W.J. Corbett, R.O. Hettel, H.-D. Nuhn

Stanford Synchrotron Radiation Laboratory
Stanford Linear Accelerator Center
Stanford University, California 94309

Abstract. As part of a program to align and stabilize the SPEAR storage ring, a switchable shunt resistor was installed on each quadrupole to bypass a small percentage of the magnet current. The impact of a quadrupole shunt is to move the electron beam orbit in proportion to the off-axis beam position at the quadrupole and to shift the betatron tune. Initially, quadrupole shunts in SPEAR were used to position the electron beam in the center of the quadrupoles. This provided readback offsets for nearby beam position monitors and helped to steer the photon beams with low-amplitude corrector currents. The shunt-induced tune shift measurements were then processed in MAD to derive a lattice model.

INTRODUCTION

In order to improve electron beam stability in SPEAR, the storage ring and photon beamlines were re-aligned in 1995 (1). After the alignment, however, the electronic offsets of the beam position monitors (BPMs) were still unknown, and beam steering remained difficult. To locate the BPM centers, we made use of an old but effective beam steering method that is now used widely and installed a shunt circuit on each quadrupole (2-7). Since a shunted quadrupole deflects the beam when the orbit passes off-axis through it, the shunt can be used to center the beam in the quadrupole.

The steering procedure involves a few simple steps. In the first step, we center the beam in quadrupoles located at each end of an insertion device straight section (see Fig. 1). With the beam centered, the offset values can be determined for BPMs located in the straight section. In some cases, the BPM offsets were up to 6mm. These values were entered into the database for use by the orbit correction program. By using the BPMs as steering fiducials, the beam could then be centered at any time without re-activating the shunts. The shunt-induced tune shifts were also numerically processed to derive a lattice model. As a result, initial machine commissioning and orbit adjustments after each fill have become routine. This paper reports on three aspects of the quadrupole shunt work at SPEAR: shunt circuit design, beam centering, and lattice diagnostics.

---

*Work supported in part by U.S. Department of Energy Contract DE-AC03-76SF00515 and Office of Basic Energy Sciences, Division of Chemical Sciences.

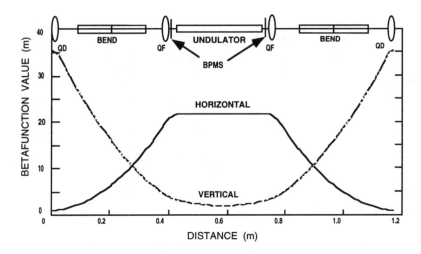

FIGURE 1. Betatron functions in SPEAR insertion device cells.

SHUNT CIRCUIT DESIGN

The design goal for the shunt circuits in SPEAR was to produce the maximum orbit shift and a betatron tune shift of order 0.01, but with the shunt resistor power load not to exceed 250W. In addition, high tolerance resistors were needed for accurate β-function measurements, and the driving circuit was needed to switch the shunts at up to a 10Hz rate. These criteria led to an optically isolated FET switch design with 250W, 1% power resistors mounted on water-cooled copper plates. Depending on the quadrupole family, the resistor impedances range from 0.1-0.3Ω, and they bypass 1-3% of the supply current to the individual quadrupoles. A multiplexer activates each FET switch separately in either AC or DC mode and reads back the bypass current. A schematic circuit diagram for a shunt is shown in Fig. 2.

To estimate the resolution for centering the beam in a quadrupole, one can take the orbit kick θ from each shunt as proportional to the change in focusing strength and proportional to the beam offset in the quadrupole,

$$\theta \propto c\,(\Delta k\,L)\,x_q, \tag{1}$$

where c is a proportionality constant, $\Delta k\ L(m^{-1})$ is the change in the integrated quadrupole strength, and x_q is the beam offset relative to the magnetic center of the quadrupole. The closed orbit perturbation Δx (suppressing the phase factor) is then

$$\Delta x \propto \theta \sqrt{\beta_k \beta_o}, \tag{2}$$

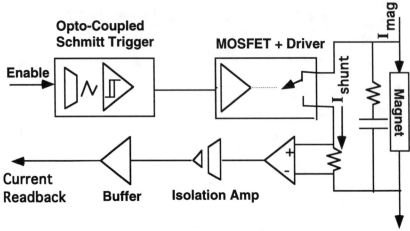

FIGURE 2. Circuit schematic for quadrupole shunt in SPEAR.

with β_k and β_o the β-function values evaluated at the kick and observation points, respectively. Substituting Eq. (1) into Eq. (2) yields the orbit offset x_q that can be detected for a specific BPM processor resolution,

$$x_q \propto \frac{\Delta x}{c \Delta k L \sqrt{\beta_k \beta_o}}. \tag{3}$$

Equation (3) shows the factors that contribute to high resolution beam centering measurements: good BPM processor resolution (small Δx), a large shunt-induced value for Δk L, and large β-functions. The fractional betatron tunes enter through the proportionality constant c. The ratio $x_q/\Delta x$ for BPMs in the beamline regions (where most SPEAR BPMs are located) and in the lattice matching cells (where β-functions are highest) are listed in Table 1 for each quadrupole family. Also listed are the sensitivities for the largest beam motion at one of the nine photon beam monitors, depending on phase.

The relative ability to resolve the electron beam offset in the different quadrupole types is found by multiplying the value listed in Table 1 by the

TABLE 1. Ratios of (Orbit Offset)/(Orbit Shift) for Quadrupole Shunts in SPEAR

	x(arc)	y(arc)	x(max)	y(max)	y(photon)
Q2	10	90	6	21	7.4
Q1	52	40	31	9	3.3
QFA	8	30	5	7	2.5
QDA	47	32	28	7	2.5
QFB	8	55	5	13	4.5
QF	8	56	5	13	4.6
QD	46	30	27	7	2.5
(arbitrary units)					

resolution of the BPM processor. The resolution for horizontally centering in a QF quadrupole, for example, is about a factor of 3 higher than the vertical resolution. As a result, we often center the beam vertically in the QD quadrupoles (higher β-function) to obtain the vertical BPM offsets in the straight sections.

The resolution is also better at the photon BPMs because the effective β-function values at these locations are typically a few hundred meters. The photon BPMs also have the advantage that the position resolution is about 1 μm, and the signal is available continuously.

BEAM CENTERING (DC)

In general, algorithms exist for steering the beam through the center of many quadrupoles simultaneously (6). These algorithms require a matrix of derivatives

$$S_{ij} = \frac{\Delta(\text{orbit shift from shunt})_i}{\Delta(\text{corrector})_j} \tag{4}$$

evaluated for each quadrupole shunt i and each corrector j. To correct a particular electron beam orbit, one measures the orbit shift induced by each shunt and multiplies by the inverse matrix S^{-1} to calculate the corrector pattern that best centers the beam in the quadrupoles (6). This matrix inversion technique, although elegant and global in extent, requires either a relatively accurate lattice model or a lengthy data acquisition process to generate the S-matrix.

Alternatively, our approach has been to find BPM offsets once the beam is centered in the quadrupoles at the ends of a straight section. With this technique, we minimize errors incurred if the orbit passes through a quadrupole at an angle. The sources of BPM calibration error are reduced to quadrupole alignment errors, the discrepancy between the mechanical and magnetic centers of a quadrupole, and the BPM processor resolution.

Two different DC shunt methods have been used at SPEAR to center the beam in quadrupoles. The first method is to sweep the beam through the quadrupole, and record both the shunt-induced orbit perturbation and absolute orbit as a function of beam position. As shown in Fig. 3(a), the horizontal center of a QF quadrupole can be determined to within about 50 μm. Although the sweep method yields good resolution for a single quadrupole, it is difficult to center the beam in two quadrupoles simultaneously with this method.

The second method is a derivative-based algorithm outlined in Table 2. Similar to the matrix inversion technique, we measure the effect of a closed bump or a single corrector on the shunt-induced orbit shift. The *change* in the orbit shift (numerator of the derivative in step IV of Table 2) can include any figure of merit: for example, rms orbit shift, peak orbit shift, or the entire orbit shift.

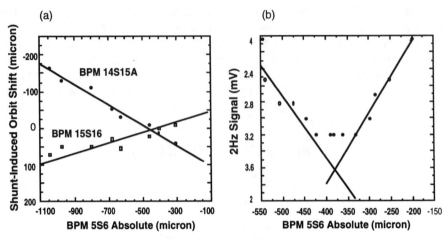

FIGURE 3. Orbit perturbation as a function of beam position read back at BPM.

TABLE 2. Algorithm for Centering Beam in SPEAR Quadrupoles

I.	Measure orbit shift from shunt.
II.	Apply closed bump over quadrupole.
III.	Re-measure orbit shift from shunt.
IV.	Form derivative $\dfrac{(\text{step III - step I})}{\text{step II}}$.
V.	Calculate requisite bump amplitude, apply bump to center beam.

We typically form the inner product of the orbit shift induced by the shunt with the vector formed for the numerator of the derivative computed in step IV. Appropriately normalized, this projection of the orbit shift yields the bump amplitude needed to center the beam. Repeated applications of the orbit adjustment servos the beam into position. Although this method is not as accurate as the sweep method, it efficiently centers the beam in two quadrupoles simultaneously.

BEAM CENTERING (AC)

Centering the beam vertically in the QF quadrupoles can be difficult because the QF magnets have small vertical β-functions. To improve resolution, the shunts are driven with a square wave of up to 10Hz (AC) and the motion synchronously detected at the photon beam BPMs. Similar examples of lock-in detection can be found at LEP (5).

As shown in Fig. 3(b), we can easily position the orbit in a quadrupole to minimize the signal power at the excitation frequency. After centering the beam with the AC technique, however, we found that the application of the shunt in the

DC mode can cause a detectable orbit shift. The discrepancy between the two techniques can give a different value for the beam position at the quadrupole center by up to 500 μm.

The source of the different measurements was traced back to the impact of the load modulation on the power supply output. This effect is shown in Fig. 4, where we plot beam position as a function of time. Each time the FET shunt switch changes state, the load modulation makes the power supply regulator overshoot before it settles to the steady state. As a result, the excitation frequency can be detected even when the beam is centered in the quadrupole. The minimum of the spectral line amplitude occurs when the AC component of the shunt-induced orbit shift best cancels the AC component of the power supply current spikes.

Beam Position Signal

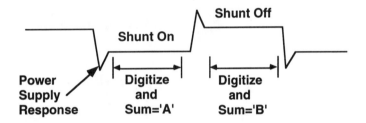

FIGURE 4. Impact of supply regulation on beam position signal.

To suppress the current spike effect, the analysis can be made on the component of the signal in the interval between spikes. Referring to Fig. 4, the beam offset is proportional to the difference between the signal evaluated during sample interval 'A' (shunt on) and during interval 'B' (shunt off). By processing on these components of the signal, the beam can be accurately centered in the quadrupoles.

LATTICE DIAGNOSTICS

Individual quadrupole magnet shunts allow the measurement of the β-functions $<\beta>$ in each plane, averaged across the length of the quadrupole, i.e.,

$$<\beta> = \frac{4\pi\Delta\nu}{\Delta kL}. \tag{5}$$

Here, $\Delta\nu$ is the observed change in betatron tune caused by the change in the integrated quadrupole strength, $\Delta k\,L$, due to the shunt activation.

In SPEAR, the shunts allow the measurement of β-functions at 50 different locations in each plane around ring. This information is sufficient to perform a numerical analysis to find the sources of deviation between the design lattice and the actual lattice; it also allows calibration of the ratio $\Delta k/\Delta I$ for each of the seven quadrupole families. Using the MAD accelerator design code (8), the seven quadrupole strengths, k, were fitted to minimize the rms difference between the measured and model β-functions and the measured and model betatron tunes.

The fitting procedure was applied to the operational configuration as well as to a bare lattice, i.e., a lattice without insertion devices. The MAD model includes the measured quadrupole positions as obtained from a global magnet survey. Before the shunt-dependent tune shift data was collected, the electron orbit was centered in the shunted quadrupole. Otherwise the shunt would shift the global closed orbit and nonlinear elements such as the sextupoles would add errors to the measurement.

For the MAD fitting routine, the measured β-function values are converted from quadrupole average $<\beta>$ to quadrupole center values β_c via

$$\frac{<\beta>}{\beta_c} = \frac{1}{L\beta_c} \int_{-L/2}^{L/2} (R_{11}(s)\beta_c - 2R_{11}(s)R_{12}(s)\alpha_c + R_{12}(s)^2\gamma_c)ds, \qquad (6)$$

where $\alpha_c = \dfrac{-1}{2}\beta'_{(s=c)}$, $\gamma_c = \dfrac{\alpha_c^2+1}{\beta_c}$, and R_{ij} are elements of the quadrupole transfer matrix. The evaluation of the integral gives

$$\frac{<\beta>}{\beta_c} = \frac{1}{2\sqrt{kL}}\left[\left(\sqrt{k}L + \sin\left(\sqrt{k}L\right)\right) + \frac{(\alpha_c^2+1)}{k\beta_c^2}\left(\sqrt{k}L - \sin\left(\sqrt{k}L\right)\right)\right] \qquad (7)$$

for the focusing plane, and $\sin(\sqrt{k}\,L)$ becomes $\sinh(\sqrt{k}\,L)$ with the second term changing sign for the defocusing plane. The second term in the bracket is much smaller than the first term, and the values for α_c and β_c can be taken as the initial model values in very good approximation.

The strength-to-current ratio, i.e., the ratio of the average quadrupole strength to its excitation current, is not well known in SPEAR. The shunt-based analysis allows us to calibrate these factors, and provides information about individual quadrupoles within the families. To determine the strength-to-current ratios, a self-consistent iterative method is used which re-calculates the average β-function from the measured shunt current and from the observed tune shifts. The fitting is done iteratively using β-function data based on the strength-to-current ratios from the previous fit, and converges quickly.

Using the updated calibration factors, new quadrupole excitation currents can be predicted to achieve the desired β-functions in the machine. The method was applied in several iterations to significantly reduce the difference between the measured and predicted β-functions. Figures 5a and 5b show the result of a fit (solid line) to the measured β-functions (square dots) for SPEAR. The error bars are based on statistical analysis of errors incurred in the data acquisition process. The residual deviations are due to variations within the quadrupole families, most notably the SPEAR QD quadrupoles which have been mechanically modified (part of the iron core was removed) to provide space for photon beam exit ports. These modifications reduced the on-axis gradient by about 1.2-1.5%. This effect has not yet been considered in the shunt data analysis.

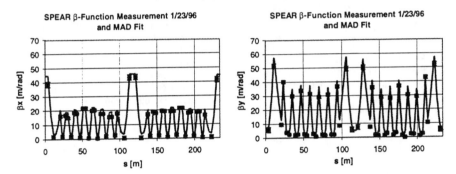

FIGURE 5. Horizontal (a) and vertical (b) β-function measurements in SPEAR.

SUMMARY

Following a re-alignment of SPEAR, we were able to use quadrupole shunts to measure horizontal BPM offset values in the straight sections. These measurements quickly led to an electron beam orbit with steered photon beams. Further application of the shunts improved the orbit in both planes throughout the storage ring. To improve resolution of beam position in quadrupoles, the shunts are modulated in an AC mode. The AC measurement system will be incorporated into a fast algorithm to automatically servo the beam position to the quadrupole center. Finally, shunt measurements to identify and correct modulations in the β-functions have proven successful. In the future, we plan to analyze the BPM readings to determine and correct individual quadrupole misalignments and to identify the dispersion component of the orbit.

ACKNOWLEDGMENTS

The authors would like to express their appreciation for the work of SSRL staff members P. Bousinna, R. Boyce, D. Ernst, R. Garrett, C. Haggart, S. Howry (SLC), L. Johnson, G. Johnson, G. Kerr, J. Montgomery, D. Mostowfi, and R.Ortiz, and for conversations with P. Röjsel of MAXLab.

REFERENCES

1. Corbett, W.J., "Beam Stabilization at SPEAR," the 1995 Synchrotron Radiation Instrumentation Conference, Argonne, Ill., Oct. 17-22, 1995, *Rev. Sci. Instrum.* **67** (9), September 1996.
2. Shunts were used at CEA in the 1960s. Private communication, A. Hofmann, H. Winick.
3. Rice, D., *IEEE Trans. Nuc. Sci.* **NS-203** (4), August 1983.
4. Barnet, I., et al., Proceedings of the Fourth International Workshop on Accelerator Alignment, IWAA95, KEK Proceedings 95-12, 1996.
5. Herb, S. et al, "A new technique to center the LEP beam in a quadrupole", in *Proc. 1st European Workshop on Beam Diagnostics and Instrumentation for Particle Accelerators*, Montreux, Switzerland, May 3-5, 1993, 120-125 (1993).
6. Röjsel, P., *Nucl. Instrum. Methods A* **343**, 374-382 (1994).
7. Portmann, G., Robin, D., Schachinger, L., "Automated Beam Based Alignment of the ALS Quadrupoles," Proc. of the 1995 Particle Accelerator Conference and Int'l Conference on High-Energy Accelerators, May 1-5, 1995, Dallas, TX, 2693-2695 (1996).
8. Grote, H., Iselin, F.C., *"The MAD Program (Methodical Accelerator Design),"* User's Reference Manual, CERN/SL/90-13, (AP), 1990.

Initial Test of a Bunch Feedback System with a Two-Tap FIR Filter Board

M. Tobiyama, E. Kikutani, T. Obina, Y. Minagawa and T. Kasuga

National Laboratory for High Energy Physics (KEK), 1-1 Oho, Tsukuba 305, Japan

Abstract.   Initial beam test of the KEKB bunch-by-bunch feedback system prototypes has been performed in the TRISTAN-AR on the longitudinal plane. A simple two-tap finite impulse response (FIR) filter system consisting of hardware logic realizes the function of the phase shift by 90°, the suppression of static components, and the delay of up to a hundred turns. With the prototype filter board and a longitudinal kicker, the feedback loop has been closed successfully. The shunt impedance of the kicker was estimated from the excitation amplitude and measured damping. The damping time of the system has been measured by using the single-bunch Robinson instability. The feedback system stabilized the coupled-bunch instability completely under 8-bunch operation.

INTRODUCTION

The rings of KEKB are designed to accumulate many bunches with huge beam current, which may cause many strong coupled-bunch instabilities in both the transverse and longitudinal planes. Even taking special care for the reduction of the sources of the instabilities, some dangerous impedance may remain high. Studies on the acceleration cavities predict that some modes have growth times on the order of a few ms in the transverse plane and a few 10's of ms in the longitudinal plane in the worst case. Therefore, the method used to analyze and suppress the instabilities is the key to achieving the designed quality of the rings. The goal of the feedback systems is to achieve damping times of 1 ms and 10 ms for the transverse and longitudinal planes, respectively.

We are now developing beam feedback systems with the bunch-by-bunch scheme and are installing the prototype systems in the TRISTAN accumulation ring (TRISTAN-AR) (1-5). In our feedback systems, we detect oscillation of each bunch individually, shift the phase of the signal by 90° of the synchrotron frequency, and then kick the bunch to damp the oscillation. We have already installed button electrodes with increased frequency response, two wideband transverse kickers, eight transverse amplifiers of 200 W each, one longitudinal prototype kicker, and two longitudinal power amplifiers of 500 W in the south straight section of the AR.

In the longitudinal signal processing, we will use a two-tap finite impulse response (FIR) filter which has the function of DC suppression and a 90° phase shift. Prior to fabrication of the full-function filter board, we made a prototype board to prove the validity of the two-tap scheme. In this paper, we show the result of the beam test of the longitudinal feedback system prototype in TRISTAN-AR. We have closed the feedback loop and measured the shunt impedance of the prototype kicker. Spontaneous instability was successfully damped at a satisfactory rate. Also we have tried to damp coupled-bunch instability under 8-bunch operation. Fabrication of the first set of full-function filter boards that can handle all the bunches with the minimum bunch spacing of 2 ns is under way. Related parameters of the KEKB accelerators as well as those of TRISTAN-AR (at beam test) are listed in Table 1.

TABLE 1. Main Parameters of KEKB and TRISTAN-AR

Ring		LER	HER	TRISTAN-AR	
Energy	E	3.5	8	2.5	GeV
Circumference	C	3016.26		377.26	m
Beam current	I	2.6	1.1	0.001	A
rf frequency	f_{RF}	508.887		508.58	MHz
Harmonic number	h	5120		640	
Particles/bunch	N	3.3×10^{10}	1.4×10^{10}	7.8×10^{9}	
Synchrotron tune	ν_s	0.01 ~ 0.02		0.02	
Longitudinal damping time	τ_e	23	23	20	ms

EXPERIMENTAL SETUP

A block diagram of the longitudinal feedback system prototype at TRISTAN-AR is shown in Fig. 1. The system consists of a position detection part, a phase shifter part, and a kicker part.

Position Detection System

The longitudinal position of a bunch is measured with the wide-band phase detection system which is capable of distinguishing individual signals from the bunches with the bunch spacing down to 2 ns. The signal from a button electrode is divided into three branches by a power combiner and summed up again by another power combiner. As the lengths of the delay cables which connect the two combiners are designed to have a time difference of $\alpha + n\lambda/c$ (n = 0,1,2), where α is constant and λ is the wavelength of the detection frequency, this system acts as

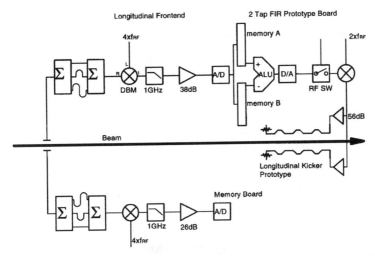

FIGURE 1. Block diagram of the longitudinal feedback system prototype installed in AR.

an FIR bandpass filter with the center frequency of nc/λ. The detection frequency is chosen to be the fourth harmonic of the rf frequency—in our case that is 2.034 GHz. The output of the filter is a burst of a sinewave-like signal less than 2 ns long.

Using a double-balanced mixer (DBM, R&K M-21), the sinewave-like burst is multiplied by the reference signal which is quadruple the rf signal. By rejecting the higher-frequency component with a low-pass filter (LPF, $f_c = 1$ GHz) from the output of the intermediate frequency (IF) of the DBM, the baseband signal of a synchrotron oscillation is detected as the form of $\sim I_b \Phi \sin(\omega_s t)$ if the amplitude Φ is small.

We have used two independent sets of detection systems; one for the feedback signal and the other for only monitoring of the oscillation. The position signal for the feedback system is amplified with three stage amplifiers for a total gain of 38 dB. For the bunch oscillation monitor, we have used two stage amplifiers with a total gain of 26 dB.

Two-Tap FIR Filter Complex

Signal processing, the functions of which are the 90° phase shift and the elimination of the static (DC) component, is performed with a two-tap FIR filter realized by a simple hardware system. The response of an FIR filter can be represented as a linear combination of the data obtained as a time series: x(1), x(2), etc. The two-tap filter has only two terms, the coefficients of which are 1 and −1, so the output has the form of

$$y(n) = x(n_1) - x(n_2)$$

and has the favorite frequency of $1/2(n_1 - n_2)$. By selecting suitable tap positions, i.e., by selecting the address-shift of the memory, we can tune the center frequency and the group delay of the filter.

Prior to the fabrication of the filter complex with full function, which will be shown in the last section of this paper, we have examined the feasibility of the filter scheme with a simple prototype board. The prototype board works only below the system clock of 6.4 MHz that corresponds to the eight equally-spaced bunch operations of AR. It has an 8-bit, 125-MHz fast analog-to-digital converter (FADC) (AD9002) for the digitizer and has the memory for 4096 turns of bunch position on a single-bunch mode. The board is packaged in a single-width CAMAC module.

For the bunch position monitor, we have used a 500-MHz FADC board (REPIC RPC-250) which is packaged in a single-width CAMAC module. It has 4096 bytes of memory per channel.

Feedback Kicker

The R&D for the KEKB longitudinal kicker has been centered on the wide-band device which is based on a series stripline structure (the series drift tube type) originally proposed by G. Lambertson (6). A combination of the beam pipe and the inner electrodes creates a coaxial structure of the characteristic impedance of 25 Ω. The rf power fed by power amplifiers propagates as the TEM wave along the structure. It has four electrodes connected with the delay lines. The carrier frequency is 1 GHz, which is double the rf frequency. The ideal shunt impedance $((R_{sh}) = V_{kick}^2/(2P_{in}))$ is 1.6 kΩ and the bandwidth will be less than 125 MHz. This kicker has two input ports of 50 Ω and two output ports.

We have prepared two wide-band amplifiers with a maximum power output up to 500 W each (R&K A0812-6057). One amplifier consists of 164 GaAs FETs (Fujitsu FLL120MK) and forms a complete A-class structure. The bandwidth is about 250 MHz (890 MHz ~ 1144 MHz) with a total gain of 60 dB. The measured time response of the amplifier from 5% to 95% of full power was about 2.5 ns.

BEAM TEST OF THE LONGITUDINAL SYSTEM ON TRISTAN-AR

Selection of the Tap Positions

The center frequency, phase shift, and the delay of the two-tap FIR filter is determined by the selection of the tap positions. As the sampling frequency was $8 \times f_{rev}$, the difference between the tap number should be $n_1 - n_2 = 8 \times i$ and should be near to the synchrotron frequency, that is $(n_1 - n_2)/8 \sim f_{rev}/f_s/2$. The position of the first tap n_1 determines the phase shifts and should be selected to coincide with the burst output of the amplifier and the beam passage. We at first searched for the best tap position to maximize the loop gain of the feedback system by exciting the synchrotron oscillation with the positive feedback loop. The best tap position was (249, 65) for positive loop for the accelerating voltage $V_c = 1$ MV. The tap difference of (249-65)/8 = 23 agrees with the measured synchrotron frequency of 19.5 kHz.

Measurement of Shunt Impedance of the Kicker

The realistic shunt impedance of the kicker is very important for the development and selection of the final kicker system for KEKB. If we know the maximum amplitude of the longitudinal oscillation excited by our feedback system in the positive feedback mode, the natural damping time of the beam, and the input power to the kicker, we can estimate the shunt impedance of the kicker by comparing the results from simulations. As the Robinson damping mechanism may strongly contribute to the longitudinal damping in the single-bunch mode, we must first measure the realistic damping time. We have measured the bunch oscillation turn by turn in the time domain just after the positive feedback has suddenly turned off. Figure 2 shows an example of the measured damping (a) and excitation (b) behavior with $V_c = 1$ MV and a beam current of 0.7 mA. The fitted damping time was 2 ms, which is much faster than the radiation damping time of about 20 ms.

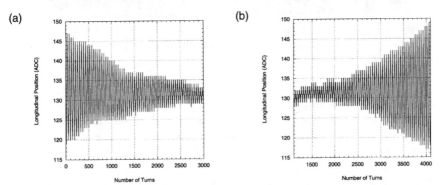

FIGURE 2. Damping (a) and excitation (b) of longitudinal oscillation just after the positive feedback loop has suddenly opened/closed. The natural damping time (a) was about 2 ms and the excitation time (b) was about 1.5 ms.

The longitudinal distribution of the zero cross point of a bunch signal from the button electrode was measured with the digitizing oscilloscope (HP 54121T). When the total input power to the kicker was about 640 W, the maximum amplitude of the longitudinal oscillation was 60 ps. To make such an oscillation in the simulation, it is necessary to have a kick voltage of about 1300 V per turn. Comparing these data, the shunt impedance R_{sh} of the kicker was estimated to be $(1300^2)/640/2 = 1.3$ kΩ, that is about 80% of the ideal one.

Damping of the Single-Bunch Oscillation

We can excite the longitudinal oscillation by intentionally shifting the resonant frequency of the cavities to induce the single-bunch Robinson instability. By tuning the resonant frequency, we controlled the growth rate of the instability. By setting the detuning angle of the accelerating cavities to be +9°, we excited a constant longitudinal oscillation without losing the beam. Figure 3(a) shows the beam spectrum with a detuning angle of +9° and a beam current of 0.7 mA. The maximum amplitude of the oscillation was about 50 ps. By closing the negative feedback loop, we succeeded in damping the oscillation completely, as shown in Fig. 3(b).

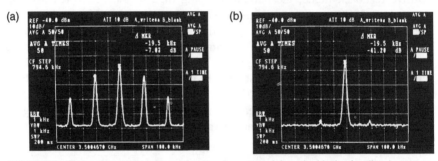

FIGURE 3. Beam spectrum without feedback system (a) and with the feedback system (b). The longitudinal oscillation was excited artificially by tuning the resonant frequency of the cavities.

The observed residual oscillation was about 4.3 ps (σ), which is attributed to the jitter of the measuring system. The input power to the kicker in the stationary state was about 16 W. Figure 4 shows the damping of the oscillation just after the feedback loop was closed. The damping time was 1.9 ms. The feedback system has damped the oscillation completely up to the detuning angle of +30°. When the detuning angle was +35°, the residual amplitude was increased to 5.9 ps. Nevertheless, we did not lose control of the oscillation. The input power to the kicker in the stationary state was about 84 W.

Damping of the Coupled-Bunch Oscillation

As the filter system can handle eight bunches with the system clock of 6.4 MHz, we tried to damp spontaneous coupled-bunch oscillations of eight equally-spaced bunch operations. Figure 5(a) shows the beam spectrum with the total beam current of 4 mA, i.e., the bunch current was about 0.5 mA. Heavy and

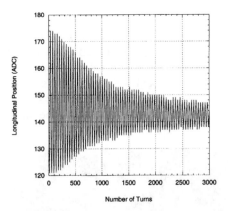

FIGURE 4. Damping of Robinson instability with the negative feedback on. The damping time is 1.9 ms.

unstable coupled-bunch oscillation was observed. This oscillation was completely damped with the feedback loop closed, as shown in the beam spectrum in Fig. 5(b). The residual oscillation was about 4.5 ps. By increasing the beam current, transverse oscillations started to grow with the beam loss even if we damped the longitudinal oscillation.

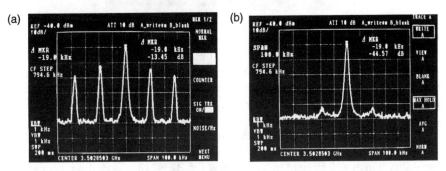

FIGURE 5. Beam spectrum without the feedback (a) and with the feedback system (b) under 8-bunch operation.

FULL-FUNCTION FILTER BOARD AND ITS APPLICATION

The fabrication of the first set of full-function boards is almost finished. Figure 6 shows a photograph of the filter board. The board is 366.71×400 mm and is controlled through a VME interface. A detailed explanation is given elsewhere (5); we will describe it briefly here. The board has one ADC daughterboard and 16 memory-ALU daughterboards on a motherboard. We adapted a MAX101 for the ADC and a TQ6122-M for the DAC. On the motherboard there are eight custom

GaAs LSIs: four fast data demultiplexers (FDMUX, Oki GHDK4211) and four fast data multiplexers (FMUX, Oki GHDK4212). The fast ECL signal from the ADC card (transfer rate of 255 Mbytes/s) is demultiplexed with the FDMUXs to a 32-channel TTL signal (16 Mbytes/s), through the memory-ALU block that forms a two-tap FIR filter, and multiplexed with the FMUXs to 255-Mbytes/s ECL signals. The memory has a depth of about 100 turns per bunch for the KEKB rings. We use FPGAs with the function of a subtracter and the bit shifter on the memory-ALU block.

FIGURE 6. Schematic view of full-function filter board.

We designed the motherboard so that it can also be a motherboard for a transient recorder by replacing the memory-ALU daughterboard with a dense memory board and by replacing the address control field-programmable gate arrays (FPGAs) on the motherboard. In our design, the maximum memory we can mount on the board will be about 40 MB. Combining the memory board and the feedback system enables us to measure the growth of the instabilities very clearly. We are now planning to use the memory board for the study of the photo electron instability which will be held in late 1996 at the BEPC ring in Beijing, China.

SUMMARY

We have examined the feasibility of the bunch-by-bunch feedback system scheme with the two-tap FIR filter board in TRISTAN-AR. The shunt impedance of the longitudinal kicker prototype has been measured with the measurement of the damping time and the maximum excitation amplitude. The estimated shunt impedance agreed with the expected value to within 80%. This encouraged us to proceed with the development of this kind of kicker.

The damping time of the feedback system has been measured with the artificially excited Robinson instability. The instability has been suppressed completely with the feedback system. Even under this heavy instability condition, our system did not lose controllability. In the 8-bunch operation, our feedback system has succeeded in suppressing the coupled bunch instability.

REFERENCES

1. E. Kikutani, T. Kasuga, Y. Minagawa, T. Obina, M. Tobiyama, and L. Ma, "Development of Bunch Feedback Systems for KEKB," in *Proc. of the 4th European Particle Accelerator Conference*, London, UK, 1613-1615 (1994).
2. E. Kikutani, T. Kasuga, Y. Minagawa, T. Obina, and M. Tobiyama, "Recent Progress in the Development of the Bunch Feedback Systems for KEKB," in *Proc. of the 1995 IEEE Particle Accelerator Conference*, Dallas Texas. USA, 2726-2728 (1996); also KEK Preprint 95-20.
3. E. Kikutani, T. Obina, T. Kasuga, Y. Minagawa, M. Tobiyama, and L. Ma, "Front-End Electronics for the Bunch Feedback Systems for KEKB," in *AIP Conference Proceedings* **333**, 363-369 (1995).
4. M. Tobiyama, E. Kikutani, and Y. Minagawa, "Longitudinal bunch feedback system with a two-tap FIR filter prototype," in *Proc. of the 10th Symposium on Accelerator Science and Technology*, Hitachinaka, Japan, 281-283 (1995); also KEK Preprint 95-103.
5. M. Tobiyama, E. Kikutani, T. Taniguchi, and S. Kurokawa, "Development of a two-tap FIR filter for bunch-by-bunch feedback systems," in *Proc. of the 10th Symposium on Accelerator Science and Technology*, Hitachinaka, Japan, 278-280 (1995); also KEK Preprint 95-104.
6. J. N. Corlett, J. Johnson, G. Johnson, G. Lambertson, F. Voelker, "Longitudinal and Transverse Feedback Kickers for the ALS," in *Proceedings of the 4th European Particle Accelerator Conference*, London, UK, 1625-1627 (1994).

Hardware Design and Implementation of the Closed-Orbit Feedback System at APS*

Dean Barr and Youngjoo Chung†

Advanced Photon Source, Argonne National Laboratory
9700 S. Cass Ave, Argonne, IL 60439

Abstract. The Advanced Photon Source (APS) storage ring will utilize a closed-orbit feedback system in order to produce a more stable beam. The specified orbit measurement resolution is 25 microns for global feedback and 1 micron for local feedback. The system will sample at 4 kHz and provide a correction bandwidth of 100 Hz. At this bandwidth, standard rf BPMs will provide a resolution of 0.7 micron, while specialized miniature BPMs positioned on either side of the insertion devices for local feedback will provide a resolution of 0.2 micron (1). The measured BPM noise floor for standard BPMs is 0.06 micron per root hertz mA. Such a system has been designed, simulated, and tested on a small scale (2). This paper covers the actual hardware design and layout of the entire closed-loop system. This includes commercial hardware components, in addition to many components designed and built in-house. The paper will investigate the large-scale workings of all these devices, as well as an overall view of each piece of hardware used.

INTRODUCTION

The Advanced Photon Source (APS) is a third-generation synchrotron light source. It is characterized by a low positron beam emittance, and hence a high-brightness x-ray beam. It is of vital importance to sustain transverse stability of the positron beam. This can be done using a closed-loop feedback system. The system at APS incorporates both global and local correction systems (2) to achieve its goal. The system has available 360 rf beam position monitors (rf BPMs), two photon position monitors for each x-ray beamline (up to 70 beamlines), and 318 dipole corrector magnets. An example of the ring with BPMs and corrector magnets is shown in Fig. 1. Miniature rf BPMs are located on each side of each insertion device (wiggler or undulator) in order to increase measurement resolution for the local feedback around these devices.

---

* Work supported by the U.S. Department of Energy, Office of Basic Energy Sciences, under Contract No. W-31-109-ENG-38.

† Present address: Kwangju Institute of Science and Technology, South Korea.

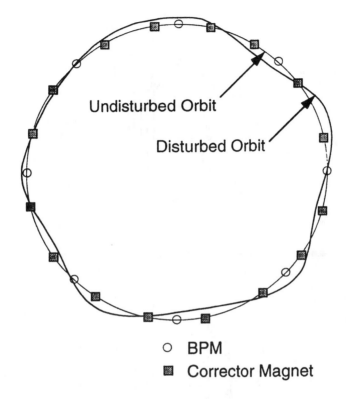

○ BPM
▦ Corrector Magnet

FIGURE 1. BPMs and corrector magnets for global beam position feedback.

The closed-orbit correction system as a whole is comprised of 20 VME crates covering 40 sectors (1104-m circumference). Each sector contains one bending magnet (for x-ray extraction) and one insertion device (either wiggler or undulator), leaving each feedback crate to cover two sectors. The design of each crate is the same thus producing an evenly distributed computing system. A diagram of a two-sectored controlled feedback system is shown in Fig. 2. Note that the two sectors are always an odd sector followed by an even sector. Figure 3 shows the layout of the feedback front panel. Figure 4 displays an aerial view of the crate. The front panel is located at the top the figure. An explanation of the various boards shown in Figs. 3 and 4 follows.

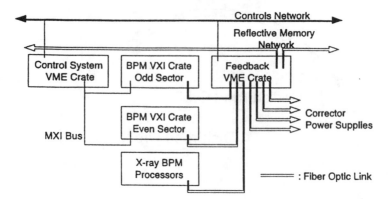

FIGURE 2. Hardware configuration of the beam position feedback system.

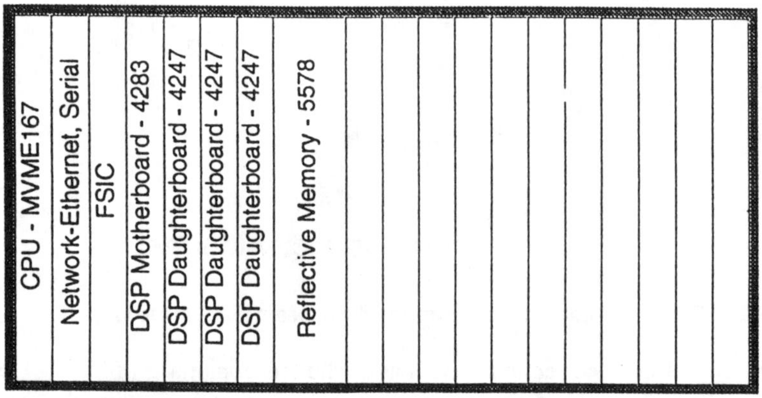

FIGURE 3. The front side of an orbit feedback crate.

FEEDBACK HARDWARE COMPONENTS

The entire system runs on a 4-kHz clock brought into the front panel of the feedback system interface controller (FSIC) (see Fig. 3). One feedback loop is thus comprised of 250 µs: the data acquisition (from the memory scanner interfaces (MSIs) and reflective memory boards) takes 50 µs, the calculation of the corrector strength by the various digital signal processors (DSPs) takes 100 µs, and the passing of the data to the power supply interface and moving of the magnets takes another 100 µs.

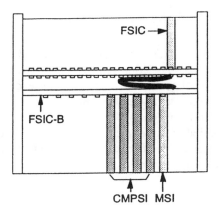

FIGURE 4. The feedback interface boards FSIC, FSIC-B, MSI, and CMPSI installed in the feedback crate (aerial view).

Each piece of hardware, whether commercially produced or manufactured and tested in house, plays a vital role in the closed-orbit feedback system. These will be covered now.

Hardware Components

MS, MSI, FSIC, and FSIC-B

Data can be acquired from either the memory scanner (MS) or the x-ray photon monitors. Currently information is only gathered from the MS as the x-ray BPM interface is not yet complete. The MS sits in a nearby VXI crate and obtains beam position data from the signal conditioning and digitizing units (SCDUs). The MS passes information at the rate of 200 Mbits/s over a fiber-optic cable to the MSI (see Fig. 4). The MSI has two channels: one for the odd sector and one for the even sector. This information is passed through the feedback system interface controller backplane (FSIC-B) (see Fig. 4) and through the FSIC (see Figs. 3 and 4) to the DSP motherboard. Each MSI is capable of handling transverse horizontal and vertical data from eight SCDUs. Since there are eighteen SCDUs in two sectors, two MSIs will usually be needed for each feedback crate.

Reflective Memory

The reflective memory is a high-speed, daisy-chained data transmission device and storage vessel for all beam position values. Each of the 20 feedback crates has one double-wide slot VME reflective memory. These commercially available boards form a network for high-speed communications between the feedback crates. APS uses the VMIC model 5578 boards which have a data transfer rate of 26 Mbytes/s in the non-redundant mode and 13 Mbytes/s in the redundant mode. Figure 5 shows a typical daisy-chained reflective memory network. Each node contains one reflective memory board.

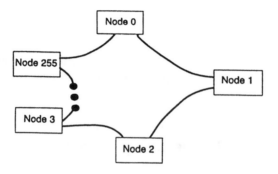

FIGURE 5. Example reflective memory network with 256 nodes.

DSP Motherboards and Daughterboards

The DSP motherboards and daughterboards are the heart of the feedback system. Each is based on the Texas Instruments' TMS320C30 floating point DSP. The motherboard has one of these processors while each daughterboard has two processors. The motherboard is capable of a peak processing power of 40 Mflops and contains 8 Mbytes of dynamic random-access memory (DRAM). This DRAM is mapped onto the VME bus and is the only way for the main CPU (the MVME167) to talk to the motherboard. The daughterboards connect to the motherboard or to each other depending on their location (see Fig. 3), and thus must use the DRAM on the motherboard for communication as well.

The DSP C30s act as the embedded controller for the entire feedback system. They handle the code for both the local and global feedback correction systems. They work on a polling system. At the end of every loop the DSPs check certain registers on the motherboard DRAM to see if they have received new instructions from the MVME167.

CPU and Ethernet Connection

The CPU (an MVME167) is a Motorola 68040-based processor running a real-time operating system called VxWorks. This processor is connected to the rest of the accelerator control system through a fiber-optic Ethernet link. The CPU can monitor status and operation of the feedback crate and send commands to the DSPs through the DRAM on each DSP motherboard.

CMPSI

The corrector magnet power supply interface (CMPSI) is the last piece of the puzzle. It is designed to take information from the DSP through the FSIC and FSIC-B and pass it over fiber-optic cables to the power supply interface which is located in another rack. The FSIC communicates with the power supply interface by sending a serial data train consisting of one start bit, 16 data bits, and 1 stop bit. For each corrector, a clock line is also run, thus making the communication synchronous. The transmission speed is 500 kHz, and therefore, it takes a minimum of 36 μsec to send data to the corrector power supplies. Each CMPSI can handle 8 channels in parallel (both data and clock included). For a complete two-sector setup using global feedback, two bending magnets, and two insertion devices, a total of four CMPSIs will be necessary for a single feedback crate.

CONCLUSION

The design and prototype of the hardware for the feedback system is completed with the exception of the x-ray BPM interface. The commercial boards have all been purchased with the exception of some of the DSP daughterboards. All in-house production boards are in various phases of ordering, assembly, and testing and should be finished by the end of September 1996. The initial local feedback system (in one sector) is being commissioned now. We hope to test the global feedback system by the end of September 1996.

ACKNOWLEDGMENTS

We wish to thank Peter Nemenyi for all his hard work in putting together the in-house boards and Frank Lenkszus, Robert Laird, and Anthony Pietryla for many hours of hardware debugging.

REFERENCES

1. Y. Chung and E. Kahana, "Resolution and Drift Measurements on the Advanced Photon Source Beam Position Monitor," AIP Conference Proceedings 333, Beam Instrumentation Workshop 1994, Vancouver, B.C., Canada, 328-334 (1995).
2. Y. Chung, "Beam Position Feedback System for the Advanced Photon Source," AIP Conference Proceedings 319, Beam Instrumentation Workshop 1993, Santa Fe, NM, 1-20 (1994).

A Closed-Orbit Suppression Circuit for a Main Ring Transversal Damper*

Hengjie Ma, Jim Steimel, John Marriner, Jim Crisp

Fermi National Accelerator Laboratory
P. O. Box 500, MS #340, Batavia, Illinois 60510

Abstract. The signals of a transversal damper pickup usually have a certain number of common-mode components due to the off-center beam at the location. For the limit in the output power and the required minimum dynamic range of the feedback system, this common-mode component must be suppressed as much as possible. An analog front end is being developed for a transversal damper of the Main Ring at Fermilab (1) for this purpose. The front end features a balanced feedforward circuit and a possible single-ended negative feedback loop. Properly set, the time constant of the feedforward circuit ensures that the slowly changing closed-orbit component will be adaptively canceled, while the betatron oscillation components will survive in the output.

CIRCUIT ANALYSIS

The simplified block diagram of the balanced feedforward circuit is shown in Figure 1. The signals a(t) and b(t) are from the A and B electrode of a damper pickup. Blocks CH.a and CH.b represent the bandpass filter and the lumped-characteristic of each channel.

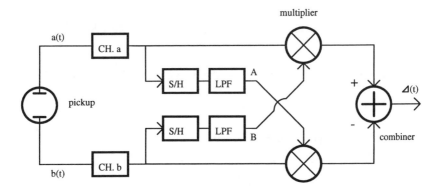

FIGURE 1. Basic scheme of the balanced feedforward circuit.

---

* Operated by University Research Association under contract with U.S. Department of Energy.

The output $\Delta(t)$ is the difference of the weighted signals, which contains the components of the betatron oscillation. To simplify the analysis, let us assume a case of simple dipole motion (2,3). Then

$$a(t) = \sum_{n=1}^{+\infty} \{c_{0n} \cos[n\omega_0 t + \theta_{0n}]$$

$$+ c_{1n} \cos[(n\omega_0 + \omega_b)t + \theta_{1n}] - c_{1n} \cos[(n\omega_0 - \omega_b)t + \theta_{1n}]\}, \qquad (1)$$

$$b(t) = \sum_{n=1}^{+\infty} \{d_{0n} \cos[n\omega_0 t + \theta_{0n}]$$

$$+ d_{1n} \cos[(n\omega_0 + \omega_b)t + \theta_{1n}] - d_{1n} \cos[(n\omega_0 - \omega_b)t + \theta_{1n}]\}, \qquad (2)$$

where ω_0 is the revolution angle frequency and ω_b is the betatron oscillation frequency. The first terms in Eqs. (1) and (2) represent the harmonics of the revolution frequency, which need to be suppressed. Sampling $a(t)$ and $b(t)$ at a rate of $\omega_0/2\pi$ causes the spectrum to repeat in a space of $\omega_0/2\pi$. Setting the cut-off frequency of the lowpass filters $\omega_c \ll \omega_b$, the surviving baseband signal at the output of the sample/hold circuit A and B will be

$$A = \sum_{n=1}^{+\infty} c_{0n} \cos(\theta_{0n}), \qquad B = \sum_{n=1}^{+\infty} d_{0n} \cos(\theta_{0n}). \qquad (3)$$

Weighting $a(t)$ with B and $b(t)$ with A, the output signal $\Delta(t)$ will be

$$\Delta(t) = B \times a(t) - A \times b(t)$$

$$= \sum_{n=1}^{+\infty} B\{c_{1n} \cos[(n\omega_0 + \omega_b)t + \theta_{1n}] - c_{1n} \cos[(n\omega_0 - \omega_b)t + \theta_{1n}]\}$$

$$- \sum_{n=1}^{+\infty} A\{d_{1n} \cos[(n\omega_0 + \omega_b)t + \theta_{1n}] - d_{1n} \cos[(n\omega_0 - \omega_b)t + \theta_{1n}]\}, \qquad (4)$$

in which the harmonics of ω_0 have been canceled out.

ERROR ANALYSIS

Although the topography of the circuit is simple, a strict and complete error analysis is not an easy task due to many unknown parameters with the components. However, three errors have been identified as the major factors affecting the effectiveness of the circuit: channel mismatch, multiplier feedthrough, and imbalance of the power combiner.

(1) Channel mismatch: The effective cancellation of the closed-orbit components completely depends on the spectral match of the two signals at the input of the combiner. Therefore, the frequency response and phase match of the two channels become critical. If the magnitude and phase frequency response of the two channels are $H_a(\omega)$, $\Theta_a(\omega)$, $H_b(\omega)$, and $\Theta_b(\omega)$, then, the rms power of the resultant common-mode components in the output P_c will be

$$P_{c1} = (1/4)\sum_{n=1}^{+\infty}\left\{\left[H_a(n\omega_0)\cdot B\cdot c_{0n}\right.\right.$$

$$\left. - H_b(n\omega_0)\cdot A\cdot d_{0n}\cos(\Theta_a(n\omega_0)-\Theta_b(n\omega_0))\right]^2$$

$$\left. + H_b(n\omega_0)\cdot A\cdot d_{0n}\sin(\Theta_a(n\omega_0)-\Theta_b(n\omega_0))\right]^2\right\}. \qquad (5)$$

(2) Multiplier feedthrough: The imperfection of the multipliers used here causes a small amount of the input signal to leak though. The feedthrough is usually specified as an attenuation from one of the input ports to the output port in dBs. Obviously, with the feedthrough of the multipliers F_a and F_b, the common-mode power component appearing in the output due to this factor will be

$$P_{c2} = (1/4)\sum_{n=1}^{+\infty}\left\{\left[10^{F_a/20}\cdot c_{0n} - 10^{F_b/20}\cdot d_{0n}\right]^2\right\}. \qquad (6)$$

(3) Imbalance of power combiner: The imbalance of a power combiner is usually specified in the amplitude and phase mismatch between its two inputs. Since it has exactly the same effect on the common-mode suppression as the channel mismatch does, the factor of the combiner imbalance can be incorporated in the transfer function of the two channels. That will yields the same result as Eq. (5).

IMPLEMENTATION

In the implementation of the circuit, some changes have to be made to adapt the limits of the available components. At the present time, the type of multiplier used here is a AD835 from Analog Devices (4). It has a 1-volt limit on the amplitude of the inputs and an output noise density of $50nV/(Hz)^{1/2}$. To meet the requirement of an effective damping, the feedback system needs quite a large amount of gain in the following stages. As a result, the noise in the output would become a problem. Because of this, it is certainly desirable to make the front end capable of handling a higher input level. One measure would be to perform Δ and Σ conversion with a 180° hybrid junction at the very front. Figure 2 shows a more detailed block diagram of the analog front end being developed.

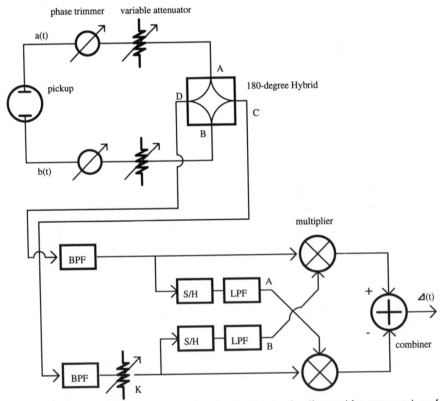

FIGURE 2. Block diagram of an analog front end using feedforward for suppression of the common-mode signal.

With the help of the mechanical variable attenuators and phase trimmers, a large part of the static common-mode components will be canceled out at the difference port of the hybrid. The following electronic circuit needs to handle only the dynamic part of the common-mode signal. During construction of the prototype, it was found that the quality of the hybrid and the matching of the attenuators and filters on the two channels appear to be critical. Normally, the sum signal from the hybrid is much stronger than the difference signal. Therefore, attenuator K is necessary to reduce its amplitude to a proper level. The sample/hold circuit consists of an ultra-fast S/H chip AD9100 cascaded with a slower AD783 (4) in order to be able to capture a beam pulse in 20 ns and hold the voltage for 21 μs.

DISCUSSION

The success of this scheme depends on the linearity of the multipliers and the match of the two channels in their frequency responses. As long as these two requirements are met, Eq. (4) will always stand regardless of the changes in the beam intensity and position. To further improve the performance, the possibility of adding a single-ended negative feedback loop has been studied. The idea is to use one more S/H circuit to sample the output signal $\Delta(t)$. The information about the residual common-mode component in $\Delta(t)$ will be fed back to the control port of the multiplier on the sum channel (lower part of Figure 3). The effect of adding this NFB loop on the characteristics including the transfer function of the subsystem is being studied.

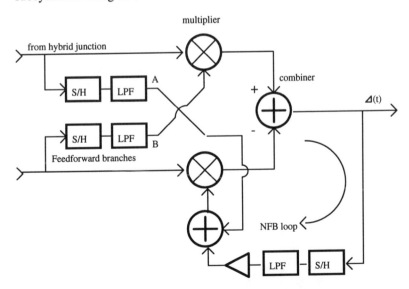

FIGURE 3. Possible incorporation of a negative feedback loop for further improvement.

ACKNOWLEDGMENT

The authors would like to thank David McGinnis for his many valuable suggestions and ideas, specially for his idea of the single-ended negative feedback.

REFERENCES

1. Steimel, J., "Overview of Coupled-Bunch Active Damper Systems at FNAL," these proceedings.
2. Pedersen, F., "Coherent Instabilities and Feedback Control," Tutorial, BIW'94.
3. Siemann, R., "Spectral Analysis of Relativistic Bunched Beams," these proceedings.
4. Data sheets AD835, AD9100, AD783, Analog Devices, 1994.

Longitudinal and Transverse Kickers for the Bunch-by-Bunch Feedback Systems of the Frascati Φ-Factory DAΦNE

R. Boni, A. Gallo, A. Ghigo, F. Marcellini,
M. Serio, B. Spataro, M. Zobov

INFN Laboratori Nazionali di Frascati, 00044 Frascati (Roma), Italy

Abstract. The development of very efficient longitudinal and transverse bunch-by-bunch feedback systems is one of the fundamental aspects for the operation of the high-current storage rings. The last part of the feedback chain is a kicker whose efficiency impacts directly on the system cost. The design and measurements of the DAΦNE longitudinal and transverse kickers are presented here. The longitudinal kicker design is based on a waveguide overloaded cavity, while a more conventional stripline structure has been chosen for the transverse kicker. The design of both devices has been optimised in order to increase the shunt impedance, i.e., their efficiency, and to lower the content of parasitic higher-order modes that could further excite coupled-bunch instabilities.

THE DAΦNE LONGITUDINAL FEEDBACK SYSTEM

The multibunch operation of DAΦNE, and in general of any "factory," calls for a very efficient feedback system to damp the coupled-bunch longitudinal instabilities. The feedback chain ends with the longitudinal kicker, an electromagnetic structure capable of transferring the proper longitudinal momentum correction to each bunch. The kicker design is optimised mainly with respect to the following parameters: shunt impedance, bandwidth, and content of higher-order modes (HOMs) that can further excite coupled-bunch instabilities. The result of the design simulations, performed with the Hewlett-Packard code HFSS (1), together with measurements on the device are presented and discussed in this paper.

Design of the Overdamped Cavity

The longitudinal kicker design is based on a waveguide overloaded cavity; a cut view of the geometry is shown in Fig. 1. The very large bandwidth required has been obtained by strongly loading a pillbox cavity with special ridged waveguides followed by broadband transitions to 7/8" standard coaxial where ceramic feedthroughs allow in-air connections to the driving amplifiers (input ports) and dummy loads (output ports).

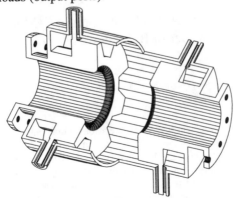

FIGURE 1. Kicker cavity cut view.

The idea of using an rf cavity as longitudinal kicker is based on some simple considerations. When all the rf buckets are filled, all possible coupled-bunch modes are present in a frequency span between nf_{RF} and $(n \pm 1/2)f_{RF}$, with n any integer and f_{RF} the bunch frequency. Therefore, the minimum bandwidth requirement for the longitudinal kicker is $f_{BW} = f_{RF}/2$, as long as the response is centered onto $f_c = (n \pm 1/4)f_{RF}$ (2). A center frequency $f_c = 3.25\ f_{RF} \approx 1197$ MHz has been chosen so that the resulting loaded quality factor of the cavity has to be set to about $Q_L = f_c\ /f_{BW} \approx 6.5$. Therefore, if the damping waveguides are symmetrically placed with respect to the fundamental mode field distribution and half of them are used as input ports while the remaining as matched terminations, the external Q values are given by

$$Q_{extinp} \approx Q_{extout} \approx 2Q_L \approx 13. \tag{1}$$

The R/Q factor of a pillbox cavity resonating around 1.2 GHz with stay-clear apertures of 88 mm is limited to about 60 Ω. The kicker shunt impedance R_s has a peak value given by

$$R_s = V_k^2/2P_{in} \approx (R/Q)\ Q_{extout} \approx 780\ \Omega. \tag{2}$$

This means that the attainable shunt impedance is about twice the value of a two-electrode stripline module (3), while no HOMs are likely to remain undamped in this structure.

The coupling waveguide is a single-ridged type waveguide with 6-mm gap to lower the TE10 cutoff frequency down to 690 MHz. The waveguide-to-coaxial transition is designed with the same criteria adopted for the main ring accelerating cavity (4). The coaxial size is the standard 50 Ω 7/8" which can withstand more than 1 kW power flow. Moreover, we can use for this coaxial standard the broadband ceramic feedthrough already developed for the transitions of the main ring cavity (5).

Computer Simulations and Experimental Results

The reflection frequency response of the transition has been computed with HFSS. The S_{11} value is lower than 0.25 along the entire frequency band up to 3 GHz, well beyond the beam pipe cut-off; the low cut-off frequency of the TE10 mode of the waveguide ($\approx$ 690 MHz) is crucial to get a good wave transmission in the low-frequency band.

From the computed transmission coefficient S_{21} for the three input ports to the three output ports for the cavity fundamental mode, we get a central frequency of about 1215 MHz and a bandwidth as large as 220 MHz.

All the HOMs (monopole and dipole modes) found up to the beam pipe cut-off frequency appear extremely damped. Then they are not considered dangerous for the longitudinal and transverse dynamics.

The most important kicker figure of merit is the shunt impedance R_s defined as

$$R_s = \frac{|V_g|^2}{2P_{fw}},$$

(3)

where V_g is the kicker gap voltage and P_{fw} is the forward power at kicker input. The gap voltage V_g may be obtained by post-processing the field solution given by the 3D simulator. In this way Eq. (3) yields an impedance peak value of about 750 Ω, in good agreement with the rough estimate from Eq. (2). Due to the transit time factor, we observe a –15-MHz frequency shift of the impedance peak value with respect to that of the transmission response.

The longitudinal kicker shunt impedance has been measured with the wire method. A 3-mm-diameter copper wire has been inserted in the cavity along the beam axis and connected to a 50-Ω line through a resistive matching network. The coaxial wire-beam tube system is a $Z_0' = 203$ Ω transmission line. The longitudinal beam coupling impedance $Z_{coup}(\omega)$, defined as the complex ratio between the cavity gap voltage and the beam current, can be calculated (6) with some approximation according to

$$Z_{coup}(\omega) \approx 2\,Z_0' \left(\frac{1}{S_{21}} - 1\right),$$

(4)

where S_{21} is the complex transmission coefficient between the two wire ports measured by a network analyzer calibrated to take into account the cable and matching network attenuations, as well as the linear phase advance due to the electrical length of the device.

The shunt impedance as defined in Eq. (3) turns out to be twice the value of the real part of the beam impedance. Then, from the plot of the function $\Re(1/S_{21})$ shown in Fig. 2, we measure a shunt impedance peak value ($\approx 620\ \Omega$), a bit lower than the expected one, at 1370 MHz (+170 MHz resonant frequency shift, due to the perturbative effect of the wire). From the measured bandwidth we find R/Q = 57 Ω, i.e., the same value obtained in the simulations. This means that the waveguide loading is underestimated by the HFSS simulations so the real loaded Q value is Q_L = 5.4 (instead of 6.5). Nevertheless, the measured shunt impedance value is still high enough, and from some preliminary experiments it seems possible to introduce some small mismatches into each coaxial line to resize the impedance curve.

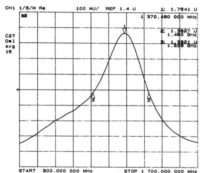

FIGURE 2. Kicker beam coupling impedance measurement (wire method).

Power Considerations

The DAΦNE operation will require a maximum longitudinal kick voltage of ≈ 400 V with 30 bunches and ≈ 1600 V with 120 bunches in order to damp an initial offset of 100 ps, a prudent estimate of the maximum injection error of the last bunch. A 200-W input power with a single kicker cavity per ring will be enough for the 30-bunch operation while two kickers per ring fed with 600 W each will eventually be required for the 120-bunch operation (7).

On the other hand, the beam current interacts with the device beam impedance, and the beam released power can be much higher than the power from the feedback system. The total power P_b released by the beam is given by

$$P_b = \sum_n \frac{1}{2}\Re e\left\{Z_{coup}\left(\omega_n\right)\right\}I_n^{\,2}, \tag{5}$$

with I_n the n-th harmonic of the Fourier expansion of the beam current. In the most critical case (120 bunches) Eq. (5) yields $P_b = 1.6$ kW per six guides.

In fact, being mainly a standing wave structure, the cavity kicker is not a directional device, and the upstream and downstream ports are almost equally coupled to the beam. Therefore, the longitudinal feedback endstage amplifiers must be protected with ferrite circulators against the backward power which can be one order of magnitude higher than the forward level. A custom circulator covering the 1- to 1.4-GHz band has been already developed and low-power tested (8).

THE DAΦNE TRANSVERSE FEEDBACK SYSTEM

The DAΦNE transverse feedback system must provide the damping of the resistive wall instability and the control of a large number of transverse coupled-bunch modes. In this system the horizontal and vertical beam position is detected by two-button-type monitors; the correction signal (in quadrature) at the kicker is obtained by summing the pick-up signals in proper proportion. Two transverse kickers, one for horizontal and one for the vertical correction, will be installed in each ring. Simulations, design criteria, efficiency and contribution to the machine impedance are reported.

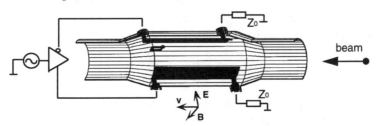

FIGURE 3. Transverse kicker: cut view and working scheme.

Design Criteria

The electromagnetic design has been carried out mainly by means of the HFSS code using a "cut and try" procedure to achieve good performances from the kicker within the machine constraints.

The kickers are based on a pair of striplines (20-cm length, 1-mm thickness). The stay-clear aperture of the device is that of the rest of the straight-section beam pipe (88 mm), and the vacuum chamber diameter in the kicker section is 120 mm. Two 50-mm-long tapers that join the kicker vacuum chamber with the rest of the beam pipe are also needed in order to minimise the losses. We found that if 80° is the stripline coverage angle, each electrode forms with the vacuum pipe a

transmission line of $Z_0 = 50\ \Omega$ characteristic impedance. Two standard 7/8"
coaxial feedthroughs connect the electrode ends with the amplifier and the
external load. The final geometry of the coax line-stripline transition minimises
the reflected power (less than 6% of the forward power in the operating
bandwidth) to the amplifier.

In order to use this device as a transverse kicker, two voltages of opposite
polarity have to drive the downstream ports with the upstream ports closed on
matched loads (see Fig. 3). The combined magnetic and electric field gives a net
deflecting Lorentz force in the transverse plane. If the particle and the wave
propagation velocities are equal we have

$$|E| = |v \times B|, \tag{6}$$

so that the transverse force is $F_\perp = 2qE$ when the kicker is driven from the
downstream ports, while the electric and magnetic deflecting forces cancel out if
the excitation is applied to the upstream ports.

Transverse Shunt Impedance

The kicker efficiency is described by the transverse shunt impedance
parameter $R_\perp$ (9), defined in the same way as R_s in Eq. (3) where V_g is now the
transverse voltage, i.e., the integral of the transverse component of the Lorentz
force per unit charge along the beam axis:

$$V_g = \int_0^l (E + v \times B)_\perp\, dz . \tag{7}$$

For a stripline transverse kicker driven in the differential mode the transverse
shunt impedance can be calculated accordingly to the following formula (3):

$$R_\perp = 2Z_0 \left(\frac{g_{trans}}{kh} \right)^2 \sin^2(kl) , \tag{8}$$

where g_{trans} is the coverage factor ($g_{trans} \approx 1$), $k = \omega/c$, h is the stay-clear radius
(44 mm), and l is the electrode length ($\approx$ 20 cm).

The results given by the previous formula can be compared to those obtained
from the integration on the kicker axis of the electric and magnetic fields as
computed by the simulation code. Figure 4 shows this comparison.

A peak value larger than 2 kΩ together with a bandwidth wider than 300 MHz
has been obtained. As a first estimate, a kicker driving power of $\approx$ 200 W seems
enough to keep the machine transverse instabilities under control (10).

Higher-Order Modes

The beam passage through the kicker structure can excite HOMs that are a potential source of coupled-bunch instability. Using the MAFIA (11) and HFSS computer codes we have found only two trapped modes resonating at 2.275 GHz and 2.530 GHz. In order to damp them, two identical rectangular loops have been introduced in the vacuum chamber near the tapers, rotated by 90 degrees with respect to the input ports. The choices of the loop shape and position have been optimised with several HFSS runs. The values of the loaded Q and R/Q factors for the two modes mentioned are no longer considered dangerous for the longitudinal beam dynamics.

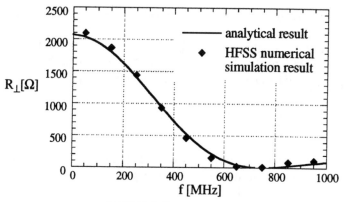

FIGURE 4. Transverse shunt impedance.

Coupling and Transfer Impedance

A simulation of the measurement, with a metallic wire placed along the kicker axis to imitate the beam, has been performed. The longitudinal coupling impedance and the transfer impedances of the kicker, defined as the ratio of the voltages at the output ports to the beam current (9), were calculated.

Coupling and transfer impedances can be easily obtained from the scattering matrix yielded by HFSS simulations according to Eq. (4) and the following formula

$$Z_{tr,i} = \frac{S_{2i}}{S_{21}} \sqrt{Z_0 Z_0'} \quad , \quad i = 3,4,5 \ , \tag{9}$$

where the subscript i refers to one of the output ports (loop, downstream, or upstream ports), the input port for the beam is port 2, and Z_0 and Z_0' are the characteristic impedance of the output ports ($\approx 50 \ \Omega$) and the wire beam tube coaxial line ($\approx 130 \ \Omega$), respectively. The real part of these functions vs. frequency is shown in Fig. 5.

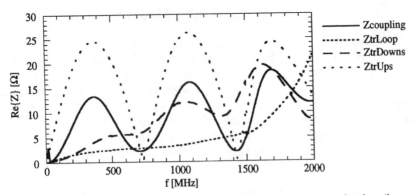

FIGURE 5. Transverse kicker coupling and transfer impedances (real part).

With a given beam current spectrum it is possible to estimate the total power P_b released by the beam to the structure (see Eq. (5)) as well as the power $P_{tr,i}$ flowing through the coaxial lines connected with loops and striplines:

$$P_{tr,i} = \sum_n \frac{1}{2Z_0} \Re\{Z_{tr,i}(\omega_n)\}^2 I_n^2 \quad , \quad i = 3,4,5 . \tag{10}$$

It must be noted that for an ideal, perfectly directional stripline device the transfer impedance and then the power flowing through the coaxial ports downstream from the beam should be zero.

The power dissipated on the kicker walls is:

$$P_d = P_b - \Sigma_i P_{tr,i} . \tag{11}$$

The wire method simulation results seem to be reliable below 2 GHz, where no significant HOMs have been found. Therefore the HOM power coupled by loops was estimated making use of the following formula:

$$P_{HOM} = \sum_k \sum_n \frac{1}{2} \frac{Q_{Lk}\left(\frac{R}{Q}\right)_k}{1 + Q_{Lk}^2 \left(\frac{\omega}{\omega_k} - \frac{\omega_k}{\omega}\right)^2} I_n^2 . \tag{12}$$

Including the effect of the roll-off due to the 3-cm DAΦNE bunch length, the previous formulas together yield the results summarised in Table 1 for two typical beam configurations.

The chosen sizes of the striplines and feedthroughs are completely sufficient to manage the power released by the beam together with the incoming power from the endstage amplifier of the transverse feedback system.

TABLE 1. Estimated Beam Power Distribution on Transverse Kicker Parts (in W)

Number of regularly spaced bunches	30	120
Total beam released power	372 + 60*	1483 + 436*
Power flowing through each loop port	9 + 30*	40 + 218*
Power flowing through each downstream port	34	135
Power flowing through each upstream port	143	555
Dissipated power on aluminum walls	≈ 0	23
Power flowing from each downstream port to each upstream port due to amplifier	100	100

*power due to HOM impedance

REFERENCES

1. Hewlett-Packard Co, "HFSS, The High Frequency Structure Simulator HP85180A™—User's Reference," December 1994.
2. Gallo, A. et al., "Simulations of the Bunch-by-Bunch Feedback Operation with a Broadband RF Cavity as Longitudinal Kicker," DAΦNE Technical Note G-31, Frascati, April 29, 1995.
3. Corlett, J. N. et al., "Longitudinal and Transverse Feedback Kickers for the ALS," in *Proceedings of the 4th EPAC*, London, 27 June-1 July, 1994, 1625-1627 (1994).
4. Boni, R. et al., "A Broadband Waveguide to Coaxial Transition for High Order Mode Damping in Particle Accelerator RF Cavities," *Particle Accelerators* **45**, (4), 195 (1994).
5. Boni, R. et al., "Update of the Broadband Waveguide to 50 Ω Coaxial Transition for Parasitic Mode Damping in the DAΦNE RF Cavities," in *Proceedings of the 4th EPAC*, London, 27 June-1 July, 1994, 2004-2006 (1994).
6. Hahn, H. and Pedersen, F., "On Coaxial Wire Measurements of the Longitudinal Coupling Impedance," BNL-50870, UC-28, April 1978.
7. Vignola, G. and the DAΦNE Project Team, "DAΦNE Status and Plans," in *Proceedings of Particle Accelerator Conference*, Dallas, TX, May 1-5, 1995, 495-499 (1996).
8. Advanced Ferrite Technology Co, Spinnerei 44, 71522 Backnang (FRG), private communication.
9. Goldberg-Lambertson, "Dynamic Devices: A Primer on Pickups and Kickers," LBL-31664, ESG-160, November 1991.
10. Serio, M., "Feedback Status," 4th DAΦNE Machine Review, Frascati, January 19-20, 1993.
11. Weiland, T., *Nucl. Instrum. Methods* **216**, 329-348 (1983).

Feedback Control and Beam Diagnostic Algorithms for a Multiprocessor DSP System

D. Teytelman, R. Claus, J. Fox, H. Hindi, I. Linscott, S. Prabhakar[*]

Stanford Linear Accelerator Center, P.O. Box 4349, Stanford, CA 94309

A. Drago

INFN - Laboratori Nazionali di Frascati, P.O. Box 13, I-00044 Frascati (Roma), Italy

G. Stover

Lawrence Berkeley Laboratory, 1 Cyclotron Road, Berkeley, CA 94563

Abstract. The multibunch longitudinal feedback system developed for use by PEP-II, ALS, and DAΦNE uses a parallel array of digital signal processors (DSPs) to calculate the feedback signals from measurements of beam motion. The system is designed with general-purpose programmable elements which allow many feedback operating modes as well as system diagnostics, calibrations, and accelerator measurements. The overall signal processing architecture of the system is illustrated. The real-time DSP algorithms and off-line postprocessing tools are presented. The problems in managing 320k samples of data collected in one beam transient measurement are discussed and our solutions are presented. Example software structures are presented showing the beam feedback process, techniques for modal analysis of beam motion (used to quantify growth and damping rates of instabilities), and diagnostic functions (such as timing adjustment of beam pick-up and kicker components). These operating techniques are illustrated with example results obtained from the system installed at the Advanced Light Source at LBL.

INTRODUCTION

A feedback system for bunch-by-bunch control of longitudinal coupled-bunch instabilities has been developed and installed at the Advanced Light Source (ALS) at Lawrence Berkeley Laboratory (LBL). The system design is described in detail in earlier publications (1,2). As shown in Fig. 1, signals from four button-type pickups are combined and fed to a stripline comb generator. The generator produces a four-cycle burst at the sixth harmonic of the ring rf frequency (2998 MHz). The resultant signal is phase detected, then digitized at the bunch crossing rate. A correction signal for each bunch is computed by a digital signal processing module and applied to the beam through a fast digital-to-analog (D/A) converter, an output modulator, a power amplifier, and a kicker structure.

---

* Work supported by U.S. Department of Energy contract DE-AC03-76SF00515.

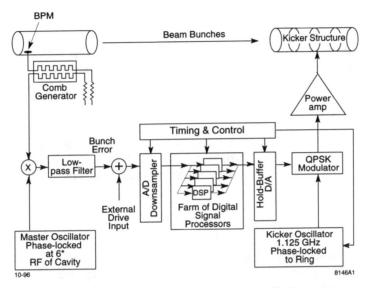

FIGURE 1. Block diagram of the longitudinal feedback system.

The correction signal for a given bunch is calculated using the past information on the motion of only that bunch. As a result, it is possible to use a highly parallel computational structure to achieve the necessary throughput. Since the system uses digital signal processors (DSPs) to process the synchrotron oscillation signals, it is possible to extend the functionality of the feedback system to include many modes of feedback operation as well as various diagnostics. A large suite of software tools has been developed to take advantage of the flexibility of the signal processing architecture.

OVERALL SOFTWARE STRUCTURE

The feedback system software consists of three major layers. On the very top level we have a graphical operator interface constructed using the Experimental Physics and Industrial Control System (EPICS) tools. This layer controls the hardware drivers and support programs of the middle layer, composed of VME and VXI single-board computers running the VxWorks (Wind River Systems) operating system. The VME crates are populated with the DSP boards while the VXI crate houses the downsampler and the holdbuffer modules. At the bottom level there are real-time programs which are coded in the native assembly language of the AT&T 1610 DSP chip (3).

DSP ALGORITHMS

All the DSP programs written to date can be divided into three classes: feedback, data collection, and feedback system support and diagnostics.

Feedback Tools

Feedback programs are the most basic of all, providing correction signal computation at the highest possible rate. The feedback kick is computed using a finite impulse response (FIR) filter with the following algorithm

$$u_i(n) = \sum_{m=0}^{N-1} c(m)\phi_i(n-m).$$ (1)

The coefficients c are chosen to maximize the gain at the synchrotron frequency and to provide the appropriate phase shift through the filter for overall phase inversion in the feedback loop. The number of taps N is variable from 2 to 12. The FIR filter supports four different coefficient sets switchable on the fly. An infinite impulse response (IIR) filter is currently being developed.

Data Collection Tools

Another class of programs are data collection tools. One of the most versatile programs currently in use is the triggerable record program. This program allows capture of the phase oscillations of all or selected bunches. It utilizes the hardware resources on the DSP board to provide uninterrupted feedback signal computation while collecting data. Each DSP is equipped with 8k words of dual-ported RAM accessible by the DSP and the VME host processor. When recording is triggered, the dual-port memory is used to store the sampled phase coordinates of the bunches. Once the recording is finished, an interrupt is sent to the VME host processor, which then reads out the data from the dual-port memory. Figure 2 shows the sequence of events for the data capture by the record program. Initially, before the trigger condition is satisfied, the program runs a normal FIR filter. The recording trigger can be generated by software (from an EPICS button) or hardware (from a logic-level signal). The software trigger is used to take a snapshot of the beam motion and may be synchronized to software events. However, in some conditions it is necessary to link the data capture to a hardware event, such as an injection into the storage ring. Such synchronization is provided by the hardware trigger. The hardware and software triggers can be combined to allow selection of one event out of a sequence. After the trigger the filter coefficients switch to an alternate set to allow the user to choose a different feedback condition, e.g., no

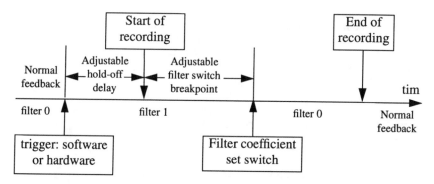

FIGURE 2. Event sequencing for the triggerable record program.

feedback or positive feedback. Before data recording, a variable-length hold off occurs after the trigger. This delay is used to optimize the recording window position for the slow transients. After the hold off we start recording data in the dual-ported RAM. At the variable breakpoint the coefficients revert to the original feedback values. After the full record is written, the DSP program interrupts the host processor and returns to the initial pre-trigger state.

The data records are stored on the Unix-based server and analyzed off-line using Matlab scripts. The most common application of the record programs is to provide grow/damp records. In these experiments the longitudinal oscillations are allowed to grow for some time and then damped. The data allows us to identify the unstable modes and calculate the growth and damping rates for those modes (4). The flowchart for the Matlab-based modal analysis code is given in Fig. 3. This code is heavily optimized to allow immediate analysis of the data in the control room.

Another application of the triggered record program is to capture the injection transients. To synchronize the recording to the event we use the hardware trigger from the logic-level signal derived from the storage ring injection pulse. Figure 4 shows part of a time-domain record of motion due to the injection of charge into bucket 311. We observe no motion in the preceding bucket (310) and a quarter-amplitude motion in the following bucket (312). That observed motion is due to the excitation by the wake fields of bucket 311, coupling due to the finite detector bandwidth, and possible spill during the injection. The initial amplitude of motion is more than 90 degrees at the detection frequency of 2998 MHz, and the detected signal "wraps around" during the first millisecond of the transient. Since we detect a product of current and phase, magnitude of motion is less than full scale due to low current in the bunch. The motion is then damped by the action of the feedback system and the natural radiation damping. The decay of recorded bunch 311 amplitude also contains effects due to decoherence of the injected and stored charge motion.

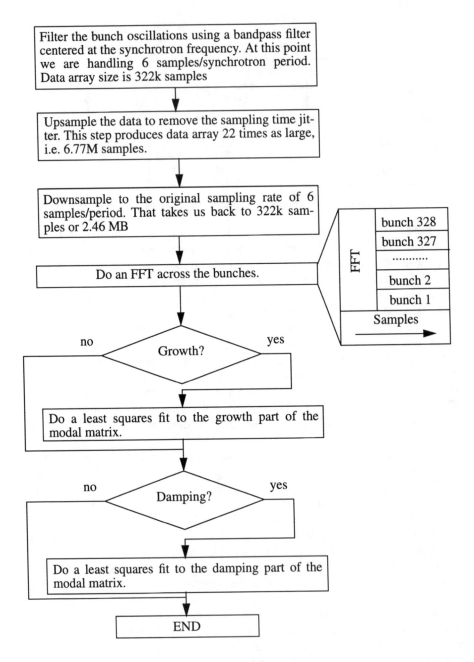

FIGURE 3. Flowchart for the modal analysis code.

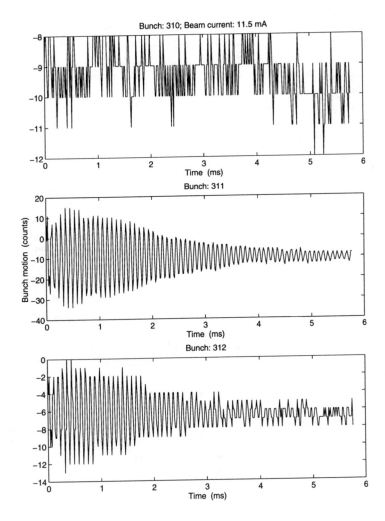

FIGURE 4. Record of an injection transient.

Diagnostic Tools

In addition to feedback filters and data recorders the software suite includes various diagnostics for setting up and calibrating the feedback system. One of the most commonly used programs is the drive program. It allows the system to drive all or selected bunches with an arbitrary excitation sequence. The sequence data is loaded in the dual-ported RAM, and the DSP program cycles through it. The sequences are only limited by the amount of the dual-ported memory available. A generator for a sine-wave sequence is available through the EPICS panel and is controlled by the amplitude and frequency desired. At the same time an interface for loading custom excitation sequences is provided where the sequence data is loaded from an ASCII file.

The drive program has many uses for feedback system setup and diagnostics. For example, driving a sinusoidal excitation at different amplitudes allows a check of the linearity of the back-end modulator and power amplifier for the operational configuration.

Another use of the drive program is to check kicker timing. For proper feedback system operation it is necessary to time align the 2-ns kick and the proper bunch. We use the drive program to excite a single bunch at the synchrotron frequency. The timing program sweeps the back-end delay line and measures the beam response at the synchrotron frequency using a GPIB-controlled spectrum analyzer connected to the A/D monitor signal. This setup is illustrated in Fig. 5. To make sure the longitudinal motion we measure is due to the kicker excitation, only one bucket in the ring is populated. Figure 6 shows the detected power at the synchrotron frequency versus delay line setting. Since the longitudinal kicker (5) is a bandpass device with a 1125-MHz center frequency, its time response has ~ 400 ps periodicity. The overall envelope of the response shows the convolution of the kicker fill time, kicker bandwidth, and amplifier rise time/bandwidth. A back-end timing program automatically takes the data and determines the best delay line setting (highest kicker gain at 8750 ps).

Another commonly used diagnostic is the front-end calibration program. To obtain a calibration factor to convert from the beam motion in A/D converter counts to beam motion in degrees of phase at the rf frequency, the DSP program first measures the DC level at the A/D by averaging a large number of samples. Then the phase of the local oscillator is shifted via program control by a known reference amount. After a delay to allow the phase shifter to settle, the input DC level is measured again. The difference between the two measurements is used to calculate the calibration factor.

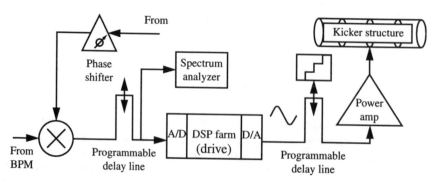

FIGURE 5. Setup for the back-end timing measurement.

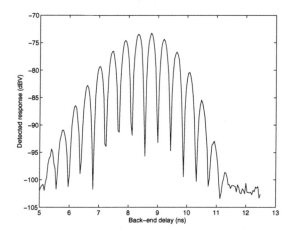

FIGURE 6. Back-end timing sweep.

SUMMARY

The programmable nature and high computational power of the longitudinal feedback system make possible a number of measurements and diagnostics in addition to the correction signal computation. The above features allow the automation of a number of otherwise tedious and slow system setup and calibration tasks. DSP-based feedback signal computation enables testing alternative feedback algorithms such as IIR filters. We are just starting to develop new techniques that take full advantage of the system's flexible architecture.

REFERENCES

1. Oxoby, G. et al., "Bunch-by-Bunch Longitudinal Feedback System for PEP-II," in *Proc. of the 4th European Particle Accelerator Conference (EPAC 94)*, London, England, June 27-July 1, 1994, 1616-1618 (1994).
2. Teytelman, D. et al., "Operation and Performance of the PEP-II Prototype Longitudinal Damping System at the ALS," in *Proc. of the 16th IEEE Particle Accelerator Conference (PAC 95) and International Conference on High Energy Accelerators*, Dallas, Texas, May 1-5, 1995, 2420-2422 (1996).
3. Claus, R. et al., "Software Architecture of the Longitudinal Feedback System for PEP-II, ALS and DAΦNE," in *Proc. of the 16th IEEE Particle Accelerator Conference (PAC 95) and International Conference on High Energy Accelerators*, Dallas, Texas, May 1-5, 1995, 2660-2662 (1996).
4. Fox, J. D. et al., "Observation, Control, and Modal Analysis of Longitudinal Coupled-Bunch Instabilities in the ALS via a Digital Feedback System," these proceedings.
5. Corlett J. N. et al., "Longitudinal and Transverse Feedback Kickers for the ALS," in *Proc. of the 4th European Particle Accelerator Conference (EPAC 94)*, London, England, June 27-July 1, 1994, 1625-1627 (1994).

Low-Intensity Beam Diagnostics with Particle Detectors

A. Rovelli, G. Ciavola, G. Cuttone, P. Finocchiaro, G. Raia

INFN-LNS, Via S. Sofia 44/A Catania, 95125 Italy

C. De Martinis, D. Giove

INFN-LASA, Via F.lli Cervi 201 Segrate (Mi), 20090 Italy

Abstract. The measure of low intensity beams at low-medium energy is one of the major challenge in beam diagnostics. This subject is of great interest for the design of accelerator-based medical and radioactive beam facilities. In this paper we discuss new developments in image-based devices to measure low-intensity beams. All the investigated devices must guarantee measurement of the total beam current and its transverse distribution.

INTRODUCTION

A radioactive ion beam (RIB) facility is now under construction at LNS in Catania (1). The Superconducting Cyclotron (CS K = 800), axially injected by a Superconducting ECR source (SERSE) (2), will deliver the heavy ion primary beams (high intensity and high energy). The secondary radioactive beams (low intensity and very low energy) will be produced by stopping the cyclotron beams on a thick target ion source complex, that provides the ions to be accelerated by means of a 15-MV HVEC Tandem (low intensity and low energy). Expected beam intensities will range from less than 10^5 pps up to 10^{10} pps with a maximum energy of some MeV/n after the Tandem. An important topic of the project is the measurement of the beam characteristics along the transport lines. This paper describes some preliminary ideas developed so far and tests that have been carried out.

The typical electrical and optical devices used in beam diagnostics can hardly operate below 10^8 pps; in these cases different devices are needed. Our goal was the development of beam detector prototypes able to work properly at low and very low intensity (i.e., below 10^8 and 10^4 pps). We are working on two different kinds of devices, an electrical one and an electro-optical one, with the aim of characterizing them in a wide range of projectiles and energies. We are investigating the possibility of their being calibrated in order to give a reliable measurement of the beam intensity and profile.

The electrical beam position monitor (BPM) is based on a microstrip gas chamber (MSGC) detector. The ionization chamber is placed transversally with respect to the beam and the strips, placed along the longitudinal axis, collect and eventually amplify the produced charges.

The electro-optical BPM consists of an electrostatic structure that conveys onto a microchannel plate (MCP) the electrons or ions produced in the residual gas along the beam pipe. After the multiplication stage, the accelerated electrons are sent onto a phosphor screen and the produced light is observed by means of a charge-coupled device (CCD) camera.

IMAGING TECHNIQUES

In the last years we carried out an extensive investigation using the light emitted by Chromox 6 alumina screens due to interaction with heavy ions (3). The experience allows us to evaluate the performance of this scintillating screen in terms of emitted light versus energy and intensity of the incident beam. The method has proven to be valuable both for the number and the quality of the simultaneous information coming from a single measurement. Sensitivity, lifetime, and spatial resolution have been carefully studied and the results have proven quite satisfactory with 10^9 pps beams. The results obtained so far, along with an ever-growing general interest in quantitative beam diagnostics by means of image analysis, has led us to investigate the use of imaging techniques in a wider range of energies and intensities. By *imaging techniques* we refer to the emitted light detection by means of a CCD camera connected to a frame grabber board fitted inside a computer-based analysis system.

The work has been concentrated in two distinct areas: a more *conservative* approach devoted to an examination of different scintillator screens directly inserted along the beam transport path (intercepting) and an *innovative* approach, aimed at integrating imaging techniques with secondary electron emission monitors (nonintercepting).

As far as the first approach is concerned, we improved the spatial resolution, the sensitivity, and the nonperturbative characteristics obtained in the above-quoted experience. Chromox 6 screens with a thickness of 0.2 mm have been used, providing a good spatial resolution (on the order of 50 μm), low-beam attenuation, and low-beam blow up. Intensity as low as 10^7 pps has been measured with such screens. To improve the sensitivity of the measure, we plan to use other scintillating screens such as ZnS(Ag), CsI(Tl), and plastic scintillators. Also the possible use of intensified CCD cameras would allow us to reduce the lower limit of the beam current.

SECONDARY EMISSION DEVICES

The nondestructive requirement for a beam detector may be met by a secondary emission device. It will measure the transversal size and intensity distribution of a particle beam by determining the position and intensity of secondary electrons or ions produced by the beam interactions with thin foils (nearly 10 μgr/cm^2) or residual gas along the beam transport line.

As far as residual gas interaction is concerned, a beam monitor prototype has been designed. It will use a microchannel plate (MCP) device coupled to a phosphor screen and to a CCD sensor (Fig. 1). Our experimental setup allows us to work in two different modes: electron collection or positive ion collection. The electrons or ions produced by the ionization of the residual gas are accelerated by a transverse electric field of suitable polarity toward the MCP that performs the electron multiplication while keeping a high degree of positional correlation between the input and the output. The output electrons are driven onto the phosphor screen placed just behind the MCP; their collision produces visible light that is collected by a CCD camera.

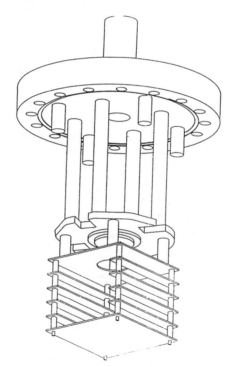

FIGURE 1. MCP setup.

We have tested an early prototype of such a detector working at a residual pressure of 10^{-6} mbar and with a sensitive length of 20 mm in order to reduce losses. Using a $^{13}C^{5+}$ beam at 86 MeV, the expected electron-ion production yield is of the order of 2×10^{10} pairs/mm/mbar/pnA; that is, 3.2 pairs/mm/mbar/particle. Neglecting the collection losses, we get a production yield of 6.4×10^{-5} pairs/particle which will give a signal of 6.4×10^{3} pairs/s with an incident beam of 10^8 pps. The obtained image is shown in Fig. 2.

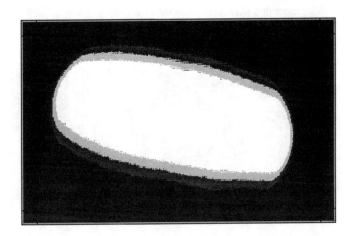

FIGURE 2. Beam trace with a $^{13}C^{5+}$ beam.

The same equipment has been tested with a $^{58}Ni^{16+}$ beam at 1.75 GeV delivered by the CS. The expected production yield was 9×10^{10} pairs/mm/mbar/pnA; that is, 14.4 pairs/mm/mbar/particle. Neglecting collection losses, we get a production yield of 2.5×10^{-3} pairs/particle which will give a signal of 10^5 pairs/s with an incident beam of 10^8 pps. Measurements have been done collecting either electrons or ions. In the second case a better sensitivity was obtained by measuring a transverse current distribution of ≈ 3.5 mm FWHM with a total beam current of $\approx 10^7$ pps, very similar to the corresponding one obtained when intercepting the beam directly on the Chromox 6. Figures 3 through 5 show the results. These preliminary results are also very interesting when considering the possibility of using CCD cameras with higher sensitivity. We evaluate that a similar setup can be used also at lower current ($< 10^7$ pps) if we increase the ionization by means of differential pumping.

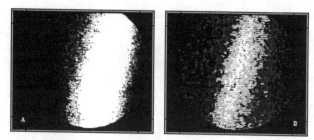

FIGURE 3. Electrons (A) and positive ions (B) collected with a $^{58}Ni^{16+}$ beam.

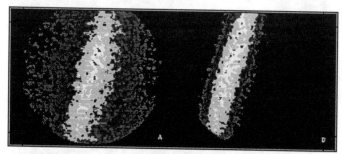

FIGURE 4. Effect of the focalization on the beam trace.

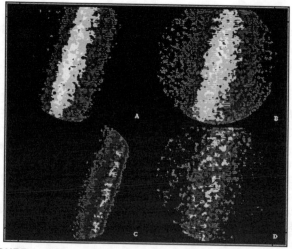

FIGURE 5. Effect of the MCP voltage change on the beam trace.

Secondary emission by means of thin foils (carbon films with 30-nm-thick or TiO$_2$ coatings) can provide a mechanism to obtain useful information from beam intensities as low as 10^3 pps with the foil transducer coupled to a device similar to the one discussed above for residual gas emission. Spatial resolutions on the order of hundreds of microns are reported in the literature (4-6). We plan to test such a transducer by measuring the level of noise available with respect to residual gas emission and studying the repeatability of the results.

LOW-INTENSITY BEAM DETECTION WITH MSGC

An alternative technique to building a residual gas-based beam monitor is to use a microstrip gas chamber (MSGC). In the last few years this type of particle detector has proven to be a good candidate for the physics at the Large Hadron Collider (LHC) accelerator. A large number of research and development projects are in progress all over the world in order to improve the features of this type of detector. The MSGC is a gas chamber with special charge-collecting electrodes: instead of the usual wires it exploits a plane of microstrips deposited onto a suitable substrate (7). A lot of discussion has arisen concerning the nature of this substrate, but the most widely accepted substrate is made of a special semiconductive glass (8,9).

We plan to develop a beam monitor by using an MSGC plate as a charge collector, exploiting an accelerating electric field (as in the MCP case) to convey the electrons produced by the beam in the gas. This plate would be placed parallel to the beam axis. A suitable choice of gas pressure in the chamber (closed by two thin plastic windows) and of voltage between anodic and cathodic strips could also give rise to a signal amplification, with a gain of 10^2 to 10^3. By sampling the signals from the strips (the usual pitch is 200 µm) one can get information about the beam profile in one direction. Therefore by placing two detectors in series, rotated 90° from each other, there should be a complete reconstruction of the XY profile.

Due to the fact that in our application the MSGC is not hit by the incident particles, and that the intensity is quite a bit lower than that foreseen by the LHC applications, our plates do not need a special substrate. Therefore we can use very cheap MSGCs deposited on normal glass to build our detectors. Our goal was the development of a prototype of the chamber equipped with the readout electronics, in order to perform a complete characterization as a function of the operating variables. At this moment, the assembling of the final prototype of the chamber with microstrips on a 5 × 5 cm D263 glass, 200-µm pitch is in progress. First tests (vacuum tightness and strips electronics) were performed with the preliminary prototype (10). Beam shifts are scheduled before the end of this year. At that time we will perform a series of response calibrations with different beam types and energies, in order to define an operating diagram suitable for our range of applications.

CONCLUSIONS

The measurement technique with the MCP allowed us to define two important characteristics of our experimental setup: dynamic range and working mode (positive or negative). In the range of 10^9 to 10^6 pps the actual configuration guarantees a very efficient technique (nondestructive) as an on-line monitor for position, size, and current distribution of the beam. The use of thin foils allows us to decrease the lower limit to 10^3 pps with the same performance. The difference we found working in the two modes (electron or positive ion collection) is that the ion collection offers a better signal-to-noise ratio and maintains the spatial information due to the lower diffusion drift of the ions. The tests on the MSGC setup are very satisfactory, and we hope to confirm this trend with the beam tests as soon as it is ready.

REFERENCES

1. R. Alba et al., "Radioactive Ion Beams Facilities at LNS," Proc. RNB III, 31-38, (Frontieres: 1993).
2. G. Ciavola et al., "The SERSE project," *Rev. Sci. Instrum.* **65**, 1057 (1994).
3. G. Cuttone et al., "Heavy Ion Beam Diagnostics at LNS," Proc. 2nd European Workshop on Beam Diagnostics and Instrumentation for Particle Accelerators, Travemunde, Germany, 1995, 84 (1995).
4. R. Jung, "Screens versus SEM Grids for Single Pass Measurements in SPS, LEP and LHC," Proc. 2nd European Workshop on Beam Diagnostics and Instrumentation for Particle Accelerators, Travemunde, Germany, 1995, 57-59 (1995).
5. P. Heeg, "Intensity Measurements of Energetic Heavy Ion Beams," Proc. 1st European Workshop on Beam Diagnostics and Instrumentation for Particle Accelerators, Montreux, Switzerland, May 3-5, 1993, CERN PS 93-35 BD, CERN SL 93-35 BI (1993).
6. R. Odom, "Non Destructive Imaging Detectors for Energetic Particle Beams," *Nucl. Instrum. Methods B* **44** (4), 465-472 (1990).
7. Proceedings of the Int'l. Workshop on Micro-Strip Gas Chamber, Padova, 1994, G. Della Mea and F. Sauli, Eds., (Progetto Padova: 1994).
8. G. Della Mea et al., Proceedings of the Int'l. Workshop on Micro-Strip Gas Chamber, Padova, 1994, G. Della Mea and F. Sauli, Eds., 30, (Progetto Padova: 1994).
9. R. Bouclier et al., Proceedings of the Int'l. Workshop on Micro-Strip Gas Chamber, Padova, 1994, G. Della Mea and F. Sauli, Eds., 39, (Progetto Padova: 1994).
10. P. Finocchiaro, private communication.

Operational Experience with the CEBAF Beam Loss Accounting System*

R. Ursic, K. Mahoney, K. Capek, J. Gordon, C. Hovater, A. Hutton,
B. Madre, M. Piller, C. Sinclair, E. Woodworth, W. Woodworth

*Continuous Electron Beam Accelerator Facility, 12000 Jefferson Ave.
Newport News, VA 23606*

M. O'Sullivan

Ford Motor Co.

Abstract. The Continuous Electron Beam Accelerating Facility (CEBAF) is a new generation particle accelerator for basic research in nuclear physics. After a successful commissioning period, we started delivering beam to first experiments in November 1995, with beam availability of better than 70%. In this paper we present specifications, design, and discussion of operational experience with the personnel safety and machine protection implementation of CEBAF's beam loss accounting system. The diagnostics section of this system uses low Q TM_{010} stainless steel cavities as beam current monitors to sample beam current throughout the beam path. The associated rf front end and analog signal conditioning electronics output signals proportional to beam current or beam loss, depending on the implementation. The personnel safety and the machine protection systems then use this current or loss information to turn off the beam when hazardous conditions occur. This system, implemented in its final configuration in January 1996, was developed in one year and is performing to specifications.

INTRODUCTION

In the CEBAF accelerator superconducting cavities operating at 1497 MHz accelerate electrons to 4 GeV energy in five passes. This unique facility is capable of delivering continuous wave (CW) electron beams with currents that span a 1-nA to 200-µA range, with energies from 800 MeV to 4 GeV, to facilitate nuclear physics research at the quark level.

An errant beam can produce a hazardous amount of radiation in areas where personnel may be present. This is the domain of the personnel safety system (PSS). An errant beam with such high power density can also damage the machine components considerably. This is the domain of the machine protection system (MPS). CEBAF's beam loss accounting (BLA) system provides protection in these two domains.

---

* Supported by U.S. Department of Energy Contract #DE-AC05-84ER40150.

PERSONNEL SAFETY SPECIFICATIONS

The PSS protects people from exposure to prompt beam-induced radiation (1). During normal operations, the PSS issues an active permission signal, which is a 625-kHz square wave. When a fault occurs, the PSS withdraws the permission, causing the removal of the beam at the electron gun. As a second level of protection, kicker magnets deflect any beam to a water-cooled aperture on PSS faults. Figure1 shows the location of the beam current monitors (BCMs) assigned to the PSS. These BCMs serve the following functions:

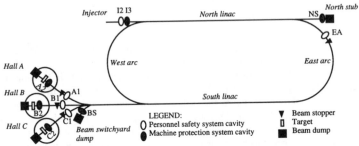

FIGURE 1. BCM cavities location along the CEBAF beamline.

1. Beam stoppers near the entrance to the experimental halls protect personnel from the beam during access to the experimental areas. When the stoppers are in place, de-energized magnets prevent the beam from going in their direction. Should this safety feature fail, it is conceivable that a beam of high power density may tunnel through the stoppers. The BCMs (EA, A1, B1, and C1) protect the stoppers by causing the termination of the beam when they detect a beam current greater than 1 µA.

2. CEBAF's contractual requirement imposes an operational limit on the beam current. The BCM (I2) ensures that the beam current in the injector does not exceed 190 µA. Hall B BCM (B1) sets a 1-µA limit to prevent the possibility of errant beam going up towards the counting house.

MACHINE PROTECTION SPECIFICATIONS

The MPS protects the accelerator and end station equipment from beam-related damage. During normal operations, the MPS issues an active permission signal, which is a 5-MHz square wave. When a fault occurs, the MPS withdraws the permission, causing the removal of the beam. Figure 1 shows the locations of the BCMs for machine protection. The BLA system performs the following functions.

1. The BLA system limits the beam current in the injector to 180 µA. This ensures that the MPS will trip before the PSS in all cases.

2. If i(I3), i(NS),.., represent the beam current values in BCMs at I3, NS,...,
the BLA system calculates the beam loss by computing i(I3)-i(NS)-i(BS)-i(A2)-
i(B2)-i(C2). If this value exceeds 2.5 mA, the system will terminate the beam.

3. In case of an instantaneous beam loss, the BLA system will terminate the
beam such that the integrated beam loss does not exceed 25000 µA-µs (2,3).

DESIGN OVERVIEW

The Cavity

The BCM is a stainless steel pillbox cavity with an intrinsic Q of 3100 that
operates in the TM_{010} mode at 1497 MHz (Fig. 2). This cavity has a larger
bandwidth and a better temperature stability than a copper or other high-Q cavity.
A micrometer stub tuner adjusts the resonant frequency of the cavity. The cavity
specifications are listed in Table 1.

FIGURE 2. The CEBAF beam current
monitor pillbox cavity, three-port
configuration.

TABLE 1. Cavity specifications

Mode	TM 010
Frequency	1497 MHz
Q_0	3100
R/Q (cavity center)	93 ½ *
Cavity Radius	7.74 cm
Cavity Length	7.50 cm
Beam Pipe Radius	1.75 cm
Material	304 Stainless Steel

* $R=V^2/2P$

The cavities in the MPS system have two couplers per cavity, whereas the
cavities in the PSS system have three couplers per cavity. The three-coupler
cavity provides a redundant output channel, which is a PSS requirement.
Magnetic field loops situated around the cavity accomplish the coupling in and
out of the cavity. The loops mount on 2.75" rotatable Conflat flanges.

In the MPS cavity, port 1 has the large area loop and port 2 has the small area
test loop. Rotating the large area loop facilitates critical coupling and the
resulting loaded cavity Q is approximately 1550. Adjustment of the small area
test loop sets the insertion loss between the two ports to 13.5 dB.

In the PSS cavity, ports 1 and 2 have the large area loops and port 2 has the
small area test loop. By carefully adjusting the large area loops, it is possible to
make the two loops identically undercoupled, while the combination results in
critical coupling and a loaded cavity Q of 1550.

Standard Electronic Modules

The 'standard' electronic modules are the downconverter, a test source, an rms-to-DC converter, an equalizer, and an integrator. Figure 3 shows the standard configuration.

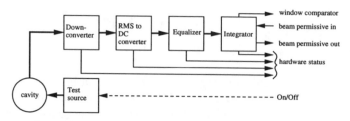

FIGURE 3. Standard modules and their configuration.

The downconverter shifts the 1497-MHz spectral component of the cavities to 1 MHz. The downconverters and the 1497-MHz test source reside near the cavity in the accelerator tunnel. Such proximity minimizes cable attenuation drifts due to temperature variation. Low loss, 10-m-long 0.5" Heliax cables connect the cavity to the downconverter and the test source. The same type of cable connects the downconverter to the electronics in the service buildings.

An rms-to-DC converter converts the bandpass filtered 1-MHz signal to a baseband signal. This module uses the AD637 rms-to-DC converter integrated circuit (IC) (4), which has excellent linearity and temperature stability.

An equalizer module compensates differences in cable losses. The output of the equalizer scales with the beam current.

All the above modules except the equalizer can be calibrated in the laboratory and interchanged with the same type in the field. The equalizer is specific to each BCM and is calibrated *in situ*.

The integrator monitors the average beam current or beam loss. When the beam current or beam loss exceeds a preset threshold, the integrator withdraws the permissions.

All modules conform to the 3U Eurocard standard. The interconnection among modules is through the back plane. Blindmate snap-on rf connectors carry the 1497-MHz signals, while 96-pin DIN connectors carry the baseband and digital signals. Keyed connectors on the modules and the backplane prevent insertion of the modules in the wrong place. Each module in the system is also equipped with a supervisory circuit to ensure safe operation. Upon detecting an error in its operation, a module will generate a fault signal, which prevents beam from leaving the injector.

Implementation of the PSS Part of the BLA System

The PSS part of the BLA system consists of four cavities at the beam stoppers and one cavity in the injector. The large-area coupler connects to a downconverter. Figure4 shows the PSS logic. The PSS permissive from the electronics associated with the BCMs at EA, A1, B1, and C1 go to the injector. At the injector, these permissives pass through logic associated with the BCM I2. The latter limit the beam current to 190 μA. For the stopper BCMs two modes of operation exist: one with beam stopper in place and the other without the stopper. For a given mode, faults occur if the beam current in any stopper BCM exceeds 1 μA when the beam stopper is in place. Occurrence of a fault results in the PSS permissive going inactive and consequent removal of the beam.

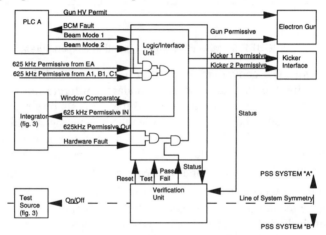

FIGURE 4. PSS BCM implementation block diagram.

The BLA system for the PSS also contains a verification unit, which is a time sequencing state machine. Through the control system interface to the verification unit, the operators can test the BLA system and remotely reset any BCM fault.

Implementation of the MPS Part of the BLA System

The MPS part of the BLA system consists of six cavities. The small-area coupler connects to the test source and the large-area coupler connects to a downconverter. Cables carry the signals from the downconverters to the beam switch yard service building (Fig. 5).

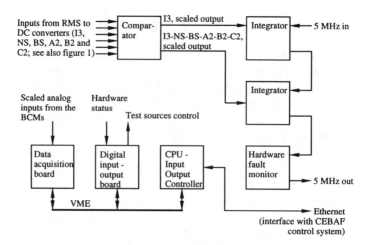

FIGURE 5. MPS BCM implementation block diagram.

The rms-to-DC converters process the signals. The MPS implementation requires a slight modification to the standard configuration of Fig. 3. This modification results in a comparator board that consists of six equalizers and a circuit that calculates the difference between the injector signal and the sum of the other five BCM signals. The comparator board has two outputs, each of which goes to an integrator. One output is used when the beam current in the injector exceeds 180μ A. The other output is used when the average beam loss exceeds $2.5\ \mu A$ or the instantaneous beam loss (above $2.5\ \mu A$) exceeds $5000\ \mu A\text{-}\mu s$. Both outputs force the integrators to withdraw MPS permissions.

A 1-inch-thick layer of thermal insulation around the cavities along with heating tapes, thermocouples, and controllers maintain the cavities' temperature at 110oF. This minimizes frequency de-tuning due to ambient temperature change.

The MPS BLA software continuously monitors the current readings, identifies failed hardware devices, and permits system testing and verification. The user interface consists of two components. The overview screen is a diagram of the accelerator with bar charts and readbacks showing the current reading on each BCM. Operators use this screen during normal operation of the machine. Additional information, namely, the hardware status readbacks, are available on the expert screen.

CALIBRATION METHOD

Calibration of the BLA system was a two-step process. The first step was to tune the cavities using a network analyzer. The second step was to calibrate the system. At the cavity end of the cable (where the cavity connects to the downconverter) we connected a 1497-MHz source of a known signal level. We then adjusted the DC signal level at the equalizer (PSS) or the comparator (MPS) to obtain an output that scales with the beam current.

PERFORMANCE

The BLA began operation in January 1996. One of the conditions to extend the contractual requirement of CEBAF's operational envelope from 120kW to 800kW was to demonstrate the reliability of the BLA system under normal operations. An independent panel of experts tested and verified the PSS system. Following is the summary of the operational experience during the first three months.

PSS System

We verified all the system requirements during acceptance testing. A Faraday cup in the injector provided a reference for absolute current measurement. Initial calibration showed an average of 13% difference between PSS BCMs and the Faraday cup measurements. After calibrating the injector PSS BCM with the beam, we investigated the cause of the discrepancy. Since we did not consider VSWR mismatch between the cavity ports and the 50-Ω transmission system during our calculations, we hypothesized that this mismatch was the source of the error. If this hypothesis is correct, then this error should be common to all PSS BCMs. In order to test this idea we applied to all PSS BCMs, the same amount of gain offset applied to the injector BCM. With these corrections, the measured maximum absolute error between the Faraday cup and the BCMs was < 1%.

MPS System

Calibration of MPS cavities showed an average of 7% difference with respect to the Faraday cup measurements. Relative accuracy, or match, between the injector and the end station BCMs is a very important characteristic of the MPS part of the BLA system. In this regard, reproducibility of the calibration results is the key issue. First calibration failed due to a yet unknown problem; the hall C BCM was off by 15%. In the second attempt we obtained a relative accuracy between the injector BCM and the three end station BCMs that was better than 1.5%. We believe that this is the maximum relative accuracy that one can obtain following the above-described calibration method. However, in order to reliably operate the MPS system above 100 μA, the relative accuracy must be less than 1%. The design team is currently evaluating an option to use the beam to "fine" calibrate the MPS BCMs after a "coarse" calibration with the above procedure. Unfortunately, this method demands confidence that we have 100% transmission between the injector BCM and the BCM under calibration.

The first two months of operational experience gave us an opportunity to estimate the long-term stability of the system. We used two methods to evaluate this. The first one relied on the operator-performed weekly confidence tests. The operator turned on a test source associated with each cavity and measured the response. The disadvantage of this method is that any drift in the test source output appears as a drift in the BCM response. The purpose of this test is not to calibrate the system but to verify that the BCM's response was within 5% of the expected performance. The second method uses archived data for CW beam runs and compares readings from different BCMs. The uncertainty associated with this method is that we are never sure if we have 100% transmission between the BCMs whose responses we want to compare.

TABLE 2. Long-term stability data for the period March 13 to April 29, 1996.

Parameter	Drift	Method
Off-set (no-beam) drift	± 0.1 µA	Beam off, test source off
PSS BCMs	± 1.5%	Regular tests using a test source
MPS BCMs	± 1%	Regular tests using a test source
MPS BCMs relative to each other	± 0.7%	CW beam

Analog signals proportional to beam current and loss are available in the control room. This diagnostic feature of the system proved to be very useful. Figure 6 shows waveforms for a typical operation of the system.

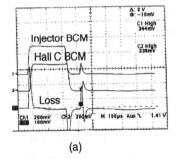 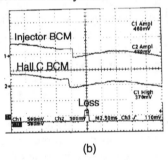

(a)   (b)

FIGURE 6. (a) Machine operators use 250-µs-long pulses to tune the machine. Spikes on the loss signal are due to delay between the BCM readings. The little pulse following the 250-µs one is a modulation pulse for multipass BPMs. (b) MPS BLA detected loss and removed the beam at the gun. The injector and hall C BCM signals pick up 60-Hz noise on their way from the electronics to the control room (500 m). Transmission of the loss signal over commercial off-the-shelf AD-serial link-DA devices is the reason for the absence of the 60-Hz noise on this signal. The scaling factor is 40 µA/V.

CONCLUSION

Both the PSS and MPS implementations of the beam loss accounting have met the performance goals required to safely operate the CEBAF accelerator and end stations at full design current. In addition, the first few months of operational experience demonstrate that the system is reproducible and stable and that it provides a useful diagnostic tool to machine operators.

ACKNOWLEDGMENTS

Many people helped shape this unique product. Realization of this complex system, within just one year, would not have been possible without the help from many persons. We thank T.Grummel, B. Smith, and J. Coleman for technical support; J. Mammoser and D.Ouimette for valuable technical input during the review and acceptance phase; H.Areti for valuable input during the development phase of this project; machine operators for excellent support and feedback during the implementation phase.

REFERENCES

1. A. Hutton et al., "Beam Current Monitors for Personnel Safety and Machine Protection," Rev. 1/10/95.
2. C. Sinclair, "Time Response Requirements for the BLM/FSD System," CEBAF -TN-92-046, October 21, 1992.
3. C. Sinclair, "Beam Loss Monitor Performance Requirements," CEBAF-TN-94-024, April 6, 1994.
4. "AD637 High Precision, Wideband RMS-to-DC Converter," Special Linear Reference Manual 1992, p. 4-23, Analog Devices, Norwood, MA.

A Multi-Batch Fast Bunch Integrator for the Fermilab Main Ring

G. Vogel, B. Fellenz, J. Utterback

Fermi National Accelerator Laboratory
P.O. Box 500, Batavia, IL 60510

Abstract. In order to support new multi-batch coalescing scenarios in the Fermilab Main Ring and Main Injector, a new fast bunch integrator system has been developed and installed. This VME-based system provides the capability to measure both the batch and central bunch intensities for up to 12 proton or 4 antiproton batches at a time in the accelerator. The system provides for variable batch lengths of up to 15 bunches and central bunch spacing down to 21 rf buckets (394 ns). A new dual-channel fast integrator circuit has been designed for the system which attains 50 dB of dynamic range with programmable integration windows to 18.8 ns in length with 2-ns rise/fall times.

INTRODUCTION

Higher-luminosity operations in colliding beam physics can be achieved in many ways including the use of higher-intensity bunches for the collisions. One of the ways this is accomplished at Fermilab is to take a group of bunches (batch) and, through a process of rf manipulation known as coalescing, convert them into a single bunch containing the majority of their previous intensity. This process is presently carried out in the Main Ring.

In order to determine the efficiency of this process and monitor where losses occur during it, a real-time intensity measuring system known as a fast bunch integrator (FBI) was developed. Utilizing a signal from the Main Ring resistive wall current monitor, this system was able to measure the intensity of both the original batch and its central bunch by using both wide-gated and narrow-gated integrators. A comparison of these two measurements after coalescing provides a measure of the coalescing efficiency. The original FBI system was capable of measurements on a single batch.

As another way of increasing luminosity the Tevatron is moving to operations using more coalesced bunches (36 on 36 vs. 6 on 6). To accomplish this, multiple batches will be coalesced in the Main Ring simultaneously for injection into the Tevatron. This multi-batch operation involves coalescing up to 12 proton or 4 antiproton (pbar) batches at one time. In support of the new multi-batch operations a new Main Ring FBI has been developed. The new FBI system supports fast time plots, snapshot plots, and datalogging of bunch intensity for all bunches as well as providing sums of bunch intensities (for use as transfer qualifiers) for both wide and narrow integration gates. The narrow gates provide the integrated intensity for individual bunches (P1-P12, A1-A4) while the wide gates provide the integrated intensity for an adjustable-sized batch centered on each narrow-gated bunch. The system has been scaled for a maximum bunch intensity of 4×10^{11} particles per bunch. The new FBI is a VME-based system utilizing a Motorola MVME-162LX embedded computer. It has been designed to provide sufficient functional flexibility to be used in the main injector with minimal software modification (only that required to adopt the gating for a machine with a different harmonic number). A typical fast time plot of coalescing studies using the new FBI system is shown in Fig. 1. Beam is seen entering the machine at approximately 1 s with the lower trace being that beam in the central bunch. The beam is accelerated up to 150 GeV with some losses and at about 3 s coalescing takes place. In this case not all the beam has been recaptured in the central bunch as is seen by the difference between the wide (M:P1IWG) and narrow (M:P1ING) gate plots.

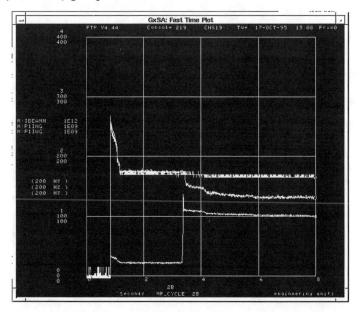

FIGURE 1. FBI plot of single-bunch coalescing.

SYSTEM HARDWARE

The system is VME based using an embedded controller with an ethernet interface to the accelerator control system. The hardware consists of a timer board (VRFT), a Tevatron clock decoder (VUCD), two digitizers (Omnibyte Comet), and two gated integrator modules along with the embedded processor (Motorola MVME-162LX) in a modified Tracewell VME chassis. A block diagram of the system is shown in Fig. 2. While most of the system hardware is either commercial or Fermilab AD/Controls general-purpose modules, the dual-channel gated integrator modules were specifically designed for this project. This was due to the high speed requirements of the analog processing. The integrator circuit needed to be capable of being gated with integrate windows as short as 18.8 ns with 2-ns maximum rise and fall times, cleared with reset pulses of 50 ns, and be ready to integrate again all within 384 ns. This module is described below.

The resistive wall current monitor is a wide-band (2 GHz), AC-coupled, low-impedance (0.16 ohms) beam current detector. A full description of these types of detectors can be found in ref. (1).

The Motorola MVME-162LX is a standard configuration, 68040-based, single-board VME computer. It serves as the system network (ACNET) interface via its ethernet connection and as the VME bus master. It is configured with VxWorks (Wind River Systems) as its operating system. Software is written in the C programming language and downloaded via the network. The MVME-162

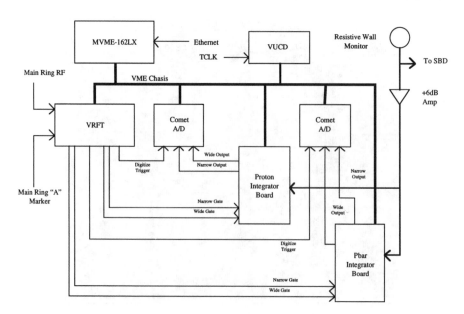

FIGURE 2. Main Ring FBI block diagram.

has 4 MByte of DRAM and 128 k of battery-backed RAM which is used to store system settings.

The VME universal clock decoder (VUCD) and VME radio frequency timer (VRFT) are standard AD/Controls VME clock decoder boards. The VUCD provides Tevatron clock decoding in order to support plotting and datalogging while the VRFT uses Main Ring rf (53 MHz) and a beam sync marker to generate the system data acquisition gates.

The Omnibyte Comet VME A/D board is a 12-bit, 5-MSa/s VME digitizer with 64 k words/channel of memory depth and an external trigger. Two channels (0,1) on each board are used to sample the wide- and narrow-gate integrated intensities based on an external trigger.

Dual High-Speed Integrator

The dual high-speed integrator board (Fig. 3) is a two-channel, fast-gated integrator circuit in a 6U VME form factor. The channels can be configured with jumpers to be either fully independent or have common signal or common trigger inputs.[*] For the FBI each board is configured for a common input but independent gates for the two integrator channels. The gate width is fixed in hardware at 18.8 ns for the first channel (via jumper) but uses the trigger gate width,

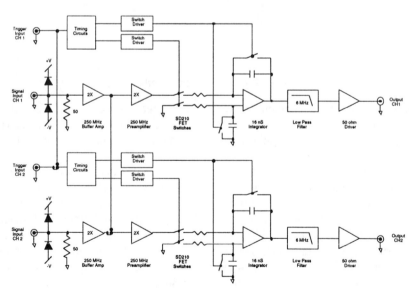

FIGURE 3. Integrator board block diagram.

---

[*] A valuable source of information on the design of high-speed integrator circuits can be found in ref. (2).

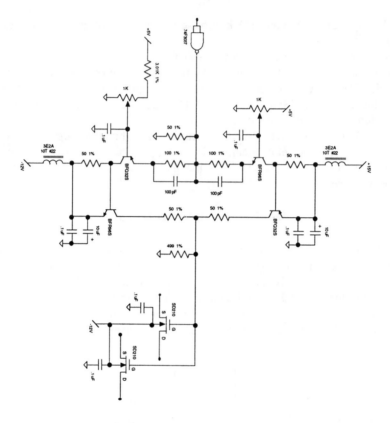

FIGURE 4. FBI switch driver schematic.

controllable from a parameter page, for the second. To accomplish the high-speed switching required for the 18.8-ns gates, Siliconix SD210 DMOS FET switches were selected and fast bipolar drivers (Fig. 4) were designed. The 1-V, full-scale output corresponds to 4×10^{11} particles.

The measured rms noise level at the output is 1.9 mV, or 0.19% of full scale and is consistent with the noise specification for the AD843 amplifier used to form the integrator. The 1/f noise below 2 kHz provides the dominant component. This part was selected for its low noise, 34-MHz bandwidth, and small bias current by virtue of the FET input. The buffer amplifier used to optimize the signal level at the integrator switch has 250 MHz of bandwidth but contributes little noise. Linearity was measured as a function of both input pulse amplitude and width. In both cases, the errors were dominated by the 0.19% of full-scale rms noise. The rise and fall time of the switches provide ± 0.5% amplitude accuracy through a 15-ns window for the 18.8-ns-wide gate.

Because the resistive wall monitor used for the FBI is AC coupled, the baseline varies depending on bunch intensities and spacing. With 12 bunches spaced by 21 rf buckets, the baseline is estimated to change by ± 0.1% between turns. To correct for this and remove constant integrator errors, the baseline is measured every other turn and subtracted from the measured intensities in software. The accuracy of this procedure depends on the distribution of bunches and where the baseline is sampled.

SYSTEM SOFTWARE

The system software is a compiled C program running in the VxWorks operating system of the embedded processor. It handles all the data, reads the digitizers, sorts data, and performs the proton and pbar summations (3). Data display is provided via standard control system fast-time and snapshot plots, parameter pages, or a local terminal.

Embedded Processor Application

A compiled C program running on the embedded processor performs high-speed data transfer across the VME backplane along with sorting and summing of the transferred data. The system reads a complete data set when a data request has been issued. The integrated intensities of the 12 proton bunches are digitized by the proton Comet board as gated externally by the VRFT. The intensity data is interleaved (P1-P11 odd, P2-P12 even) in the digitizer FIFO memory. Channel 0 carries the narrow-gate data while channel 1 holds the wide-gate data. The integrated intensity data for the 4-pbar bunches reside similarly in the pbar digitizer. This data is read out and sorted and used to calculate narrow and wide gate sums for protons and pbars. It then returns only those bunch intensities or sums that have been requested.

When a change of VRFT settings is initiated by a console page, data acquisition stops, the VRFT is disabled, the new memory pattern is configured and downloaded to the VRFT, and it is re-enabled. The pattern is triggered on receipt of a request from ACNET or a local terminal and plays over two turns. The pattern is configured with an integrator prepulse (to clear any integrator drift since the last measurement) and odd bunch gates on the first turn followed by even bunch gates and background sample on the second. A partial sample of the gating scheme for protons is shown in Fig. 5.

Console Control

The ability to change the 12-proton gate delays, the 4-pbar gate delays, the proton and pbar zero sample times, as well as the proton and pbar wide-gate

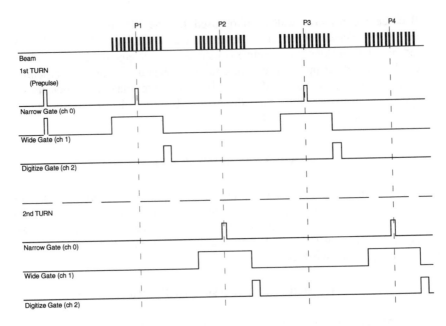

FIGURE 5. FBI system timing for bunches P1-P4.

widths is provided via a console page application. Bunch delay timing for the system is controlled via the digital-to-analog (D/A) setting of the narrow-gate intensity parameter for each bunch and zero sample. This setting is defined in rf buckets. Default bunch spacing is 21 buckets. The standard wide-gate setting for protons and pbars is 13 buckets. As each setting is changed it is sent to the local processor for use in reconfiguring the VRFT. Data can be displayed using standard console plotting packages (fast time plot and snapshot plot) as well as by monitoring the parameter page. Fast time plot rates of up to 720 Hz are supported and a maximum sample rate of 23 kHz is available on the snapshot plots.

CONCLUSIONS

The system has been operational in the Fermilab Main Ring since October 1995. It has provided all the Main Ring FBI data for both the 36 on 36 operation studies and the last months of Collider Run Ib. It has proven a reliable, easy-to-use, and easy-to-maintain system for real-time individual bunch intensity monitoring.

ACKNOWLEDGEMENTS

We would like to thank J. Crisp for his assistance with the design and analysis of the high-speed gated integrator, C. Briegel for his assistance with the embedded processor software, and S. Moua for his assistance in assembling the system.

REFERENCES

1. Webber, R.C., "Longitudinal Emittance: An Introduction to the Concept and Survey of Measurement Techniques Including Design of a Wall Current Monitor," *AIP Conference Proceedings* **212**, 85-126 (1989).
2. Wang, X., "Ultrafast, High Precision Gated Integrator," *AIP Conference Proceedings* **333**, 260-266 (1994).
3. Utterback, J., and Vogel, G., "Fast Bunch Integrator, A System for Measuring Bunch Intensities in Fermilab's Main Ring," in *Proc. of ICALEPCS 95*, Chicago, IL, October 29, 1995, to be published.

Intensity Measurement of High-Energy
Heavy Ions at the GSI Facility

P. Forck, P. Heeg*, A. Peters

Gesellschaft für SchwerIonenforschung, Planckstrasse 1, D-64291 Darmstadt, Germany

Abstract. The intensity of the high-energy heavy ion beam, extracted from the heavy ion synchrotron (SIS) at the GSI, Darmstadt, is determined with three types of detectors: scintillator counters, ionisation chambers (ICs), and secondary electron monitors (SEMs). The accuracy of the ICs, compared to a general formula, is about 25%; for the SEMs it is within a factor of two. Intensity limits for the IC are measured. A SQUID-based current comparator is in preparation for an absolute determination of the primary ion current.

INTENSITIES OF SLOWLY EXTRACTED IONS FROM SIS

The Gesellschaft für SchwerIonenforschung (GSI) heavy ion facility can deliver the whole spectrum of heavy ions in a wide energy range. Using the heavy ion synchrotron (SIS), ions from p to U can be accelerated to energies from 50 MeV/u to 2 GeV/u. The number of stored particles in the SIS lies between 10^5 for rare isotopes and a few times 10^{10} for Ne, for example. The future upgrade of the accelerators will deliver higher currents up to the space charge limit of the SIS. This is, for example, about 3×10^{11} Ne ions or 5×10^{10} Xe ions (1), see Fig. 1. Most of the experiments use the slow extraction mode with a spill length between 0.5 s and 10 s. This is equivalent to about 10^4 pps (particles per second) and 10^{11} pps, or, expressed in electrical current between 0.1 pA and 100 nA. These low, quasi-DC currents cannot be measured with standard current transformers. Instead, other techniques are used, such as directly counting low numbers of extracted ions with plastic scintillators. For an intermediate range of particles, gas-filled ionisation chambers are used. For several decades this device has been used mainly for proton beams at CERN (2) and BNL (3) and for heavy ions at LBNL (4) and KEK (5). As shown below, this device has some saturation behaviour for particle rates higher than about 10^9 pps for heavy ions (depending on the ion species). Secondary electron monitors are also used to cover the higher intensity range. Again, these devices have been used a long time, e.g., at CERN (6); however, their absolute precision is not as high as for the ionisation chambers.

---

*Kernphysikalische Detektoren, Niedergartenweg 15, D-64331 Weiterstadt, Germany.

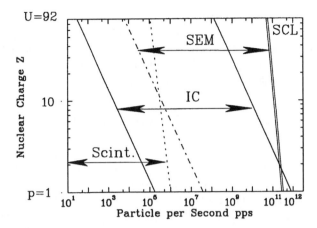

FIGURE 1. The three types (scintillator (Scint.), ionisation chamber (IC), and secondary electron monitor (SEM)) of currently used intensity detectors are compared for ions with nuclear charge Z, a kinetic energy E_{kin} = 1 GeV/u, and a beam spot size of 1 cm^2. The lower limit of the IC and SEM correspond to 1 pA for the secondary current, and the upper limit of the IC corresponds to a dose rate of 200 Gy/s. The space-charge limit (SCL) of particles inside the SIS is also drawn.

As shown in Fig. 1, the overlap between the three detectors is large enough to perform an absolute calibration. The aim is to get a precision of 5% in the intensity measurement for all ions beams. As it is cumbersome to do a calibration for each ion beam, an automatic algorithm is desirable.

We developed a compact array of these three detectors, mounted on one 200-mm-diameter flange. These arrays are now installed at seven different locations; six additional devices are in preparation.

The highest currents of extracted ions, e.g., 10 nA of Ne^{10+} (corresponding to about 10^{10} ions), will be measured in the future with a beam transformer using the cryogenic current comparator (CCC) principle. This completely different detector is currently being investigated to get absolutely calibrated results with a nondestructive measurement method.

THE IONISATION CHAMBER

Outline of the IC

The IC is mounted inside a cylindrical pocket, which can be driven to the beam path. It is separated from the vacuum by 100-μm stainless steel. The electrodes of the chamber are made of 1.5-μm Mylar and are coated with 100-μg/cm^2 silver. They are separated by 5 mm and an electric field between 1 kV/cm and 4 kV/cm is applied. The size is 64 × 64 mm^2 and a mixture of 90% Ar and 10% CO_2 is used.

The ions transverse the chamber perpendicular to the surface. The advantage of this geometry is the homogenous electric field and the short drift length of the secondary electrons and ions; therefore, it is also suitable for single particle counting. One important disadvantage is the higher probability of recombination of electrons and ions due to the mutual penetration of the charge clods.

Calculation of the Secondary Charge

The output of secondary charge Q is proportional to the energy loss of the ion inside the the active gas volume of 5 mm. This energy loss is calculated according to a formalism described by Ziegler (7), based on the Bethe-Bloch equation. The energy loss of a particle with kinetic energy per amu E_{kin} and the nuclear charge Z scales approximately with $\Delta E \propto Z^2/E_{kin} \cdot \ln (E_{kin}/W)$. Ziegler modifies this mainly by using an effective nuclear charge, fitted from a large set of data. Using the average energy for the production of one ion-electron pair W, which is known from measurements with lower energy particles (8), the output charge Q of the IC for one ion is expressed as $Q = e \cdot \Delta E/W$ with e being the elementary charge.

By using a 90% Ar + 10% CO_2 gas mixture ($W_{Ar} = 26$ eV and $W_{CO_2} = 33$ eV) a high count rate and a good timing should be possible. The recombination probability depends on the ionisation density, and thus on the beam intensity, the ion velocity, and the chamber length. A distorted field due to space charges amplifies the effect. Previous investigations (4,9) showed that recombination effects become important at dose rates above 20 Gy/s in air or nitrogen, whereas in gases which do not form negative ions, the limit is expected at about 200 Gy/s, for an electric field of about 2 kV/cm (9).

Measurements

Calibration measurements against the scintillator have been performed with a lot of different ion species. A typical beam diameter was 10 mm FWHM. The region of linearity between IC and scintillator extends up to about 10^6 pps. For higher count rates the scintillator starts to saturate.

To get the experimental (and in some sense the theoretical) precision of the IC, a comparison of the calculated and measured calibration factors f is shown in Fig. 2. First, it shows that light particles give more than two orders of magnitude lower output charge than heavy particles. The measured and calculated calibration factors are linear over three orders of magnitude, i.e., there is no systematical deviation visible. In addition, it proves that one can use the W values previously determined from lower energy measurement. The deviation, defined as $(f_{ex} - f_{theo})/f_{theo}$, is shown in the upper part of the figure. One can see a deviation of about 25%. There might be some experimental problems, like the uncertainty of the exact pressure and mixing ratio of the gas, or some offset problems of the current-to-frequency converter. To reach the desired accuracy of 5%, these parameters

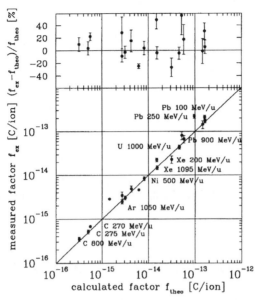

FIGURE 2. Comparison between measured f_{ex} (i.e., calibrated against the scintillators) and calculated f_{theo} calibration factors for different ions. Some representative ion species are labelled. The error bars are due to uncertainties in the fit. In the upper part the deviation between the calculated and measured factors is shown.

have to be controlled more accurately; e.g., the counting gas will be changed to pure Ar as soon as the technique is possible.

As an example of the high intensity limits, the output of the IC is shown in Fig. 3 for $^{40}Ca^{11+}$ and $^{20}Ne^{10+}$. The output charge, converted to particles using the calibration, is plotted as a function of the number of ions stored in the SIS using the reading of the DC transformer before the extraction starts. The number of circulating ions is varied with some quadrupoles behind the ion source, so that the extraction efficiency is not changed; this can be proven by the non-saturable output of the SEM (see below). The electrode voltage is varied between 500 V and 1900 V. For low voltage values (i.e., 500 V), saturation starts for the 500-MeV/u Ca beam at about 3×10^8 pps of detected particles, corresponding to a flux of 5×10^8 s^{-1}cm^{-2}; the deposed charge is 2.2×10^{-6} C/s and the average dose rate is 50 Gy/s. For the 300-MeV/u Ne beam this occurs at about 2×10^8 pps or a flux of 1×10^8 s^{-1}cm^{-2}; the deposed charge is 10^{-6} C/s, and the average dose rate is only 10 Gy/s. This behaviour is due to recombination of the secondary ions and the electrons. The shielding of the applied voltage due to the high charge density amplifies this effect, because the ions are not accelerated fast enough to the electrode. The saturation dose rate is relative low, but one has to take into account that the rate of extracted ions can vary within a factor of 5 on a 100-μs time scale. These saturation effects have to be considered in more detail in the future. For higher electric fields of about 3 kV/cm, the losses are small for the presently used number of ions.

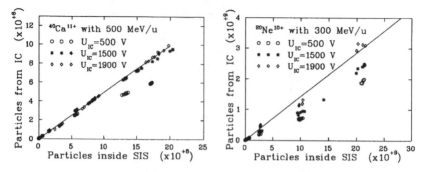

FIGURE 3. Two examples for the saturation behaviour of the IC. The applied voltage is varied; each point is one spill. The IC output is converted into particle numbers (note the different axis for both plots).

THE SECONDARY ELECTRON MONITOR

Outline and Calculation of Secondary Charge of the SEM

The SEM is mounted on the same pocket as the IC, but inside the vacuum. It consists of three slightly curved 100-μm Al plates to increase their mechanical strength and to prevent microphonic pickup. The size of the electrodes is 80 × 80 mm^2 and 100 V is applied.

The energy loss of the ion inside the Al material is calculated using again the Ziegler formalism (7). As only electrons within a certain depth can escape from the surface, a proportionality to the *specific* energy loss dE/ρdx is expected (6), ρ being the mass density of Al. The fitted value for the amount of detectable secondary charge per unit energy loss from the data in Fig. 4 is Y = 23 e$^-$/(MeV/mg/cm^2). A slight dependence on the nuclear charge of the incoming ion is reported (10), but is not used in our calculation yet.

Measurements

Calibration measurements have been performed where the SEM secondary charge output is compared to the scintillator or to the IC (itself calibrated against the scintillator). The output of the SEM is about three orders of magnitude lower than from the IC; both detectors' outputs are well linear. Particle intensities from about 10^6 pps, depending on the ion species, can be measured.

A comparison between the calculated and measured calibration factors for various ions is plotted in Fig. 4. The linearity is reasonable; no Z-dependence is used here. The deviation $(f_{theo} - f_{ex})/f_{theo}$ is larger than for the IC; here we are only

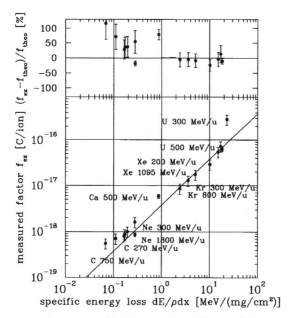

FIGURE 4. Comparison between measured f_{ex} (i.e., calibrated against the scintillators or IC) and calculated specific energy loss $\frac{dE}{\rho dx}$ for different ions. Some representative ion species are labelled. From this data a fit is done to get the fraction of emitted electrons Y. The upper part shows the deviation between the calculated $\left(f_{theo} = Y \cdot \frac{dE}{\rho dx}\right)$ and measured factors.

reaching about a factor of two. The main reason is the fact that here we are dealing with a surface effect, i.e., the amount of secondary charge is very sensitive to any surface modification. In addition, fast electrons ('δ-electrons') can be generated and emitted from the surface, an effect which is dependent on the ion velocity and charge. It is reported that due to aging effects, the efficiency can decrease considerably (10,11) due to radiation damage for high numbers of transmitted ions. This should be avoidable by using gold as a coating (11).

THE CRYOGENIC CURRENT COMPARATOR

The main part of this detector is a toroidal superconducting flux coupling coil as the antenna and a coupled DC superconducting quantum interference device (SQUID) system as the magnetic flux sensor (12). These parts are wrapped in a superconducting magnetic shielding with a small gap where only the azimuthal magnetic field component, which is proportional to the ion current, can enter with small attenuation. All other field components are strongly weakened. The detector system itself is mounted in a special bath-cryostat with a so-called "warm hole" (100 mm in diameter) for the passing ion beam.

After testing and assembling all parts of the detector (13), measurements concerning the current resolution and the noise limitations were done. To simulate the ion beam a simple wire loop (one winding) around the flux transducer was installed. Using a calibrated picoampere current source (Keithley 261), the current sensitivity of the detector system was determined to 175 nA/ϕ_0 (1 ϕ_0 corresponds to an output signal of 2.5 V in the most sensitive range of the SQUID system). Figure 5 shows a series of 10-nA test pulses taken with a bandwidth of 100 Hz. During the measurements the vacuum pumps are switched off to minimise microphonic effects. To determine the current resolution of the system, a noise spectrum was taken. The measured noise level of 0.43 mV$_{RMS}$ corresponds to a current resolution of 0.24 nA/$\sqrt{Hz}$.

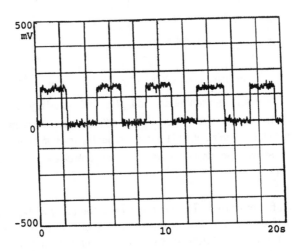

FIGURE 5. 10-nA pulses, 2.2-s pulse width from a wire detected with the CCC.

But there are still some effects that influence the detector performance:
- The output signal shows a strong zero drift at the beginning of each test. After a cooling time of about 100 hours this drift drops to acceptable values under 0.5 mV/s ($\cong$ 35 pA/s current drift).
- Measurements were carried out to study the influence of external magnetic fields; a field of 10^{-5} T yields a current of 3.3 nA ($\vec{B} \parallel \vec{I}$) and 22 nA ($\vec{B} \perp \vec{I}$). These values are small enough to allow further tests under ion beam conditions.
- Investigations were carried out to study the microphonic sensitivity of the detector system. Beside the vacuum pumps, the noise spectrum shows several characteristic lines. The results of the measurements are taken into account for the design of the vibration-isolated mounting of the detector.

CONCLUSION

The three presently used detectors can serve as a cheap and reliable standard system for all high-energy beams. The IC and the SEM can be calibrated against the scintillators. The accuracy of the IC compared to a general formula is about 25% for the full range of ions. It is expected to get an accuracy of about 5% after solving some technical problems. The high-intensity beams, where the IC saturates, can be measured with the SEM. This device shows very good linearity, i.e., a good relative precision, but calibration measurements seem to be necessary.

First results of the cryogenic current comparator are expected during a high-current test phase of the SIS in May 1996 where a $^{20}\text{Ne}^{10+}$ ion beam will be accelerated with intensities of some 10^{10} particles per machine cycle. The extracted beam currents of 30 to 50 nA will deliver good measurable signals with a possible resolution of 5%.

REFERENCES

1. K. Blasche, B. Franzke, in *Proceedings of the Fourth European Particle Accelerator Conference*, London 1994, 133 (World Scientific, Singapore: 1994).
2. V. Agoritsas, A Sealed Metal Argon Ionisation Chamber, CERN/PS/EI 81-7 (1981).
3. C.E. Swartz, G.S. Levine, R.L. Carmen, *Rev. Sci. Instr.* **34**, 1398 (1963).
4. T.R. Renner, W.T. Chu, B.A. Ludewigt, M.A. Nyman, R. Stradtner, *Nucl. Instrum. Methods A* **281**, 640 (1989).
5. Y. Sugaya et al., *Nuc. Instr. Meth.* **A 368**, 635 (1996).
6. V. Agoritsas, in *Symposium on Beam Intensity Measurements*, 117 (Daresbury: 1968).
7. F.J. Ziegler, *The Stopping Power and Ranges of Ions in Matter*, Vol. 5, (Pergamnon Press: 1980).
8. E.g., F. Sauli, "Principles of Operation of Multiwire Proportional and Drift Chambers," CERN 77-09 (1977).
9. W.T. Chu, B.A. Ludewigt, T.R. Renner, *Rev. Sci. Instrum.* **64**, 2055 (1993).
10. A. Junghans, K. Sümmerer, GSI, private communication.
11. J. Camas, G. Ferioli, J.J. Gras, R. Jung, in *Proceedings of the Second European Workshop on Beam Diagnostics and Instrumentation for Particle Accelerators*, Travemünde, 1995, (DESY M 9507), 57 (1995).
12. I.K. Harvey, *Rev. Sci. Instrum.* **43**, 1626 (1972).
13. A. Peters, W. Vodel et al., *Proceedings of the Fourth European Particle Accelerator Conference*, London 1994, 290 (World Scientific, Singapore: 1994).

On-line Luminosity Measurements at LEP

P. Castro, B. Dehning, G.P. Ferri, P. Puzo

CERN, CH-1211 Geneva 23, Switzerland

Abstract. At each LEP interaction point, the luminosity is monitored on-line by small angle Bhabha detectors. These detectors are optimized to observe in all bunches relative luminosity changes in a few seconds. The description of the detectors is given, together with the method used to calculate the luminosity after background correction. Optimization of the LEP performances was done with beam separation scans using the luminosity measurements. Those scans also provide a unique measurement of the vertical beam size at the interaction point.

INTRODUCTION

The aim of the on-line LEP luminosity monitors is to provide precise relative measurements of the luminosity at the four interaction points. These detectors measure the rate of elastic scattering of electron-positron pairs (Bhabha scattering). In order to get high Bhabha scattering rates, the detectors are placed at the very forward regions of the interaction points, very close to the circulating beams. For scattering angles θ less than 50 mrad, the Bhabha scattering is dominated by the Coulomb contribution (1). The differential cross section can be approximated by

$$\frac{d\sigma}{d\Omega} \cong \frac{4\alpha^2}{E^2\theta^4} \, , \tag{1}$$

where E is the beam energy and α the fine structure constant.

The setup is shown schematically in Fig. 1. Each luminosity monitor consists of two calorimeters located at opposite sides of the interaction point, one internal and one external to the orbit in the horizontal plane. Two separate monitors are used at each interaction point: monitor 1 detects coincidences of e^+e^- pairs with the positron outside and the electron inside as it is shown in Fig. 1, while monitor 2 detects e^+e^- pairs with the positron inside and the electron outside. Using both detectors allows us to increase the statistical precision on the measurement and decrease systematic effects.

In order to measure the high flux of Bhabha scattered e^+e^- pairs at very small scattering angles, the detectors are housed inside horizontal collimators placed inside the vacuum chamber 8.3 m away from both sides of each interaction point.

430

Straight Section   (~500 m)

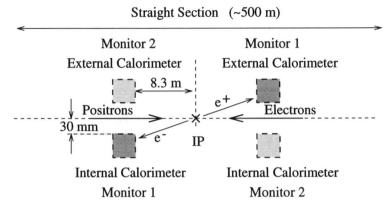

FIGURE 1. Top view (not to scale) of four calorimeters installed at each of the four LEP interaction points.

As part of the horizontal collimators, the detectors can be moved to within 30 cm of the circulating beams. The vertically focusing superconducting quadrupole QS0, located between the interaction point and the detectors, decreases the minimum scattered angle accessible on the detector in the horizontal plane. Finally, the detectors intercept Bhabha e^+e^- pairs scattered at angles between 2 and 5 mrad.

The Bhabha rate is typically about 60 Hz for a luminosity of $10^{31}cm^{-2}s^{-1}$. Thus, luminosity measurements with 5% statistical error are obtained in a few seconds. The total cross section of each monitor is about 3.8 ± 0.2 µb in data taking conditions.

In 1995, LEP was mainly running with 12 bunches per beam. Three bunches are grouped together with 250 ns separation (*families*) making four groups (*trains*) separated by 22 µs. The luminosity monitors are able to measure luminosity for all 12 bunch crossings per revolution.

GLOBAL SYSTEM

The system of all LEP luminosity monitors is shown schematically in Fig. 2. In each interaction point, the signals from both pairs of calorimeters are treated separately by a microprocessor in a shielded gallery 40 m away from the detectors. This microprocessor performs some first-stage calculations and sends data every 9 s over the network to a main workstation in the control room. The global computations are done there, allowing on-line display of the luminosity every 9 s.

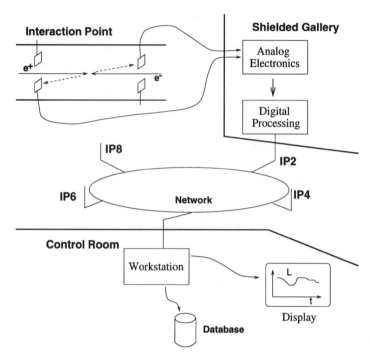

FIGURE 2. Schematic view of the measuring system of the LEP luminosity monitors.

DETECTORS

The detectors used for luminosity monitoring are calorimeters made of tungsten and thin layers of silicon semiconductors. The calorimeters are 86 mm long and 41 mm wide to fit in the limited size of the collimators. The description of the detectors and their associated front-end electronics are reported in detail in (2).

A silicon detector is mounted after five radiation lengths, close to the location of the maximum number of e^+e^- pairs in the longitudinal profile. When the particles created in the tungsten during the cascade development enter the silicon detector, they create electron-hole pairs that are collected by the reversed bias voltage. One single output channel collects the signal of the 40×40 mm^2 sensitive area of the detector. The total charge collected by the silicon detector is integrated and digitized by a 12-bit ADC.

Figure 3 shows the measured amplitude spectrum on one detector for 45-GeV (a) and 68-GeV (b) beam energies. There is a peak for values above 100, mainly due to off-momentum particles hitting the detector. Their energy is about 1% lower than the real Bhabhas and the resolution of the calorimeter is not good enough to discriminate them.

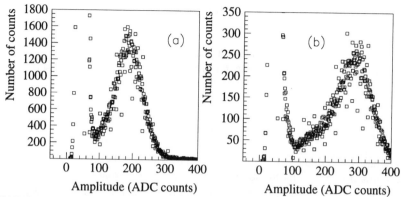

FIGURE 3. Amplitude spectrum of 45-GeV (a) and 68-GeV (b) positrons measured in IP8. The threshold for high-energy particle discrimination is at 100.

As we are only using one silicon detector, the fluctuations of the longitudinal development of the shower are large. Moreover, a fraction of the hits is close to the inner edge of the calorimeter, where part of the shower is not contained in the tungsten and remains undetected. This explains the broad amplitude spectra of Fig. 3.

By going from 45 GeV to 68 GeV, the peak position is increased by the ratio of the energies, but the resolution[†] of the peaks does not change significantly.

The peak observed for amplitudes lower than 100 is the noise due to the combination of the noise in the silicon detector and the noise in the 40-m cable between the detector and the electronics. In this example, the threshold for high-energy particles tagging was set to 100.

DATA ACQUISITION

At each bunch crossing the digitized signals from both internal and external calorimeters are read by a digital signal processor (DSP) (3). This DSP compares the signals of both calorimeters to the threshold values used to discriminate high-energy particles. When the signals from both calorimeters are above their respective thresholds, a counter for *pairs* N_{pair} is incremented by one. If only the signal from the internal detector is higher than the threshold, a counter for *internals* N_{int} is incremented by one. The same rule applies to a counter for the *externals* N_{ext}. Finally, the three counters N_{pair}, N_{int}, and N_{ext} are transferred to a microprocessor every 3 s. Three of those basic acquisitions are performed before their average is sent to the control room.

[†] Defined as the ratio between the FWHM of the peak and its position.

LUMINOSITY

The Bhabha rate and the background estimation are obtained from N_{pair}, N_{int}, and N_{ext}, thus allowing the complete determination of the luminosity and its statistical error.

Background

The luminosity detectors (see Fig. 1) are protected on their back side (opposite to the interaction point) by the tungsten block of the collimator to which they are linked. This represents more than 30 radiation lengths, so no high-energy particle can be detected from the back side.

However, from the front side, the detector can be hit by off-momentum particles. Even if a devoted set of collimators in the arcs and at the entrance of the straight section allows us to reduce their rate, there are still some off-momentum particles hitting the detector, either coming from the arcs directly or from interaction with residual gas molecules or thermal photons in the straight section.

Synchrotron radiation photons have only several tens of keV. They do not induce any background, because their complete electromagnetic shower is absorbed in the first five radiation length of tungsten and do not reach the silicon detector.

The only background which affects the detectors is then due to off-momentum particles. Table 1 shows the typical normalized background rates measured for the two different running energies of LEP in 1995 for each monitor. A typical current was about 300 µA per bunch. The current normalized Bhabha rate is about 15 Hz/mA at 45 GeV, which is two orders of magnitude smaller than the external background rate.

The background rate is much higher in the detectors located on the external side of the ring than on the internal one.

TABLE 1. Typical Normalized Background Rates Measured in 1995 in Each Monitor

	Internal calorimeter (Hz/mA)	External calorimeter (Hz/mA)
45 GeV	0 to 120	60 to 1800
68 GeV	0 to 220	1800 to 3600

Luminosity Calculation

The accidental coincidence of background particles hitting both the internal and external detectors at the same bunch crossing affects the coincidence pair rate $\dot{N}_{pair}$ computed from the counter N_{pair}. The Bhabha scattering rate $\dot{N}_b$ is

$$\dot{N}_n = \dot{N}_{pair} - \dot{N}_{acc} \, , \qquad (2)$$

where $\dot{N}_{acc}$ is the *accidental pair* rate. With a statistical method (4) assuming uncorrelated background, the accidental pair rate $\dot{N}_{acc}$ is calculated from the *internal* and *external* rates $\dot{N}_{int}$ and $\dot{N}_{ext}$. Finally, we obtain for the Bhabha rate $\dot{N}_b$ and its statistical error $\sigma(\dot{N}_b)$:

$$\dot{N}_b = \dot{N}_{pair} - \frac{\dot{N}_{int}\dot{N}_{ext}}{kf_{rev} - \dot{N}_{int} - \dot{N}_{ext} - \dot{N}_{pair}} \, , \qquad (3)$$

$$\sigma(\dot{N}_b) \approx \sigma(\dot{N}_{pair}) \, , \qquad (4)$$

where $\sigma(\dot{N}_{pair})$ is the statistical error on $\dot{N}_{pair}$, f_{rev} is the revolution frequency, and k is the number of bunches per beam. The correction term to $\dot{N}_{pair}$ in Eq. (3) is on the order of 50% at 45 GeV.

It is worthwhile noting that the often-used delayed coincidence method has a much higher error (4).

Statistical Errors

The statistical error σ is computed by Eq. (4). An estimate of the quality of the background subtraction is given by the fluctuations of the luminosity. If one considers two consecutive luminosity measurements $\mathcal{L}_i$ and $\mathcal{L}_{i+1}$, the rms of the difference $\mathcal{L}_{i+1} - \mathcal{L}_i$ divided by its error $\sqrt{\sigma_{i+1}^2 + \sigma_i^2}$ must be 1 when the luminosity variations are only due to statistical fluctuations. This difference is plotted in Fig. 4, where the line is a Gaussian fit to the data.

To ensure that the luminosity was not affected by sources other than statistical fluctuations, the data used for Fig. 4 are registered during stable data-taking conditions in beam separation scans (see next section). The acquisition time for each point is 9 s, giving a statistical uncertainty on each measurement of about 5%. The shape of the histogram is well described by a Gaussian with an rms of 1.016 $\pm$ 0.04. This demonstrates that the statistical errors are estimated correctly.

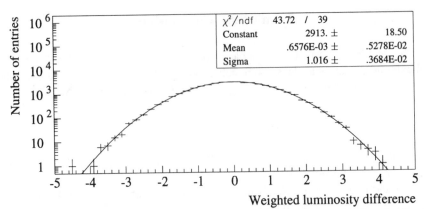

FIGURE 4. Histogram of the difference between two consecutive luminosity measurements, divided by their combined error. The line is a Gaussian fit to the data.

BEAM SEPARATION SCANS

The optimization of the luminosity at LEP is done by measuring the luminosity for different separations of the electron and the positron beams. These separations are controlled by vertical electrostatic separators located on both sides of each interaction point.

For each train crossing, the luminosity is recorded for the three families and plotted versus the beam separation. Figure 5 shows such a separation scan performed in one interaction point. The origin of the separation corresponds to the theoretical beam axis. Each point corresponds to an acquisition time of 27 s. The curves are Gaussian fits to the luminosity recorded for each family, the maximum of the Gaussian being reached when the vertical separation between the electron and the positron bunches is zero for this family.

This method allows us to find the optimum overlap between each family with a typical precision of 0.2 μm. The three families do not exactly overlap for the same value of the separator field, which is in agreement with simulations (5). The maximization of the luminosity is done by weighting the three individual optima.

The width σ_{scan} of these individual Gaussian curves is related to the vertical beam sizes σ_{e^+} and σ_{e^-} at the interaction point:

$$\sigma_{scan} = \sqrt{\sigma_{e^+}^2 + \sigma_{e^-}^2} \ . \tag{5}$$

For the example given in Fig. 5, assuming that both electron and positron beam sizes are equal, the measured vertical beam sizes are 4.9 ± 0.3 μm, 4.6 ± 0.2 μm, and 3.7 ± 0.2 μm for families A, B, and C, respectively.

This method is a unique way to measure the vertical beam sizes at the interaction point, with a typical precision of 0.2 μm.

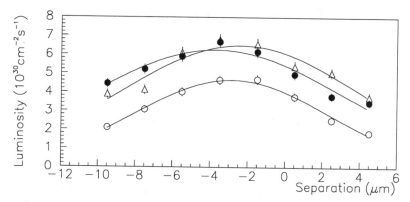

FIGURE 5. Separation scan #631 performed in IP2 for the three families: Family A (•), Family B (△) and Family C (o). The lines are Gaussian fits to the data. The origin of the separation is the theoretical beam axis.

CONCLUSION

The LEP luminosity monitors provide reliable on-line luminosity measurements for all the colliding bunches. They are intensively used for luminosity optimization and allow us to measure the vertical beam sizes at the interaction point with very good accuracy.

ACKNOWLEDGMENTS

This work was performed in the CERN SL/BI group. The authors wish to thank M. Lamont and J. Wenninger for their implementation of the separation scan procedure.

REFERENCES

1. G. Altarelli, R. Kleiss, and C. Verzegnassi, "Z Physics at LEP 1," CERN Yellow report 89-08, Vol. 1 (1989).
2. G.P. Ferri, M. Glaser, G. von Holtey, and F. Lemeilleur, "Silicon Detectors Used for Beam Diagnostics in the LEP Collider," *Nucl. Phys. B*, Supplements Section (1990).
3. P. Castro, L. Knudsen, R. Schmidt, "The Use of Digital Signal Processors in LEP Beam Instrumentation," Proceedings of the 3rd Annual Accelerator Instrumentation Workshop, CEBAF, Newport News, USA, October, 1991, *AIP Conference Proceedings* **252**, 253-259 (1992).
4. P. Castro, "Luminosity and beta function measurement at the electron-positron collider ring LEP," PhD thesis at the Valencia University, Spain (17th May 1996).
5. E. Keil, "Truly Self-consistent Treatment of the Side Effects with Bunch Trains," CERN SL/ 95-75 (1995).

Transmission Monitoring in the DESY Accelerator Chains

H. Burfeindt, W. Radloff

Deutsches Elektronen Synchrotron (DESY)
Notkestraße 85, D-22603 Hamburg, Germany

Abstract. Standardized beam current transformers are installed in each of the accelerators and in the transport lines, close to the injection/ejection elements. Their signals are proportional to the charge in a single bunch and are suitable for both direct visualization on digital oscilloscopes and for single-pass data acquisition. Corresponding bunch current signals or signal trains to be compared "at a glance" are arranged in perceptible tandems. We report on the technical design of the pickup electrodes, optical-fiber-based analog delay line techniques, and calibration procedures.

INTRODUCTION

The HERA e/p collider (27 GeV e^+/e^-, 820 GeV p) and the DORIS synchrotron radiation facility (4.6 GeV e^+) are currently in operation at the DESY laboratory in Hamburg, Germany. Pre-accelerator chains are required to boost the particle energies to inject into these machines. A schematic of the electron and proton accelerator chains is shown in Fig. 1, and a sketch of the DESY complex is shown in Fig. 2. The storage ring PETRA II is used as booster for both electrons and protons; the stored beams orbit in opposite directions in these injection modes.

Efficient transfer through the injection chains is needed to minimize the filling times. Particle losses can also lead to local radiation heating. To achieve and maintain maximum transfer efficiency, the accelerator operators require diagnostics throughout the chain, with easily selected signal combinations for quick com-

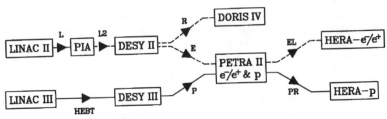

FIGURE 1. Schematic of the accelerator chains at DESY.

parisons.   Standardized measurement setups and simplicity of calibration are important design requirements.

In 1995 the achieved transfers in quantities of particles per bunch were:

$e^{\pm}$: PIA/DESY II $1 \times 6 \cdot 10^9/6.25$ Hz $\overset{0.5h}{\Rightarrow}$ HERA-e $189 \times 2.5 \cdot 10^{10}/\text{run}$

p: DESY III $1 \times 1 \cdot 10^{11}/0.223$ Hz $\overset{0.4h}{\Rightarrow}$ HERA-p $180 \times 6 \cdot 10^{10}/\text{run}$

BASIC CONCEPT

The beam current monitor unit consists of a broad-band transformer pickup directly connected to a pulse-forming low-pass filter for integration, followed by a remote-controlled attenuator (measurement range extension) and a low-noise, wide-band amplifier (normalization).   Twenty such devices have been installed in the accelerator chain (see Fig. 2).

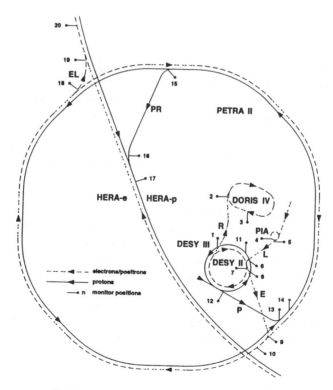

FIGURE 2. HERA and the preaccelerators.

For comparison of signals and signal groups, all these monitors should have the same electrical characteristics, although they are different in their mechanical dimensions and apertures. In addition, all monitors avail on one basic measurement range with concordant calibration factors per direction. The chosen time constants of the monitors allow their signals to be digitized without exceeding the minimum bunch interval of 96 ns. Individual cable or fiber insertions for analog signal delaying, electronic adder devices, together with the properties of dedicated "nailed" long transfer lines including optical TX/RX converters become an integral part of signal quality and of the absolute calibration. An advantage of this concept is that no timing or gate settings are required up to the multiplexer (MPX) input ports in the control room (BKR).

DESIGN CONSIDERATIONS

Mechanical Layout of the Pickup

Transformers with toroidal cores made of *permalloy*[*] tape with a thickness of 0.025 mm and with various apertures in gradings between 60 and 230 mm in diameter serve as the sources for the beam current signals. For practical reasons we decided to arrange the pickup electrode entirely outside the vacuum. This requires a vacuum-sealed isolating gap close to the setup. This gap consists of a small ring of radiation-resistant Al_2O_3 ceramic. In the case of very short bunches the inner surface of the gap may additionally be bridged for wave tapering by interleaving metallic fingers or combs.

This technical solution has significant advantages because the transformer can be cut into halfrings. By this measure, the pickup electrode can be completely finished in the lab and quickly mounted, modified, or exchanged in situ even if the vacuum system is closed. The necessary shieldings are made of engineering steel or aluminum and can also be cut into halves.

The fully equipped halfcore is embedded in a shielding halfcover. The mounting on the vacuum vessel is done simply by clamping both halfcovers together on their mechanical seats with screws. Long strips of flat springs inserted between the halfcovers and the framework of the ringcore keep the gaps of the cores tightly closed.

All electrical connections are made pluggable immediately outside the shielding cover using BNC- or SMA-type chassis receptacles. The first active electronics may be located up to 2 meters away from the beam pipe.

---

[*] Permenorm 360, K1, Trafoperm, etc. are trade marks of *VAC Vakuum Schmelze*, Hanau, Germany.

Electrical Layout of the Pickup

The electrical layout of the pickup is defined by the requirements listed below.

1. The bunch lengths range from 10^{-10} to 10^{-8} seconds. The signal heights are expected to be proportional to the charge in a single bunch and independent of the real bunch shapes.
2. On one hand, the signals should decay to zero within 2/3 of the available 96-ns bunch-to-bunch interval. On the other hand, the peak of the signal has to be flat and long enough to allow a single pass for data acquisition.
3. High-frequency components produced by very short bunches should not reach the active electronics.
4. Overall calibration must be feasible at any time.

On the basis of these design goals and in combination with our experience with magnetic beam position monitors (1), in particular in DESY III, we have established the following electrical configuration.

Permalloy tape as core material with its circular magnetic texture was preferred, because its permanence is not strongly changed if exposed to stray fields (2,3). The transformer core is furnished with four single pickup loops symmetrically spaced over the circumference. The primary loop consists of a solid, flat copper strip tightly bent around the core. A small, ferrite-loaded rf transformer for impedance matching is mechanically integrated into each loop. The secondary side carries 15 turns of enameled wire; the ends are guided directly to the outside by means of chassis receptacles. In principle all four outputs are connected consecutively in parallel: the two outputs from each halfcover are joined together by short pieces of coaxial cable of equal length. These points are then joined with equal cable lengths to the inputs of a balun transformer, which is part of the low-pass filter (see Fig. 3).

A separate winding, wrapped around the core and positioned between two pickup loops, serves for overall calibration and test purposes.

ELECTRONIC COMPONENTS

Low-pass Filter Module

A strongly damped, twin LC low-pass filter inserted next to the pickup forms the final shape of the signals to be displayed and digitized. The slope of the response function softly decays passing the upper cut-off at 20 MHz. Because undesired rf components in the spectra of the bunches may be picked up through the device, the rejection band of the filter must be carefully investigated. The filter is assembled with SMDs and is separately housed in a tin box. After the filter all

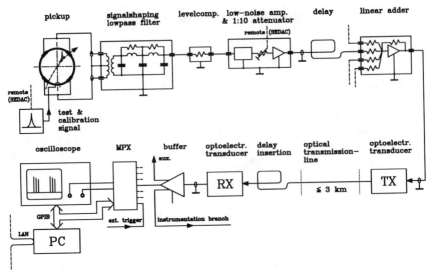

FIGURE 3. Overview of a transmission monitoring setup.

bunch current signals have rise and fall times of ≈ 16 ns. The total length of the signal occupies 70% of the available 96-ns bunch spacing.

Attenuator and Amplifier

Attenuators inserted between the filter and amplifier serve for final calibration adjustments. Variations in losses and signal levels due to differences in pickup dimensions require gains of 5 to 20 in order to attain standard measurements of 2×10^{10} particles/Volt for electrons and 1×10^{11} particles/Volt for protons. A relay-driven attenuator in the input stage extends the measurement range by a factor of 10.

Adder

Corresponding signals from e- or p-monitors to be displayed in tandem arrangements are subjected to suitable delays by means of cable insertions. Up to four pickup sources may be non-reactively combined at the ports of an operational transimpedance amplifier before they are sent via cable or fiber to the accelerator control room.

Signal Transfer and Delay Techniques

The length of transmission links ranges from 60 m to over 2 km, including the required delay insertions. For short distances (up to 400 m) and in irradiated areas we exclusively use air-dielectric rigid coaxial rf cable, mainly of 5/8" or 7/8" type. For longer distances (from 400 m to over 2 km) we make use of 50/125 gradient glass fiber cables with excellent transmission properties such as low attenuation losses and neglectable dispersion. With a specific delay of 5 μs/km and attenuation of 2.3 dB km, analog delays of several microseconds can be easily established. So even longer signal trains can be set serially on one trace or set congruently on parallel traces.

Optoelectronic Conversions

The optoelectronic TX and RX transducers that are used are in-house developments, utilizing commercial laser diode modules which operate in the 890-nm window. The achieved device data for a TX/RX set linked with 1 km of 50/125 fiber cable are listed in Table 1.

TABLE 1. TX/RX Device Data

transmission band	10 Hz - 40 MHz
dynamic range	> 36 dB
modulation capability	± 1.5 Volt /50 Ω
adjustable gain	−6 dB
connectors	FSMA, FC-PC, ST

Signal Display

Finally the signals end in standard setups located in the control room. A typical setup consists of a PC-driven coaxial multiplexer MPX (HP3488A) wired with a digital oscilloscope (Tektronix TDS 540). Menu buttons integrated in the front panel frame of the oscilloscope are used to select the various signal combinations which are updated with the relevant transfer trigger.

CALIBRATION AND TEST DEVICE

Because of the symmetric configuration of the pickup electrode we eliminated frequency-dependent couplings within their usable apertures. Finally a reliable calibration became feasible based on the following measures:

- All pickup transformers contain a test loop for recalibration.
- The pickup stations are calibrated by means of a stub antenna device before their installation.
- We use very short pulses with 2-ns FWHM; overshoots or reflections of the calibration current must decay to zero within the rise time of the system.
- Each station is measured twice with same charge: first using the antenna in the test device, then using the integrated test loop.
- The measured relative deviation of both outputs must be taken into account as a correction factor for future recalibrations.

Deviations of < 3% typically have been measured.

The generation of appropriate δ-pulses with zero baseline shift is based on the Avalanche effect (4). We only need to measure the DC portion of the periodic δ-pulses if its period is quartz controlled. With, e.g., $f = 100$ kHz and $i_{DC} = 1$ A, we would have to establish $n = i/(e\,f) = 62.5 \cdot 10^{12}$ particles per pulse; but at an available yield of 3.3×10^9 particles/pulse, a current of 52.8 µA has to be precisely measured!

A front-end active probe directly connected to the stub or loop connection is used to measure the voltage drop of the calibration current across a 50-Ω feedthrough resistor by means of an rf-blocked amplifier OP07. A pulse generator, DVM, and power supply are housed in a portable backend cabinet. Accuracies attained with this method have been repeatedly confirmed in good agreement to a commercial precision parametric current transformer[†]. A simplified δ-pulser module remains plugged into the test loop port of each pickup station. For test purposes the pulses may be remotely activated at any time.

PERFORMANCE

Figures 4 through 7 show selected signal arrangements taken as printouts from the scope stations. The attained transmission efficiency appears clearly by visual comparison of the pulses or pulse trains.

CONCLUSION

The design of this beam current transmission monitoring is tailored to given preconditions in the accelerator chains at DESY. The direct display of arranged but unprocessed signals is reliable and useful for daily routine operation, though the drop in the signal may be a little inconvenient. Although the length of bunch signals are far off the real bunch length, misalignments in the injection lead immediately to striking signal distortions.

---

† PCT from *Bergoz*

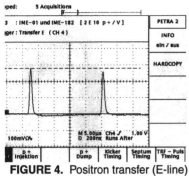

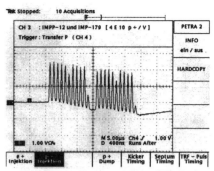

FIGURE 4. Positron transfer (E-line)
DESY II - PETRA II

FIGURE 5. Proton transfer (PR-line)
DESY III - PETRA II

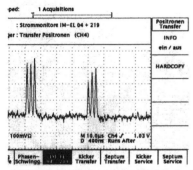

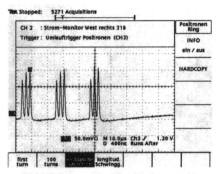

FIGURE 6. Positron transfer (EL-line)
PETRA II - HERA-e ...

FIGURE 7. ... and finally deposed in
HERA-e.

Over a separately kept signal path we have investigated a "state of the art" fast
data acquisition module for single-pass recordings (5). This commercial ADC has
a bandwidth of 33 MHz with a resolution of 12 bits and has been successfully
tested for well over a year during different injection conditions for PETRA II.

REFERENCES

1. S. Battisti, M. Le Gras, J. M. Roux, B. Szeless, D. J. Williams, "Magnetic Beam Position Mon-
    itors for LEP PRE-Injector," Proc. of the 1987 IEEE Particle Accelerator Conference, March
    16-19, 1987, 605-607 (1987).
2. W. Radloff, "The Intensity Monitor Device for PETRA," DESY report M-79/14, March 1979.
3. W. Kriens, W. Radloff, "Fast Lifetime Measurements of Stored e^+/e^- Single Bunches in
    PETRA and DORIS II Utilizing the AC-Signals of Simple Beam Current Transformers," Proc.
    of the IEEE Particle Accelerator Conference, March 21-23, 1983, 2193-2195 (1983).
4. M. J. Browne, J. E. Clendenien, P. L. Corredura, R. K. Jobe, R. F. Koontz, J. Sodja, "A Multi-
    channel Pulser for the SLC Thermionic Electron Source," SLAC-PUB-3546, January 1985.
5. R. Neumann and M. Wendt, private communication.

LANSCE Beam Current Limiter[*]

Floyd R. Gallegos

*Los Alamos Neutron Scattering Center, AOT Division, Los Alamos National Laboratory,
Los Alamos, NM 87545 USA*

Abstract. The Radiation Security System (RSS) at the Los Alamos Neutron Science Center (LANSCE) provides personnel protection from prompt radiation due to accelerated beam. Active instrumentation, such as the beam current limiter, is a component of the RSS. The current limiter is designed to limit the average current in a beamline below a specific level, thus minimizing the maximum current available for a beam spill accident.

The beam current limiter is a self-contained, electrically isolated toroidal beam transformer which continuously monitors beam current. It is designed as fail-safe instrumentation. The design philosophy, hardware design, operation, and limitations of the device are described.

INTRODUCTION

The current limiter (XL) was designed to be an integral part of an instrumentation-based, engineered personnel protection system at the Los Alamos Neutron Scattering Center (LANSCE). Other components of the Radiation Security System (RSS) include the personnel access control systems, fail-safe ion chamber systems, safety system logic and wiring, and safety system beam transmission mitigation devices (beam plugs or stoppers).

As part of a limited-scope probabilistic safety analysis of selected safety systems at LANSCE, the device underwent a reliability analysis by an experienced team of Los Alamos National Laboratory safety analysts. The analysis provided estimates of system unavailability (ratio of average downtime to uptime in the interval between testing) of 3.7×10^{-3} with an estimated error factor of 2.2. Annual testing and operation for half a year per year were assumed (1,2).

DESIGN PHILOSOPHY

The XL is instrumentation that provides fail-safe beam current limiting protection. "Fail-safe" is defined as functioning as specified or if a single failure occurs the device will 1) function as intended (due to the redundant circuitry), 2) function with a more sensitive fault threshold, or 3) generate a fault condition.

The device specifications and design requirements are listed below. Refer to Fig. 1 for functional block information.

---

[*] Work supported by the U.S. Department of Energy.

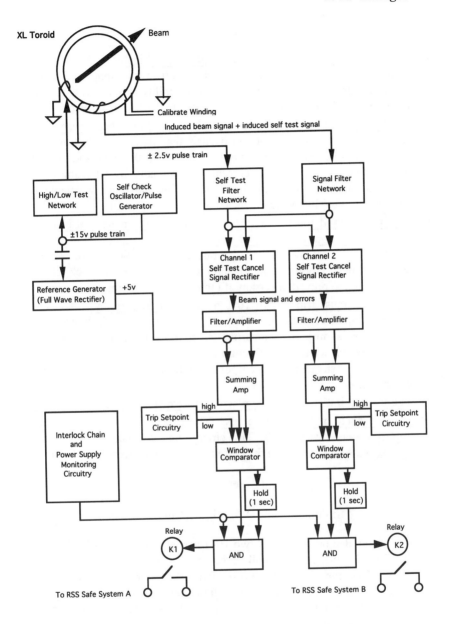

FIGURE 1. Functional block diagram of the LANSCE beam current limiter (XL).

Specifications

- Current detection must be nonintercepting and bipolar. The device is a current transformer with a signal winding, self-test winding, and calibrate winding. The calibrate winding provides the ability to inject test signals to verify the trip level settings without removing the XL cover plates.

- The XL must detect average current with an input peak beam current up to 20 mA. The circuitry is designed such that an input of 20-mA peak current will not saturate the input amplifiers (10 volts) and they will respond linearly. Higher peak currents are attainable by adjusting the gain of the current-to-voltage amplifier circuitry. The maximum detectable peak current is limited by the magnetic saturation of the toroid.

- The trip setpoint must be adjustable to address the maximum beam spill concerns of various experimental areas. Trip level setpoint adjustment (1, 3, 10, 30, 100, 300 mA) is made by physically changing the filter rectifier and amplifier setpoint printed circuit boards. An interlock chain is incorporated to prevent insertion of the incorrect combination.

- The device must trip within 10% of the desired setpoint. To achieve required trip tolerances, grounding of the device is critical. An internal electrical isolation joint is incorporated to prevent signal interference by currents which would flow through the XL via the beam pipe. Multiple layers of EM shielding are incorporated into the XL design to reduce the effect of noise.

- The time to initiate a 3-μA trip with an average beam current of 1.2 mA must be equal or less than 25 ms. The overall circuit time constant is approximately five seconds. The circuit time constant equation can be approximated by (assuming the current to voltage conversion is linear):

$$i_{trip} = i_{final}(1 - e^{-trip\ time/time\ constant})$$

trip time $= -\ln[1 - (i_{trip}/i_{final})]*$time constant

It has been demonstrated that two 10-mA peak, 1-ms pulses 8.33 ms apart (120 Hz, 1.2 mA average) will cause a fault condition (3 μA trip) within 22 ms. Therefore three beam pulses will pass prior to the initiation of a fault condition. Experimental operations required that the circuitry be designed to permit a single 10-mA peak, 1-ms-wide pulse pass without causing a fault condition.

- The XL must be self-contained and essentially isolated. The device is only dependent on incoming AC power for its operation. Trip information supplied by the unit is electrically isolated from internal circuitry and power supply. The introduction of external effects such as electrical loading or problems due to operator error are minimized. External cabling consists of the AC power cable and two armored cables which transmit the XL fault state to the RSS logic.

- It must be active at all times (no dead-time). The XL is an AC-coupled device with full-wave rectification. As such it does not require a means of DC restoration to determine average current levels. Monitoring is active at all times and will provide protection regardless of the beam timing. External timing gates are not required for proper operation.

- Fail-safe operation is critical. The XL was designed to be as fail-safe as practical (limited by information available on component failure modes). In the design of the XL, an attempt was made to ensure that the unit would fail in a condition that would not compromise the safe operation of the accelerator. Certain parameters are monitored within the unit such that if an out-of-tolerance condition is detected, the XL will fault. The various design schemes used are:

  a. Asynchronous test signal generation for self checking of circuitry and toroid connectivity --- Self checking pulses are generated in the XL. These pulses are directed along two paths. One set of pulses is sent directly through circuitry on the filter-rectifier card. The other set of pulses goes through the toroid via a test winding. Current due to the pulses is induced on the signal winding and processed in the filter-rectifier card. The voltages due to the currents from these two paths are filtered and compared. If they are identical, the effect is canceled. If they are different in magnitude or phase, an output is generated and rectified. This output is an error signal. If the signal wiring or test wiring was disconnected, the error would be sufficient to cause a fault of the unit. If there was a problem with the gains of the signal path or the test path, the error generated would effectively reduce the trip setpoint since the error would manifest itself as a current level that is always present.

  b. Full-wave rectification --- By full-wave rectifying, an absolute error is generated thus eliminating the possibility of cancellation of error signals due to opposite polarities. For example, if bipolar noise is introduced into the system both polarity components would manifest themselves (additive) as a current offset level that is always present.

  c. Self-test signal --- To ensure that the self-test signal is present, the signal is full-wave rectified and used to generate a reference level that is summed with the error signal output. If the reference level is not present, the XL will fault since it is outside the trip band.

  d. Power monitoring --- The power supply voltages are monitored by self-check circuitry. The power supply voltages ($\pm$ 15) are used to generate the current pulses directed to the test winding. The voltages used to generate the current pulses sent to the self-check circuitry are derived from ± 2.5 volt reference diodes. If the power supply voltages drift from their required values, an error signal is generated since cancellation is not achieved in the filter-rectifier card. The components used by the XL are commercial. The reliability of their electrical characteristics is decreased if stated absolute maximum ratings are exceeded or if values fall below minimum ratings. The power supply voltages are internally monitored by additional circuitry to ensure that maintenance personnel are aware that the voltages remain greater than 12 volts and less than 18 volts; otherwise the XL will fault.

e.   Redundant circuitry --- Certain sections of the XL circuitry are not easily checked using a self-test method. Where self checking of the circuitry cannot be performed, redundant fault channel circuitry has been added. Redundant fault outputs (relay contacts) are supplied to the RSS.

f.   Complete loss of power --- On a complete loss of power, the two redundant fault contacts will open.

g.   Interlocking --- The system has been designed with an interlock chain incorporated on the printed circuit board. If the proper combination of printed circuit cards is not inserted in the correct card locations, the unit will fault. The printed circuit cards are also keyed to prevent insertion into the incorrect slot.

DEVICE DESCRIPTION

The XL is 18 in. high by 17.5 in. wide and 14 in. long (beam axis). The current transformer consists of a 2-mil, tape-wound supermalloy toroid, 10 in. O.D., 5 in. I.D., and 1 in. length. It is wound with a 100-turn copper signal winding and two single-turn (calibrate and self-check) windings. The beam pipe is electrically broken internal to the device to prevent return currents from flowing through the toroid.

The device and power supply are designed to minimize electronic and acoustic noise pickup. The processing and fault generating electronics are contained within the unit chassis to minimize noise introduction. The power supply is externally mounted. Single-point grounding concepts in circuit and shielding implementation are used to reduce introduction of noise (see Fig. 2 and Fig. 3).

FIGURE 2. Unassembled current limiter hardware. Clockwise from top left: outer casing and beam pipe, current transformer with multiple layers of shielding, and power supply and multiple shielding casings.

FIGURE 3. Front view of current limiter with cover plates removed. Card cage with associated printed circuit boards, mounted power supply, and outer hardware casing.

OPERATION

The XL requires periodic checkout and operational verification. The unit will trip and remain faulted only as long as the fault condition exists. Latching of the fault is accomplished in the external RSS logic.

Since it is located in the beamline, access to the XL during beam operations is restricted. Diagnostic indication, test points, and switches are located on the top panel of the XL. Light emitting diodes are illuminated to indicate the fault condition, the trip level setpoint, and the condition of the DC power. To verify proper fault operation of the unit, two test switches (high/low) are used to increase or reduce the magnitude of the pulses directed to the self-test winding. The test points can be used to determine the power supply voltage levels. Remote test signal injection capability via the calibrate winding has recently been added. This allows frequent testing of the fault capability without requiring access to the beamline.

VULNERABILITIES

- The XL is an AC device. The location of the XL must be selected such that beams of opposite polarity are not simultaneously present or signal cancellation will occur. The XL is also ineffective with neutral beams.

- Internal electronics are susceptible to radiation damage. Since the electronics must be located as close as possible to the current transformer to minimize noise effects, they are exposed to ionizing radiation fields. This may cause the XL to fail due to radiation damage to the solid-state electronics. Location of the XL with this constraint in mind is critical.

- Signal levels induced by prompt radiation. During the 1990 operating period the average current delivered to the primary neutron scattering target was significantly increased. An XL which was located on a beamline adjacent to the Proton Storage Ring (beam compressor) extraction line faulted when exposed to a large radiation pulse produced by beam spill from the occasional misfiring of the ring extraction kickers. An identical device placed upstream, shielded from the extraction line, did not exhibit the coupling phenomena.

SUMMARY

LANSCE current limiters have been in service since 1989 with satisfactory results. All failures (three) were fail-safe in nature. The first failure was due to radiation damage to the XL electronics since the device was located in immediate proximity to a beam stop. The components were highly activated and postmortem analysis was not possible. The other two failures (single leg of redundant circuitry) were due to component failure of a FET-controlled relay. The FET developed leakage current that was sufficient to maintain an energized relay regardless of the FET input state. This failure is being addressed by redesigning the circuitry with robust components.

XLs have demonstrated high reliability and exhibited their ability, as part of the safety system, to provide effective protection from exceeding the operating envelope.

ACKNOWLEDGMENTS

The author is indebted to Andrew Browman for his support and advice in the development and implementation of the LANSCE current limiter.

REFERENCES

1. Sharirli, M. et al., "Limited-Scope Probabilistic Safety Analysis for the Los Alamos Meson Physics Facility (LAMPF)," presented at the Probabilistic Safety Assessment International Topical Meeting, Clearwater Beach, Florida, January 26-29, 1993, 554-558; also LANL Report LA-UR-92-3438 (1993).
2. Macek, R.J., "Beam-Limiting and Radiation-Limiting Interlocks," presented at 9th World Congress of the International Radiation Protection Association, Vienna, Austria, April 14-19, 1996; also LANL Report LA-UR-96-443 (1996).

The APS Machine Protection System (MPS)*

R. Fuja, W. Berg, N. Arnold, G. Decker, R. Dortwegt, M. Ferguson,
N. Friedman, J. Gagliano A. Lumpkin, G. Nawrocki, X. Wang

Argonne National Laboratory, 9700 South Cass Ave., Argonne, IL 60439 USA

Abstract. The machine protection system (MPS) that protects the APS storage ring
vacuum chamber from x-ray beams, is active. There are over 650 sensors monitored and
networked through the MPS system. About the same number of other process variables
are monitored by the much slower EPICS control system, which also has an input to the rf
abort chain. The MPS network is still growing with the beam position limits detection
system coming on-line. The network configuration, along with a limited description of
individual subsystems, is presented.

INTRODUCTION

The APS storage ring has a circumference of 1104 meters and is divided into
40 sectors. Each sector has a bending magnet (BM) x-ray source and there will be
35 insertion device (ID) sources. At 100 mA of 7-GeV stored beam, the vacuum
chambers are passively safe to bending magnet radiation under normal cooling
water flow conditions. For closed gap insertion devices, 10 kW x-ray beams
generated by a missteered beam pose an immediate danger to certain parts of the
vacuum chamber, on the order of a few ms. The machine protection system
(MPS) uses many different sensors to monitor storage ring conditions and beam
steering parameters (1,2). If an unsafe condition arises, a fast beam abort is
generated by removing the rf drive to the klystrons. The beam then coasts inward
to the scrapers and is lost in about 300 ms. All fault conditions are latched and, in
an improved version of the logic, all faults are time stamped.

---

* Work supported by U.S. Department of Energy, Office of Basic Energy Sciences, under
Contract, No. W-31-109-ENG-38.

THE MPS SYSTEM

The MPS system is divided into three functional categories:
1. Different fault condition sensors check everything from water and vacuum to beam missteering.
2. Local MPS cards collect signals from various fault sensors and report to a main MPS control card. There is one local MPS card for every two sectors.
3. A main MPS card with decision-making capabilities acts as the interface to the rf system. This network carries 1-MHz heartbeat signals over fiber cables to a main MPS logic circuit that controls rf switches, removing the drive to the klystrons. This logic circuit also receives input from beam current monitoring circuits, giving it decision-making capabilities.

The network comprises 20 cells, called local MPS cards, located in odd-numbered sectors around the storage ring. Two types of cells exist: the earlier version has inputs for 8 heartbeat signals, and an improved version has 12 inputs and provides a time stamp of requested aborts. Input to an unused cell is manually disabled. When the heartbeats to all active inputs are received, a cell generates an output heartbeat that goes to collection points located near the main MPS control logic card. Outputs from collection points activate the main logic card. Inputs to a local MPS cell come from individual sensors, such as beam position limits detector electronics, or from summation cards that receive information from many sensors, such as water flow, etc. Heartbeats from sensors monitoring each sector are fed into the inputs of local a MPS cell. There are 70 cell inputs reserved for experimental area beamline front ends; 35 for BM beamlines, and 35 for ID beamlines. The EPICS control system also has an input to the main heartbeat collection point. Activating a button on the control screen will dump the beam.

THE MAIN MPS CARD

The main MPS card acts as the interface to the low-level rf system and is triggered by the loss of a heartbeat signal, the removal of one of the beam current information signals, or the beam current exceeding a maximum preset threshold. If stored beam is above 0.5 mA, loss of a heartbeat signal will remove the low-level rf signal to the klystron drivers for 100 ms. If beam is not below 0.5 mA, the klystron high voltage power supplies are tripped after 100 ms and, if beam is below 0.5 mA after 100 ms, the low-level rf is reapplied. From the time an abort is requested until the time the beam decayed to 0.8 mA was measured at 700 microseconds. As long as the heartbeat signal remains missing, the main MPS card will remove the rf for 100 ms every time beam exceeds 0.5 mA. The loss of one of the beam current signals will remove rf until the signal returns.

THE SENSORS

A list of the sensors in use around the ring appears below.

Sensor	Quantity
Absorber Water Flow	320
Vacuum Chamber Water Flow	76
Storage Ring Flags	10
Isolation Valves	80
Beam Position Limits Detector	7
Experiment Hall Front Ends	14 BMs
Experiment Hall Front Ends	9 IDs

A variety of sensors verify that machine systems are within their limits. The cooling water for the absorbers and the vacuum chamber as well as the vacuum gate valves are monitored through a series of VME-based latch cards. The status of storage ring flags and beamline exit port vacuum valves is input to the latch cards. Two latch cards are required to collect one sector's worth of signals; 80 latch cards cover the ring. When all inputs to the latch cards are present, a 1-MHz heartbeat is sent to a local MPS cell card.

The threat of a missteered ID beam or a BM beam greater than 100 mA has resulted in a series of devices which protect the vacuum chamber. Beam position limits detector (BPLD) cards (3) are VME based and receive signals from the rf BPMs. These cards can average beam passes in the ring and have offset capabilities. The card functions are controlled from an EPICS page. If stored beam is within limits, a heartbeat signal is sent to a local MPS cell card. Each insertion device requires its own BPLD electronics, and wire monitors also check for beam missteering and blow up. There are eight beam position wire monitors installed and monitored by the EPICS control system but not connected to the abort system. A total of 320 vacuum chamber surface resistance temperature monitors (RTMs) are planned. Output data from all the RTM sensors in a sector will be combined via electronics to produce a single heartbeat signal per sector; the same will be true for the wire monitor sensors. Wire monitor and RTM sensors will have separate inputs to the MPS system.

Experiment hall front ends are stand-alone in nature and generate their own heartbeat signal when satisfied. Each beamline has its own protection system and an input to the MPS system. Figure 1 shows a block diagram of a typical local MPS cell and how it is connected to the main machine protection logic card.

A machine protection system cell for two sectors is shown in Figure 1, and 80 cells cover the storage ring. Heartbeat signals are carried over fiber optic cables, from latch cards and stand alone chassis to local MPS cells. When all activated inputs on a local MPS cell are present, the cell generates and transmits its heartbeat over fiber cables to a collection cell located next to the main MPS card.

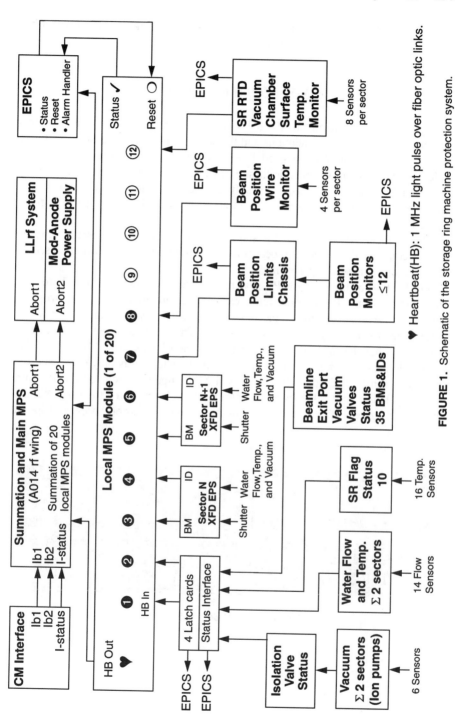

FIGURE 1. Schematic of the storage ring machine protection system.

The main MPS card has digital outputs to the low-level rf system and fiber outputs to the klystron power supplies.

THE MPS CONTROL SCREEN

Figure 2 shows a layout of the MPS control screen (4). This screen depicts the network and its interconnections and the status of individual sensors. Software beam limit conditions are imposed here. When the beam current transformer indicates beam conditions outside these limits, the EPICS input to the main MPS card is activated and latched. A software alarm handler can monitor any parameter in the control system and can activate the EPICS output to the main MPS card.

The 20 local MPS cells and the status of their inputs are displayed. As new systems come on-line, inputs to local cells are hardwired active. From this screen a postmortem of aborts is performed.

Verification of the entire MPS system is performed after a major shutdown, while verification of sector systems can be performed once sector work is complete. Every system connected to the MPS system must verify its performance to the next higher-level system. The beam position limits system must verify that the loss of its heartbeat is detected by its local MPS cell. The water and vacuum systems verify that they trip the latch cards they drive, and the latch cards must verify that they trip local MPS inputs. Each local MPS card input is cycled to verify that the main MPS card has been cycled to verify that the low-level rf goes off and the klystron high voltage power supplies trip on command.

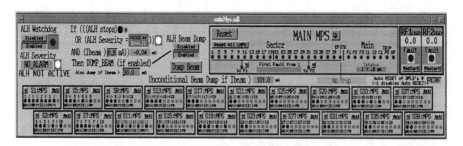

FIGURE 2. MPS Control Screen.

SUMMARY

The MPS is active when beam is stored. The hardwired beam current limit can be set to a maximum of 300 mA. Validation of beam position limits cards occurs before each run with beam current under 0.5 mA. The wire monitor system and the RTM system are planned for future installation. The machine protection system as configured works well; there are very few nuisance aborts and they are being corrected. An upgrade to the main MPS card, now in progress, will allow for automatic testing and troubleshooting of this card.

REFERENCES

1. A. Lumpkin et al., "Overall Design Concepts for the APS Storage Ring Machine Protection System," Proceedings of the 1995 Particle Accelerator Conference, 2467-2469 (1995).
2. G. Decker, "Abort Interlock Diagnostic for Protection of APS Vacuum Chamber," Proceedings of the 1993 Particle Accelerator Conference, 2196-2198 (1993).
3. X. Wang, private communication.
4. N. Arnold, private communication.

Initial Commissioning Results from the APS Loss Monitor System[*]

Donald R. Patterson

Advanced Photon Source, Argonne National Laboratory, Argonne, IL 60439

Abstract. The design of the beam loss monitor system for the Argonne National Laboratory Advanced Photon Source is based on using a number of air dielectric coaxial cables as long ionization chambers. Results to date show that the loss monitor is useful in helping to determine the cause of injection losses and losses large enough to limit circulating currents in the storage ring to short lifetimes. Sensitivities ranging from 13 to 240 pC of charge collected in the injector BTS (booster-to-storage-ring) loss monitor per picocoulomb of loss have been measured, depending on the loss location. These results have been used to predict that the storage ring loss monitor leakage current limit of 10 pA per cable should allow detection of losses resulting in beam lifetimes of 100 hours or less with 100 mA stored beam. Significant DC bias levels associated with the presence of stored beam have been observed. These large bias levels are most likely caused by the loss monitor responding to hard x-ray synchrotron radiation. No such response to synchrotron radiation was observed during earlier tests at SSRL. However, the loss monitor response to average stored beam current in APS has provided a reasonable alternative to the DC current transformer (DCCT) for measuring beam lifetimes.

INTRODUCTION

The APS loss monitor system provides a relative measurement of beam loss rates throughout the entire APS accelerator, from the linac gun through the storage ring. This is done by detecting the high energy photons given off by the accelerated particles as they collide with the vacuum chamber wall, residual gas molecules, or some other obstruction. The loss monitor system provides

---

[*] Work supported by the U. S. Department of Energy, Office of Basic Energy Sciences, under Contract No. W-31-109-ENG-38.

diagnostic information only. It does not provide an input to the personnel safety system or the machine protection system.

SYSTEM DESCRIPTION

The loss monitor system is based on using a 7/8-inch air dielectric coaxial cable as an ionization chamber. It is similar in many ways to loss monitor systems in use at Brookhaven (1) and SLAC (2). Five hundred volts DC is applied to the center conductor of the cable and the cable shield is grounded. An ionization gas consisting of a mixture of 95% argon and 5% carbon dioxide at 8 psig is passed through the cable. The average current flowing from the center conductor through the ionization gas to the cable shield is measured and is proportional to the average beam loss rate along the cable. The system design is described in more detail elsewhere (3,4,5).

The loss monitor cable is installed parallel to the vacuum chamber subject to the restrictions caused by mechanical interference with other equipment and the need for personnel movement and access to equipment. When possible, the cable was installed below the plane of the rings or inside the rings in an attempt to minimize exposure to synchrotron radiation. Air dielectric coaxial cables are placed along the entire length of each of the major machine components.

The current signals from the coaxial cable ionization chambers are multiplexed into one of nine electronics packages. Each electronics package can accept inputs from up to seven coaxial cables. Accelerator physics personnel determined the number and length of the coaxial cables used to cover the accelerator components, taking into account the need for spatial resolution and cost considerations. A total of 35 individual cables are multiplexed into the nine loss monitor electronics packages. The electronics packages include a DC-coupled current amplifier that measures the total ionization current flowing through the selected coaxial cable. Personnel safety considerations require that the cable shields be grounded and the high voltage be applied to the cable center conductor. This requires that the current amplifier be floating on top of the high voltage.

The loss monitor system is controlled by the EPICS-based (6) accelerator control system. Data can be presented to the operators and physicists numerically or graphically and can be recorded for later analysis and archival purposes using a series of software tools.

RESULTS

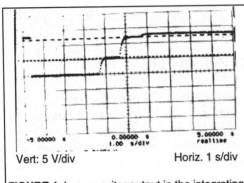

Vert: 5 V/div   Horiz. 1 s/div

FIGURE 1. Loss monitor output in the integrating mode.

Figure 1 shows an oscilloscope trace taken of the loss monitor output voltage during testing in the storage ring. For these tests, a 0.5-nC charge bunch was accelerated to 7.0 GeV and transported into sector 40 of the storage ring. The entire charge bunch was intentionally forced into the vacuum chamber wall in sector 40. This process was repeated at a 1-Hz rate. The loss monitor electronics was set to its integrating mode, so that the two steps in the oscilloscope trace show the collection of the charge induced in the loss monitor cable from the loss of two 0.5-nC bunches. The voltage step indicated by the cursors is 6.72 volts in amplitude, and corresponds to 6.72 nC of collected charge, since the loss monitor electronics was in the 10-nC range. This results in a sensitivity of about 13.5 pC of charge collected from the loss monitor cable per picocoulomb of charge lost in the storage ring at 7 GeV at this location.

A similar technique was used to measure sensitivities in the BTS, as shown in Figure 2. In this case, the location of a loss of 1.1 nC at 7 GeV was adjusted to eight different locations along the BTS. As shown in Table 1, the loss monitor

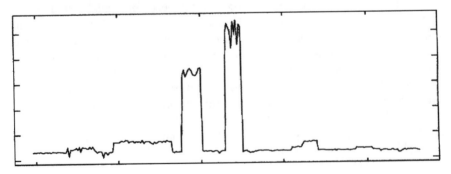

FIGURE 2. Loss amplitude as a function of location in the BTS.

sensitivity varies by more than a factor of 10 depending of the location of the loss. This is probably due to the fact that losses at some locations see relatively little shielding (primarily the aluminum vacuum chamber wall) between the loss point and the loss monitor cable, while losses as other locations may see several feet of shielding (steel and copper in large magnets) between the loss point and the loss monitor cable. The conclusion is that the loss monitor reading should be only taken as a relative reading. The readings can be used to tune for minimum loss at one fixed location, but great caution should be exercised when using the loss monitor to compare the amplitudes of losses in different locations.

TABLE 1. Loss Monitor Sensitivities at Various Locations in the BTS

Loss Location	Loss Monitor Reading, μC	Calculated Sensitivity, pC Collected per pC of Beam Loss
Thick Septum	0.016	15
Thin Septum	0.026	24
First Vertical Corrector	0.036	33
Second Vertical Corrector	0.166	150
Third Vertical Corrector	0.26	240
First Horizontal Corrector	0.022	20
First Flag	0.020	18
Second Flag	0.014	13

Leakage currents in the loss monitor electronics, air dielectric coaxial cables, and signal lead-in cables determine the minimum collected charge that can be measured. Leakage currents in the storage ring are usually less than 10 pA. In the booster synchrotron tunnel, where temperature and humidity are uncontrolled during the summer, leakage currents can occasionally approach 100 pA but are usually below 10 pA. These leakage currents result in a minimum detectable charge of less than 5 pC in the storage ring and 50 pC in the booster for a 0.5-s integration time. With measured sensitivities ranging from roughly 13 to 240 pC collected per picocoulombs lost, beam current losses of 0.021 to 0.38 pC are expected to be detectable in the storage ring and losses of 0.21 to 3.8 pC or better are expected to be detectable in the booster and BTS (booster-to-storage-ring transport line), depending on location and humidity.

The expected sensitivities to losses with stored beam in the storage ring can also be calculated from the above measurements. With stored beam, the average beam current follows an exponential decay,

$$I = I_o e^{-t/\tau},$$

(1)

where I is the DC circulating current in the storage ring as a function of time, I_0 is the DC circulating current at time $t = 0$, and τ is the lifetime of the stored beam.

If Q_c is the circulating charge, Q_{co} is the circulating charge at time $t = 0$, and T is the time required for the charge to make one pass around the storage ring, then

$$I = \frac{Q_c}{T}, \quad I_0 = \frac{Q_{co}}{T}, \quad \text{and} \quad Q_c = Q_{co}e^{-t/\tau}. \tag{2}$$

Then

$$\frac{dQ_c}{dt} = -\frac{Q_{co}}{\tau}e^{-t/\tau} = -\frac{dQ_L}{dt} = -\frac{1}{k}\frac{dQ_{lm}}{dt}, \tag{3}$$

where Q_L is the charge lost from the stored beam, Q_{lm} is the charge collected by the loss monitor, and k is the sensitivity factor in pC of collected charge per pC of charge lost from the beam current. From this,

$$\frac{I_0Te^{-t/\tau}}{\tau} = \frac{1}{k}\frac{dQ_{lm}}{dt} = \frac{1}{k}I_{lm}, \tag{4}$$

where I_{lm} is the current collected by the loss monitor and

$$I_{lm} = \frac{I_0Tke^{-t/\tau}}{\tau} = -\frac{dI}{dt}kT. \tag{5}$$

Since the leakage currents in the storage ring are less than 10 pA in each of the ten cables, a uniformly distributed loss that generates 10 pA in each of the ten loss monitor cables for a total of 100 pA should be detectable. Using Eq. (5) above with a charge revolution time T of 3.68 μs and an effective k value (averaged over all locations) of about 100, losses in the storage ring associated with lifetimes in the range of $\tau = 100$ hours or less with $I_0 = 100$ mA circulating beam should be detectable. However, actual measurements taken on April 3, 1996, with 20 mA of stored beam and 17 hours lifetime show loss monitor readings of about 1.1 μA in each of the ten cables. These readings are about five orders of magnitude larger than expected from the above equations using the loss sensitivities obtained with single-bunch data. Apparently the loss monitors are responding to scattered synchrotron radiation despite the attempt to locate the cables below the plane of the storage ring. No such response to synchrotron radiation was observed during earlier tests at SSRL (3). This apparent response to synchrotron radiation makes the storage ring loss monitors of little value in measuring losses that may exist with reasonable beam lifetimes.

It has been observed, however, that the loss monitor readings are proportional to beam current (see Figure 3). This has allowed the loss monitor signal to be used as a low-noise input signal for the calculation of beam lifetimes.

The response of the loss monitor to beam obstructions can be further

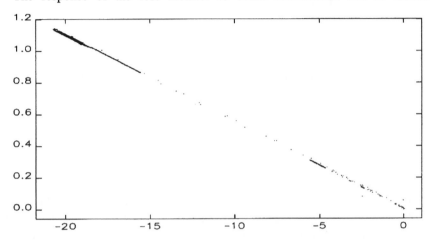

FIGURE 3. Loss monitor readout (vertical axis, µA) versus beam current (horizontal axis, mA).

illustrated by observing the response of the booster-to-storage-ring loss monitor as the scrapers are inserted into the beam. Figure 4 shows the loss monitor signal as the second top and bottom scrapers in the BTS are inserted, then removed. The initial loss monitor signal is caused by the beam hitting the septum. The first peak in the signal is caused by the beam striking the edge of the top scraper. The first valley is the loss monitor signal caused by the beam striking the body of the top scraper. The second peak is again generated by the edge of the top scraper as it is withdrawn. The second valley is the loss monitor signal caused by the beam striking the bottom scraper. Note the lack of signal peaking as the bottom scraper grazes the beam. It appears that the beam strikes the edge of the top scraper as it grazes the beam and the resulting radiation shower is directed downward toward the loss monitor cable, resulting in a signal peak. Similarly, the beam strikes the edge of the bottom scraper as it grazes the beam and the resulting radiation shower is directed upwards away from the loss monitor cable, resulting in no loss monitor signal peak.

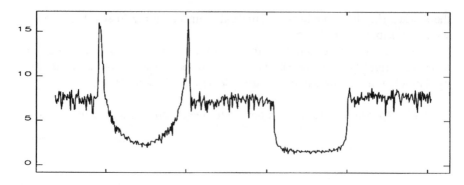

FIGURE 4. Loss monitor signal as scrapers are inserted into the beam.

FURTHER WORK

Further work is required to verify that the predominant loss monitor signal with stored beam is due to response to synchrotron radiation. It may be possible to partially shield the loss monitor cable by placing it inside a metal conduit so as to attenuate the softer synchrotron radiation while allowing the harder Bremsstrahlung radiation to pass into the loss monitor cable. It might also be possible to reduce the sensitivity to synchrotron radiation by relocating the loss monitor cable. However, this seems unlikely to be productive because of mechanical and access restrictions on the cable location and the likelihood that the synchrotron radiation is scattered sufficiently to almost uniformly bathe the storage ring tunnel.

The loss monitor electronics package has been designed to allow measurement of the voltage pulses that arise in the loss monitor cables from large, localized losses. Preliminary measurements indicate that it may be possible to measure the timing of these pulses relative to the beam timing to determine the location of the loss within a cable. The feasibility of automatically making these timing measurements will depend largely on the noise pickup in the cabling. This will be studied in more detail in the future.

REFERENCES

1. Witkover, R. L, "Beam Instrumentation in the AGS Booster," Proceedings of the Third Annual Workshop on Accelerator Instrumentation, Newport News, VA, *AIP Conference Proceedings* **252**, American Institute of Physics, NY, 188-202 (1991).
2. McCormick, D., "Fast Ion Chambers for SLC," Proceedings of the 1991 IEEE Particle Accelerator Conference, San Francisco, CA, May 6-9, 1991, 1240-1242 (1991).
3. Patterson, D. R., "Design and Performance of the Beam Loss Monitor System for the Advanced Photon Source," Proceedings of the Beam Instrumentation Workshop, Vancouver, Canada, October 2-6, 1994, *AIP Conference Proceedings* **333**, 300-306 (1995).
4. Patterson, D. R., "Preliminary Design of the Beam Loss Monitor System for the Advanced Photon Source," Proceedings of the Fourth Accelerator Instrumentation Workshop, LBL, Berkeley, CA, October 27-30, 1992, *AIP Conference Proceedings* **281**, 150-157 (1993).
5. Lumpkin, A., Patterson, D., Wang, X., Kahana, E., Sellyey, W., Votaw, A., Yang, B., Fuja, R., Berg, W., Borland, M., Emery, L., Decker, G., and Milton, S., "Initial Diagnostics Commissioning Results for the Advanced Photon Source (APS)," Proceedings of the 1995 IEEE Particle Accelerator Conference, Dallas, Texas, May 1-5, 1995, 2473-2475 (1996).
6. Documentation available on the World Wide Web at http://www.aps.anl.gov /asd/controls/epics/EpicsDocumentation/WWWPages/EpicsFrames.html.

Position and Collision Point Measurement System for Fermilab's Interaction Regions

M. Olson, A. A. Hahn

Fermi National Accelerator Laboratory, Box 500, Batavia IL 60510

Abstract. A higher resolution beam position monitor (BPM) system has been developed at FNAL to measure the transverse position of the beam at opposite ends of the Collision Hall. A secondary function is to measure the longitudinal location of the collision point. This system is called the Collision Point Monitor (CPM). The transverse positions are determined by software rectification and integration of a BPM signal obtained from a sampling oscilloscope. A difference over sum calculation of the A and B signals yields the position. The longitudinal location is obtained by measuring the difference in time between the proton and antiproton bunches at both ends of the Collision Hall. The downstream difference is then subtracted from the upstream difference and the result is multiplied by half the speed of light to yield the collision point error.

INTRODUCTION

Fermi National Accelerator Laboratory (FNAL) has two regions in the Tevatron (B0 and D0) where the proton and antiproton beams collide. These collision regions are bounded upstream and downstream by horizontal and vertical beam position monitors (BPMs) as drawn in Fig. 1. These BPMs are striplines with the same cross section as the standard Tevatron detector but are only half their length (8.3 cm) (1,2). The detector signals are split and shared with the Tevatron BPM system. To conserve space, the BPMs were built as an integral part of the cryogenic quadrupole magnet. This required four meters of cable inside the magnet to transport the signal to the accessible end. The stripline directionality was defeated by placing shorts on the cables attached to one end of the detector. This allowed both proton and antiproton signals to be measured using the same cables and electronics, reducing systematic errors.

A typical store consists of six proton and six antiproton bunches. The proton bunch intensities are three times larger than the antiproton. To reduce the beam-beam tune shift and increase luminosity, separators were installed in the Tevatron to move the protons and antiprotons onto helical orbits that intersect only at B0 and D0. The location of the collision point can be determined by measuring the proton-antiproton position difference upstream and downstream of the Collision Hall and projecting straight lines. This procedure ignores any beam-beam steering. The 150-micron resolution of the existing Tevatron BPM system was not sufficient. The Collision Point Monitor (CPM) provides 20-micron resolution.

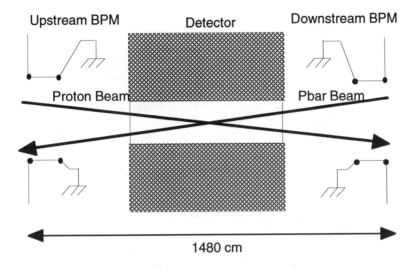

FIGURE 1. Tunnel geometry of CPM pickup plates.

A secondary function of the system is to accurately determine the longitudinal collision point by measuring the arrival time difference between the proton and antiproton bunches at each end of the Collision Hall. A resolution of about 50 ps, or 1.5 cm, has been achieved.

CONFIGURATION

The CPM is composed of a LabVIEW application program operating on a Macintosh computer utilizing a Tektronix TDS520 oscilloscope with a Keithley rf multiplexer for data acquisition. The Collision Point Monitor is diagrammed in Fig. 2. The interface between the Macintosh and the accelerator control system is provided by a token ring link (3). The computer communicates with the oscilloscope and the multiplexer through a GPIB interface. The four position detectors are sequentially connected to the oscilloscope's inputs through the Keithley rf multiplexer. The scope's main sweep is triggered by a beam sync pulse generated by a Camac 279 module. The 279 module has a resolution of 7 rf buckets, so the oscilloscope must be operated in the delay trigger mode to obtain the finer timing resolution required to capture the 20-ns bunch signal. The oscilloscope's trigger delay must be changed for each BPM to compensate for differences in cable lengths and beam flight times. After the beam signal has been acquired, a GPIB read is performed to transfer this information to the computer for processing.

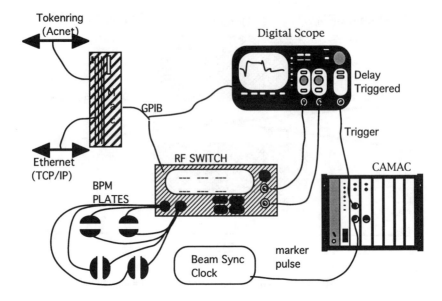

FIGURE 2. The Collision Point Monitor configuration.

TRANSVERSE POSITIONS

At both interaction regions there are six proton-antiproton bunch collisions each turn. One of the six collisions is selected with the Acnet parameters T:B0TRIG or T:D0TRIG. These parameters are referenced to the Tevatron beam sync "A-marker" and are in units of rf buckets.

A beam position resolution of 20 microns can be achieved by setting the oscilloscope to single-sweep averaging mode with 16 averages, 5 ns per division, delayed trigger mode, and full bandwidth.

During each data collection cycle the CPM program performs four iterations through a loop, one iteration for each of the BPMs. Nested within this outer loop is a second loop that iterates twice, once to obtain the proton position data and again to obtain the antiproton position data. The vertical gain of the scope is set to 2 volts per division on the proton pass and 500 millivolts per division on the antiproton pass. Improved performance on lower intensity stores can be achieved by setting the oscilloscope's vertical scale to a more sensitive setting. It is planned to have these scales automatically set in a future revision of the program. The delay for the oscilloscope trigger is set automatically from a look-up table depending upon which bunch is selected and which BPM is being read. The look-up table values have been empirically determined to trigger the oscilloscope about 10 ns before the bunch arrives.

When the beam traverses the detector, it generates a doublet signal such as the one illustrated in Fig. 3. Two signals, one from the A plate and one from the B plate, are transferred to the computer through the GPIB interface.

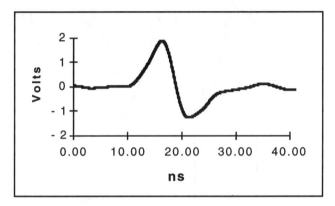

FIGURE 3. A typical BPM doublet signal.

The procedure used to calculate the position is listed below.

1. The digitized signal is passed through a software 5-90 MHz bandpass filter to remove unwanted frequency components and optimize the signal-to-noise ratio.

2. The most positive and most negative points of the trace are found and the points between them are fit to a cubic polynomial. The zero crossing of the polynomial is used as the zero crossing for the signal.

3. The signal is rectified by multiplying all points after the zero crossing by minus 1.

   The advantage of steps 2 and 3 over a simple addition of the absolute values is that any offset, noise, ringing, or satellite bunch signals outside the central bunch averages to zero.

4. The signal strength of the A and B doublets is then determined by digitally integrating the rectified signals.

5. A difference-over-sum calculation is performed to obtain a position reading.

This process repeats until all four proton and all four pbar positions have been obtained. A sample datalog plot of the <u>D0</u> <u>D</u>ownstream <u>H</u>orizontal <u>P</u>roton <u>P</u>osition signal is displayed in Fig. 4.

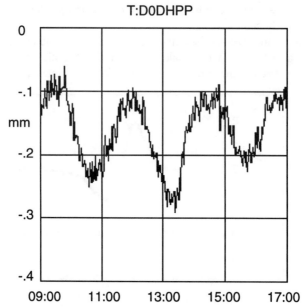

FIGURE 4. Sample datalog plot of a transverse position reading.

LONGITUDINAL POSITIONS

The longitudinal location of the collision point is measured using another routine that obtains a signal from one plate at each of the four BPM locations. The scope is set to 20 ns per division, single-channel acquisition, 16 averages, and an appropriate trigger delay. Waveforms similar to those in Fig. 5 are sent to the computer and processed to determine the zero crossing times of the proton and antiproton doublets. From these zero crossings, the time-of-flight differences (Delta-t) is determined. A sample datalog plot of the longitudinal positions covering a 16-hour time span is shown in Fig. 6.

Consider the case diagrammed in Fig. 7 where the collision point is shifted downstream. At T1, the proton bunch crosses the upstream BPM but the antiproton bunch still has an x amount of time before it crosses the downstream BPM. Later, at T4, the proton bunch crosses the downstream BPM but the antiproton bunch still has an x amount of time before it crosses the upstream BPM. This results in Delta-t upstream being larger than Delta-t downstream by a factor of 2x. Therefore

$$\text{Cogging Error} = (\text{Delta-t upstream} - \text{Delta-t downstream})/2*C. \qquad (1)$$

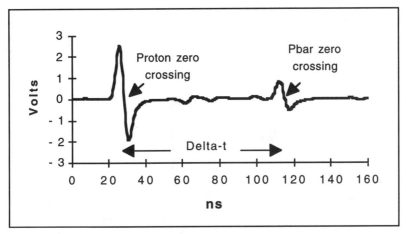

FIGURE 5.   A typical Delta-t waveform.

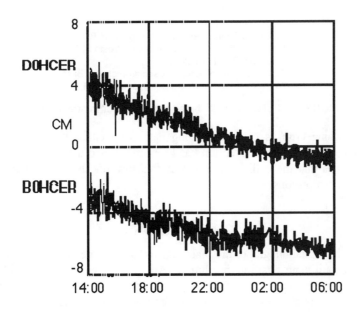

FIGURE 6. Sample datalog plot of the longitudinal positions.

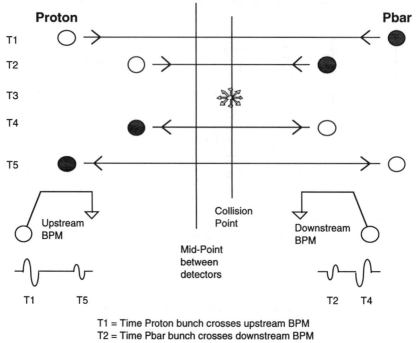

T1 = Time Proton bunch crosses upstream BPM
T2 = Time Pbar bunch crosses downstream BPM
T3 = Time Bunches collide
T4 = Time Proton bunch crosses downstream BPM
T5 = Time Pbar bunch crosses upstream BPM

FIGURE 7. Collision point shifted toward the downstream BPM.

The cogging error is calculated using data from either the vertical or the horizontal detectors. The convention has been established that a positive cogging error indicates that the collision point is shifted towards the downstream BPM, with respect to protons.

CONCLUSIONS

Using the digital integral of the plate signals to calculate position was thought to offer the ultimate position resolution. Any analog equivalent would clearly suffer from at best the same difficulties as this digital system. An annoyance of the system is the slow cycle rate that makes it awkward to use during separator scans. Approximately 90 seconds is required for a complete set of transverse and longitudinal measurements. The majority of time is used to perform averaging which could be reduced by using a faster scope or a digitizer.

The success of the system was the automatic measurement of the absolute longitudinal beam position. The rms resolutions obtained were about 50 ps or 1.5 cm. A less-than-successful result was the absolute transverse measurement.

Statistically the transverse measurements are good to the 20-micron level. The absolute accuracy is difficult to verify. With the separators turned off, proton and antiproton orbits should be identical, but the measured positions disagreed by up to 0.69 millimeters. Figure 5 illustrates one source of error. Between the proton and antiproton doublets there are reflected signals caused by the shorted end of the pickup delayed by the internal cable. Some ringing is corrupting the antiproton reading. Compensation for this problem in software is possible, but further analysis is required. Rebuilding the BPM with an internally shorted plate to eliminate the internal cable run would reduce this effect but cost may be prohibitive.

REFERENCES

1. Gerig, R.E., "Fermilab Energy Doubler Beam Position Monitor System," Operations Bulletin 888, 1982.
2. Shafer, R.E., Gerig, R.E., Baumbaugh, A.E., Wegner, C.E., "The Tevatron Beam Position and Beam Loss Monitoring Systems," Proc. of the 12th International Conference on High Energy Accelerators, Fermilab, August 11-16, 1983, 609-615 (1983).
3. Blokland, W., "An Interface From LabVIEW to the Accelerator Controls Network," Proc. of the Accelerator Instrumentation Workshop, Berkeley, CA, October 27-30, 1992, *AIP Conference Proceedings* **281**, 320-329 (1993).

Trigger Delay Compensation for Beam Synchronous Sampling*

James Steimel

Fermi National Accelerator Laboratory
P.O. Box 500, Batavia, IL 60510

Abstract. One of the problems of providing beam feedback in a large accelerator is the lack of beam synchronous trigger signals far from the rf signal source. If single-bucket resolutions are required, a cable extending from the rf source to the other side of the accelerator will not provide a synchronous signal if the rf frequency changes significantly with respect to the cable delay. This paper offers a solution to this problem by locking to the rf, at the remote location, using a digital phase-locked loop. Then, the digitized frequency value is used to calculate the phase shift required to remain synchronized to the beam. Results are shown for phase lock to the Fermilab Main Ring rf.

INTRODUCTION

Most fast, wide-band beam instrumentation and control systems require some kind of reference signal which remains stable and in phase with the beam. For storage rings or rings with highly relativistic beam, a stable fixed oscillator would provide the necessary signal. For lower energy accelerators, however, one needs a source which tracks the changing velocity of the beam. The Fermilab Main Ring is such an accelerator, and it derives beam synchronous triggers from the rf acceleration system. These triggers are then used to control dampers (1), kickers, and diagnostics equipment. Although the position of the beam relative to these triggers remains constant from cycle to cycle at a given time in the cycle, they do not remain in the same position relative to beam throughout one cycle. This paper discusses the reason for this phenomenon as well as offering solutions to counteract the problem.

---

* Operated by the University Research Association, Inc. under contract with the U.S. Department of Energy.

MAIN RING RF TIMING

The Main Ring rf system is used to accelerate the beam; therefore it must remain phase locked to the beam in order to remain effective. The phase of the voltage-controlled oscillator (VCO) must lead the phase of the beam by the same amount of time it takes for the signal to get to the amplifiers, through the cavities, and across the gap at the correct phase for the beam. By providing a delay in the feedback loop of the VCO, the phase error between the beam and the VCO can be reduced by a factor of the open loop gain of the system. This delay must be short enough, however, to maintain the stability of the loop. Figure 1 shows a simplified diagram of the Main Ring phase loop. With the configuration shown, the phase error at the phase detector will equal the phase error at the accelerating cavity. Thus, the signal coming out of the VCO is not synchronous with the beam until 500 ns after it is created.

The triggers used to control synchronous events are derived from the VCO signal. The Main Ring Beam Synch (MRBS) system divides the VCO frequency by seven and distributes this clock signal around the ring. Encoded on this clock are different timing critical events such as the event for firing booster extraction kickers, the event for firing Main Ring extraction kickers, and events marking the first bunch every revolution for a partially filled machine. The events for the

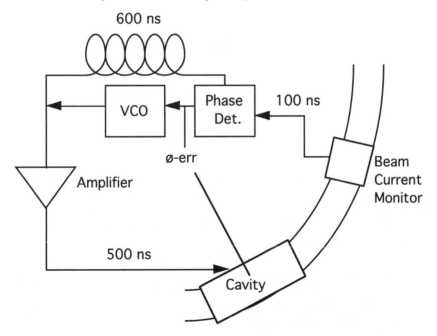

FIGURE 1. Block diagram of the Main Ring beam phase feedback system. Notice that the phase error at the phase detector will match the phase difference between the VCO and the beam at the cavity.

kickers can be calibrated at a particular point in the acceleration cycle (usually the beginning and end of the cycle) and remain precise to a beam bucket on each cycle. The events for marking a bunch, however, only remain precise at short distances from the VCO.

Effects of Delay Errors

Using the MRBS system as a beam synchronous trigger requires careful monitoring to ensure that the trigger is delayed properly with respect to the beam. If the signal is not delayed properly, phase errors between the trigger and the beam will develop as the velocity of the beam increases. One way to simulate the effect is to drive two cables of unequal delay with an rf source. At any one frequency, the phase difference between the two cables is fixed, but as the frequency changes, the phase difference between the two cables changes according to Eq. (1). The last approximation in Eq. (1) assumes that we perform the integration where the frequency slew appears relatively flat over a time span equal to the difference in the delays.

$$
\begin{aligned}
\Delta\phi &= \int_0^t \left(\omega_{rf}(t') - \omega_{rf}(t' - \Delta t) \right) dt' \\
&= \int_{t-\Delta t}^t \omega_{rf}(t') dt' - \int_{-\Delta t}^0 \omega_{rf}(t') dt' \\
&\cong \left(\omega_{rf}(t) - \omega_{rf}(0) \right) \Delta t
\end{aligned} \tag{1}
$$

The Main Ring rf changes its frequency by about 300 kHz over the entire cycle. For triggering applications close to the VCO, the delay error will be on the order of the total delay of the acceleration system. This corresponds to a maximum phase error of about 300 kHz * 600 ns = 65°, which is too large a phase error to ignore even close to the VCO.

A real problem exists for synchronization at any location other than close to the rf system. About two miles of cable is require to send the VCO signal to the other side of the ring. This results in a delay of about 16 μs, which corresponds to a phase error of almost five complete cycles. This means that a trigger set up to be synchronous with the start of every batch at injection will be off by five buckets at extraction.

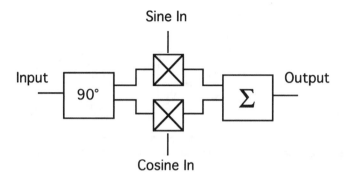

FIGURE 2. I & Q phase modulator. The cosine and sine inputs are driven by a voltage which is proportional to the real and imaginary parts of the output signal desired.

DELAY DIFFERENCE COMPENSATION

For any application which relies on a synchronous trigger signal to sample or kick the beam throughout the Main Ring cycle, the phase errors caused by differences in delays are unacceptable. One way to correct the problem is to delay the beam signal by an amount equal to the extra delay of the trigger signal from the VCO. Unfortunately, this could be as much as two miles of cable at a place in the ring far from the VCO. Compensations of this type would be impossible for kickers because an equal negative delay would be required.

Open Loop Compensation

Another possibility for compensating the delay difference is to use an I & Q phase modulator and two arbitrary function generators. Figure 2 shows a block diagram of the phase modulator. The function generators drive the multipliers which control the ratio of the real and imaginary components of the rf drive. These function generators are programmed to provide a phase shift equal and opposite to the phase shift induced by the difference in delay. One drawback to this method is that the generators must be reprogrammed every time the acceleration cycle program is changed. The generators must also be programmed accurately relative to each other, or they could cause phase and amplitude modulations on the output of the modulator.

A more versatile method of correcting the phase errors is to replace the function generators with a signal derived from the VCO frequency. If a table of digital frequency values that tracks the VCO can be created, then the configuration shown in Fig. 3 will compensate for delay differences. The frequency value is multiplied by a delay value (positive or negative), and a phase

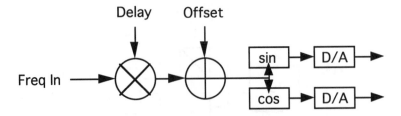

FIGURE 3. I & Q modulator driver circuit. A digital frequency input will produce two analog signals which are proportional to the real and imaginary parts of the phase shift required.

offset is added. Then this phase value is passed through sine and cosine look-up tables and converted to analog signals to drive the I & Q phase modulator mentioned above. The advantage of this method over the function generators is that only one function needs to be programmed (frequency), and the sine and cosine look-up tables are calibrated to reduce phase and amplitude modulations. Unfortunately, this method will still require a different frequency table every time the acceleration cycle program is changed.

Phase-Locked Loop Compensation

Ideally, there should be a way to measure the VCO frequency throughout the cycle and send the value to the phase shifters without some kind of fixed table. A commercial frequency counter could provide the needed digital frequency value. The value would drive the frequency input of phase shifter, and the VCO itself would drive the other phase shifter input. This works fine if the actual VCO signal is distributed around the ring.

Another way to solve the problem is by using a digital phase-locked loop that drives a direct digital synthesizer (DDS). Figure 4 shows a block diagram of the phase-locked loop. The value of the rf wave is sampled by a digitizer which is triggered by the MRBS clock. The digital value is scaled and summed with an integrated signal. Both the scalar and the integration constants are controlled digitally, so the poles of the phase-locked loop filter can be adjusted easily. The primary feature of this system is that the circuit provides a filtered digital value equal to the VCO frequency. Another feature of this system is that it provides a new rf signal, which makes it unnecessary to distribute the actual VCO signal.

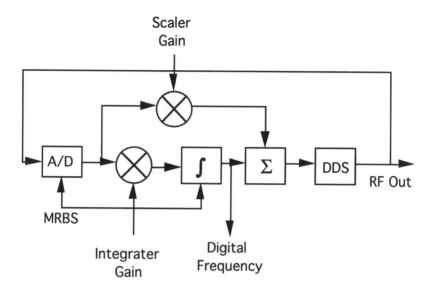

FIGURE 4. An rf regeneration circuit. The rf phase locks to the MRBS and produces the digital frequency signal necessary to drive the phase shifter.

Measurement Results

The entire circuit including the rf regeneration circuit, the I & Q modulator driver, and the I & Q modulator was constructed. A Sciteq® ADS-431-203 was used as the DDS. It is capable of being clocked at 1600 MHz and can output frequencies up to 400 MHz. The digitizer is 8 bits wide and drives an Altera® erasable programmable logic device (EPLD) which acts as the digital filter and delay processing unit. For a clock frequency of 803 MHz and an input frequency of 53 MHz/7, the scalar value of the filter is 3.91×10^5 s^{-1}, and the integrator value is 3.61×10^8 s^{-2}. With these settings, the delay precision is about 40 ns, and the maximum possible delay is about 41 μs.

The system was then tested on the bench using a network analyzer and a divide-by-seven rf counter. The network analyzer swept through the Main Ring operating frequencies from 52.8 MHz to 53.1 MHz, and the sweep time was adjusted to provide the maximum frequency slew rate that the Main Ring requires (about 2 MHz/s). Tracking errors were less than half of a degree at this slew rate. However, the system does have repeatable errors in the phase shift circuitry. As the programmed delay causes the phase shifter to shift the signal through 360°, there is a ± 0.2 dB error in amplitude and a ± 3° error in phase. These errors can be subtracted by reprogramming the sine/cosine look-up tables.

Random errors are much more detrimental to the operation of the system than the fixed errors. Phase modulation noise was measured, and it was usually about −110 dBc/Hz out to 15 MHz. Amplitude modulation noise measured about −120 dBc/Hz out to 400 kHz, but the amplitude modulation spectrum also contained spurs at about −57 dBc. These values are typical values. One can measure the worst-case noise properties by setting the input frequency to a point where 21 bits on the DDS are toggling to sustain lock. In this case, there are phase modulation spurs at -54 dBc and amplitude modulation spurs at −49 dBc. Also, the AM noise floor is increased to −80 dBc/Hz.

Future Plans

This system will certainly be applied to the operation of the Main Ring dampers, and it may also be applied to other beam synchronous applications. Hopefully, though, the days of the rf regeneration circuit are numbered. The entire accelerator will be converting its analog VCO and accelerating system into a digital DDS system. When this happens, the value of the DDS frequency could be distributed around the ring with a carrier at the DDS frequency itself. Once the rf signal and frequency value are extracted, the only necessary components are the I & Q modulator and the I & Q modulator driver. Plans are currently being discussed for maintaining ring-wide beam synchronous signals in the Main Injector.

ACKNOWLEDGMENTS

The author would like to thank Ken Koch for his efforts in laying out and constructing the circuit boards for this project. The author would also like to thank Keith Meisner, Brian Chase, and Bob Webber for their advice on diagnosing problems in phase-locked loops.

REFERENCES

1.   James M. Steimel, Jr., "Fast Digital Dampers for the Fermilab Booster," in *Proceedings of the 1995 Particle Accelerator Conference*, May 1-5, 1996, Dallas, TX, 2384-2388 (1996).

Design and Commissioning of the Photon Monitors and Optical Transport Lines for the Advanced Photon Source Positron Accumulator Ring*

W. Berg, B. Yang, A. Lumpkin, and J. Jones

Advanced Photon Source, Argonne National Laboratory
9700 South Cass Avenue, Argonne, IL 60439

Abstract. Two photon monitors have been designed and installed in the positron accumulator ring (PAR) of the Advanced Photon Source. The photon monitors characterize the beam's transverse profile, bunch length, emittance, and energy spread in a nonintrusive manner. An optical transport line delivers synchrotron light from the PAR out of a high radiation environment. Both charge-coupled device and fast-gated, intensified cameras are used to measure the transverse beam profile (0.11 to 1 mm for damped beam) with a resolution of 0.06 mm. A streak camera ($\sigma_\tau = $ 1 ps) is used to measure the bunch length which is in the range of 0.3 to 1 ns. The design of the various transport components and commissioning results of the photon monitors will be discussed.

INTRODUCTION

The Advanced Photon Source (APS) positron accumulator ring is designed to be filled by a 450-MeV positron linac at 60 Hz (1). The positron accumulator ring (PAR) is 30.667 meters in circumference and operates with a fundamental frequency of 9.776 MHz. However, the addition of power at the 12th harmonic (117 MHz) produces further compression resulting in a shorter bunch for efficient injection into the booster synchrotron. High field dipoles (1.48 Tesla) are used to bend the positron beam around the ring (2). Synchrotron light emitted from the source points within the dipoles allow measurement of the transverse beam size, bunch length, emittance, and energy spread via the photon monitors.

PHOTON MONITOR TRANSPORT LINES

Two optical transport lines have been designed and installed in order to deliver synchrotron light out of the machine's high radiation environment into a user-safe area. Some of the key considerations for the design were light transport efficiency,

---

* Work supported by U. S. Department of Energy, Office of Basic Energy Sciences under Contract No. W-31-109-ENG-38.

vibration, and reduction of air/thermal turbulence. Evacuated vacuum chambers, which house the optical components, are used to reduce the effect of air currents. The transport system overview is shown in Fig. 1(a). Figure 1(b) shows a diagram of a typical source point located in a dipole vacuum chamber. A retractable fiducial target inserted at the source point will provide a reference to verify the system alignment, focus, resolution, and magnification.

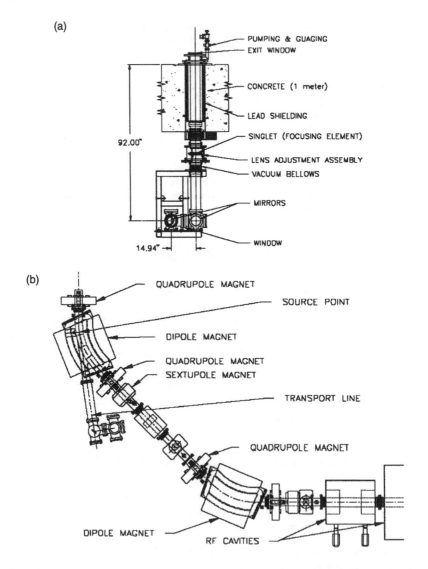

FIGURE 1. (a) Transport line components. (b) One quadrant of the PAR showing the location of the source point.

The main imaging component of the transport line is a double convex quartz lens. The lens projects an image to the center of the optics table where the cameras are mounted. Table 1 shows the design parameters of the transport system. The performance of the transport optics were modeled using the commercial ray-tracing program ZEMAX (3). The results are given in Table 2.

TABLE 1. Design Parameters for the Transport Line

Lens material / diameter	quartz / 150 mm
Focal length	1.65 m
Distance from the source	3.00 m
Distance from the image	3.67 m
Magnification	1.22

TABLE 2. Spatial Resolution for the Transport Line
(λ=532 nm, $\Delta\lambda$=10nm, 8 mrad horizontal acceptance)

Diffraction contribution	14 µm
Depth of source contribution	5 µm
Geometric and chromatic aberration	12 µm
Imperfect optics distortion (estimated)	12 µm
Combined resolution	23 µm
Beam size	600 - 1110 µm

The visible synchrotron light from the source point is intercepted by a 4" moly mirror (radiation cooled). The light is reflected through a quartz window which isolates the ring from the transport system. This isolation enables adjustment and maintenance of the optics without venting the ring. A number of quartz mirrors direct the light through a fused silica lens to an optics table. Considerable effort was put into design and selection of components for a vacuum-compatible lens adjustment housing. The lens can be adjusted with three degrees of freedom for final alignment.

The x-ray shielding specifications required that a significant amount of lead be placed in and around the transport penetrations that were cast into the concrete. A 1.5" thick lead sleeve poured between two steel tubes with support flanges welded to the ends provide the bulk of the shielding. The lead liner, pictured in Fig. 2, shows the lead as it is being installed into a penetration. Vibration isolation pads located under the supporting base of the exit vacuum chamber help reduce the low-frequency vibrations generated from nearby machinery. Soft bellows were incorporated into the transport line on either side of the lens housing to reduce vibration transfer from the mezzanine floor above the ring. The light exits the transport vacuum chamber through a quartz window and is directed onto an optics table for imaging. Several views of the finished installation are shown in Figs. 3(a), (b), and (c).

FIGURE 2. Installation of lead shielding.

CAMERAS AND IMAGING RESULTS

A gated, intensified camera (Stanford Computer Optics: Quik05) was placed at the focal point of the transport system to readout beam images (4). The fast gate (minimum of 5 ns) allows the acquisition of beam images in a single turn/single pass. Such capability is useful in studying transient phenomena at the time of injection, extraction, and compression (turn-on of 12th harmonic power).

Two charge-coupled device (CCD) cameras with secondary optics provide routine observation of the beam. Since the two photon ports have very different dispersions, the transverse beam profile data for stored beam yields information on both beam emittance and energy spread. The images from the above cameras are combined with a video quad (American Dynamics AD1476) for routine viewing by the machine operators (5). Table 3 lists the measured beam profile and Fig. 4 shows a typical image and the transverse beam profiles.

TABLE 3. Lattice Function and Measured Beam Profiles (RF12 off)

Direction	x		y	
Port	No. 1	No. 2	No. 1	No. 2
Beta function (m)	4.1	3.2	13.5	3.4
Dispersion (m)	0.0	-2.6	0.0	0.0
Beam size (mm)	0.91	1.01	0.18	0.11

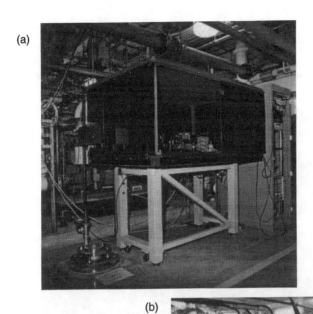

FIGURE 3. (a) Exit window and optics table. (b) Source point transport section. (c) Transport and lens housing.

A streak camera is used to observe the bunch length. Figure 5(a) is a streak image which shows the effect of the bunch compression on stored beam. After the 12th harmonic power is turned on, the bunch starts to lengthen from 0.8 ns but eventually damps down to 0.3 ns (Fig. 5(b)). The damping time constant is

(a)

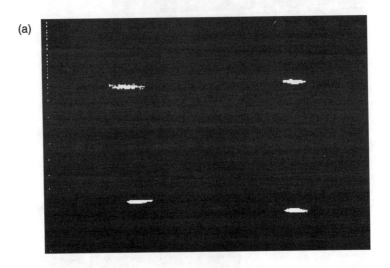

(b)

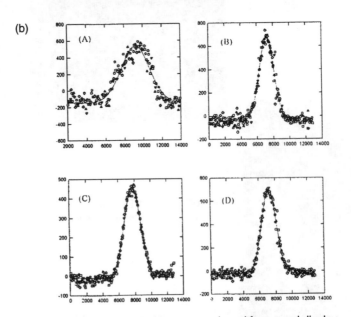

FIGURE 4.   (a) Typical images as viewed from quad display.
(b) Typical plot profiles, Intensity vs. Size (μm).

deduced from the progressive changes of the profiles. Table 4 summarizes the measured beam parameters which compare favorably with the design values. The longer bunch lengths are due to the lower rf voltage used (17 kV applied vs 30 kV designed) during these measurements.

(a)

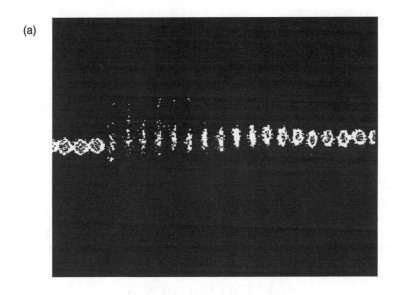

(b)

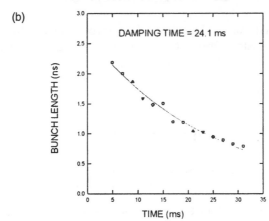

FIGURE 5. (a) Dual sweep streak image with 12th harmonic turned on. The vertical span of the image is 10 ns and the horizontal is 50 ms. The vertical streak was triggered every 2 ms. (b) Bunch length as a function of time demonstrating the damping effects with 12th harmonic applied.

TABLE 4. Transverse and Longitudinal Beam Profiles (E=375 MeV)

Quantity	Measured	Design (6)
Emittance (mm·mrad)		
e_x	0.20	0.25
e_y	0.0024	< 0.025
Vertical coupling	0.012	< 0.1
Energy spread	2.5×10^{-3}	2.8×10^{-3}
Bunch length		
RF12 off	0.84 ns	0.67 ns
RF12 on	0.31 ns	0.21 ns
Longitudinal damping time		
RF12 on	24.1 ms	25.4 ms

CONCLUSION

There are now two transport lines installed in the PAR, and commissioning of the photon monitors is underway. Initial measurements of the emittance, vertical coupling, bunch length, and longitudinal damping time have been performed. The results are found to be consistent with the design objectives. Further refinement and integration of the photon monitors into the APS control system is in progress.

ACKNOWLEDGMENTS

The authors would like to acknowledge the extra efforts put in by CAD designers Dave Fallin and Anatoly Oberfeld. Without their special attention to detail and nurturing of this project installation efforts would have been much more painful, to say the least. Also many thanks go to Mike Borland for his advice and, more importantly, his patience.

REFERENCES

1. A. Lumpkin et al., "Initial Diagnostics Commissioning Results for the Advanced Photon Source (APS)," Proceedings of the 1995 Particle Accelerator Conference, Dallas, Texas, May 1-5, 1995, 2473-2475 (1996).
2. Annex to 7-GeV Advanced Photon Source Conceptual Design Report, ANL-87-15 Annex, p. II. 10-10, May 1988.
3. Focusoft Inc., ZEMAX Optical Design Program, Pleasanton, CA.
4. Stanford Computer Optics, Palo Alto, CA.
5. American Dynamics, Orangeburg, NY.
6. M. Borland, private communication.

Particle-Beam Profiling Techniques on the APS Storage Ring[*]

B. X. Yang and A. H. Lumpkin

Advanced Photon Source, Argonne National Laboratory, 9700 South Cass Avenue, Argonne, IL 60439 USA

Abstract. Characterization of the Advanced Photon Source storage ring particle beams includes transverse and longitudinal profile measurements using synchrotron radiation-based techniques. Both optical (OSR) and x-ray synchrotron radiation (XSR) stations are now installed. Spatial resolution of about $\sigma = 55$ μm was obtained at low current in the visible field initially. This is expected to improve during its commissioning. UV/visible light from the storage ring bending magnet was employed to measure the particle beam with a resolution of $\sigma \sim 80$ μm and allow operation at 100 mA with the initial x-ray pinhole setup. Early OSR measurements of beam size are consistent with 8.2 nm·rad emittance and 2-3% vertical coupling. Early results with the x-ray pinhole camera are also presented.

INTRODUCTION

The 7-GeV storage ring at the Advanced Photon Source (APS) began commissioning in early 1995. Measurements of the transverse sizes and the bunch length of the stored beam provide important information in characterizing transverse and longitudinal emittance, respectively. Imaging techniques with optical synchrotron radiation were used first due to their simplicity. But OSR spatial resolution is limited mostly due to the diffraction of the synchrotron light beam and depth of the source (Table 1). The optimum resolution is given approximately by $\sigma \approx \left(\rho \lambdabar^2 \right)^{1/3}$ when the full acceptance angle of the imaging system is around $\sigma \approx \left(\lambdabar / \rho \right)^{1/3}$, where $\rho = 39.8$ m is the radius of the particle trajectory.

It is evident that the resolution of the optical imaging system will not be adequate for the low emittance (low vertical coupling) mode of operation, and imaging with shorter wavelength radiation is needed. In this report we discuss the design and characterization of the APS UV/visible and x-ray pinhole camera beamlines and present initial results obtained from their operation.

[*] Work supported by U.S. Department of Energy, Office of Basic Energy Sciences under Contract No. W-31-109-ENG-38.

TABLE 1. The APS Storage Ring Beam Parameter and Optical Imaging Resolution

particle beam natural emittance	8.2 nm·rad	
energy spread	~ 0.1%	
dispersion	92 mm	
vertical coupling	10 %	1 %
rms betatron oscillation (μm)	114 (H) × 117 (V)	114 (H) × 37 (V)
total beam size (μm)	146 (H) × 117 (V)	146 (H) × 37 (V)
optimal acceptance angle (λ = 400 nm)	3.5 mrad	
optimal resolution	54 μm	

UV/VISIBLE BEAMLINE RESULT

The design details of the UV/visible beamline have been reported elsewhere (1,2). Its main challenge is diffraction-limited imaging (angular resolution ~ 2 μrad) under high power load. The light transport uses a spherical imaging mirror as the only imaging element to avoid chromatic aberrations. This feature allows convenient alignment and calibration and high transport efficiency which allowed streak and CCD cameras to operate at low beam current. A CCD camera (1/2" format) was placed at the focal point of the mirror to read out the beam image.

Another critical component is the first mirror which, by virtue of high angle reflection, separates the UV/visible radiation from higher energy (x-ray) photons. To achieve 1 μrad or less wavefront distortion, the mirror needs to maintain rms flatness to within 60 nm, and the rms temperature difference over the entire mirror surface to less than 1°C. A molybdenum mirror mounted on a water-cooled copper plate was used in the commissioning stage with satisfactory results for stored beam current up to 20 mA. At higher current, we used a water-cooled slotted mirror which allows the high power x-ray beam to pass through and only intercept UV/visible beams at high angles from the orbit plane (2). Unfortunately it distorted slightly under the water/vacuum differential pressure due to a manufacturing defect and could not meet our specifications. An alternative solution, using a tube absorber in front of the pick-up mirror to shield it from the high-power x-ray beam, is currently under development.

Due to its limited spatial resolution, our OSR station is mainly used for streak camera measurements of beam bunch length and longitudinal dynamics since the setup of the x-ray pinhole camera. Figure 1 shows a typical dual sweep streak camera image taken with the Mo-mirror at low current (7 mA single bunch). The vertical scan is driven by a synchroscan unit operating at 117.3 MHz. Since the frequency is one-third of that of the storage ring rf frequency, the streak camera acquires the bunch data every three buckets. Figure 2 shows dual sweep images with different horizontal time scale and signatures of phase oscillations at 40 kHz and 360 Hz, which are likely due to output ripples of the rf power amplifier power supply. The amplitude of these signatures is checked periodically such that any abnormal conditions of the rf system should be spotted early.

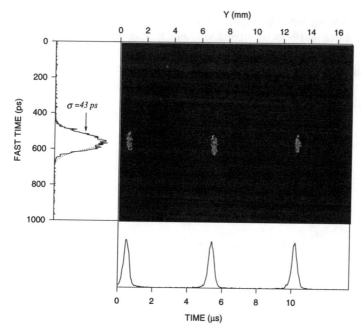

FIGURE 1. A typical dual-sweep streak camera image, a side view of the beam bunch with the fast-time axis in the vertical direction, and slow-time in the horizontal. In setting up this experiment, the beam image was rotated 90 degrees. Hence the horizontal width of the image actually shows the vertical size of the beam (top scale), while the vertical dimension of each bunch image shows the beam length in time ($\sigma \sim 45$ ps). To read the time between passes, the lower horizontal scale should be used.

(a) (b)

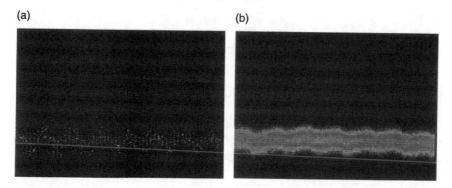

FIGURE 2. Dual-sweep streak camera images of a single bunch. The stored beam is 4.9 mA. (a) The full vertical scale is 0.96 ns and the full horizontal scale 200 µs. The bunch exhibits a 40-kHz phase oscillation with a peak-to-peak amplitude of about 80 ps. (b) The full vertical scale is 0.96 ns and the full horizontal scale 20 ms. The bunch exhibits a 360-Hz phase oscillation with a peak-to-peak amplitude of about 30 ps.

Figure 3 shows the dependence of bunch length on the single-bunch current. In the run dated 8-19-95, a significant bunch lengthening (slope change) was observed before the maximum bucket charge was reached, indicating the onset of a longitudinal instability at high bunch current. It can also be seen that the measured values are well within the APS design goal of a 100-ps bunch length at 5-mA single bunch current.

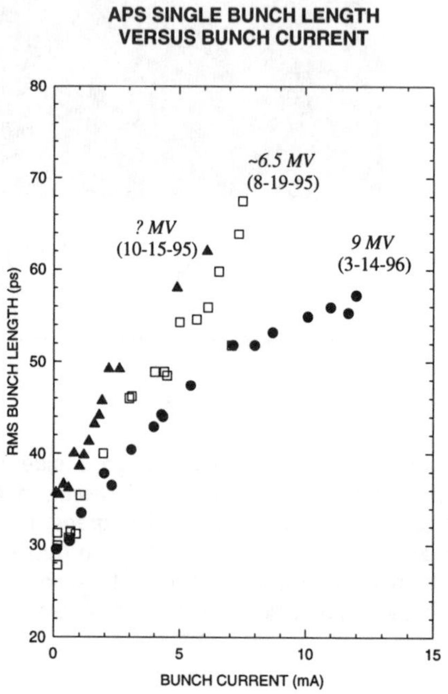

**APS SINGLE BUNCH LENGTH
VERSUS BUNCH CURRENT**

FIGURE 3. Single-bunch current versus bunch length for different rf gap voltages.

X-RAY PINHOLE BEAMLINE

The initial setup of the pinhole camera is entirely inside the storage ring tunnel and is shown in Fig. 4. It is similar to those used elsewhere (3,4). The design parameters are given in Table 2. The magnification of the pinhole camera is

$$M = \frac{\text{detector - pinhole distance}}{\text{source - pinhole distance}} = 0.442 \ .$$

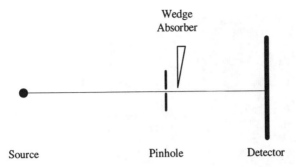

FIGURE 4. Experimental setup of the x-ray pinhole camera.

TABLE 2. Design Parameter of the Pinhole Camera (in-tunnel setup)

Component	Location (from source)	Comments
Aluminum windows	13.30 m	2 mm (H) × 12 mm (V), water-cooled, two windows and a pumping station interlocked with the photon shutter are used for vacuum safety
Pinhole	13.35 m	25 µm × 25 µm square aperture, formed with two sets of perpendicular tungsten blades 1 mm thick
S/S wedge	13.50 m	stainless wedge, adjustable attenuator (0 to 10 mm thick)
Scintillator	19.26 m	10 mm × 10 mm CdWO$_4$ crystal mounted on a grid; light is collected from the back side of the crystal
Read-out optics/camera	19.26 m	a telemicroscope (Questar SZ-FR1) with a ½" CCD camera, resolution (1-σ) = 8 µm at the scintillator, 18 µm at the bending magnet source

A typical pinhole image is shown in Fig. 5. The beam profile fits well with Gaussian form and the profile widths are 157 µm horizontally and 115 µm vertically. These values are consistent with the design goal of the APS storage ring with 8.2 nm·rad emittance and 10% vertical coupling. However while the horizontal width is the same as measured with OSR imaging, the vertical size is significantly higher than that measured with OSR or with an x-ray pinhole camera with higher magnification (4). It is likely that the point spread function of the scintillation detector dominates the total resolution due to the low magnification in the present setup.

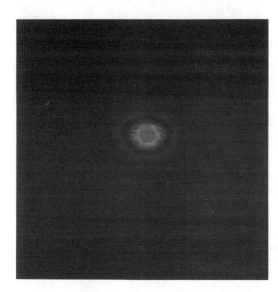

FIGURE 5. X-ray pinhole image of stored beam in the APS storage ring (100 mA @ 7 GeV, May 6, 1996). The camera field of view is 2.7 mm × 2.7 mm, and the rms beam size is 157 μm horizontally and 115 μm vertically.

One significant advantage of the x-ray pinhole camera over the UV/visible imaging is the stability of its image, since the latter is susceptible to mirror vibrations in the long transport line. This feature can be used to measure the dispersion function at the source point. Figure 6 shows a set of beam profiles taken with slightly different rf frequencies, or different beam energies. From the displacement of beam centroid, we deduce the dispersion at the source to be 74 μm per 0.1%ΔE/E. This value can be compared with the calculated value of 92 μm per 0.1%ΔE/E at the design lattice.

SUMMARY

Optical and x-ray synchrotron radiation imaging techniques have been used to characterize the Advanced Photon Source storage ring particle beams. The longitudinal measurement shows that the bunch length is within the design specifications of 100 ps, and the x-ray pinhole measurement shows that the transverse beam size is consistent with the design emittance of 8.2 nm·rad and less than 10% vertical coupling. The dispersion function at the bending magnet source point has also been measured to be 74 μm per 0.1%ΔE/E, less than the design value of 92 μm per 0.1%ΔE/E.

The resolution of our current x-ray pinhole setup is not sufficient for measuring the APS storage ring beam at vertical couplings lower than 10%. Design and construction of a beamline with higher magnification and resolution is in progress, which will improve the resolution for characterizing low emittance beam.

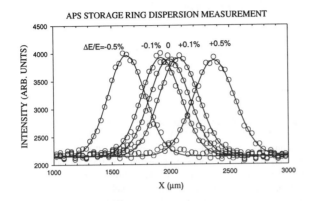

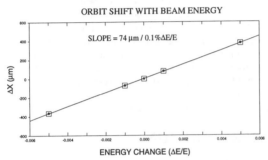

FIGURE 6. Top: horizontal beam profiles at different beam energies. Bottom: the horizontal coordinates of the beam centroid as a function of beam energy change.

REFERENCES

1. Yang, B. X., and Lumpkin, A. H., "The Planned Photon Diagnostics Beamlines at the Advanced Photon Source," Beam Instrumentation Workshop 1994, *AIP Conference Proceedings* **333**, 252-258 (1995).
2. Yang, B. X., and Lumpkin, A. H., "Initial Time-Resolved Particle Beam Profile Measurements at the Advanced Photon Source," Proceedings of the Confeence on Synchrotron Radiation Instrumentation 1995, *Rev. Sci. Instrum.* **67** (9), September 1996.
3. Elleaume, P., Fortgang, C., Penel, C., and Tarazona, E., "Measuring Ultra Small Electron Emittance Using an X-Ray Pinhole Camera," ESRF/MACH ID 95/25, 1995.
4. Cai, Z., Yun, W., Lai, B., Gluskin, E., and Legnini, D., "Beam Size Measurement of the Stored Electron Beam at the APS Storage Ring Using Pinhole Optics," Proceedings of the Conference on Synchrotron Radiation Instrumentation 1995, Rev. Sci. Instrum., 67 (9), September 1996.

Beam Diagnostics for the DIAMOND Project at Daresbury

M.T. Heron, G. Mrotzek, R.J. Smith

CCLRC Daresbury Laboratory, Warrington, Cheshire, WA4 4AD, UK

Abstract. Beam stability and orbit control have become of prime importance for all third-generation light sources. The proposed DIAMOND project, at Daresbury, offers the chance to incorporate the latest ideas and designs now required to provide significant improvements in beam diagnostics. These are essential to ensure the most accurate control of the beam position and give maximum stability of the source for users. Systems employed will include highly sensitive electronics and electron beam position monitoring (BPM) pickups, the performance of which will rely heavily upon the mechanical stability of the BPM vessel. Photon beam monitors, particularly in insertion device lines, will be required to accommodate the small beam size and the potential high heat loads this implies. The large number of photon and electron BPMs, when coupled with fast position servoing systems, will require complex data transfer networks. These are necessary to ensure adequate data rates and capacity to fully realise the beam stability specified. This paper gives a representative overview of the systems and solutions likely to be employed, building on the experience gained with presently installed systems on the SRS at Daresbury.

INTRODUCTION

The proposed DIAMOND Synchrotron Radiation Source (SRS) is currently in the early stages of design, and as such very little detailed engineering design has been undertaken. Hence this paper presents the state of the project and the general issues concerning electron beam position monitors (EBPMs), photon beam position monitors (PBPMs), and feedback systems. Naturally, considerable account will be taken of results and techniques from other third-generation rings now coming on line, before any designs are finalised. The proposals presented here represent developments of ongoing work on the beam diagnostic systems of the SRS at Daresbury.

DIAMOND Project Background

The SRS at Daresbury is a national facility for synchrotron radiation that supports over 40 experimental stations. Although the quality of synchrotron radiation is currently world class, two issues have drawn attention to the need for a new design: 1) space limitations that preclude the use of modern insertion devices and 2) beam emittance (120 nm rad) that will soon be too high to conduct state-of-the-art research. The most recent review of synchrotron radiation facilities for the UK recognised the requirement to replace the SRS by a new medium energy x-ray source. The new source, DIAMOND (1), should provide high-brilliance photons in the soft x-ray region (0.2 to 5 keV) and high fluxes in the medium x-ray region (3 to 30 keV), with this radiation being generated by undulators and multipole wigglers.

Specifications

The DIAMOND project will consist of three accelerators, a microtron producing electrons at 50 MeV into a booster synchrotron which will feed electrons into the storage ring at full energy, 3 GeV. Various options for the lattice of the DIAMOND storage ring were considered, and a recent decision was made to use a Double Bend Achromat (DBA) cell. The standard straights, 5 m long, will provide for undulators, multipole wigglers, and wavelength shifters. There are two super-long straights (20 m) which will enable novel insertion devices to be accommodated in the long-term future. An option for the inclusion of superconducting dipoles as a substitution for some of the normal dipoles has been included. The major parameters for the DIAMOND storage ring are shown in Table 1.

TABLE 1. DIAMOND Storage Ring Major Parameters

Energy	3 GeV
Circumference	346 m
Natural Emittance	15 nm.rad
Cell Type	DBA
Dipole Field	1.4T
No of Cells	16
Straight Length	14 × 5m; 2 × 20m
Beam Current	300 mA
Max. Beam Dimensions (H/V)	0.45/0.04 mm
rf	500 MHz

Status

A costed feasibility study for the DIAMOND project was produced during the Autumn of 1995. Initial thoughts were favourable for funding to be available in 1997. However, changes to the organisation of the scientific funding bodies have meant that the funding mechanism is at present unclear. Work on improvements to the lattice design and the specifications for major systems is ongoing.

ELECTRON BEAM POSITION MONITORS

Storage Ring Injection and Associated
Beam Transport Systems

Requirements for electron beam diagnostics on the booster and its accompanying beam transport system are far less stringent than those for a modern storage ring. Based on the most likely scenario of a pulsed 50-MeV microtron injecting into a cycling booster synchrotron to give full-injection energy into the storage ring, a basic scheme to implement this has been produced. Table 2 summarises some of the relevant operating parameters that affect the system design.

TABLE 2. Injection System Major Parameters

	Booster	Microtron-to-Booster Flight Path	Booster-to-Storage Ring Flight Path
rf frequency	499.654 MHz	499.654 MHz	499.654 MHz
Pulse rep. rate	2 or 10 Hz	2 or 10 Hz	2 or 10 Hz
Pulse duration	0.5 µS	1 µS	0.5 µS
Beam current	< 10 mA	10-20 mA	< 10 mA
Resolution	≈ 100 µm	100 µm	100 µm
No. of detectors	20	5	5

The pulsed nature of these systems limits the options available for processing the EBPM signals. To detect such fast repetitive signals, conventional button-type BPMs will be adequate. Simple on-axis BPM vessels will be suitable for the transport system, but off-axis detectors will be required in the booster due to synchrotron radiation production. The required resolution of these BPMs is not stringent compared to those of the storage ring, and measurements accurate to ≈ 100 µm peak to peak will be adequate in all injector systems. Having had considerable experience with 180° preprocessing hybrid-based systems on the SRS, a similar setup has been proposed to fulfill diagnostic requirements. One pair of horizontal and vertical plane synchronous detectors for each section of the injector will be used. These will be based around the ANZAC MD-149 rf mixer, which is highly suitable for single-pass systems. These detectors will be multiplexed to each EBPM using semiconductor GaAs rf switches. The main challenge for development will be to produce a windowing sample and hold digitiser to remove any unwanted modulator high-voltage pickup from the signals. Oscilloscope outputs will be available to give real-time commissioning signals and an instrumented tune measurement system will be included for the booster.

Storage Ring EBPMs

The proposed storage ring EBPM system is the most extensive and complex of all DIAMOND diagnostics. Each of the sixteen cells will be instrumented by seven two-plane monitors. It is proposed to add a further four EBPMs in the extra-long straights where large insertion devices will be installed. Of the seven EBPMs in each cell, two will be of the high performance, high stability type. Mounted in such a way as to minimise any mechanical movement, they will be isolated by bellows and fixed to individual water-cooled stands. Continuous physical position measurements on these vessels will also be made to discern beam from vessel movements. The remaining EBPMs in the cell will be fitted to the vacuum vessel wherever possible, providing lower accuracy but usable position information. Figure 1 shows such a BPM with its accompanying theoretical electrical calibration curves.

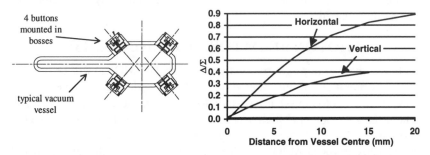

FIGURE 1. Unoptimised quadrupole vessel EBPM with its electrical calibration curves.

The intention is to maintain the vessel geometry constant throughout as much of the machine as possible. The simple addition of buttons as shown produces an unoptimised monitor geometry that has an adequate prediction of 12.5 for the horizontal electrical calibration but is particularly insensitive in the vertical plane, giving a large calibration factor of 27.5. Considerable further work will be required to develop the high-performance BPM geometry. With these monitors, it is planned that calibration factors below 10 will be achieved in both planes to permit global position control to within 10% of beam size. It is unlikely in this case that a standard vessel geometry similar to the rest of the storage ring will be possible.

Much work has been done on the SRS to develop new EBPM processing electronics based around a 180° hybrid preprocessing system and down-converting detector. This detector produces a high stability output with an rms noise figure of 1 micron. Work is in hand to develop the processing electronics further to improve its specification for installation in DIAMOND. Details of this design have been reported extensively elsewhere (2). Recent advances in commercial diagnostic systems in conjunction with the Advanced Light Source

(ALS) at Berkeley, California have resulted in the development of a commercial Eurocard-based detector system (3). The card uses the popular scanning technique whereby the amplitude from each button from a two-plane monitor is selectively sampled and detected. Each card produces DC outputs for X, Y, and Σ that can be detected using a local analog-to-digital converter (ADC) system. Button scan rates at up to 2 kHz are adequate to ensure sufficient data rates for both local and global beam position control. Consideration is being given to the possible adoption of this system based on the results reported from its performance in the ALS. In addition, a limited installation of this commercial system will be made on an extra EBPM location within the SRS to allow comparative tests of the two systems to be made. In either case, these detectors must be compatible with the high performance and general EBPM monitors in order to standardise spares and reduce cost. Both systems are adaptable to provide some first-turn capability at the high performance monitor locations. With the provision of a separate tune measurement system, normal operational and accelerator physics requirements will be easily satisfied.

PHOTON BEAM POSITION MONITORS

In line with other third-generation light sources, DIAMOND will be optimised for the incorporation of insertion devices. For wiggler and dipole lines the employment of photo-emission-type monitors presents little or no problem. For undulator lines the potential high heat loads and gap-dependent position feedback can pose difficulties. It is here that the greatest care in the design will have to be taken. The design for DIAMOND is however still in an evolutionary stage and it is difficult to project exact final PBPM configurations. In general, two vertical PBPMs per beamline will be employed with the second monitor on undulator lines being a combined horizontal/vertical unit. The present vertical servo steering magnet configuration on the SRS consists of a three-magnet corrector bump which does not allow independent position and angle correction. For DIAMOND the independent adjustment of angle and position will be possible.

At present, the translation stage motion resolution on the SRS PBPMs is around 6 μm. The position encoder is a rotary device mounted on the motor/gearbox and thus does not provide direct information on the blade position. For DIAMOND a design with a linear encoder that is directly coupled to the vane mounting in conjunction with a closed feedback loop will largely reduce backlash. This will provide better position accuracy and repeatability than the present system. A finer motion resolution through use of a smaller pitch lead screw and possible higher motor/gearbox ratio is envisaged with a motion resolution of 1 μm or better.

As with the EBPMs, the stability of the PBPM vessel will be important. The high sensitivity of the monitors will make it necessary to have adequate insulation against vibration and movement caused by thermal variations. The PBPMs at the

SRS are mounted on various ports of the vacuum vessels of the beamlines. For DIAMOND the PBPMs will be mounted (at least for the insertion device lines) in their own vessel, isolated from the beamline via bellows to reduce vibration (from operating valves, pumps, etc.). Measures to guard against movement (insulation, temperature control) caused by thermal fluctuations will be implemented. Survey spheres will allow for more accurate alignment of the beam vessel.

The biggest problem to date for the employment of PBPMs in undulator beamlines is the apparent change in beam position when the insertion device gap is changed. At the moment there seems to be no definite solution to overcome this effect. The monitor or combination of monitors (photoemission, imaging) that is finally implemented will depend in part on the degree of the dipole contamination that is seen by the photon monitors on DIAMOND and on advances that are being made with new or existing monitors.

The generic insertion devices presently proposed for DIAMOND (1,4) should not pose a problem in terms of heat load on the photon monitor blades. The inclusion of the super-long straights could conceivably lead to the incorporation of novel insertion devices that generate high heat loads which might require novel solutions (5). For most beamlines, however, a conventional tungsten blade intercepting the beam at grazing angle should be sufficient to distribute the generated heat load.

At present the processing ADCs for the PBPM position output are not local to the amplifiers, and low-pass filters (cut-off frequency 0.6 Hz) are employed to reduce the noise seen by the ADCs. At the moment only the slow drift caused by the thermal (day-to-day refill) cycling of the machine is corrected by the feedback system at a sample time of 30 s. For DIAMOND a much higher closed-loop bandwidth of around 50 Hz is envisaged, and the ADC will be placed as close as possible to the PBPMs. To support such a bandwidth, the sampling frequency will have to be in the kHz range. To handle the large amount of data from the EBPMs and PBPMs, a sophisticated high-speed data communication system will be required. This is discussed in more detail in the following section.

FEEDBACK SYSTEMS

The requirement to maintain the electron beam on the desired orbit by feedback is fundamental to third-generation light sources because of the small beam size. The global feedback system will maintain the electron beam orbit to the 1-μm level by correcting for mechanical vibration and thermal movements in accelerator components to a frequency of 1 Hz. The local feedback will maintain the electron beam stability in insertion devices and photon beam stability on the photon beamlines. This will correct for local disturbances and electrical effects to a frequency of 50 Hz.

A single feedback system incorporating global and local feedback loops (6) would provide control over the coupling between local and global but the higher data rate required for local would complicate hardware design and increase cost. The proposed system for DIAMOND uses local feedback at a sample frequency of 2 kHz and global feedback with a sample frequency of 10 Hz. By operating the global system at the lower data rate, standard network components can be used to move the data around. The global system would feed information into the local system at 10 Hz to integrate the global changes into the local control loops. The global feedback system is integral to the machine control system which at the low level consists of distributed microprocessors performing front-end processing interconnected by a network.

For each DIAMOND cell there are seven EBPMs which would be accessed by a processing crate consisting of VME processor ADCs and a network interface. This will enable the EBPMs and PBPMs to be monitored at a 10-Hz rate with a 16-bit resolution while performing limited digital filtering and averaging. The beam position information will be available at 10 Hz to a dedicated global feedback processor in a VME crate to apply the global correction algorithm, the results of which would be written over the network to steering magnets.

The local feedback system cannot be realised at the 2-kHz sample rate over a conventional network and would use a dedicated VME processing unit for each local loop. These would receive the analogue position values from the required EBPMs and PBPMs and the 10-Hz global information over a network connection. They would operate a PID algorithm at a 2-4 kHz sample rate to give angle and position control that would be applied to a compensated bump, based on four magnets, by direct analogue output or digital connection to the steering magnet supplies. Information would be available to the control system at 10 Hz. Figure 2 shows the structure of the feedback system for DIAMOND consisting of global feedback and only four local feedback loops.

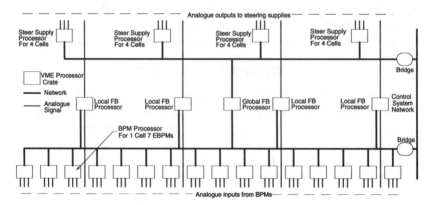

FIGURE 2. Proposed feedback structure for DIAMOND only showing four local feedback loops.

ACKNOWLEDGMENTS

Special thanks to Paul Quinn for additional information and to Mike Collier and the mechanical drawing office for technical drawing support.

REFERENCES

1. Suller, V. P et al., "Updated Plans for DIAMOND, a New X-Ray Light Source for the UK," *Proceedings of the 1995 IEEE Particle Accelerator Conference*, Dallas, Texas May 1-5, 1995, 180-182 (1996).
2. Smith, R. J., McIntosh P. A. and Ring T., "The Implementation of a Down Conversion Orbit Measurement technique on the Daresbury SRS," *Proceedings of the 4th European Particle Accelerator Conference*, (World Scientific Publishing Singapore), London, 1994, 1542-1544 (1994).
3. Hinkson, J. A., and Unser, K. B., "Precision Analog Signal Processor for Beam Measurements in Electron Storage Rings," *Proceedings of the 2nd European Workshop on Beam Diagnostics and Instrumentation for Particle Accelerators* , (DESY M 95-07), Travemunde, Germany, May 1995, 124-126 (1995).
4. Queralt, X., DPG/95/14 , CCLRC Daresbury Laboratory , Warrington WA4 4AD , U.K, 1995.
5. Shu, D. et al., "The APS X-ray undulator photon beam position monitor and tests at CHESS and NSLS," *Nucl. Instr. and Meth.* A **319**, 56-62 (1992).
6. Chung, Y., "A Unified Approach to Global and Local Beam Position Feedback," *Proceedings of the 4th European Particle Accelerator Conference*, (World Scientific Publishing Singapore), London, 1994, 1595-1597 (1994).

Beam Profile Monitors for High-Intensity Proton Beams[*]

Gianni Tassotto

Fermi National Accelerator Laboratory
P. O. Box 500, Batavia, Illinois 60510

Abstract. A description of a secondary emission electron detector (SEED) is given. This device has been designed and constructed to monitor the beam position and profile of two fixed-target beamlines, namely, KTeV and NuTeV. Both beams have an energy of 800 GeV. KTeV will take beam at an intensity of 5×10^{12} protons over a 20-s spill and NuTeV at an intensity of 2×10^{12} protons over a 2-ms fast spill.

INTRODUCTION

KTeV and NuTeV are two of Fermilab's fixed-target beamlines that are scheduled to take 800-GeV proton beam early in the summer of 1996. The beamlines will be instrumented with beam position monitors (BPMs), a beam current transformer, segmented wire ion chambers (SWICs), loss monitors, and secondary emission electron detectors (SEEDs). The SEEDs are being built to measure precisely the beam position and profile. They can also display beam intensity by adding an intensity section.

BEAM PARAMETERS

The NuTeV beam will be measured at two stations before the target. At the upstream (US) station the beam has a FWHM of 2.6 mm horizontal and 1.9 mm vertical and at the downstream (DS) station 0.7 mm horizontal and 2.4 mm vertical. The beam is delivered in 2-ms "pings." There will be a total of five pings; the first four will have 3×10^{12} protons and the fifth ping will have 10^{13} protons. The SEEDs will also be used in the NuTeV alignment run. In this short run the beam will be delivered in 2-µs pulses of about 10^{12} protons at 150 GeV.

---

[*] Operated by Universities Research Association Inc., under contract No. DE-AC02-76CHO300 with the United States Department of Energy.

The KTeV beam will also be measured at two stations. At the upstream station the beam has a FWHM of 0.42 mm horizontally and 0.85 mm vertically, while at the DS (target) station the beam size is 0.11 mm horizontally and 0.23 mm vertically. The beam intensity will be 5×10^{12} protons over 20 seconds.

DESIGN PARAMETERS

A detailed analysis of SEED design parameters has been made (1).   A 0.003-in gold-plated wire was selected for mechanical and signal strength and to allow soldering of the wire to the pads on the ceramic boards. A tension of 80 grams was chosen in order to keep the wire from sagging during beam heating.

MECHANICAL ASSEMBLY

The Research Division mechanical group designed the vacuum assembly to achieve a positioning repeatability of 0.002 inches in x, y for KTeV and 0.010 inches for NuTeV. The main detector assembly drawing is shown in Fig. 1.

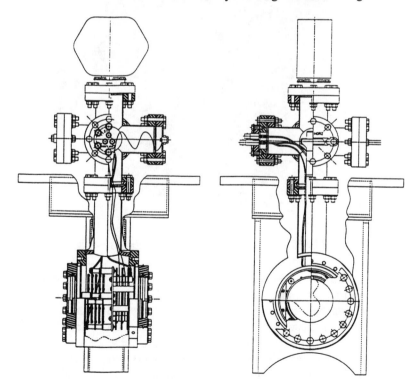

FIGURE 1. Main detector assembly.

PROFILE SECTION

The profile section contains three bias foils made with grade 1145-0 aluminum, free of holes and defects, and two signal planes. In order to satisfy the beamline requirements two types of ceramic boards have been designed and built.

1. A single-sided board with 0.5-mm spacing between pads and, therefore, a wire pitch of 0.5 mm.
2. A double-sided board with 0.25-mm spacing between pads. The offset between top-to-bottom pads is 0.125 mm. This enables selection of either a 0.25-mm or a 0.125-mm wire pitch. Figure 2 shows the profile and intensity sections of the first detector.

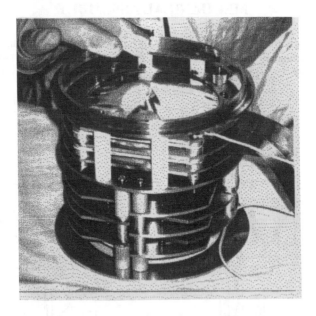

FIGURE 2. Assembly of profile and intensity sections.

A 0.003-in gold-plated tungsten wire was pre-wound at 80 grams of tension on a transfer frame at a pitch of 0.5 or 0.25 mm. The wire frame was then positioned over the ceramic board and the wires lined up with the center of the pads. A Dow Inc. epoxy resin, type DER331-12DETA-4-Cab-O-Sil, was then used to adhere the wires to the ceramic substrate, thereby relieving the pads of the tension. The epoxy was allowed to cure for 16 hours and post-cure for 4 hours at 66 degrees C. Under tension tests, the wires broke around 900 grams without slipping in the epoxy. The wires were then soldered to the pads using solder paste (62MA-A made by Solder Plus Inc.). A flex tape was designed to take the

48 signals (and two grounds) from the ceramic board to the vacuum feedthrough. This tape, manufactured by Litchfield Inc., consists of a Kapton AP8525 underlay and LF0110 overlay. The ceramic board/flex tape assembly was tested to better than 10^{-7} Torr.

INTENSITY SECTION

If an intensity section is needed it can be added to the detector by replacing appropriate spacers. This section contains three bias foils made with grade 1145-0 aluminum and two signal foils that have been gold plated to a thickness of 0.2 mm. Two ceramic-coated wires carry the bias voltage and signal to vacuum feed-throughs.

ELECTRONICS

The readout electronics for the profile section consists of a 1091 timing module and a 96-channel charge integrator scanner (2) that interfaces to a VAX computer via a 1032 data transfer module. The electronics for the intensity section consist of a current digitizer and a timing module.

CONCLUSIONS

As of this writing the NuTeV detectors are being assembled. The expectation is to have the detectors pumped down, tested, and installed in the beamline in early June 1996. Some issues are still of concern, such as the failure of the wires or the epoxy due to radiation damage. The NuTeV beam will deposit a significant amount of heat to the wires with temperatures exceeding 500 degrees C.

REFERENCES

1. Bob Drucker, "A Review of SEED Design," to be published as a Fermilab Technical Memo.
2. W. Higgins, "User's Manual for Microprocessor Based SWIC Scanner," Fermilab report TM-868A, April 1979.

Noninterceptive Beam Diagnostics
Based on Diffraction Radiation[*]

D. W. Rule and R. B. Fiorito

Naval Surface Warfare Center, Carderock Division
10901 New Hampshire Avenue, Silver Spring, MD 20903-5640

W. D. Kimura

STI Optronics, Inc., 2755 Northrup Way, Bellvue, WA 98004

Abstract. Diffraction radiation (DR) can be described as the scattering or diffraction of pseudo-photons associated with the transverse electromagnetic field of a relativistic charged particle, when the particle passes through an aperture or near an object. DR is closely related to transition radiation (TR) which is produced when a charged particle crosses a boundary between media with different dielectric constants. DR is also associated with "beam loading" which occurs when beams pass through discontinuties in their transport lines. DR is very similar in its properties to TR, so that much of what has been developed for TR beam diagnostics can readily be adapted to DR diagnostics. We will present concepts for using DR as a compact, nondestructive diagnostic that can measure the beam divergence, beam profile, and thus emittance in real time. These concepts, together with the previously demonstrated use of DR for longitudinal bunch length measurement, make an attractive suite of noninterceptive, time-resolved diagnostics which parallel those based on TR.

INTRODUCTION

A wide range of high-energy accelerators exist around the world, with more recent ones featuring exceptionally low emittance and high-brightness beams (1). A full characterization of beam parameters: emittance, beam profile, and bunch length, etc. is required to optimize accelerator performance and often needed by users. In particular experiments, single beam pulse diagnostics are desireable, e.g., in laser acceleration studies. A compact, nondestructive, real-time diagnostic system capable of measuring beam parameters—divergence, profile, emittance, energy, and bunch length—in real time does not yet exist. We are developing simple methodolgies based on diffraction radiation (DR) to noninterceptively diagnose multiple beam parameters.

The phenomena of DR is perhaps more familiar to the accelerator physicist as the reaction radiation associated with "beam loading" or loss of electron energy of a beam as it passes through an accelerator cavity or other discontinuity in the accelerator structure (2). DR has been studied theoretically since the early 1960s

---

* This work was funded in part by the NSWC-CD Technology Transfer Office.

(2-5). However, experimental observations of the DR and the verification of the unique properties predicted for DR have been performed only very recently (6).

In 1984 the potential of DR as a noninterceptive diagnostic for relativistic beams was theoretically evaluated at the Naval Surface Warfare Center (NSWC) (7,8). In particular it was determined and proposed (7) that an analysis of the frequency and angular distribution of the DR from a charged particle beam could be used as a noninterceptive diagnostic of the beam centroid position (e.g., relative to the center of a circular aperture), current density, divergence, emittance, and energy. In addition it was recognized that coherent transition radiation (TR) as well as DR would be produced at wavelengths comparable to *l*, the beam bunch length, and that the spectrum of the radiation could in principle be used to measure *l*.

In the last ten years these predicted diagnostic capabilities of TR have been fully realized (9-11). However, this is not the case with DR, which has only recently been studied in any detail. In the first experimental observation of DR reported (6), the spectrum of coherent DR in the far infrared (FIR) and mm regimes was quantitatively measured to determine bunch length. The method employed in this study was identical to that used to measure bunch length with coherent TR (12-16).

The measurements of ref. (6) also showed that the angular distribution of coherent DR was qualitatively similar to that of TR (13). These studies observed a combination of forward (0 degree) DR from a circular aperture and backward TR produced by a mirror used to optically transfer the DR to the detector. The mirror intercepted the beam, eliminating the advantage of the noninterceptive nature of DR as a diagnostic.

We are developing noninterceptive DR diagnostic methods which use simple apertures (irises) inclined at 45° to the beam axis. This scheme is easy to implement and is capable of diagnosing multiple beam properties, including divergence, transverse profile, energy, emittance as well as the longitudinal beam distribution. These DR diagnostics are capable of operating over a wide range of beam energies (e.g., $\gamma \sim 100$ to >1000), and, as we will show, the measurement of DR becomes easier as the beam energy increases.

COHERENT AND INCOHERENT DR FROM BUNCHES OF PARTICLES

The intensity of DR per unit frequency ω, per unit solid angle Ω produced by a bunch of N electrons passing through a hole in a conducting screen as shown in Fig. 1 is given by

$$\frac{d^2 I}{d\omega d\Omega} = (e^2 / 4\pi^2 c) R(\theta, \Psi) \left[\frac{\theta^2}{(\gamma^{-2} + \theta^2)^2} \right] B(\mathbf{k}, l, \sigma) G(q, a) , \qquad (1)$$

where e is the electron charge and γ is the Lorentz factor. The reflectance factor of the screen $R(\theta,\Psi) \cong 1$ for backward-directed radiation from a good metallic reflector and is identically unity for forward-directed DR. $B(\mathbf{k},l,\sigma) = N + N(N - 1)F(\mathbf{k},l,\sigma)$ is the bunch form factor for a bunch of N particles with wave vector $k = 2\pi/\lambda$.

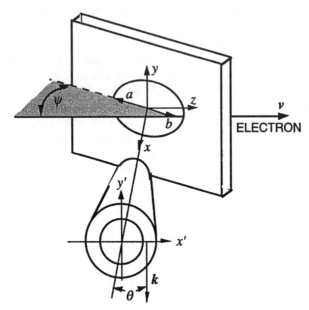

FIGURE 1. Diffraction radiation of wave vector $\mathbf{k}$ produced by a particle with velocity v transiting a distance b from the center of an aperture of radius a in a foil inclined at an angle ψ relative to the particle trajectory. Angle θ is measured from the direction of specular reflection.

The term in B proportional to N is the ordinary incoherent radiation by the bunch; the term $N(N-1) \propto N^2$ for $N \gg 1$ is the coherent radiation from the bunch; and the last term, $F(\mathbf{k},l,\sigma)$, is the squared modulus of the Fourier transform of $S(\mathbf{r})$, the normalized bunch distribution. The longitudinal dimension parameter l and transverse dimension parameter σ, respectively, select the range of wavelengths that will radiate coherently, i.e., the range of λ's for which $F(\mathbf{k},l,\sigma) \sim 1$.

The last factor in Eq. (1), $G(q,a)$, is a geometrical factor determined by the diffraction of the relativistic particle's field by the aperture, where $q = ka \sin\theta$. G has been derived (4) for a circular hole in which the particle's trajectory is displaced from the center of the aperture by an amount b. Assuming that $a \ll \gamma\lambda$ and $b \ll a$, $G(q,a) = J_0^2(qa) + (b/a)^2 J_1^2(qa)$. This expression consists of the first two terms in an infinite series expansion. Equation (1) shows that the DR intensity will always be less than TR. For the case when $a = \gamma\lambda / 2\pi$ and $\theta \approx 1/\gamma$, then $qa = 1$ and $J_0^2(1) = 0.6$, which means the intensity of the DR is $\sim 60\%$ of TR.

DR FROM A CIRCULAR APERTURE

To understand how DR scales, consider DR from a circular aperture of radius *a*. Appreciable DR is generated when $\gamma\lambda \geq 2\pi a$. As an example, let $a = 1$ cm. The DR threshold wavelength can be defined as $\lambda \geq 5/\gamma$ (cm), where we have made the simplifying approximation $2\pi \sim 5$. These wavelengths are tabulated in Table 1 for a wide range of beam energies. Note that λ becomes shorter as γ increases and reaches the optical[†] regime for electron and proton energies of ~ 1-50 GeV and ~ 5-100 TeV, respectively. Of course, there is a trade-off between hole size and wavelength; e.g., a 50-MeV beam could be focused through a 0.4-mm-diameter aperture to produce DR at 10 μm.

TABLE 1. Comparison of Diffraction Radiation (DR) Parameters for Various Representative Accelerators

E(GeV)	γ	λ^1	L_v^2	a^3	a^4	Linac[5]
ELECTRONS						
0.05	10^2	500 μm	5 m	20 μm	8 mm	ATF
0.5	10^3	50 μm	50 m	0.2 mm	8 cm	
5	10^4	5 μm	500 m	2 mm	80 cm	CEBAF
50	10^5	500 nm	5 km	2 cm	8 m	SLAC
250	5×10^5	125 nm	30 km	8 cm	40 m	NLC
PROTONS						
50	50	1 mm	2.5 m	8 μm	4 mm	
500	500	100 μm	25 m	80 μm	40 cm	
10^3	10^3	50 μm	50 m	160 μm	80 cm	Fermi
10^4	10^4	5 μm	500 m	1.6 mm	8 m	0.5 SSC

[1]Threshold DR wavelength for incoherent DR assuming an aperture radius of $a = 1$ cm.
[2]DR maximum coherence length $\sim \gamma^2\lambda/2\pi$.
[3]Maximum aperture radius for coherent bunch length measurements of 1-μm-long bunches.
[4]Maximum aperture radius for coherent bunch length measurements of 500-μm bunches.
[5]Representative accelerators within the energy range.

The emission wavelength during coherent DR will tend to peak at the bunch length *l*, (i.e., $l \sim \lambda \sim 2\pi a/\gamma$). This means the maximum aperture size for a given bunch length is $a \sim \gamma\lambda/2\pi$. Values of *a* are listed in Table 1 for 1-μm and 500-μm bunch lengths. Note that for the longer bunches, the aperture sizes are quite large even for moderate energy beams.

† The term "optical" here means the spectral range from visible light to far infrared.

REPRESENTATIVE INTENSITY CALCULATION OF DR

We have estimated the number of DR photons produced per bunch in the mm and the visible regimes for Stanford Linear Collider (SLC) parameters. The assumed parameters are: beam energy = 47 GeV, beam radius = 1 μm, bunch length ~ 1 mm, and number of electrons per bunch $N = 3 \times 10^{10}$. For a bunch-length measurement $\lambda = 6.28$ mm, $B \approx N^2 = 9 \times 10^{20}$ and we have coherent DR. If we integrate Eq. (1) over $\theta = 0$ to $\theta = 10^3/\gamma$ and assume a 0.1% bandwidth (BW), we obtain 3×10^{-5} photons/electron. An aperture radius of $a \le 5$ cm will produce this result. The total number of photons per bunch $dN/d\Omega \approx 1.6 \times 10^{16}$ per 0.1% BW per 3.7×10^{-4} sr for $\lambda = 6.28$ mm.

For the case of visible radiation, consider $\lambda = 0.5$ μm. Then $B_N \sim N = 3 \times 10^{10}$, and the radiation is incoherent. We find that for θ integrated over 0 to $10/\gamma$ and a 1% bandwidth, Eq. (1) gives $dN/d\Omega$ (bunch) $\approx 6 \times 10^5$ photons/bunch per 1% BW into 3.7×10^{-8} sr. This result requires $a \le 1$ mm, giving $J_0(qa) \sim 0.5$.

DIFFRACTION RADIATION DIAGNOSTICS

Angular Distribution of Backward Reflection DR from a Foil Inclined at 45 Degrees

In direct analogy to TR from a single foil inclined at 45° with respect to the beam velocity, backward-reflected DR from a similarly inclined foil will be observed at 90° from the beam axis. The angular distribution of DR will be modified by the reflectivity of the foil and the effects of diffraction. Such a geometry has the advantage of diagnosing the beam well out of the direction of interfering radiation directed along the beam's trajectory and, more importantly, allows the measurement to be completely noninterceptive.

The dominant angular dependence of $I_{DR}(\theta)$ is given by the first term of Eq. (1), which is identical to that of TR. Our work on OTR divergence and emittance diagnostics (11) has already shown how the expressions for $I_{TR}^{\parallel}(\theta)$ and $I_{TR}^{\perp}(\theta)$ — the parallel and perpendicular components of TR, respectively—are influenced by beam divergences s_x and s_y associated with the two orthogonal directions in a plane perpendicular to the velocity vector of the beam. Assuming a Gaussian distribution for the projected trajectory angles α_x and α_y for the beam particles, expressions for the convolution $I^{\parallel,\perp}(\theta)*P(\alpha_x, \alpha_y)$ were derived and fitted to angular distribution data to produce values for s_x and s_y. This can also be done in a similar manner with the parallel and perpendicular components of DR.

Alternatively, in principle one can determine $P(\alpha_x, \alpha_y)$ and the divergences from the measured intensities $I^{\parallel,\perp}(\theta)$, of TR or DR by inverse Fourier transformation:

$$P(\alpha_x, \alpha_y) = F^{-1} \left\{ \frac{F([I^{\|,\perp}(\theta) * P(\alpha_x, \alpha_y)]_{meas})}{F[I^{\|,\perp}(\theta)]} \right\}. \tag{2}$$

Beam Radial Distribution and Beam Size

If we have a radial bunch distribution $NS_\perp(\rho)$, then we can estimate the <u>incoherent</u> DR production from the term proportional to N in Eq. (1):

$$\left. \frac{d^2 I_{DR}}{d\omega d\Omega} \right|_{incoh} \propto N \int S_\perp(\rho) G(q, \rho) \rho d\rho = N[J_0^2(qa) + (\langle \rho^2 \rangle / a^2) J_1^2(qa)] , \tag{3}$$

where $\langle \rho^2 \rangle$ is the mean square beam radius. This expression shows how the angular/spectral density of incoherent DR depends on the beam radius and indicates how DR can be used to determine rms beam size.

Consider the situation when $qa \approx 2.4$, which is the first zero of $J_0(qa)$, for a particular λ and θ. In this case the first term in brackets in Eq. (3) is zero. Then by varying a (by changing the size of the iris) or q (by changing wavelength or angle) while keeping $qa \approx 2.4$, one obtains the greatest sensitivity in measuring $\langle \rho^2 \rangle$.

If one images the diffracting screen directly, it may be possible to determine the surface current density distribution producing the image, and consequently the beam distribution. This is distinct from the determination of $G(q,a)$ from the angular radiation pattern discussed above. $G(q,a)$ appears in the Fraunhofer diffraction pattern, whereas the image of the aperture is more directly related to the spatial bunch distribution. This is analogous to determining beam profiles from an image of TR produced by the beam at the position of the TR screen.

One method of analyzing diffraction of ordinary light is to consider it as the result of the interference of "boundary diffraction waves" around the edge of the aperture and the geometric radiation passing directly through the aperture. In the case of DR, we have only the boundary diffraction waves, since the fields in the aperture associated with the relativistic particles are <u>not</u> radiation fields.

Multifoil Diffraction Radiation

In direct analogy to TR, DR from two or more radiators separated by a distance less than or of the order of the coherence length of the radiation will produce an interference pattern which is very sensitive to a change in beam parameters (e.g., divergence) (10,11). A double-aperture DR interferometer arrangement is shown in Fig. 2.

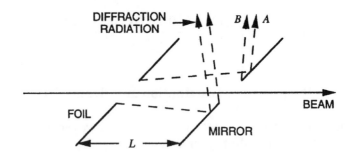

FIGURE 2. Two-aperture DR interferometer.

The intensity of DR per unit frequency per unit solid angle for the 2-foil arrangement of Fig. 2 can be expressed as

$$\frac{d^2I}{d\omega d\Omega} = (e^2/4\pi^2c)\left[\frac{\theta^2}{(\gamma^{-2}+\theta^2)^2}\right]B(\mathbf{k}, l, \sigma)[G_F(q, a) - G_B(q, a)e^{-i\phi}] , \quad (4)$$

where the terms on the right are the same as those appearing in Eq. (1), except that G_F and G_B are modified geometrical factors that take into account DR produced in the forward and backward directions, respectively, by a screen at 45° with respect to the beam direction. G_F also includes the effect of the reflection off the back of the second screen, i.e., a mirror with a hole in it.

The phase difference in the forward and backward DR rays A and B in Fig. 2 is given by $\phi = L/L_v$, where L is the path length between foils and L_v is the vacuum coherence length[‡] of DR and TR, which is defined as the length in which the phase of the forward photon produced at the first foil exceeds that of the electron's field by 1 radian. For a highly relativistic particle and at small angles θ with respect to the beam trajectory, $L_v \approx (\lambda/\pi)(\gamma^{-2} + \theta^2)$. If $\theta \sim 1/\gamma$ in this approximation, then $L_v \sim \gamma^2\lambdabar$. Note that the reflection at the second foil (mirror) introduces a phase change of π. As seen in Table 1, the coherence lengths for highly relativistic energies are all very long, which implies that a system utilizing DR would not require keeping the foil separation distances very short.

By a suitable choice of L and λ for a given γ, an interference pattern in angle or wavelength space can be produced whose fringe visibility is highly sensitive to the beam divergence (10,11). Incidentally, the relative size of the fringes as a function of λ yields information about the bunch coherence factor B.

[‡] The term "coherence length" used here refers to the distance over which one can have coherent addition of DR and/or TR from <u>two</u> or more separate foils. The radiation fields coherently adding can be produced by a single charge or a bunch. This coherence should be clearly distinguished from coherent radiation produced from a bunch acting as a single charge, which occurs at wavelengths $\lambda \gtrsim l$, the bunch dimension.

SUMMARY

To date, DR has been utilized as a beam diagnostic for only one parameter—bunch length—and the geometry of the method used compromised the inherent noninterceptive nature of DR. We have devised a completely non-interceptive methodology which uses incoherent or coherent DR from one or more apertures inclined at 45° with respect to the direction of the beam. These diagnostic methods can measure the beam divergence in a manner very similar to TR; the beam spatial distribution, by analyzing the angular and spectral characteristics of DR; the beam emittance in two orthogonal directions; and bunch length.

Single-shot measurements may be possible with this device by taking advantage of coherent DR, where the intensity scales as N^2 (N = number of electrons in bunch) and/or n multiple foils, where the intensity increases as n^2.

Variations on our basic measurement strategies, such as changing the aperture size and/or using noncircular or other aperture geometries, are possible and may provide more information about the beam. We are considering such variations in the design of DR diagnostics.

REFERENCES

1. X. J. Wang et al., in Proceedings of 1995 Particle Accelerator Conference, May 1-5, 1995, 890-892 (1996).
2. E. Keil, *Nucl. Instrum. Methods* **100**, 419 (1972).
3. F. G. Bass, V. M. Yakovenko, *Sov. Phys. Uspecki* **8** (3), 420 (1965).
4. M. L. Ter-Mikaelian, *High Energy Electromagnetic Processes in Condensed Media*, (J. Wiley-Interscience, New York, 1972).
5. N. J. Maresca and R. L. Liboff, *Can. J. Phys.* **53**, 62 (1975).
6. Y. Shibata, S. Hasebe, K. Ishi, T. Takahashi, T. Ohsaka, M. Ikezawa, T. Nakazato, M. Oyamada, S. Urasawa, T. Yamakawa, and Y. Kondo, *Phys. Rev. E* **52**, 6787 (1995).
7. D. W. Rule and R. B. Fiorito, Naval Surface Weapons Center, Tech. Report No. TR 84-134 (1984).
8. D. W. Rule and R. H. Ritchie, in Proc. of Werner Brandt Workshop on Penetration Phenomena, Oak Ridge National Laboratory, 15-16 April 1985.
9. L. Wartski et al., *IEEE Trans. Nuc. Sci.* **20**, 544 (1973).
10. L. Wartski et al., *J. Appl. Phys.* **46**, 3644 (1975).
11. See R. B. Fiorito and D. W. Rule, "Optical Transition Radiation Beam Emittance Diagnostics," Beam Instrumentation Workshop, Santa Fe, NM, October 1993, *AIP Conference Proceedings* **319**, 21-37 (1994) and references therein.
12. U. Happek, A. J. Sievers, and E. B. Blum, *Phys. Rev. Lett.* **67**, 2962 (1991); R. Lai, U. Happek, and A. J. Sievers, *Phys. Rev. E* **50**, 4294 (1994).
13. Y. Shibata et al., *Phys. Rev. E* **49**, 785 (1994) and *Phys. Rev. E* **50**, 1479 (1994).
14. P. H. Kung et al., *Phys. Rev. Lett.* **73**, 967 (1994).
15. Walter Barry, in Proceedings Workshop on Advanced Beam Instrumentation, KEK, Tsukuba, Japan, April 1991, 224-235 (1991).
16. J. C. Schwartz et al., *Phys. Rev. E* **52**, 5416 (1995).

Longitudinal Bunch Deformation of a Multi-Bunched Beam in the TRISTAN Accumulation Ring

T. Obina*, K. Satoh, T. Kasuga, Y. Funakoshi, and M. Tobiyama

KEK, National Laboratory for High Energy Physics, 1-1 Oho, Tsukuba, 305 Japan

Abstract. A remarkable bunch deformation has been observed in the TRISTAN Accumulation Ring (AR) during multi-bunch operations. When two bunches that have different populations are stored in the ring, the bunch length of the weaker bunch is longer than that of the stronger one. The phenomenon can be explained as an effect of wake fields due to higher-order modes (HOMs) of accelerating cavities. We tried to find out the frequency of the mode and the strength of the wake field, and introduced a new technique called a test bunch measurement. The estimated field strength from the experiment shows a reasonable agreement with the calculation of HOM impedance.

INTRODUCTION

When an electron storage ring is operated in a single-bunch mode, the current dependence of the bunch length is explained by the potential-well distortion and the microwave instabilities (1-3). These theories were successfully applied to the bunch lengthening in the TRISTAN Accumulation Ring (AR) at KEK, National Laboratory for High Energy Physics (4). On the other hand, a remarkable bunch deformation was observed when several bunches were stored in the AR (5). In some cases the stationary longitudinal distribution of electrons, which we call the longitudinal bunch shape hereafter, exhibits two peaks or a trapezoidal shape in four-bunch or eight-bunch modes. These phenomena were explained qualitatively with the potential-well distortion due to the long-range wake field, although the source of the wake field had not yet been identified. The main purpose of the present study is to find out the sources and the mechanism of the anomalous bunch-shape deformation.

OBSERVATION OF BUNCH SHAPE

First of all, we carefully measured the bunch shape with a streak camera (Hamamatsu Photonics, C1587 with M1955 synchroscan unit and M2887 dual time base extender). The synchroscan unit scans the photoelectrons in the vertical

---

* The Graduate University for Advanced Studies, Department of Accelerator Science, 1-1 Oho, Tsukuba, 305 Japan

direction with a frequency of $f_{RF}/4$ = 127 MHz. Once trigged, a sawtooth wave is applied to the horizontal deflector. The horizontal sweep time can be changed from 10 μs to 10 ms, much slower than a typical vertical sweep time of several picoseconds. We can, therefore, neglect the contribution of the horizontal sweep to the beam profile and can observe the bunch shape in successive revolutions. Figure 1(a) shows an example of the bunch profile measured by the streak camera during the two-bunch operation. The longitudinal bunch shape of each bunch is shown in Fig. 1(b). At first we measured the current dependence of the bunch length in the

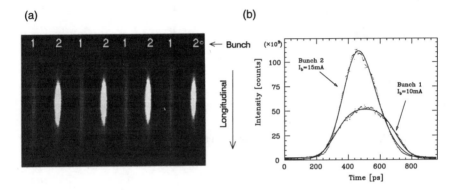

FIGURE 1. Longitudinal bunch profile measured by a streak camera (a) and the longitudinal bunch shape (b). The picture shows the bunch shape of successive revolutions in the two-bunch operation. The bunch current of number 1 and number 2 is 10 mA and 15 mA, respectively.

single-bunch operation. We also measured in the two-bunch condition, where two bunches were injected into opposite sides of the ring and their currents are equal or unequal with each other. Figure 2 summarizes the bunch length both in the single-bunch and the two-bunch operations as a function of the scaling parameter defined by $\xi = I_b/V_c$ [mA/MV]. We hereafter call the smaller-current bunch a weak bunch and the larger-current bunch a strong bunch. When two bunches that have different currents are stored in the ring, the weak bunch becomes quite a bit longer than the strong bunch although, in standard bunch lengthening theory, the bunch is lengthened with the increase of the bunch current.

MODEL OF BUNCH DEFORMATION

We consider a simple model in which the bunch shape is determined by the accelerating rf voltage and a higher-order mode (HOM) of a cavity. The voltage seen by the beam is expressed as

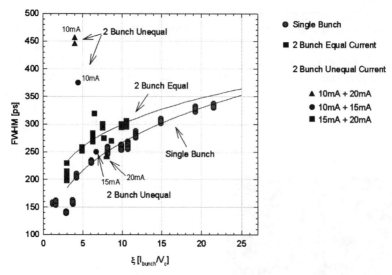

FIGURE 2. Bunch length in single-bunch and two-bunch operations.

$$V(\phi) = V_c \left\{ \sin(\phi + \phi_s) + k\cos\left(\frac{m}{h}\phi + \phi_m\right) \right\} , \qquad (1)$$

where h is the harmonic number of the ring, m is an integer which represents the revolution harmonics, ϕ_s is the synchronous phase, kV_c is the peak voltage of the HOM, and ϕ_m is the phase of the HOM which has the frequency mf_{rev}. For any rf waveform, the formula for the electron line density is given by Haissinski in (6). In our model, the bunch shape is expressed as

$$\rho(\phi) = \rho(0) \exp\left[\frac{A}{V_c} \int_0^\phi \left\{ \sin(\phi' + \phi_s) + k\cos\left(\frac{m}{h}\phi' + \phi_m\right) \right\} d\phi' \right] , \qquad (2)$$

where A is a constant.

The effect of the HOM fields for the even and odd harmonics of the revolution frequency are discussed separately. The even harmonics have the same effect on the two bunches because the HOM field has the same slope at the opposite side of the ring. If the HOM field lengthens one bunch, it also lengthens another bunch in the opposite side. Similarly, if the HOM field shortens one bunch, it shortens another bunch. On the other hand, the odd harmonics have the opposite effect on two bunches. The HOM field lengthening one bunch shortens the other bunch and vice versa. The anomalous bunch deformation, therefore, arises from the odd harmonics the effective strength of which is proportional to the difference of the bunch currents.

ESTIMATION OF HIGHER-ORDER MODES

Test Bunch Method

We developed a test-bunch measurement method to estimate the HOMs in the rf cavity which deform the longitudinal bunch shape (7). We inject a strong bunch which induces a wake field and, at the same time, inject a sufficiently weak bunch whose wake field is negligibly small. From the fact that the shape of the weak bunch is deformed due to the wake field induced by the strong bunch, we can estimate the frequency and the strength of the HOM by changing the distance between the two bunches. (In this paper, we call the strong bunch a main bunch and the weak bunch a test bunch.)

To simulate the test-bunch measurement, we choose the TM012 mode as a HOM field deforming the bunch shape. The frequency of the mode f_{HOM} is 1465 MHz with $m = 1844$. We assume $\phi_m = 0$ and $k = 0.3$ in the example. The energy loss per turn is set to be zero for simplification and the synchronous phase is equal to π. Because the phase advance of the rf and the TM012 mode is different, the longitudinal bunch shape of the test bunch varies depending on the distance from the main bunch. Figure 3 shows the calculated longitudinal bunch shapes of the test bunch in 20 successive rf buckets behind the main bunch. We used Eq. (2) in the calculation and normalized so that $\int i(\phi) d\phi = 1$. Bucket number 0 means the main bunch, bucket number 1 means the bucket just after the main bunch, and so on.

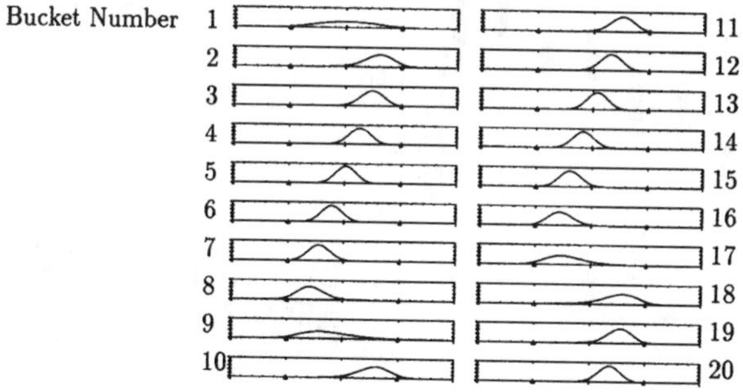

FIGURE 3. Simulation of the bunch shapes in successive buckets.

From the data shown in Fig. 3, there are two methods to estimate the HOM frequency. One is to measure the peak position of the test bunch and the other is to measure the bunch length. With both methods, the Fourier transform of the obtained data gives us information on the frequency of the mode. According to the sampling theorem, frequencies higher than half of the sampling frequency cause an aliasing. Therefore, we have to determine the real frequency by other means.

We decided to measure the peak position of each bunch because the effect of the HOM appears more sensitive in the peak position than in the bunch length.

Furthermore, the detection of the peak has the advantage that, even if the test bunch performs the coherent synchrotron oscillation, the averaged center of the test bunch represents the bunch center which would be observed without the oscillation. On the other hand, if we measured the FWHM of the test bunch, we could not distinguish the contribution of the wake field from the coherent synchrotron oscillation.

We used a high-speed PIN photodiode (Antronics Research, AR-S2) to detect the bunch center. The output of the photodiode was fed to a digitizing sampling oscilloscope (Hewlett Packard, HP54120B and HP54121A sampling head) which has an analog bandwidth of 20 GHz. Because it is very time-consuming to measure all buckets in the AR, we measured only the successive 21 buckets in nine parts in the ring. We selected eight equally-spaced places in the ring and measured 10 buckets just before and after each place. Because we wanted to measure the detail of the wake field around the main bunch, we selected 20 buckets before and after the main one.

Figure 4(a) shows the peak position of the test bunches. The abscissa shows the bucket number from 1 to 640 and the ordinate represents the peak position in units of picoseconds. There is some offset component in the figure due to the change of the synchronous phase by the transient beam loading effect of the main bunch. The calculated difference in the synchronous phases between just after and before the main bunch is 128 ps, that agrees well with the measurement.

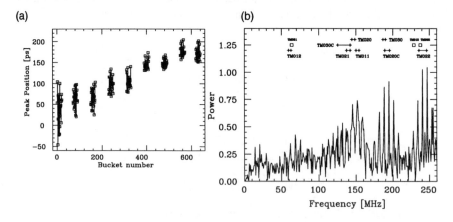

FIGURE 4. (a) Peak position of the test bunch. (b) Fourier transform of the test-bunch measurement and the aliasing frequency of the measured HOMs.

We subtracted the offset due to beam loading and calculated the Fourier transform of the variations of the peak position. Peak positions of the unmeasured buckets were set to zero. The result is shown in Fig. 4(b). There are sharp peaks every 6.36 MHz. These peaks arise from the fact that we measured selected buckets in only eight locations around the ring rather than all buckets. Three large peaks are seen in the figure. The Fourier transform in each part suggests that the 150-MHz component appears at buckets numbered 0 to 100, and the other two large peaks—around 190 MHz and 240 MHz—appear in the whole ring. We can therefore conclude that the 150-MHz component arises from the short-range wake field that covers about 100 buckets after the main bunch and has less contribution to the bunch deformation at the two-bunch operation. Figure 4(b) also shows the aliasing frequency of the measured HOMs of the rf cavity.

Cavity HOMs

We estimate the induced voltage of the HOMs by impedance calculations. There are two rf sections in the AR. Each section contains four alternating periodic structure (APS)-type cavities, each of which has 11 accelerating cells. The imped-ance of an APS cavity is calculated with the computer code SUPERFISH and is shown in Table 1. In the actual cavity, TM011 and TM021 modes are damped with HOM dampers. Due to the multi-cell structure, each mode has several resonant frequencies.

TABLE 1. Higher-order Modes and Shunt Impedances of an APS Cavity.

Mode	Frequency (MHz)	Aliasing Freq. (MHz)	Q Factor	Impedance (MΩ)
TM011	865.38	150.78	38000	3.4 (damped)
TM020	1163.38	146.22	50000	0.10
TM020C*	1211.06	193.90	15000	1.0
TM021	1385.88	139.86	45000	5.2 (damped)
TM012	1465.55	60.19	50000	1.7
TM022	1793.54	240.78	54000	1.9
TM030	1845.19	189.13	72000	1.0
TM030C*	1901.62	132.70	23000	2.2
TM031[†]	1967	67.3	46000	0.71
TM032[†]	2305	237.9	55000	0.60
TM013[†]	2314	228.9	76000	0.30

* 'C' means that the field is concentrated in the coupling cell of the cavity.
[†] not measured

When the AR was operated in the single-bunch mode, we observed the beam-induced signal with a pickup at the end-plate of the cavities because all longitudi-nal modes have field components normal to the end-plate. Because the diameters of the accelerating cells and the coupling cells are different, the TM020 and TM020C modes have different resonant frequencies. We did not measure the three modes in the table marked with [†] because it was difficult to separate these modes from the others. Furthermore, these modes have low shunt impedance relative to the other modes. From the measurement of cavity HOMs and the test-bunch measurements, modes TM020C, TM030, and TM022 are candidates that play an important role in the bunch deformation.

DISCUSSION

Three modes are candidates for the source of the deformation. It is difficult, however, to include three modes into the model simultaneously. Because the TM020C mode has the highest beam spectrum and the largest R/Q value among the three modes, we considered only the TM020C mode, and introduced $m = 1517$ in Eq. (2).

In the actual accelerating cavities, the tuning angle of the HOM (ϕ_m) cannot be fully measured. If the tuning angle is equal to $0°$, there is no bunch deformation. We adopted $\pm 90°$ as the tuning angle because the HOM acts effectively on the bunch shape at that angle. The solid line in Fig. 1(b) shows the fitted results. Because the field strength of the HOM, expressed with the value k, is proportional to the current difference between two bunches, we estimated the k value per 1 mA. The result is shown as k_{fit} in Table 2.

TABLE 2. The k values obtained from fitting bunch shape (k_{fit}), impedance calculation (k_{calc}), and test-bunch measurement ($k_{testbunch}$). The value is normalized with 1 mA.

Total Current (mA)	Bunch Current (mA)	k_{fit}	k_{calc}	$k_{testbunch}$
25	10	4.4×10^{-3}	3.0×10^{-3}	
	15	2.4×10^{-3}		
35	15	1.6×10^{-3}	2.5×10^{-3}	1.8×10^{-3}
	20	1.8×10^{-3}		
30	10	4.0×10^{-3}	2.7×10^{-3}	
	20	3.0×10^{-3}		

The HOM tuning angle is unknown in the actual cavities. To estimate the highest cavity impedance, we assumed the HOM tuning angle to be $45°$ and also assumed the eight cavities had the same HOM tuning angle. Therefore k, expressed as k_{calc} in Table 2, might be overestimated.

From the Fourier transform of the test bunch measurement, the strength of the TM020C mode is calculated to be 49.6 kV from the peak value of the imaginary part. The mode has negative slope at the main bunch which shortens the main bunch itself and lengthens the bunch at the opposite side. The k value per 1 mA is calculated to be $k_{testbunch} = 1.8 \times 10^{-3}$. Table 2 summarizes the fitted and calculated k values.

SUMMARY

We have studied the bunch deformation of the multi-bunched beam in the TRISTAN AR. We carefully measured the bunch shape in the single- and two-bunch operations and found that the shape of bunches are deformed remarkably when two bunches having different currents are stored in the ring.

To identify the source of the deformation, we studied the effectiveness of the test-bunch measurement with the numerical calculations and adopted the method. The HOM wake fields measured with this method are classified into two categories: one is the short-range HOM wake field that lasts until about 100 buckets after the main bunch and the other is the long-range wake field that affects the bunch shape in the two-bunch operation. We measured the beam-induced signal in the cavity with the pickup at the end-plate of the rf cavity and concluded that three modes, TM020C, TM022, and TM030, are candidates for the source of the deformation.

We estimated the strength of the wake fields in three approaches. The first approach was the calculation of the HOM impedance with the computer code SUPERFISH. The calculated resonant frequencies were compared with the measurement of the beam-induced signal. The second approach was the bunch shape under the two-bunch operation. We fitted the measured bunch shape with the Haissinski formula and estimated the strength of the HOM. The last approach used the Fourier transform of the test-bunch measurement to determine the strength of the modes.

Estimated k parameters were different among the three approaches by a factor of 2 to 3. The difference is attributed to the fact that there are eight cavities in the AR and the resonant frequency and shunt impedance of HOMs may be different among cavities.

We have also made preliminary bunch-shape measurements in four-bunch and eight-bunch operations and observed anomalous bunch deformation. A future project will study these cases in detail. The APS cavities in the AR will be temporarily replaced with another type of cavity for a test of large-current storage simulating the KEKB operation. It would be very interesting to measure the bunch-shape deformation after removing the APS cavities and minimizing long-range HOM fields.

ACKNOWLEDGMENTS

The authors wish to thank Dr. K. Akai and Dr. Y. Morozumi who have kindly supported the experiment and the calculations of the cavity HOMs.

REFERENCES

1. B. Zotter, CERN/ISR-TH/78-16, (1978).
2. A. W. Chao and J. Gareyte, SPEAR-197/PEP-224 (1976).
3. H. Yonehara, T. Kasuga, M. Hasumoto, and T. Kinoshita, *Jpn. J. Appl. Phys.* **27**, 2160 (1988).
4. K. Nakajima et. al., in *Proc. 5th Symposium on Acc. Sci. and Tech.*, Tsukuba, 297 (1984).
5. Y. Funakoshi et. al., in *Proc. 7th Symposium on Acc. Sci. and Tech.*, Osaka, 270 (1989).
6. J. Haissinski, *IL NUOVO CIMENTO*, **18B**, 72 (1973).
7. T. Obina et. al., in *Proc. 10th Symposium on Acc. Sci. and Tech.*, Hitachinaka, Japan, 347 (1995).

New Generation Electronics Applied to Beam Position Monitors

Klaus B. Unser

BERGOZ Precision Beam Instrumentation
F-01170 Crozet, France

Abstract. Cellular telephones and global positioning system (GPS) satellite receivers are examples of modern rf engineering. Taking some inspiration from those designs, a precision signal-processor module for beam position monitors was developed. It features a heterodyne receiver (100 MHz to 1 GHz) with more than 90 dB dynamic range. Four multiplexed input channels are able to resolve signal differences lower than 0.0005 dB with good long-term stability. This corresponds to sub-micron resolution when used with a beam position pick-up with 40 mm free aperture. The paper concentrates on circuit design and modern dynamic testing methods, used first during development and later for production tests. The frequency synthesizer of the local oscillator, the phase-locked synchronous detector, and the low-noise preamplifier with automatic gain control are discussed. Other topics are design for immunity to electromagnetic interference to ensure reliable operation in an accelerator environment.

INTRODUCTION

Beam position monitors (BPMs) are essential diagnostic instruments in a particle accelerator and every accelerator project usually includes the development of a specific beam position monitoring system. Conventional beam positions pick-ups use two or four electrodes (directional couplers or buttons) as position sensors. Beam position is a function of the amplitude difference of these electrodes' signals. The fundamental frequencies are specific to each particular accelerator and the power spectrum may extend to many GHz. Many methods for processing the beam position pick-up signals have been used or proposed in the past (1). Modern circular machines and storage rings, for example synchrotron light sources, have very high demands on the precision of these measurements. The objective is now to measure beam position with a resolution better than 1 µm. This requires measurement of the ratio of four electrical signals with a precision better than 0.0005 dB. The amplitude of the signals themselves could change over a range of more than three decades (> 60 dB) for different operating modes.

A UNIVERSAL BPM SIGNAL PROCESSOR

We have taken the initiative to develop a universal BPM signal processor that would adapt easily to the frequency patterns of different accelerators. The starting point was a new BPM system designed by J. Hinkson (2) that is optimized to measure the average closed-orbit position with high resolution. His primary application was transverse feedback and beam interlock systems near insertion devices in the Advanced Light Source (ALS) at Lawrence Berkeley National Laboratory.

Our BPM signal processor uses the same basic concept and architecture, but its center frequency is programmable anywhere between 100 MHz and 1 GHz. It tracks input signals over a limited frequency range ($\pm$ 200 kHz, extension possible up to $\pm$ 1 MHz). Optional features are fast signal gating and a limited capability for single-turn measurements. Low cost, reliability, and simplicity of production were the other key requirements.

The central function of the signal processor is a receiver in the VHF/UHF frequency range. It was therefore a good idea to have a closer look at modern telecommunication equipment before starting our project. High-technology consumer products like cellular telephones and global positioning system (GPS) satellite receivers have started a revolution in modern rf technology. Complex systems are now being built on very small circuit boards, using novel integrated circuits functions and miniature passive components. They do not need heavy shielding enclosures.

Taking inspiration and components from those designs allowed us to shrink the processor module to a single-height Eurocard (100 $\times$ 160 mm). A built-in, pre-programmed frequency synthesizer for the down-converter makes the unit completely self-contained and independent of any external control or timing system. In its default operating mode, the user has only to connect the pick-up signals and the power supplies to obtain the normalized X and Y outputs. All intermediate coaxial connectors were eliminated by using on-board miniaturized bandpass filters, micro-strip low-pass filters, and adjustable attenuators. This improves the reliability of critical interconnections and lowers the component cost significantly.

Operating Principle

The BPM signal processor has four parallel input channels for the four input signals A to D. The input multiplexer switches sequentially from one input channel to the next. A single receiver measures the signal for each input channel in turn and stores its value in four corresponding analog memories. The voltages (V_A to V_D) in each of these memories are therefore proportional to the power levels of the original input signals. To normalize the readings (to make them independent of changing beam current), automatic gain control (AGC) is used to keep the sum $V_A + V_B + V_C + V_D$ constant.

Sequential scanning and the use of a single receiver simplifies the circuits and has as major advantage, i.e., the gain for all channels is always identical. Gain changes due to temperature effects or component aging are eliminated by the automatic gain control. The measurements are only valid if input conditions do not change during the scan. In practice, this is a very good method for beam position measurements of the average closed orbit.

The X and Y coordinates are calculated with a matrix of sum and difference analog circuits. They use the following algorithms for a typical position pick-up (with the four electrodes rotated at 45° with respect to the vertical axis).

$$X = K_x \frac{V_A + V_B - V_C - V_D}{V_A + V_B + V_C + V_D} \quad (V/mm)$$

$$Y = K_y \frac{V_A - V_B - V_C + V_D}{V_A + V_B + V_C + V_D} \quad (V/mm)$$

The calibration constants Kx and Ky of the beam position pick-up are an approximation and apply only to the center of the vacuum chamber.

The basic operating principle, as described above, has been used for the first time at the NSLS storage ring (3). In the following chapters, design criteria for certain elements of the system will be discussed. A more complete description of the prototype for the ALS can be found in an earlier report (2).

Input Multiplexer

The block diagram (Fig. 1) shows 1-GHz low-pass filters in each of the four input channels. They are necessary to protect the GaAs switches from the very short, high-voltage pulses which are typical of wide-band pick-up signals. Stripline filters have been used before (4) for this function, but in our design they are folded up to occupy a very small area on the circuit board. Not shown on the diagram are the high-pass filters (7 MHz) which protect the switches from an accidental DC or low-frequency AC input. The attenuators (if required) reduce the signal power to the maximum level which the amplifier can accept. A 1-dB variable adjustment is used to equalize insertion loss in all four channels. The output impedance of the four channels, as seen by the switch, have to be matched to very tight tolerances.

Low-Noise Amplifier

This is a key element for the performance of the BPM signal processor. It has the dual function of low-noise amplifier and attenuator with a 50-dB gain

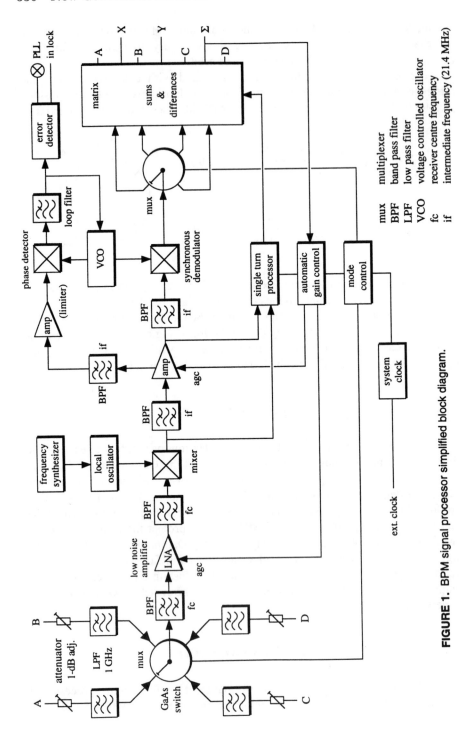

FIGURE 1. BPM signal processor simplified block diagram.

control range. More than 100-dB dynamic range is required, without gain compression, because gain compression anywhere in the signal path would change the calibration of the BPM signal processor as a function of input signal power.

Amplifiers with gain control have a tendency to change input admittance when changing gain. Considering the unavoidable matching errors of the input channels, this could be the cause of an unbalance in channel gain. This has to be avoided, because the resulting effect for the BPM signal processor is a zero offset which changes as a function of input signal power.

The requirements above are very different from the usual specifications of telecommunication equipment. It was therefore necessary to build a special amplifier with discrete transistors to cover the 100-MHz to 1-GHz frequency range and reach an acceptable compromise between conflicting demands.

Local Oscillator and Frequency Converter

An active mixer is used as down-converter to the 21.4-MHz intermediate frequency (IF). This is another critical element in the signal chain, where gain compression is likely to limit the available dynamic range. Thanks to the AGC in the preamplifier, the signal amplitude changes are already reduced at this point.

The local oscillator should not contribute any significant amplitude or phase noise to the processed signals. A frequency synthesizer (Fig. 2) is used as a source. It consists of a voltage-controlled oscillator (VCO) with programmable divider (5) phase locked to a reference frequency (f_{ref}). A 17-bit serial word from a plug-in F-key module selects the divider ratio for a specific receiver center frequency. Spectral purity of the local oscillator depends on the choice of the reference frequency and the dynamic response of the phase-locked loop (PLL).

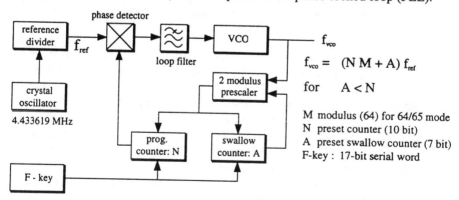

FIGURE 2. Frequency synthesizer.

IF Amplifier and Demodulator

The integrated circuit[*] used as an IF amplifier and demodulator is designed for television applications. We had to add discrete transistor stages at input and output of the IF amplifier to match the IF filters and improve the gain compression limits. The IF bandwidth (Fig. 3) can be tuned from 200 kHz to 2MHz to suit any particular application. The bandwidth is large enough to allow a fast step response when switching from one input channel to the next. Sidebands at the revolution frequency are rejected in order to make the detector see the same kind of signal independently of the accelerator operating mode: multibunch or single bunch.

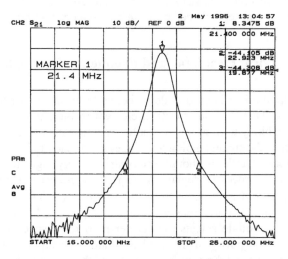

FIGURE 3. Frequency response of IF amplifier (BPM processor for ALS). Bandwidth (–3 dB) = 434 kHz. Markers indicate the sidebands for a revolution frequency of 1523.4 kHz.

Synchronous detection of the IF signal requires a reference frequency. This reference is generated by a VCO. A phase detector compares the reference with a filtered sample of the carrier frequency and controls the VCO in a PLL. The dynamic response of the PLL is controlled by the loop filter.

If the pick-up signals applied to the BPM signal processor inputs are not in phase, switching from one input channel to the next would produce sudden phase jumps of up to 180°. Even a fast PLL would encounter difficulties, and this would seriously limit the maximum scanning frequency.

Many of the BPM systems presently in use require precise phase matching on all signal cables coming from the pick-ups. We preferred to solve this problem

---

[*]LM1823, National Semiconductor Corporation

in a different way. Our loop filter has an electronic switch (not shown in Fig. 1) which reduces the time constants during the 10 µs after channel switching. This allows the receiver to settle on a new signal level; as a result, phase errors up to 180° are tolerated.

ELECTROMAGNETIC INTERFERENCE

The BPM signal processor will be used in machine protection interlocks, where reliable operation is extremely important. In addition to shielding the sensitive parts of the board, a number of special precautions have been incorporated in the design to improve immunity to electromagnetic interference.

- The coaxial inputs pass via an uninterrupted ground plane before connecting to the circuit board.
- All other input and output connections have low-pass filters to keep the interference outside and limit the frequency response to what is absolutely necessary.
- All control inputs have 10-ms low-pass-filter time constants.
- Each circuit function on the board is specifically designed to be insensitive to frequencies outside its normal operating range. This is particularly important for operational amplifiers which have a tendency to demodulate rf signals and, as a result, produce erroneous DC output.

DYNAMIC TESTING

The different rf circuit and filter functions of the BPM signal processor require alignment and dynamic testing during development and in the final production phase. Conventional rf circuits have special coaxial test connectors. Considering the high packing density on our printed circuit, this was not a desirable solution. Instead, after some initial experimenting, we developed a special miniature rf test set. On the board, this required simply two plated-through holes at a distance of 0.1 inch, with one of them connected to the ground plane. The coaxial test probes have two corresponding test pins, consisting of gold-plated, spring-loaded contacts. For use with a network analyzer (Hewlett Packard HP8753D), a matching calibration set (open, short, and 50-Ω load) permits us to establish the correction constants, which take care of the small residual discontinuity of these connections. The test set is used for measuring impedance and transfer functions (Fig. 3) for frequencies up to 3 GHz.

Testing the transient response of feedback (AGC) and phase-locked loops uses the same type of test connections, but requires a different instrument set-up. Generally speaking, the method consists of injecting a square wave modulated stimulus, most often a DC current, in a suitable summing junction of the loop under test. The transient response is observed with an oscilloscope at another test point and can be optimized by tuning the different elements of the loop filters or phase correction networks.

A spectrum analyzer (Anritsu MS2601B) is used to measure the signal at different points of the signal path. It can determine the 1-dB gain compression point and the saturation limit for every location along the signal processing chain. Spectral purity and stability of the local oscillator is verified with the spectrum analyzer.

A special test generator has been built to simulate fast pick-up signal pulses with repetition rates up to 15 MHz. This allows testing of the BPM signal processor by simulating different filling patterns and single-bunch mode. The peak amplitude of the generator is at present limited to 22 V and is therefore insufficient to test the upper limit of the dynamic range in these modes.

An automated test bench for final testing and certification of the BPM signal processor is presently under development. It uses a programmable continuous wave (CW) signal generator (Marconi Instruments 2022D) as a signal source. A precision 4-way power splitter and four specially designed precision attenuators with remote switching control prepare the four input signals. A Macintosh computer with a 16-bit ADC, multiplexer, and control interface (National instruments NB-MIO-16X) runs a specially developed LabVIEW application program (6). The test bench establishes a test report, recording noise and offset errors in all quadrants over the entire dynamic range (> 80dB)

The noise spectrum of X and Y outputs, as a function of input signal power, is recorded with an FFT spectrum analyzer (Stanford Research Systems SR760).

Performance Figures

The following dimensional data refer to calibration with a "reference" pick-up: Sensitivity in X and Y plane = 5%/mm (40-mm aperture).

Noise   → X and Y signal (for the first 40 dB of the dynamic range):
            less than 1.5 μm (rms) for a bandwidth of 0 to 1 kHz
            less than 0.5 μm (rms) for a bandwidth of 0 to 100 Hz

Zero stability → X and Y zero drift less than 0.5 μm during 10 hours

Linearity   → Over 60 dB of input signal range:
            less than ± 5 μm error within ± 0.5mm of center
            less than ± 10μm error within ± 3mm of center

REFERENCES

1. R. E. Shafer, "Beam position monitoring," *AIP Conference Proceedings* **212**, 27-58 (1989).
2. J. A. Hinkson, K. B. Unser, "Precision Analog Signal Processor for Beam Position Measurements in Electron Storage Rings," Proc. of the European Workshop on Beam Diagnostics and Instrumentation for Particle Accelerators (DIPAC), Travemünde, Germany, DESY M 95-07, 124-126, May 1995.
3. R. Biscardi, J. W. Bitter, "Switched Detector for Beam Position Monitor," Proc. of the 1989 IEEE Particle Accelerator Conference, Chicago, IL, **3**, 1516-1518 (1989).
4. S. Brinker, R. Heisterhagen, K. Wille, "DELTA Beam Position Monitor," Proc. of the 1991 IEEE Particle Accelerator Conference, San Francisco CA, **3**, 1154-1156 (1991).
5. "The Technique of Direct Programming by using a Two-Modulus Prescaler," Application Note AN-827, Motorola Semiconductor.
6. K. Scott, "BPM (test bench) LabVIEW instrument," private communication.

Bunched-Beam Measurements of Very Small Currents at ASTRID

F. Abildskov and S.P. Møller

ISA, University of Aarhus, DK 8000 Aarhus C, Denmark

Abstract. Stored currents in low-energy ion storage rings, like ASTRID, are often very small. Absolute current measurements are nevertheless important for absolute measurements of cross sections and also for machine operation purposes. Experimental results, using a beam charge monitor (BCM) from Bergoz, are shown for both light ions (H⁻) and heavy ions (N_2^+). The velocities are low, $\beta \sim 0.001$ to 0.05, and the detected currents are in the 0.1- to 2-μA range. The storage ring ASTRID, where the measurements are made, will be described. The principle of the BCM will be briefly mentioned, and the obtained performance (resolution, stability, noise, etc.) will be given.

THE ASTRID STORAGE RING

ASTRID is the first facility which combines a storage ring for ions with a synchrotron radiation source. The motivation for this was to make a relatively expensive piece of equipment available to a wider user community. The layout of the storage ring with injectors is shown in Fig. 1, and Table 1 lists ASTRID's operational parameters.

Ions are preaccelerated in an isotope separator using a very stable (rms < 1 V) 200-kV high-voltage supply. A variety of ion sources for both positive and negative ions can be used with the separator to produce singly-charged ions and molecules of almost any type. A charge exchange cell has been installed after the separator magnet to produce negative ions by electron capture in a Na, K, or Cs vapour. Differential pumping in the injection beamline separates the high-pressure ion source (10^{-2} Torr) from the ring vacuum (10^{-11} Torr).

A pulsed (10 Hz) race-track microtron has been built to produce the 100-MeV electrons for the storage ring. The "ring" is a square as formed by two 45° bending magnets, excited by a common coil, in each corner. The 16 quadrupoles are grouped in four families. Two families of eight sextupoles are available for chromaticity corrections. Superimposed on the air-core sextupoles are eight horizontal and eight vertical correction dipoles. Furthermore, four horizontal correctors are available as back-leg windings on the main dipoles.

Two different rf systems are used. For the ions, a ferrite-loaded cavity operating in the region of 0.5-3.3 MHz is available, giving a maximum voltage of 200 V.

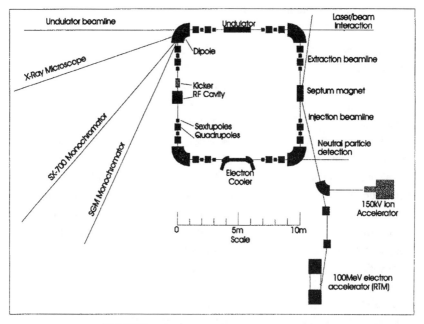

FIGURE 1. Layout of ASTRID and injectors.

For the electrons, a capacitively loaded coaxial TEM cavity operating at 104.9 MHz is used. This cavity was fabricated in steel, which was then copper plated. The obtained Q is around 9000.

Ions and electrons are injected with a magnetic septum (DC), and a kicker is placed diametrically opposite. For the ions, the electrostatic kicker excited by a square pulse injects one turn. For the electrons, a magnetic kicker excited by a half-sine pulse is used to accumulate electrons. Since the start-up of the facility many different ions have been stored in the ring; a list is given in Table 2.

The stored ion beams had rigidities between 7 MeV/c for 6 keV $^4He^-$ and 300 MeV/c for 45 MeV H_2^+. Many of the ions were accelerated to around 6 MeV. The most complicated operation procedure of ASTRID has been acceleration of H_2^+ from 0.15 to 4 MeV with harmonic 6, debunching and rebunching at harmonic 1, followed by acceleration from 4 to 45 MeV, storage for laser interaction, and then deceleration from 45 MeV to 4 MeV for recombination with the electron beam.

The lifetime at injection energy of stored beams of positive ions was limited by the vacuum, typically a few 10^{-11} Torr, giving lifetimes of around 10 seconds. Injected currents for the positive ion beams were in the range of 1 to 10 μA.

The lifetime of a negative ion beam is determined by rest gas stripping, intra-beam stripping, and field stripping (in the bending magnets). A new stripping mechanism has been identified for loosely bound ions. Black-body radiation can ionize ions with small electron affinities like Ca^- and He^-. Furthermore, some ions are metastable and autoionize on timescales around a millisecond.

TABLE 1. Operational Parameters of ASTRID

General:		
Magnetic rigidity	1.87 Tm	
Circumference	40 m	
	Electrons:	**Ions:**
Injection energy	100 MeV	5 – 150 keV
Hor./vert. tune	2.208/2.640	2.29/2.63
Hor./vert. chromaticity	–4.3/–7.1	–3.4/–7.5
Corrected chromaticity	+1/+1	
Momentum compaction	0.068	0.053
Injected current	< 265 mA	1 pA – 10 µA
Accelerated current	< 235 mA	0.1 – 3 µA
Max. energy	580 MeV	45 MeV (H_2^+)
Horizontal emittance	0.14 mm mrad	
Critical energy	0.36 keV	
Critical wavelength	35Å	
Energy loss/turn	8.3 keV	
Beam lifetime (Touschek)	16 hours (at 150 mA)	
Number of bunches	14	1 – 28
rf frequency	104.9 MHz	0.5 – 3.3 MHz
rf voltage	125 kV	200 V

TABLE 2. Atomic and Molecular Ions Stored in ASTRID

$^2D^+$	$^4He^+$	$^6Li^+$	$^7Li^+$	$^{16}O^+$	$^{19}F^+$	$^{20}Ne^+$	$^{24}Mg^+$	$^{40}Ar^{++}$	$^{151}Eu^+$	$^{166}Er^+$
H_2^+	HD^+	H_3O^+	CH_3^+	$^{13}CO^+$	^{14}N	$^{15}N^+$	$^{12}CO^{++}$	$^{13}CO^{++}$	O_2^+	CS_2^{++}
C_{48-60}^+	C_{60}^{++}	C_{70}^+								
H^-	D^-	$^3He^-$	$^4He^-$	$^9Be^-$	$^{11}B^-$	$^{12}C^-$	$^{16}O^-$	$^{19}F^-$	$^{28}Si^-$	$^{32}S^-$
$^{35}Cl^-$	$^{40}Ca^-$	$^{56}Fe^-$	$^{88}Sr^-$	$^{138}Ba^-$	$^{174}Yb^-$					
$^4He_2^-$	$^{12}C_{2-8}^-$	OH^-	Al_{1-12}^-	C_{48-60}^-	C_{70}^-					

Many negative ions can only be produced in small quantities, leading to currents in the pA to nA range. Hence they can only be observed with 'neutral' detectors monitoring the decay of the stored beam by counting neutralized ions at the end of the straight sections.

THE BEAM CHARGE MONITOR

For ions with intensities in the 0.1- to 10-µA range a beam charge monitor (1) from Bergoz (Fig. 2) with a resolution ≤ 10 nA rms is used. This monitor has been developed for LEP (2-4) and HERA (5). DC current transformers do not have the required sensitivity (resolution ≤ 800 nA rms) (5).

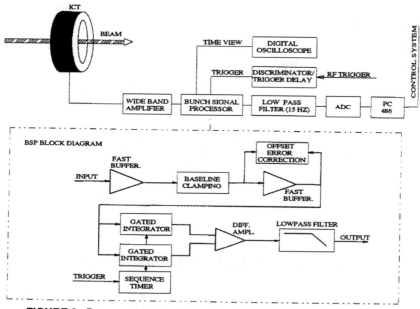

FIGURE 2. Beam current monitoring system of the ASTRID storage ring.

The beam current sensor in the beam charge monitor (BCM) is the integrating current transformer (ICT). The ICT is followed by a wideband amplifier, a bunch signal processor (BSP), and at ASTRID a 15-Hz low-pass filter. The ICT is a linear integrator for the entire frequency spectrum of the beam current signal. The BSP is controlled by an external trigger synchronous to the rf frequency. A sequence timer in the BSP creates two identical time intervals in cascade after an external trigger event. The intervals can be set to any required value within the limits of 50 ns and 1600 ns. The time intervals are used to gate the input signal in two independent integrator channels. One channel integrates the portion of the signal containing the selected bunch pulse and the other channel samples the baseline close to the bunch pulse in the same way, see Fig. 3. The result is subtracted from the output of the first channel. The output of the differential amplifier is therefore proportional to the charge of the bunch signal. The sampling integrators give a high degree of noise suppression.

The ASTRID beam charge monitor is the continuous averaging version. It measures the average charge over time, of repetitive selected bunches and therefore, measures currents.

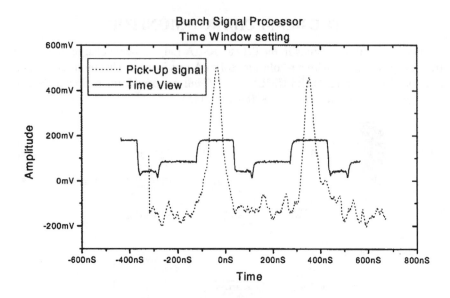

FIGURE 3. Setting of timing windows around bunches and between bunches (HD⁺ beam).

BEAM CURRENT MEASUREMENTS

Experimental results are shown for both a light ion H⁻ (Fig. 4) and a heavy ion N_2^+ (Fig. 5). The velocities are low, $\beta \sim 0.001$ to 0.05, and the detected currents are in the range of 0.1 to 2 μA.

For precise current measurements the zero-offset is deducted from the beam measurements as shown in Fig. 4. For the continuous averaging version BCM the zero-offset is trigger frequency-dependent. At ASTRID the zero-offsets for the individual current measurements are measured before and after the beam current measurements. It is done under actual operating conditions—the same trigger frequency (rf-cavity running), magnets ramping, etc.—but without circulating beam in the ring.

To measure the resolution of the BCM, the noise and zero drift are measured. The data collection system has recorded the BCM output without circulating beam in the ring. In the zero stability measurement, shown in Fig. 6, each point is the average of 20 individual measurements taken at 2.5-s intervals. In the noise measurement (Fig. 7) each point is the standard deviation σ of the same data set.

The resolution is measured to be < 10 nA rms. The absolute calibration depends on the external timing and the position of the bunch in the integration window and must be adjusted accurately.

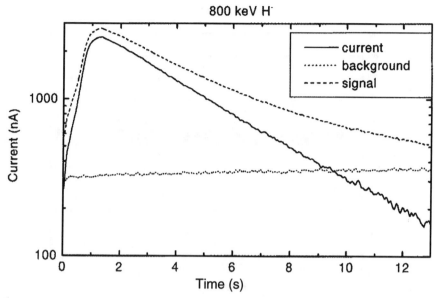

FIGURE 4. H⁻ current for one acceleration cycle showing the detected signal with and without beam and the corresponding 'true' beam current.

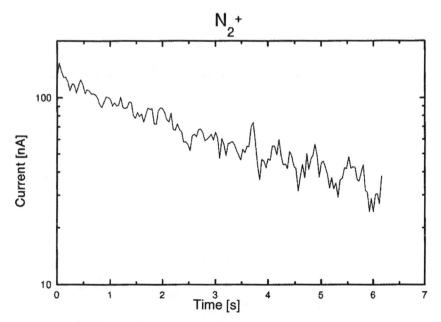

FIGURE 5. N₂⁺ current at 5.5 MeV for one cycle (no averaging).

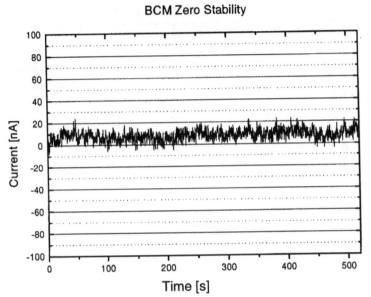

FIGURE 6. Zero stability measurement (BCM).

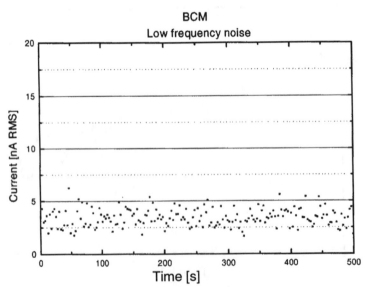

FIGURE 7. Measurement of low frequency noise (BCM).

REFERENCES

1. Bergoz, J., "Beam Charge Monitor, Continuous Averaging User's Manual," Julien Bergoz, Crozet, France, October 1995.
2. Unser, K.B., "Design and Preliminary Tests of a Beam Monitor for LEP," Proc. of the 1989 Particle Accelerator Conference, Chicago, Illinois, March 1989, 71-73 (1989).
3. Unser, K.B., "Measuring bunch intensity, beam loss and bunch lifetime in LEP," Proc. of the 2nd European Particle Accelerator Conference EPAC 90, Nice, France, June 1990, 786-788 (1990).
4. Burtin, G., Colchester, R.J., Fischer, C., Hermery, J.Y., Jung, R., Vanden Eyden, M., Vouillot, J.M., "Mechanical design, signal processing and operator interface of the LEP beam current transformers," Proc. of the 2nd European Particle Accelerator Conference EPAC 90, Nice, France, June 1990, 794-796 (1990).
5. Shütte, W. and Unser, K.B., "Beam current and beam lifetime measurements at the HERA proton storage ring," Proc. of the 1992 Accelerator Instrumentation Workshop, Lawrence Berkeley Laboratory, Berkeley, CA, October 27, 1992, *AIP Conference Proceedings* **281**, 225-233 (1993).

Beam Diagnostics in the Indiana University Cooler Injector Synchrotron[*]

M. Ball, V. Derenchuk, B. J. Hamilton

Indiana University Cyclotron Facility
Milo B. Sampson Lane, Bloomington, IN 47405

Abstract. Diagnostic systems are being designed for the new Cooler Injector Synchrotron (CIS) being built at the Indiana University Cyclotron Facility (IUCF). There will be two separate beam position monitor (BPM) systems used in the CIS project. In the 7-MeV injection beamline, where the beam will be a DC pulse 200 to 300 μs long with a 1-s period, the pickups will be 4-quadrant electrostatic type used with high input impedance, front-end amplifiers. In the CIS ring, elliptically shaped, 4-quadrant electrostatic pickups will be located at the entrance and exit to each bending magnet. Logarithmic amplifiers will be used in the front end to detect the rf bunched beam. The 200-MeV extraction beamline will have a 150-ns-pulse beam with a 0.2- to 1-s period and use 4-quadrant electrostatic pickups and logarithmic amps with a peak detector circuit. The output of the BPMs will include intensity as well as horizontal and vertical position. Harps are being designed for use in the injection beamline, the CIS ring, and the high-energy extraction beamline. An emittance measurement system, incorporating a harp and movable slits, will be used in the 25-keV beam transport line between the H⁻ source and the RFQ and the 7-MeV injection beamline.

BEAM POSITION MONITORS

The 7-MeV injection beamline of the CIS ring will contain a beam position monitor (BPM) system and two harps. The beam in this section will be a DC pulse, 200 μs long, and have a 0.2- to 1-s period. The BPM system will use 4-quadrant electrostatic pickups in the beamline. A chassis will be attached to each of four pickups that will house the electronics used in signal detection and amplification. The detector amplifiers used will be field-effect transistor (FET), unity gain, buffer amplifiers. FETs are used for their high input impedance which will provide the long time constant that is needed to detect such a slow signal. Low-noise, low-offset voltage, high-gain op-amps will follow the detectors. The left/right, up/down (L/R, U/D) signals will be summed and subtracted, then passed through a sample/hold (S/H amp) amplifier. The L+R signal will also provide the trigger to the S/H amp. The four held signals—L+R, L-R, U+D, U-D—will exit

[*]Work supported by the National Science Foundation (Grant No. NSF PHY-9413383).

the circuit board to an ADC. The signals will be normalized in the computer, providing the position and intensity to be displayed as the operator wishes. The system has an operating range of 40 dB. The minimum detectable signal is approximately 10 µA. The predicted range in which the injection beamline will operate is from 10 µA to 400 µA. The signals provided will be used for initial tuning, but could also be used for feedback in steering loops.

The CIS ring will use elliptically shaped, 4-quadrant electrostatic pickups in the beamline. They will be positioned at the entrance and exit of each of four bending magnets. Each pickup will have a dedicated set of electronics attached to it. The detector used will be an Analog Device AD640 logarithmic amplifier (1). The log amp detector design (2) was selected for its wide dynamic range (65 dB) and ease of use (no local oscillators or mixers required). The log amps will be followed by low-noise op-amps with 30-kHz active filters. The difference of the log amps, L-R and U-D, gives the normalized position in the x and y planes. The sum of the four signals will provide the intensity. The signals will go to dedicated, high-speed ADC modules with on-board memory. This type of signal buffering will allow the operator to view the position and intensity history of the beam during injection and ramping in great detail.

BEAM PROFILE MONITOR

A secondary emission system (harp) to be used for beam profile and emittance measurement, has been designed for the injection beamline. The harp design was patterned after the design used at Fermilab. The harp uses a 5-printed-circuit-board stack; each board has 24 wires and is spaced 1 mm apart. The first, third, and fifth boards will be biased with a small voltage to collect scattered electrons. The second and fourth boards will be aligned in the horizontal and vertical planes as detector grids. The signal will be carried from the vacuum chamber on a ribbon cable and through a vacuum connector designed and built in house. A 100-ft folded and shielded ribbon cable will be used to send the currents to the electronics board, which is housed in a NIM crate.

The front end of the circuit board electronics will have 5-MΩ resistors to ground in parallel with 0.22-μF capacitors providing a time constant of 1.0 s. The signals will then be sequentially multiplexed and integrated. The integrator has a shorting switch in parallel with the feedback resistor which will be used to zero the amp between each channel actuation of the multiplexer. After a buffer amp, the signal will go to a fast ADC with built-in memory. Using this type of ADC allows the user to acquire data at a fast rate and display it whenever it is wanted.

The timing for the harp will be on the same circuit board as the multiplexer and integrator. A trigger from the master timing clock will be all that is needed to start the data acquisition process. During commissioning a stand-alone system will be used incorporating a PC and a graphical programming package, HP VEE, from Hewlett-Packard (3). In the operational system the signals will be displayed at the operations console through the controls computer.

Emittance Measurement System

Commissioning of the IUCF CIS project will take place using a pulsed beam of negative hydrogen ions with a pulse length of up to 300 µs and period of up to 5 Hz. The 25-keV H⁻ beam will be produced by a duoplasmatron and transported to a radiofrequency quadrupole drift-tube linac (RFQ/DTL) through a 1.5-m beamline consisting of three Einzel lenses and one x, y steerer assembly. The ion source x-y position and angles are also adjustable; together with the lenses and steerers they allow control of all degrees of freedom necessary for matching into the RFQ/DTL.

To insure proper matching to the RFQ during commissioning and run-to-run reproducibility, the 25-keV beamline is being equipped with an emittance scanner, a horizontal beam profile harp, a Faraday cup, and a viewer. The viewer consists of a BeO scintillator, 2.54 cm × 2.54 cm square, mounted on a holder oriented 45° to the beam axis. A television camera can view the pulses through an acrylic window. The emittance scanner will consist of a 48-wire harp 10 cm downstream of a 0.25-cm slit mounted on a stepping-motor-driven platform.

The 48-wire harp is constructed using standard wire chamber winding techniques. It consists of three planes of 50-µm wires: one set of 48 signal wires on a 0.5-mm spacing and two planes of 24 wires spaced 4 mm from the front and back of the signal plane. The two 24-wire planes are biased to a positive voltage greater than about + 90 V in order to collect electrons knocked off the signal wires. The harp wires on the emittance scanner are oriented parallel to the slit jaws. The 48-wire harp has been tested with a peak H⁻ current of 200 µA and pulse length of 300 µs at 1 Hz. Preliminary emittance measurements of the pulsed beam using a pepperpot with 0.5-mm apertures on a 3-mm spacing upstream of the harp agree well with a two-slit scanning measurement of DC beam from the same source.

BETATRON PING TUNE MEASUREMENT SYSTEM FOR THE IUCF CIS RING

The betatron ping tune system (4) measures the horizontal and vertical beam position on a turn-by-turn basis for beam currents in the range from < 1 µA to > 1 mA. A fast Fourier transform of this position data is performed by a PC-based DSP, yielding the betatron fractional tune.

System Hardware and Software

A logarithmic amplifier (LOG amp) has been incorporated into the IUCF cooler synchrotron ring tune measurement system (Fig. 1), greatly improving performance and reducing cost. This new design will be used in the new CIS ring. The LOG amp design provides the required dynamic range, eliminating the need for automatic level control. The output of a two-stage LOG amp is filtered and amplified, then synchronously digitized using a transit recorder (TR) ADC clock derived from the buncher rf. Detecting in this manner eliminates the need for a phase-locked loop and sample/hold amplifiers. The TRs reside in the PC and are connected to the DSP board via a high-speed digital bus.

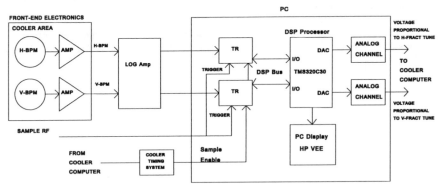

FIGURE 1. Betatron ping tune measurement system utilizing logarithmic amplifiers and HP VEE grahical programming operator display software.

Operator displays are now implemented using HP VEE. This allows rapid development and modification of displays such as FFT spectrums, tune vs time during an energy ramp, etc. The FFTs are performed by the DSP board with a background task moving the tune information from the DSP board to the HP VEE software running on the PC. The system is automatically initialized upon power-up with no operator interaction required. This is a display system only, system variables such as tune kicker amplitude and timing must be set using the control computer.

System Performance

System sensitivity has been greatly improved using the LOG amps design. This allows the tune kickers to be run a factor of four less in amplitude than when using the old system. This is sufficiently low that there is no noticeable kicker-induced beam loss. The synchronous nature of this detector design makes it less susceptible to phase noise during an energy ramp, making this mode of operation much more reproducible.

ACKNOWLEDGMENTS

We would like to thank Jim Lacky and Keith Solberg and the Wire Chamber Group for their help in developing the harp system. Thanks also to the Operations Group for their support and patience while helping us build and test the new systems.

REFERENCES

1. Analog Devices, "Linear Products Data Book," pp. 7.7-22, 1990/91.
2. Wells, F. D., Gilpatrick, J. D., Shafer, R. E., and Shurer, R. B., "Log-ratio Circuit For Beam Position Monitoring," Proc. of the 1991 Particle Accelerator Conference, San Francisco, CA, May 6-9, 1991, 1139-1141 (1991).
3. Hewlett-Packard, "Test and Measurement Catalog," 47-49 (1996).
4. Hamilton, Brett J., Ball, Mark S., Ellison, Timothy J.P., "Betatron ping tune measurement system for the IUCF cooler synchrotron/storage ring," *Nucl. Instrum. Methods A* **342**, 314-318 (1994).

A Multiwire Ionization Chamber Readout Circuit Using Current Mirrors

W. R. Rawnsley, D. Smith, T. Moskven

TRIUMF, 4004 Wesbrook Mall, Vancouver, B.C., Canada, V6T 2A3

Abstract. A circuit which utilizes current mirrors has been used to apply high voltage bias to the wires of a multiwire ionization chamber (MWIC) profile monitor while still allowing measurement of the beam-induced ion-electron currents collected on the wires. Bias voltages of up to 250 V have been used while wire currents over a range of 0.5 nA to 50 nA have been measured. The circuit is unipolar but can be designed for positive or negative bias. The mirrors also provide a current gain of 10, reducing the effects of transistor leakage and extending the useful range of the circuit to lower signal levels. A module containing 32 Wilson current mirrors has been constructed and is used with a MWIC monitor in TRIUMF's Parity experiment beamline.

INTRODUCTION

Safety considerations motivated the use of current mirror circuits in this application. The MWIC monitor used in an experiment (Parity) consists of a 16 by 16 wire grid with 3-mm spacing inside an Al case with 0.010-inch-thick Al windows. The 0.005-inch-diameter Au-plated Mo wires operate in an Ar/CO_2 (90/10) mixture at 1 atm. Negative ions and electrons created by the passage of a 230-MeV proton beam are attracted to the wires by an electrostatic field. Normally a high negative voltage is applied to the case which is insulated from ground, and the wire currents are measured by a scanning current amplifier whose input potential is near ground. In this experiment the monitor is located in a vacuum vessel which also contains a liquid hydrogen target. The failure of a series of vacuum seals could allow the formation of an explosive mixture of hydrogen and air. In order to provide an additional layer of safety, it was decided to ground the monitor's metallic case to eliminate a possible source of sparks and instead apply a high positive voltage to the wires inside.

THE CURRENT MIRROR WITH GAIN

The use of a current mirror circuit to allow the measurement of beam-induced currents from a biased intercepting electrode has been described in ref. (1). The concept has been extended here to a system requiring more measurement channels and lower signal currents. Two versions of the circuit have been used, and simplified diagrams are shown in Figs. 1 and 2.

Referring to Fig. 1, the collector of transistor Q1 holds the MWIC wire at a voltage close to +200 V and a current I_{IN} flows into the wire. If V_{GAIN} is zero, the high-gain transistors Q1 and Q2 share the same base-to-emitter voltage and have similar collector currents. Then

$$\frac{I_{C2}}{I_{IN}} = \frac{\beta}{\beta+2}\left(1 + \frac{V_{CB2}}{VAF}\right), \tag{1}$$

where I_{C2} is the collector current of Q2, V_{CB2} is its collector-to-base voltage, β is the current gain of the transistors (and is typically 475 for the devices used here), and VAF is their Early voltage of 45.7 V. The term in brackets is due to basewidth modulation.

Q1 and Q2 are in a quad pack but do not share the same die. They are isothermal and have similar characteristics but are not identically matched. Their collector breakdown voltage is only 60 V. Q3, however, is a high-voltage transistor and it drops the majority of the 200 V so that Q2 need only withstand 15 V. Q3 is connected in a common base configuration and passes the current from Q2 to the circuit output with only a small loss due to its base current. To reduce the effects of transistor leakage, a 60-mV bias has been added to the emitter of Q1. The effect is to create current gain in the mirror circuit. Then

$$\frac{I_{C2}}{I_{IN}} = \frac{\beta G}{\beta+G+1}\left(1 + \frac{V_{CB2}}{VAF}\right), \tag{2}$$

where $G = e^{qV_{GAIN}/kT}$ and q is the electron charge, k is Boltzmann's constant, T is the temperature, and $kT/q = 25.9$ mV at 300°K.

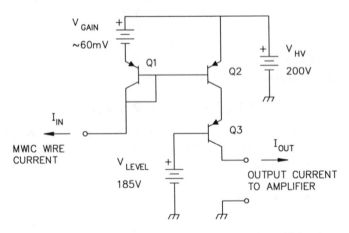

FIGURE 1. A schematic of a simple current mirror with gain plus a high-voltage common base output stage.

The current gain extends the useful range of the circuit to lower signal levels. At room temperature a gain of 10 was found to increase the leakage current from the mirror from about 0.05 nA to 0.1 nA. The majority of the leakage in the circuit output, typically up to 5 nA, arises from Q3, however. The overall effect is to reduce the equivalent leakage at the circuit input to 5 nA/10 = 0.5 nA.

A Wilson current mirror circuit was used in the second version of the circuit, Fig. 2, and adds one more transistor, Q4. This circuit has a much higher output impedance than that of the simple mirror and its gain, governed by Eq. (3), is fairly independent of basewidth modulation:

$$\frac{I_{C2}}{I_{IN}} = \frac{\beta^2 G + \beta G + \beta + G + 1}{(\beta + G + 1)\beta}. \tag{3}$$

The Wilson configuration is commonly used in circuits with $V_{GAIN} = 0$ as it provides near unity gain.

SPICE SIMULATION

Figure 3 shows a PSpice simulation of the Wilson circuit in Fig. 2 (2). Models for the Motorola MPQ3799 and 2N6519 at 27°C were used. The simulation indicates that the current gain (I_{OUT}/I_{IN}) of 10 is accurate to 20% from an I_{IN} of 0.93 nA to 6.1 mA and to 2% from an I_{IN} of 6.7 nA to 0.43 mA. The accuracy is limited at the low end of the range by leakage and at the upper end by the effects of high-level injection and the transistor internal base and collector resistances.

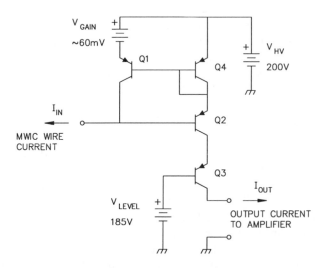

FIGURE 2. Wilson current mirror.

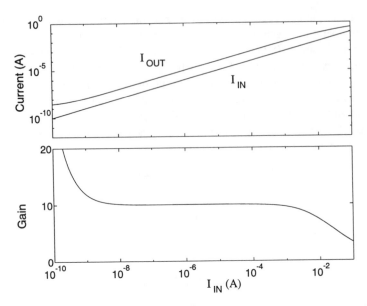

FIGURE 3. A PSpice simulation of the circuit in Fig. 2 showing the currents and the gain.

FULL SCHEMATIC

The full schematic of a single channel is shown in Fig. 4. The gain control voltage is provided by a voltage divider consisting of P1, R7, R2, and R3. The gain of each channel is individually set using its potentiometer. The common gain control voltage V_{DIST} is distributed to the cards at a high level, about −7.1 V with respect to V_{HV} and is reduced by a factor of about 100 in each channel to form the final voltage V_{GAIN}. This prevents gain errors due to ground voltage differences between cards. The signal current also flows through R3 and adds to the gain control voltage. Though this change is only significant at very high signal currents, it is canceled by adding R4. R4 is made 10 times less than R3 but carries 10 times the signal current and generates the same offset voltage.

Equation (3) indicates that the gain is temperature sensitive. In our case, β is large compared to the gain over much of the useful signal current range, and the gain approaches $e^{qV_{GAIN}/kT}$. This implies that for a gain of 10 at 27°C the gain drops by 0.76%/°C. In this application, the current mirrors are used to measure beam profiles, and it is only necessary that all channels have the same gain, not a specific gain. Since all 32 channels are in close proximity inside a metal enclosure and there is very little heat produced inside it, the channel gains were not individually compensated for temperature. Overall compensation to first order was provided, however, by making the common gain control voltage V_{DIST} a function of temperature. A National Semiconductor LM334 temperature-sensitive current source was incorporated in the bias supply for this purpose. Equation (3) and PSpice modeling both indicate that V_{GAIN} should be increased by 0.2 mV/°C for proper compensation; however, it was found empirically that 0.45 mV/°C was required. This difference is probably due to a decrease in β with temperature which was ignored when simplifying Eq. (3) and which is not fully included in the PSpice model of the MPQ3799. R1, R5, and R6 provide current limiting in case a MWIC wire shorts to ground. Diodes D1, D2, and D3 are made by Siliconix and have a maximum leakage of 50 pA. D1 discharges the pick-up wire and signal cable through the 200-V supply when it is turned off. D2 and D3 limit the output voltage and protect the current amplifier which follows.

A photograph of the unit is shown in Fig. 5. The current mirrors and power supplies are mounted in a 19-inch rack module 3.5 inches high by 16.5 inches deep. Each 4.5 × 6" card holds eight channels. Commercial open-frame power supply modules were used. A zener diode shunt regulator was used to smooth the high-voltage supply. A Motorola 2N6519 with a breakdown voltage of 300 V was used for Q3. Other suitable high-voltage transistors are discussed in ref. (1). Q1, Q2, and Q4 share the same package in a Motorola MPQ3799 transistor array. Table 1 shows suitable PNP, NPN, and mixed arrays that have high gain, low leakage, are inexpensive, and readily available. The CA3046A contains both NPN and PNP transistors on a single die. Fortunately, the typical values of I_{CBO} (the collector cutoff current) are orders of magnitude less than the listed maximums.

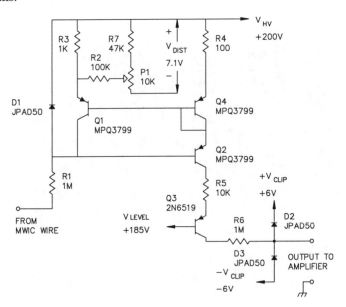

FIGURE 4. The full schematic of one channel.

TABLE 1. Transistor Arrays

Device	Source	Type	β Min.	I_{CBO} Max. (nA)
MPQ3799	Motorola	Quad PNP	225	10
MPQ2484	Motorola	Quad NPN	200	20
CA3096A	Harris	3 NPN	130	40
"	"	2 PNP	40	40

β min. at $I_c = 0.1$ mA, I_{CBO} max. at 25°C

The wire currents are brought out of the vessel on a vacuum feedthrough. The signals are then brought from the beamline to the electronics on RG-174U coaxial cables about 50 feet long. The connectors used on the air side are Amp Coaxicon pins and sockets that are intended for signals of less than 80 V. Though no problems were encountered with sparking in the air, it was found that leakage in the Coaxicon connectors was sometimes worse than the leakage of the current mirrors. No entirely satisfactory method of cleaning the connectors was found. In addition, the polypropylene dielectric used in the connectors has very poor radiation hardness. Connectors better suited to this application are needed.

EXPERIMENTAL RESULTS

Figure 6 shows profiles of a 0.4-nA beam. Curves A and B were recorded using the simple current mirror circuit with a high-voltage bias of +250 V and +150 V, respectively, on the wires. Curve C was recorded with the wires

FIGURE 5. A photograph of the completed rack mount module.

connected directly to the scanning current amplifier without the current mirror circuit. For the latter test the wires were at ground potential and a bias of –100 V was applied to the monitor's case (the maximum allowed by the Safety Group for this test). In Fig. 7 curve A was recorded using the Wilson current mirror circuit with a bias of +200 V on the wires. Curve B was recorded without the current mirror circuit, the wires at ground potential, and a bias of –100 V applied to the monitor's case.

Background (beam off) subtraction was used to produce the profiles. The background correction for each wire was approximately 0.5 nA with the current mirrors; this correction would not usually be required. Beam steering indicated that the apparent increase in signal on wire #17 was not due to a localized peak in the otherwise smooth profile. This is the lower-most wire of the vertical profile plane and it is thought that it was collecting extra charge, though the reason is not known. The height of the profiles increased with increasing bias voltage indicating that a plateau in the sensitivity had not been reached. The similarity of the shape of the profiles indicates that no adverse physical effects in the monitor were caused by changing the biasing scheme and that the new scheme provides correct profile measurements.

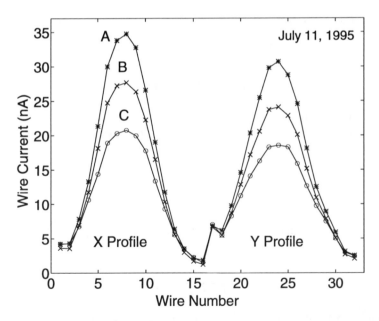

FIGURE 6. Profiles of a 0.4-nA beam. Curves A and B were recorded using the simple current mirror circuit with a bias of +250 V and +150 V, respectively, on the wires. Curve C was recorded without the current mirror circuit, the wires at ground potential, and a bias of –100 V applied to the monitor's case.

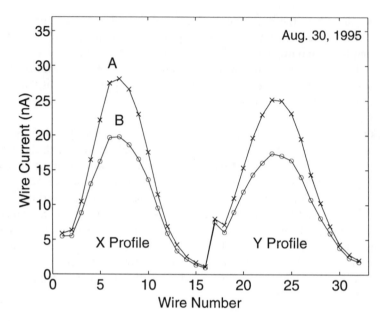

FIGURE 7. Curve A was recorded using the Wilson current mirror circuit with a bias of +200 V on the wires. Curve B was recorded without the current mirror circuit, the wires at ground potential and a bias of −100 V applied to the monitor's case.

CONCLUSION

The current mirror circuit has been successfully applied to allow biasing and readout of multiwire ionization chamber wires. It provides an extra margin of safety in the Parity experiment.

ACKNOWLEDGMENTS

We would like to thank Richard Helmer who suggested this application and Charles Davis and Shelly Page who facilitated testing during a tight beam schedule.

REFERENCES

1. Rawnsley, W.R., Moskven, T., "A Current Mirror Circuit to Bias Beam Intercepting Pick Up Electrodes," Beam Instrumentation Workshop, Sante Fe, NM, October 20-23, 1993, *AIP Conference Proceedings* **319**, 249-254 (1993).
2. MicroSim Corporation, PSpice A/D (1995).

A Very Sensitive Nonintercepting Beam Average Velocity Monitoring System for the TRIUMF 300-keV Injection Line

Yan Yin, Robert E. Laxdal, Anotoli Zelenski

TRIUMF, 4004 Wesbrook Mall, Vancouver, B.C., Canada V6T 2A3

Peter Ostroumov

INR, 117312 Moscow, Russia

Abstract. A nonintercepting beam velocity monitoring system has been installed in the 300-keV injection line of the TRIUMF cyclotron to reproduce the injection energy for beam from different ion sources and to monitor any beam energy fluctuations. By using a programmable beam signal leveling method the system can work with a beam current dynamic range of 50 dB. Using synchronous detection, the system can detect 0.5 eV peak-to-peak energy modulation of the beam, sensitivity is 1.7×10^{-6}. The paper will describe the principle and beam measurement results.

INTRODUCTION

A nonintercepting beam average velocity monitoring system has been developed and installed at TRIUMF for about a year (1). The system is being used to match the energy of different ion sources in order to avoid retuning the injection beamline optics and the center region of the cyclotron when switching the sources. In addition, the monitor can be used to stabilize the input energy during high-resolution operation. The nuclear physics experiments carried out with the TRIUMF cyclotron beam also require a very sensitive energy fluctuation monitoring system.

THE ENERGY MONITORING SYSTEM

The system (see Fig. 1) consists of a resonant pickup, that operates at 46 MHz, and a phase detector to compare the phase between the beam signal picked up by the resonator and the main rf (23 MHz) taken from the first buncher. The second harmonic is chosen to both increase the sensitivity and reduce the interference from the main rf. The distance between the pick-up and buncher is 3.25 meters. In order to make the system independent of the beam intensity, which varies from 1 μA to 300 μA, a programmable attenuator with a pre-amplifier and an HP8347A rf amplifier with a leveling function are used to keep the signal input into the phase detector constant in both amplitude and phase. The main rf signal is doubled in frequency for the phase comparison with the beam signal.

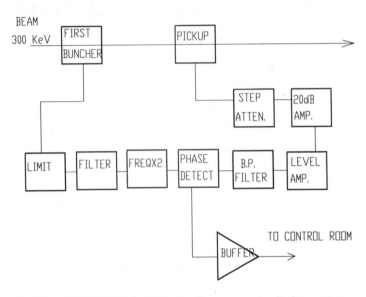

FIGURE 1. TRIUMF 300-keV injection line energy monitoring system.

Measurements have shown that when the beam intensity changes from 300 μA to 0.1 μA the total phase shift is only 1.7 degrees (see Fig. 2). The step in the curve is due to a noncontinuous step attenuator adjustment. With an ordinary 100-MHz oscilloscope, the system can easily match the energy to 60 eV, which is 2×10^{-4}. The system has been routinely used for about a year. During that time it has been running reliably, and there is no obvious temperature drift observed.

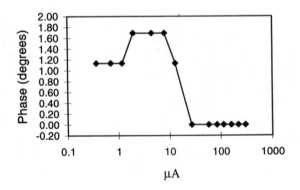

FIGURE 2. The phase shift due to the beam intensity change.

THE PRINCIPLE OF THE MEASUREMENT

Time-of-flight (TOF) methods are useful for low-energy, bunched heavy ion beams, since at a low velocity an energy change produces a relatively large TOF variation over a short distance. For bunched beams, the TOF can be measured directly with two wide-band frequency pickups over a given distance and give not only the absolute energy, but also the energy spread. However, the TRIUMF cyclotron injection line beam is not bunched until it reaches the injection point, so it is only a beam modulated with the cyclotron 23-MHz rf signal. In this case we measure the phase difference of one component of the beam signal between a pickup and the main rf signal taken at the first buncher. Therefore it only gives the average velocity of the beam, which can be converted into beam average energy.

Given the measured rf frequency $f = c/\lambda$, where c is the velocity of light and λ is the wavelength, and assuming the beam flight time Δt over the separation distance Δd, we can obtain the phase shift: $\Delta \varphi = \varpi \Delta t = 2\pi f \Delta t$ between the first buncher and the pickup. Since $\Delta t = \Delta d/\beta c$, where βc is the normalized beam velocity, then $\varphi = 2\pi \Delta d f/\beta c$. When the beam energy fluctuates, $\beta = \beta_0 + \Delta \beta$, $\varphi = \varphi_0 + \Delta \varphi$, and $\Delta E/E = -2\Delta \varphi/\varphi$; hence, $\Delta \varphi = -\pi \Delta d \Delta E/\lambda \beta E$. In this application $\Delta d = 3.25$ m and E is ~ 300 keV, so a 1-degree phase change is then equivalent to a beam energy change of 83 eV.

ENERGY MEASUREMENT WITH SYNCHRONOUS DETECTION

The Parity experimental group has developed a data acquisition system to do beam spin parity measurements. A synchronous detection method is used to suppress noise and reduce statistical errors (2). This data acquisition system was used to determine the sensitivity of this energy monitoring system. A 25-Hz square wave with a timing pattern that allows cancellation of the linear drift of the signals was used to modulate the ion source beam. The output signal of the phase detector, which represents the energy fluctuation of the beam, was first converted to a frequency and then counted synchronously with the modulation. The v/f conversion ratio is 4×10^5 counts/volt.

Let S_{up} and S_d represent the counts at the up and down of the square wave (see Fig. 3). The difference between the up and the down values normalized by their sum values $\dfrac{S_{up} - S_d}{S_{up} + S_d}$ corresponds to the x-axis in Fig. 4 and Fig. 5. The shift of the event count histogram along the x-axis corresponds to the amplitude of the

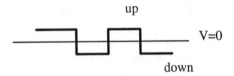

FIGURE 3. Energy modulation waveform.

energy modulation of the beam, which is $E_{up} - E_{down}$. The counts are integrated for a time sufficient to reduce the statistical error. When there is no modulation, as in Fig. 4, the shift value is only –0.002, but with a modulation of $(E_{up} - E_{down}) = 0.5$ eV it is $(1.8 \pm 0.3) \times 10^{-5}$ (in Fig. 5 this is not printed on the plot). This means that the system can actually measure 0.1 eV of synchronous energy change in a 300-keV beam. The measurement was made with a 3.4-µA polarized beam. It has also been verified that the measurement is independent of beam current, which has been varied from 0.1 to 30 µA.

Measurements have also been made for 1 eV and 2 eV. Figure 6 shows the results of these measurements. The right Y axis is in phase (degree), and the dashed line shows the relationship of phase shift and modulation voltages of 0, 0.5, 1, and 2 volts.

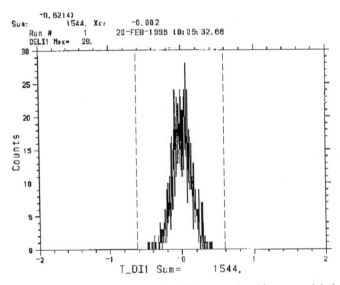

FIGURE 4. Synchronous detection difference signal for no modulation.

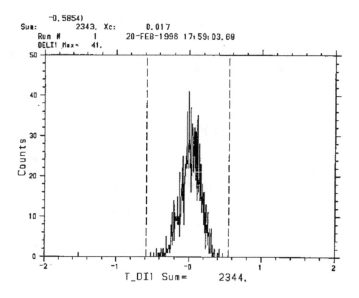

FIGURE 5. Synchronous detection difference for a 0.5-eV energy modulation. Note small shift in centroid of distribution.

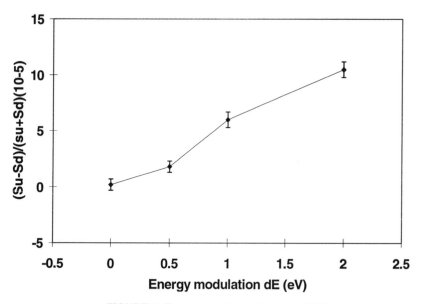

FIGURE 6. Energy modulation (1 eV = 0.012°).

ACKNOWLEDGMENTS

The authors would like to thank the rf control group for their help in using the existing TRIUMF rf electronics system.

REFERENCE

1. Yin, Y., Laxdal, R.E., Ostroumov, P., "TRIUMF injection line nonintercepting beam energy monitoring system through phase measurement," Proc. of 6th International Conference on Ion Sources, Whistler, Canada, Sept. 10-16, 1995, *Rev. Sci. Instrum.* **67** (3), 1255-1257, March 1996.
2. Page, S.A., Birchall, J., Van Oers W.T.H. et al, "TRIUMF Experimental Proposal E497," (1987).

APPENDICES

Summaries of Discussion Groups and Closeout

As has been customary, discussion sessions were held on Monday and Tuesday afternoons with three topics each day running in parallel. A session chairman for each group was selected from the BIW '96 Program Committee. These sessions were intended to provide a forum for informal exchange of ideas, problems, and solutions across the laboratories. Five of the discussion topics were chosen on the basis of interest questionnaires returned by participants with registration materials. The sixth topic, on "Measurements Needed for Fourth-Generation Light Sources," was identified by Hettel and Lumpkin as an evolving area for diagnostics development in new parameter spaces.

The session chairs gave a brief summary of the discussion in each of their groups on Wednesday and Thursday. The Fourth-Generation Light source topic ended up as part of the closeout, along with R. Witkover's leading the "Workshop Feedback" session and A. Lumpkin's providing closing comments.

Precision Beam Position Monitors

Jim Hinkson, Lawrence Berkeley National Laboratory

The discussion began with the discussion leader showing some tranparencies from an earlier talk. In view of the demanding requirements of new and improved light sources and linear colliders, BPMs are needed with higher resolution and repeatibility with lower intensity dependence and cost. Also, for active global and local beam orbit control, the BPMs must have good bandwidth.

The speaker wanted opinions on 180° hybrids and their suitability for high-precision beam position measurements in view of the rather poor null seen at the difference port with equal input signals. M. Kahana stated he believed hybrids are capable of delivering the desired performance if the null errors are calibrated out.

The speaker asked for presentations of prepared material. T. Shintaki showed results with a three-cavity, TM010 BPM centered at 6 GHz. He claimed about 100-nm resolution, single shot. The cavities were small with an approximate 10-mm beam bore. Accurate placement of the cavity probes was difficult. The discussion moderator suggested slot-coupled TEM pickups might be easier to construct with good precision. Coupling to the beam may be controlled with accurately machined slots.

The speaker asked for opinions on beam-based BPM alignment vs. the requirement for comphrensive BPM lab measurements and accurate survey. No one argued against beam-based alignment. Also, no one argued with the speaker's opinion that beam-based BPM alignment should be considered an integral part of any modern BPM system.

Beam Profiling

R. Witkover, Brookhaven National Laboratory

General Comments on the form of the session. The discussion session was well attended with 25-30 people present. Although the meeting lasted for nearly the full assigned time, there was some hesitancy to discuss material that would be presented later in the Workshop. This might be avoided in the future by delaying the discussion groups until the last day or dedicating each day's oral and posters to a specific area which would be the subject of the discussion as well. The room was set up for classroom seating, which I felt inhibited discussion as compared to a face-to-face "roundtable" setting. A number of hardware issues were presented for discussion.

Wire Scanners and Multi-wire (HARP) Monitors. The question was asked about the possibility of using wires of different thickness on a single drive (for a slow stepping or crawling wire device in a transfer line) for measuring beams of different intensity. It was commented that stepping devices at CEBAF were said to have this capability. George Mackenzie mentioned that at TRIUMF they use a scanning blade to get more sensitivity than a wire because of the larger surface area.

There were also a number of questions about wire burnout, which affects both stepping and flying wire scanners. Most devices use tungsten, silicon carbide, or carbon wires, all of which have failed. Keith Jobe of SLAC said that the 4-micron carbon wires used at the final focus burn out. Roland Jung of CERN said that they burn out carbon wires at LEP but that quartz filaments look promising and referred to his paper to be presented on the subject.

Many of these devices use secondary emission of electrons from the surface as the signal source. It has long been known that the condition of the surface has a strong influence on the emission coefficient. Some samples show a short-term change of the coefficient if the surface has not been well cleaned before exposure to beam, but a significant change has also been observed after long-term exposure at levels of the order of 10^{18}/cm^2 (1,2). R. Jung also reported seeing such an effect with as much as a 35% change in the coefficient for aluminum foils and presented data on various materials as well. This effect is seen in secondary emission chambers used to measure beam intensity by moving the beam away from its normal spot on the device to a relatively unexposed area with a resulting change in intensity. It will also show up as a change in the beam shape (and area under the curve) on profile monitors. At present there is no known material or surface process which entirely eliminates the problem.

A similar effect occurs in microchannel plates which are often installed in ionization profile monitor (IPM) devices used for high-vacuum or low-intensity beam monitoring. There is a well-documented (3) deterioration of the gain with total charge extracted (50% at 0.1 Coulomb/cm^2). Arnold Stillman of BNL described a procedure using an ultraviolet lamp to generate charge in the microchannel plate, which can then be used to calibrate the response and compensate for the change.

Transition and Diffraction Radiation. Don Rule described work that he and Ralph Fiorito are doing on using diffraction radiation to monitor beam profile. He stated that it was similar to transition radiation in that a foil is placed in the beam path, but in this case there is an aperture through which the beam passes and it is radiation from the aperture that is detected. The advantage is that the beam does not go through the foil material, eliminating scattering, beam loss, and radiation and foil burnout in high intensity beams. The disadvantage is that the aperture size may not be practical for all beams. Shibata and group in Japan have apparently observed coherent diffraction radiation and measured bunch length, but Fiorito and Rule are analyzing the case for incoherent radiation and feel that the analogy to transition radiation is close. The preliminary results indicate it might be interesting for SLAC where at 50 GeV the radiation is in the optical range. At $\gamma\lambda/2\pi \sim 1$ the response is about 50% that for transition radiation. On the practical side, Keith Jobe pointed out that it may be difficult to untangle unwanted reflections.

Phillipe Piot of CEBAF reported that they had made measurements that confirmed the resolution limit using transition radiation was basically the diffration limited resolution, as predicted by Fiorito and Rule.

References.
1. E. L. Garwin and N. Dean, "Methods of Stabilizing High Current Secondary Emission Monitors," Proc. Symp. On Beam Intensity Measurement, Daresbury, England, 22-26, April 1968.
2. V. Agoritsas and R. L. Witkover, "Tests of SEC Stability in High Flux Proton Beams," *IEEE Trans. Nucl. Sci.* **NS-26** (3), 3355-3357 (1979).
3. Technical Data Sheet for Rectangular MCP and Assembly Series, Hamamatus Corp., 360 Foothill Rd., P. O. Box 6910, Bridgewater, NJ 08807, August 1986.

Closed Orbit, Tune, and Beta(s) Measurement and Correction

R. Webber, Fermi National Accelerator Laboratory

Approximately 20 persons participated in the Closed Orbit, Tune, and Beta(s) Measurement and Correction discussion group. The following laboratories, sections, and/or projects were represented: Argonne APS, Brookhaven AGS, Brookhaven RHIC, CERN PS, Cornell CESR, Fermilab RF & Instrumentation,

Fermilab Booster, Fermilab Antiproton Source, Indiana University Cyclotron Facility, Lawrence Berkeley Lab, and Stanford SSRL.

Closed Orbit Discussion. The discussion began with a description and discussion of the local and global orbit control systems planned for APS. A goal of orbit stabilization to 17 microns horizontally and 4.6 microns vertically was stated. The APS system relies on digital signal processor technology and reflective memory packaged in 20 VME crates to process position information and to adjust corrector magnets for the local and global orbit control sysems. Position information is presently sampled at 1 kHz. Separate magnets of the same design are used for the two systems. These magnets have been measured to demonstrate magnetic field frequency response to greater than 200 Hz, although the bandwidth is reduced by eddy current effects in the beam pipes which are different for the global and local magnets. The global orbit control feedback loop bandwidth is designed to be 100 Hz. This is achieved by using a stainless steel beam tube for the global loop magnets. Position information from up to 320 BPMs is used to control 38 magnets in the complete global orbit control system. The local orbit control loop is designed to operate at a reduced bandwidth, and the standard aluminum beam chamber is used for these magnets.

SSRL reported on their global closed orbit stabilization feedback system which reduces, by an order of magnitude, millimeter-scale orbit fluctuations caused by day to night temperature changes. A microVAX and a CAMAC-based control system uses the information from about 32 BPMs to control about 40 correctors. SSRL has non-laminated corrector magnets and corrections are made at 6-second update intervals. Magnet hysteresis has presented some problems. Local corrections with a bandwidth of up to nearly 100 Hz has been accomplished using strictly analog feedback.

The question of orbit stability in the Fermilab Tevatron was raised. It was reported that there is no automatic feedback on the orbit in the Tevatron (other than beam energy control). Orbit measurements are manually initiated hourly during a colliding beam store. If the rms distortion is > 1 mm, a program to calculate new corrector magnet settings is manually run and those settings are downloaded. Since the protons and antiprotons see the same magnetic fields and react identically, magnetic perturbations do not affect collision; both beams will travel identical orbits (in opposite directions). Only the electrostatic separators affect the beams differentially.

Tune Measurement. IUCF solicited thoughts about tune measurement methods useful for their ring. Tune control is important for them in order to avoid polarization resonances; their measurement problem is complicated by a large increase in particle velocity during acceleration.

Stimulation of the beam by white noise was suggested, but it was noted that the power spectral density required to stimulate the beam sufficiently might be achieved only with significant total noise power and that the signal analysis problem remains. It was suggested that a narrow-band FM stimulus signal around the expected tune frequency might be more efficient and effective. This idea was expanded to include use of a direct digital synthesizer programmed to track the changing beam revolution frequency as part of both the FM signal source and the signal processing electronics.

AGS reported attempts to use similar methods with mixed results. They ultimately rely on the brute force method of impulse beam stimulus (kicks), turn-by-turn beam position response meauserment, and FFT signal analysis to obtain tune information. This can be done multiple times during an acceleration cycle.

DSPs. APS's extensive use of digital signal processors (DSPs) prompted some discussion of the state of that technology and the promises and problems it offers in the realm of beam instrumentation.

Concern was expressed by several participants regarding the rapid evolution and often obsolecense of these types of products. SSRL had considered use of an AMD microcontroller chip until suddenly the product line was dropped. The RHIC BPM system makes use of Motorola DSPs. APS utilizes DSP boards based on the Texas Instruments C030 DSP chips. It was generally agreed that TI is the "Intel" of DSPs since its products are in widespread applications, and a broad base of good support, including software development tools, is provided. Some discussion, without conclusion, concerned the merits of using DSP-specific chips versus general-purpose computer chips such as Intel 486's or Motorola 680xx's. The general-purpose chips seem to offer extensive support, widespread application, and generational compatability, stability and continuity; whereas the DSPs offer significant performance advantages in specific applications. It was noted that, whenever seeking the highest performance, one is often testing the waters of chip development and subject to the whims and limitations of product support by the manufacturer.

It was mentioned that programmable logic array chip suppliers such as Altera are now providing (free?) libraries of DSP-type functions which can be directly programmed into their devices. This may be advantageous when a limited number of fixed functions can satisfy one's requirements, and the dynamic programming that a full-fledged DSP can provide is not necessary.

Beta(s) and Beam Transfer Function Measurements. There was no discussion on this topic.

Feedback, Damping, Drivers

Ralph J. Pasquinelli, Fermi National Accelerator Laboratory

This discussion group began slowly with some specific questions concerning TWT operation at the ALS in Berkeley, wide dynamic range multipliers required for damper systems at Fermilab, and 150-MHz bandwidth fast attenuators. This was about the first 10 minutes. After some prodding on my part, I could not get anyone else to discuss anything about feedback systems.

At this point I changed the discussion session into a different form of feedback discussion, what the perception of the BIW '96 was to that time. Very quickly the discussion picked up. The following points were brought to the floor. The tutorials were thought to be very good and seemed appreciated. Some people would like to see the topics expanded, one thought they should be more like the accelerator school. I indicated that is more the role of the accelerator schools as opposed to the BIW.

- The Instrumentation bulletin board is underutilized.
- Posters, vendors, and discussion sessions could be in parallel as opposed to having multiple discussions in parallel. This year, there was too long a session for posters and vendors. It could be organized in a way such that people could arrange to meet interested parties concerning their poster.

The topics then returned to beam feedback systems with questions on resistive wall instabilities in the newly proposed Fermilab Recycler Ring and SLAC B-Factory secondary electron problems. There was no one in attendance that was able to address these questions. Other comments were brought up about nonlinear feedback performance such as "bang bang" feedback, pickup and kicker response, and damping times.

Instrumentation for Low-Current and Radioactive Beams

George Mackenzie, TRIUMF

The discussion started with two groups requesting advice regarding noninter-cepting monitoring of low current beams that are set up initially using intercepting diagnostics. A new slow extraction line from the MIT-Bates recirculator (400- to 800-MeV electrons) delivers 1 to 15 μA quasi-DC. Suggestions were a) to use residual gas profile monitors such as those developed for DESY by Hornstra, b) to observe OTR from a very thin minimally intercepting carbon foil (CEBAF estimated that 50 nA of 850-MeV electrons should give a usable signal), and c) to install at sensitive places thin foil halo monitors with an aperture slightly larger than the normal beam size. A therapy facility is being designed to deliver 60- to 250-MeV protons, 0.1 to 20 nA, in a 250-ms spill at 2 Hz. The therapists will use ion chambers to monitor dose, the group wishes to monitor the primary beam also. TRIUMF uses 0.0003-inch (2 mg/cm^2) aluminum foil secondary emission monitors for current and position. The beam degradation from these is small compared with the therapist's beam preparation equipment. A well-shielded, vibration-free current transformer with a resolution of ~ 1 nA was described at the 1991 BIW by F. R. Gallegos, LANL (AIP Conf. Proc. 252).

Heavy ions produce copious secondary electrons even at very low, keV/amu, energies. Many laboratories collect these in a channeltron, microchannel plate, or other multiplier to obtain timing, current, or position signals from very weak beams. CERN-ISOLDE has measured profiles of 2- or 3-mm-diameter beams as low as 6×10^4 ions/s. GANIL and Catania measure higher energy beams (MeV/amu) using the ionization in the residual gas; see papers by Anne and Rovelli in these proceedings. Alberto Rovelli believes that ~10^6 C^{5+} may be imaged. MSU-NSCL has successfully imaged transverse distributions using a scintillation screen made up of a thin layer of ZnS sprayed onto a metal backing. Longitudinal distributions are measured with a resolution of ~ 100 ps by counting pulses from channeltrons, etc.; ~ 10 ps is achieved with bunch length monitors (Feschenko, BIW '92). Energy distributions are measured using spectrometers, time of flight (~ 0.1% at ANL), and bunch-length velocity detector (~ 0.2%, Feschenko and Ostroumov, BIW '94). Very low currents are measured by counting, intermediate currents in Faraday cups; electronic noise sets the lower limit of the CERN system at 10 to 100 pA. A nonintercepting squid-based current comparator is being developed at GSI; 10-nA resolution has been achieved on the bench, and it will be installed in a beamline shortly.

Accelerators and beamlines are usually tuned for very weak beams by first setting up a more abundant analog beam with a similar q/m. Settings are then changed by a precalculated amount to suit the desired ion species. Radioactive species are often accompanied by isobars both stable and unstable. The accumulation of longer-lived species or daughters on a beamstop can give rise to a current from β or α decay which masks the incoming beam current. To determine the species present and to monitor performance, it is becoming common to accumulate the beam on a tape or wheel which is periodically advanced to a GeLi detector which identifies species by their γ-ray spectrum. Once identified, β detectors, which are cheaper and more convenient, can be used downstream. A new requirement, to confirm stable operation of an ion source and mass spectrometer, is to monitor the production of off-mass species at the focal plane of the spectrometer.

The discussion ended by creating a list of equipment manufacturers, references, and other resources. The latter included an article on beam diagnosis by Peter Strehl in the *Handbook of Ion Sources*, ed. B. Wolf, CRC Press, 1995 and the simulation program SRIM, available from J. F. Ziegler, IBM-Research, Yorktown, NY.

Measurements Needed for Fourth-Generation Light Sources

R. Hettel, Stanford Linear Accelerator Center

The discussion group on instrumentation needs for fourth-generation synchrotron light sources began with a review by the group chairman of photon beam properties that characterize such sources and with a description of storage ring and FEL-based fourth-generation implementations that are envisioned. This was followed by a discussion of measurement methods and instrumentation that could be used for these facilities. The conclusion of the group was that most instrumentation needs can likely be met with existing technology developed for advanced synchrotron radiation storage ring and electron/positron linear accelerator facilities (such as the SLAC SLC), although challenges remain, including measuring and maintaining the true micron beam alignment required for these extremely bright facilities.

Fourth-Generation Source Properties. Properties that characterize synchrotron radiation facility performance and that can be used to define fourth-generation light sources were delineated at the 10th ICFA Beam Dynamics Panel Workshop on 4th Generation Light Sources held at the ESRF in January, 1996. They include:
- high coherence
- high brightness
- emittance on the order of the diffraction limit (i.e., emittance less than ~ 8% of photon wavelength)
- short bunch length (~ 1 ps or less for time-resolved measurements)
- high spectral purity (transform limit)
- high peak power
- polarization options

High transverse coherence and brightness are both obtained by reducing beam emittance towards the diffraction limit. Coherence is also enhanced in laser-like FEL sources. Spectral purity and peak power are increased when the radiation from low-emittance sources is concentrated in a narrow band(s), as is the case with undulator and FEL generators. Peak power from these sources is maximized when the bunch length is short; short bunches also permit high time resolution in measuring dynamic properties of material. Beam polarization is determined by the magnet structure generating the particular synchrotron radiation beam (e.g., bending magnet, wiggler, or undulator) and is improved in general by reducing field errors in these structures and by controlling the trajectory of the charged particle beam through them with an accuracy that in some cases may be a fraction of the beam size.

For VUV photons (of order 100 nm) and x-ray photons (of order 0.1 nm), the diffraction limited emittances are about 10 nm-rad and 0.01 nm-rad, respectively. While the diffraction limit for low-energy VUV light has already been reached in third-generation rings, these rings are more than two orders of magnitude away from the x-ray diffraction limit, except in cases where the horizontal-vertical lattice coupling has been reduced to a few parts in a thousand (e.g., at the ESRF). With further reduction of the horizontal emittance in x-ray rings to a few nm-rad, this reduced coupling will result in a diffraction-limited source in the vertical plane only. Intrabeam scattering prevents reaching the limit in both planes. In addition, the bunch length in storage rings can only be reduced to a few picoseconds due to turbulent bunch lengthening from the microwave instability. These factors limit the obtainable average brightness from storage rings to the order of 10^{22}, about a factor of 10^4 higher than that specified for the third generation. Corresponding beam power densities will also increase by this factor. This higher brightness will likely be reached by improvements made to existing third-generation sources, creating "generation 3.5."

The ESRF Workshop identified two other source types that can truly be termed "fourth generation": storage ring FELs for UV photons, and linac-driven FELs for UV and x-rays. FELs on third-generation UV rings can produce coherent beams with average brightness on the order of 10^{26}. Linac-driven x-ray FELs, such as the 1.5-Å Linear Coherent Light Source (LCLS) FEL driven by a 15-GeV beam now being studied at SLAC, may produce average brightnesses comparable to the generation 3.5 storage rings. However, because they are diffraction-limited in both planes and have bunch lengths of less than a picosecond, they may produce fully coherent beams with peak brightnesses on the order of 10^{33}, with corresponding peak powers on the order of 10^{10} W. This is many orders of magnitude greater than those obtainable with improved storage rings. These phenomenal numbers will only be reached if the underlying FEL process, self-amplified spontaneous emission (SASE), truly works as predicted by calculation. Several projects are now underway to demonstrate SASE in the UV regime (e.g., TESLA at DESY and DUV at BNL). The future of more ambitious projects like the LCLS will depend on the success of the UV projects.

Fourth-Generation Instrumentation Needs. Examples of beam parameters for an improved storage ring that were considered by the discussion group come from the ESRF, where reducing the horizontal emittance to 3 nm-rad together with a reduced coupling of 3×10^{-3} will result in a vertical emittance of 0.01 nm-rad, near the diffraction limit for 1-Å photons. The corresponding nominal electron beam size and divergence will be 170 μm rms and 20 μrad rms horizontally, and 10 μm rms and 1 μrad rms vertically. Bunch length will be a few picoseconds. Beam properties will most likely be measured in much the same way as they are for third-generation sources, using rf BPMs for orbit, photon BPMs for beamline trajectory, x-ray imaging for transverse profile and emittance, streak camera with sub-picosecond resolution for longitudinal profiles, and x-ray polarimeters.

One significant challenge for these storage ring sources identified by the group is to develop BPM electrode assemblies that have true micron stability (10% of the beam size) in the presence of changing thermal gradients in the vacuum chamber so as to match the micron resolution already obtainable with modern BPM processing electronics. This is perhaps a factor of ten better mechanical stability then is presently realized in third-generation rings. Another challenge is to develop similarly stable photon BPMs and other x-ray optical devices that can handle the much higher power densities from the improved sources. Sufficient resolution in emittance measurements (< 10 μm size resolution, < 1 μrad divergence resolution) is already forseeable using x-rays.   Anticipated beam properties for the SLAC LCLS FEL were presented and the group discussed the demanding instrumentation needs for this facility. The planned 15-GeV beam emittance is ~ 0.03 nm-rad in both planes, the diffraction limit for 4-Å photons; the fundamental photon wavelength will be tunable in this regime, down to 1.5 Å. A photocathode electron gun will produce a 1-nC short bunch (a few ps) or train of bunches at a linac repetition rate of up to 120 pulses/s. Each bunch will be compressed to the order of 0.1 ps (5000-A peak current) before entering a permanent magnet undulator (planar or helical) that will be between 30 m and 80 m long. The undulator aperture is only a few millimeters over this length. Beam size and divergence are of order 10 μm rms and 1 μrad rms, respectively. The electron beam and FEL photon beam must be colinear to ~ 10% of the beam size (the order of 1 μm rms) over a few gain lengths to maximize the SASE process. The generated 100-fs photon pulse will have a peak brilliance on the order of 10^{33} and peak power of order 10^{11} W. Nobody is quite sure what effect the extreme electric fields created by this photon pulse will have on materials; it will certainly destroy many, requiring a constant replenishing of experimental sample material on a pulse-by-pulse basis.

A significant instrumentation challenge for the LCLS is to monitor and control beam position within the undulator to the micron level. While it may be possible to install stripline BPMs having single-shot micron resolution (this resolution is obtained for the SLAC SLC using 10-cm-long electrodes), it has yet to be determined whether the wake fields from these electrodes will lengthen the passing bunch or increase its energy spread; both would be deleterious to the SASE process. It will also be difficult to machine long striplines with the necessary tolerances for micron resolution in a 5-mm-diameter chamber. Electrodes plated on ceramic may be required. Alternatives to stripline BPMs were discussed, including resonant cavity monitors that have 0.1-μm resolution but which may not fit in the undulator space, and a variation of the "laser wire" monitor developed for the SLC, where beam position is determined by measuring the Compton-scattered high-energy photons created when the electron beam interacts with a transverse thin filament of optical light created by a high-power laser. This method would require a system of mirrors and windows along the undulator. It is likely that position measurement at the far end of the FEL of either the electron beam as a function of undulator quadrupole strength modulation or of incoherent synchrotron radiation caused by different sections of the undulator can be used to establish initial beam alignment and to calibrate embedded position monitor assemblies.

Insertable transition radiation or wire/laser wire monitors along the length of the undulator may also be considered for determining absolute beam alignment.

Other instruments discussed included those for LCLS profile, emittance, and bunch length measurements. It is likely that a conventional carbon wire scanner can be used for integrated profile and emittance measurement, although the more elaborate laser wire was also suggested for this. Single-shot profile and emittance measurements can be obtained in the usual way using transition radiation foils; however, it was proposed that the interference of diffraction radiation from multiple foils may provide higher resolution and be less disruptive or subject to damage from the electron beam since it passes through holes in the foils. Wake field effects from these foils have yet to be determined. Coherent interferometry of transition or diffraction radiation can be used to measure bunch length with ~ 10-fs resolution; the frequency resolved optical gating (FROG) method can be used to provide unambiguous single-shot bunch length since both amplitude and phase interfering radiation is measured. Another method that was suggested is to sample the longitudinal electron beam profile with a femtosecond laser pulse. Finally, it was noted that streak cameras having resolutions of < 100 fs are now becoming available and may be usable for single-shot bunch length measurement.

Closeout

Alex Lumpkin, Argonne National Laboratory

Following Bob Hettel's summary of the "Measurement Needs for Fourth-Generation Light Sources," Richard Witkover (BNL) led the traditional "Workshop Feedback" session. He reminded us that the workshop format had evolved during this series that was first held at Brookhaven National Laboratory in October 1989. At that time tutorials were emphasized with 2-3 hours allotted for each of them. The present mix and duration of tutorial, invited, contributed oral, and contributed poster has been developed through the BIW Program Committee's use of input from these "feedback" sessions. A specific query and poll on this mix generated no specific comment, so it is assumed as acceptable.

There were suggestions and comments about the parallel discussion sessions. Of course, some persons would have liked to go to more of them so suggestions of serial or other days came up. The roundtable room setup was preferred to encourage discussion as compared to the two rooms with classroom seating. This was partly a logistic issue this year since the discussion groups were so well attended in Robert Shafer's view (perhaps attributable to the captive audience that was bused to the workshop site). These suggestions for roundtable format and serial discussion groups seem counter to a suggestion to shorten the workshop to three days. More discussion of engineering problems encountered, solved, understood or not was suggested. This depends on the participants themselves. It is anticipated that the Program Committee will continue this discussion prior to BIW '98.

Finally, as we close this Workshop, I believe we are actually entering another interesting period for beam diagnostics developments. As pointed out during this Workshop, the PEP-II B factory is adding high current beam aspects to the lepton colliders, the Relativistic Heavy Ion Collider (RHIC) facility is being commissioned, and fourth-generation light source research is pushing both storage ring and linac-driven FEL diagnostics closer to collider beam specifications. The fourth-generation light source R&D already has SASE research activities at UCLA/LANL for 10 µm, at BNL at 0.9 µm, DESY at 0.2 µm, and SLAC at 0.1 µm. APS will be utilizing an injector linac for 0.1- to 0.2-µm radiation SASE tests. The short bunches needed for FELs and colliding beams is supported by the development in coherent transition radiation- and diffraction radiation-based techniques. Submicron beam position resolution capability exists in storage rings now and may be extended to single-shot linacs by laser-based probes or rf cavity-based techniques. It is my view that the diagnostics developments are ahead of the accelerators and thus can be in place for the critical commissioning phases of the new machines. The next few years will be evidently quite active, and it is most appropriate that SLAC/SSRL will host the next workshop with their involvement in many of these issues.

List of Participants

FINN ABILDSKOV
Institute for Storage Ring Facilities
University of Aarhus
Ny Munkegode
DK-8000 Aarhus C
DENMARK

Phone: 45-8942-3776
Fax: 45-8612-0740
E-Mail: fabild@dfi.aau.dk

ROBERTO G. AIELLO
Controls Department
SLAC
P.O. Box 4349
Stanford, CA   94309-4349

Phone: 415/926-3552
Fax: 415/926-3882
E-Mail: aiello@slac.stanford.edu

MICHEL R. ANNE
GANIL
B^d H. Becquerel
BP 5027
14021 Caen-Cedex
FRANCE

Phone: 33-31-45-45-03
Fax: 33-31-45-46-65
E-Mail: anne@ganac4.in2p3.fr

JEFFREY A. ARTHUR
Accelerator Division/RFI
Fermi National Accelerator Laboratory
P.O. Box 500, MS 308
Batavia, IL 60510

Phone: 630/840-3162
Fax: 630/840-3754
E-Mail:

PATRICK AUSSET
Laboratoire National Saturne
CEA
91191 Gif.Sur Yvette Cedex
FRANCE

Phone: 33-1-6908 29 16
Fax: 33-1-6908 48 58
E-Mail: ausset@
             chatelet.saclay.cea.fr

ROBERT J. AVERILL
Bates Laboratory
Massachusetts Institute of Technology
P.O. Box 846
Middleton, MA 01949

Phone: 617/253-9254
Fax: 617/253-9599
E-Mail: averill@bates.mit.edu

MARK S. BALL
Department of Beam Dynamics
Indiana University Cyclotron
2401 Milo Sampson Lane
Bloomington, IN 47405

Phone: 812/855-5162
Fax: 812/855-6645
E-Mail: ball@iucf.indiana.edu

BARRY W. BARNES
Accelerator Division/RFI
Fermi National Accelerator Laboratory
P.O. Box 500
Batavia, IL 60510

Phone: 630/840-4110
Fax: 630/840-3754
E-Mail: barry_barnes@
              admail.fnal.gov

DEAN S. BARR
Accelerator Systems Division/APS
Argonne National Laboratory
Building 401
9700 South Cass Avenue
Argonne, IL 60439

Phone: 630/252-9146
Fax: 630/252-4732
E-Mail: dbarr@anl.gov

WALTER C. BARRY
CBP Department
Lawrence Berkeley National Laboratory
MS 71-259
Berkeley, CA 94720

Phone: 510/486-6705
Fax: 510/486-7981
E-Mail: Walter_Barry@
              Macmail.lbl.gov

EDWARD L. BARSOTTI
Accelerator Division/RFI
Fermi National Accelerator Laboratory
P.O. Box 500, MS 308
Batavia, IL 60510

Phone: 630/840-8104
Fax: 630/840-3754
E-Mail: ebarsotti@adcalc.fnal.gov

WILLIAM J. BERG
Accelerator Systems Division/APS
Argonne National Laboratory
Building 401
9700 South Cass Avenue
Argonne, IL 60439

Phone: 630/252-6929
Fax: 630/252-4732
E-Mail: berg@aps.anl.gov

JULIEN BERGOZ
Bergoz Beam Instrumentation
01170 Crozet
FRANCE

Phone: 33-50-41-0089
Fax: 33-50-41-0199
E-Mail: bergoz@bergoz.com

CHANDRA M. BHAT
Accelerator Division/MI
Fermi National Accelerator Laboratory
P.O. Box 500
Batavia, IL 60510

Phone: 630/840-4821
Fax:
E-Mail: cbhat@fnal.gov

WILLEM BLOKLAND
Accelerator Division/RFI
Fermi National Accelerator Laboratory
P.O. Box 500
Batavia, IL 60510

Phone: 630/840-2681
Fax: 630/840-3754
E-Mail: blokland@adcalc.fnal.gov

PETER CAMERON
Dept. RHIC, Beam Instrumentation
Brookhaven National Laboratory
Building 1005
Upton, NY 11973

Phone: 516/344-7657
Fax: 516/344-2588
E-Mail: cameron@bnl.gov

JOHN A. CARWARDINE
Accelerator Systems Division/APS
Argonne National Laboratory
Building 401
9700 South Cass Avenue
Argonne, IL 60439

Phone: 630/252-6041
Fax: 630/252-5291
E-Mail: carwar@aps.anl.gov

BRIAN E. CHASE
Accelerator Division/RFI
Fermi National Accelerator Laboratory
P.O. Box 500
Batavia, IL 60510

Phone: 630/840-3040
Fax: 630/840-3754
E-Mail: chase@adcalc.fnal.gov

DONG CHEN
Accelerator Division/RFI
Fermi National Accelerator Laboratory
P.O. Box 500, MS 308
Batavia, IL 60510

Phone: 630/840-5056
Fax: 630/840-3754
E-Mail: dongchen@fnal.gov

MICHAEL J. CHIN
Engineering Division
Lawrence Berkeley National Laboratory
Building 46/125
1 Cyclotron Road
Berkeley, CA 94720

Phone: 510/486-5143
Fax: 510/486-5775
E-Mail:

ROGER C. CONNOLLY
Dept. RHIC, Beam Instrumentation
Brookhaven National Laboratory
Building 1005
Upton, NY 11973

Phone: 516/344-4698
Fax: 516/344-5729
E-Mail: connolly@bnl.gov

JOHN N. CORLETT
Center for Beam Physics
Lawrence Berkeley National Laboratory
1 Cyclotron Road
Berkeley, CA 94720

Phone: 510/486-5228
Fax: 510/486-7981
E-Mail: jns@bc1.lbl.gov

JAMES L. CRISP
Accelerator Division/RFI
Fermi National Accelerator Laboratory
P.O. Box 500, MS 308
Batavia, IL 60510

Phone: 630/840-4460
Fax: 630/840-3754
E-Mail: crisp@admail.fnal.gov

JEAN-CLAUDE DENARD
Accelerator Development Group
CEBAF
MS 58B
12000 Jefferson Avenue
Newport News, VA 23608

Phone: 804/249-7555
Fax: 804/249-7658
E-Mail: denard@cebaf.gov

JOSEPH E. DEY
Accelerator Division/RFI
Fermi National Accelerator Laboratory
P.O. Box 500, MS 306
Batavia, IL 60510

Phone: 630/840-8380
Fax: 630/840-4552
E-Mail: dey@fnal.gov

BRIAN J. FELLENZ
Accelerator Division/RFI
Fermi National Accelerator Laboratory
P.O. Box 500, MS 308
Batavia, IL 60510

Phone: 630/840-2512
Fax: 630/840-3754
E-Mail: fellenz@admail.fnal.gov

THOMAS J. FESSENDEN
Department 47/112 Accelerator
Fusion & Energy
Lawrence Berkeley National Laboratory
1 Cyclotron Road
Berkeley, CA 94720

Phone: 510/486-5356
Fax: 510/486-5392
E-Mail: tjfessenden@lbl.gov

RALPH B. FIORITO
Department of Weapons Materials
NSWCCD
10901 New Hampshire Avenue
MC 684
Silver Spring, MD 20903

Phone: 301/394-1908
Fax: 301/394-5135
E-Mail: fiorito@oasys.dt.navy.mil

ALAN S. FISHER
PEP-II
SLAC
P.O. Box 4349, MS 17
Stanford, CA 94309

Phone: 415/926-2436
Fax: 415/926-3882
E-Mail: afisher@slac.stanford.edu

ROGER FLOOD
Accelerator Development Department
CEBAF
MS 58B
12000 Jefferson Avenue
Newport News, VA 23608

Phone: 804/249-7083
Fax: 804/249-7658
E-Mail: flood@cebaf.gov

PETER FORCK
Beam Diagnostic Department
Gesellschaft fuer
Schwerionenforschung GSI
Planckstrasse 1
D-64291 Darmstadt
GERMANY

Phone: 49-6159-712313
Fax: 49-6159-712104
E-Mail: p.forck@gsi.de

JOHN D. FOX
Department PEP-II
SLAC
Bin 18, P.O. Box 4349
Stanford, CA   94025

Phone: 415/926-2789
Fax: 415/926-8533
E-Mail: jdfox@slac.stanford.edu

HORST W. FRIEDSAM
Accelerator Systems Division/APS
Argonne National Laboratory
Building 401
9700 South Cass Avenue
Argonne, IL 60439

Phone: 630/252-9169
Fax: 630/252-5948
E-Mail: horst@aps.anl.gov

RAYMOND E. FUJA
Accelerator Systems Division/APS
Argonne National Laboratory
Building 401
9700 South Cass Avenue
Argonne, IL 60439

Phone: 630/252-6442
Fax: 630/252-4732
E-Mail: fuja@aps.anl.gov

JOHN N. GALAYDA
Accelerator Systems Division/APS
Argonne National Laboratory
Building 401
9700 South Cass Avenue
Argonne, IL 60439

Phone: 630/252-7796
Fax: 630/252-1512
E-Mail: galayda@aps.anl.gov

FLOYD R. GALLEGOS
Department AOT-6
Los Alamos National Laboratory
MS H812
Los Alamos, NM 87545

Phone: 505/667-6848
Fax: 505/665-0046
E-Mail: frg@lanl.gov

DAVID M. GASSNER
AGS Department
Brookhaven National Laboratory
Building 911A
Upton, NY 11973

Phone: 516/282-7870
Fax: 516/282-5676
E-Mail:

JOSÉ LUIS GONZALEZ
Proton Synchrotron Division
CERN
1211 Geneve 23
SWITZERLAND

Phone: 41-22-767-3928
Fax: 41-22-767-9145
E-Mail: gonzalez@ps.msm.cern.ch

SHLOMO GREENWALD
Department LNS
Cornell University
Wilson Laboratory
Ithaca, NY 14853

Phone: 607/255-4882
Fax: 607/255-8062
E-Mail:

BRETT J. HAMILTON
Department of Beam Diagnostics
Indiana University Cyclotron
2401 Milo Sampson
Bloomington, IN 47405

Phone: 812/855-5162
Fax: 812/855-6645
E-Mail: brett@iucf.indiana.edu

M. T. HERON
Synchrotron Radiation Department
CCLRC Daresbury Laboratory
MS B86
Keckwick Lane, Daresbury
Warrington, Cheshire WA4 4AD
UNITED KINGDOM

Phone: 44-1925-603345
Fax: 44-1925-603124
E-Mail: m.t.heron@dl.ac.uk

ROBERT O. HETTEL
Accelerator Systems Department
SSRL
P.O. Box 4349, MS 69
Stanford, CA 94309

Phone: 415/926-3489
Fax: 415/926-4100
E-Mail: hettel@slac.stanford.edu

JAMES A. HINKSON
Engineering Division
Lawrence Berkeley National Laboratory
Building 46/125
1 Cyclotron Road
Berkeley, CA 94720

Phone: 510/486-4194
Fax: 510/486-5775
E-Mail:

KEITH JOBE
Accelerator Department
SLAC
4349 Sandhill Road
Menlo Park, CA 94025

Phone: 415/926-2084
Fax: 415/926-2407
E-Mail: rkj@slac.stanford.edu

RONALD G. JOHNSON
Controls Department
SLAC
P.O. Box 4349, MS 17
Stanford, CA 94309

Phone: 415/926-8520
Fax: 415/926-3882
E-Mail: ron_johnson@
          slac.stanford.edu

KEVIN JORDAN
FEL Department
CEBAF
12000 Jefferson Avenue
Newport News, VA 23606

Phone: 804/249-7644
Fax: 804/249-6355
E-Mail: jordan@cebaf.gov

ROLAND R. JUNG
SL Division
CERN
CH-1211 Geneva 23
SWITZERLAND

Phone: 41-22-767-3295
Fax: 41-22-767-9560
E-Mail: rjung@sl2.msm.cern.ch

EMANUEL KAHANA
Advanced Photon Source
Argonne National Laboratory
Building 401
9700 South Cass Avenue
Argonne, IL 60439

Phone: 630/252-7383
Fax:
E-Mail: kahana@aps.anl.gov

RODERICH KELLER
Advanced Light Source
Lawrence Berkeley National Laboratory
1 Cyclotron Road
Berkeley, CA 94720

Phone: 510/486-5223
Fax: 510/486-4960
E-Mail: r_keller@lbl.gov

ANTHONY H. KERSHAW
Isis Accelerator Division
Rutherford Appleton Laboratory
Building R2, Ral
Chilton, Didcot, Oxon 0X110QX
UNITED KINGDOM

Phone: 44-0-1235-446646
Fax: 44-0-1235-445720
E-Mail:

KEVIN J. KLEMAN
Synchrotron Radiation Center
University of Wisconsin
3731 Schneider Drive
Stoughton, WI 53589

Phone: 608/877-2147
Fax: 608/877-2001
E-Mail: kevin@src.wisc.edu

MARTIN J. KNOTT
Accelerator Systems Division/APS
Argonne National Laboratory
Building 401
9700 South Cass Avenue
Argonne, IL 60439

Phone: 630/252-6609
Fax: 630/252-1512
E-Mail: mjk@aps.anl.gov

MICHAEL KOGAN
LNS Bates Linac
Massachusetts Institute of Technology
21 Manning Road
Middleton, MA 01949

Phone: 617/253-9533
Fax: 627/253-9799
E-Mail: kogan@bates.mit.edu

ERICH KUGLER
ISOLDE, PPE Division
CERN
1211 Geneve 23
SWITZERLAND

Phone: 41-22-7673189
Fax: 41-22-7678990
E-Mail: ekug@cernvm.cern.ch

PETER M. KUSKE
Department of Beam Diagnostics
BESSY
Lentzeallee 100
14195 Berlin
GERMANY

Phone: 0049-30-82004188
Fax: 0049-30-82004103
E-Mail:

ANA MARIA LABRADOR GARCIA
Laboratori del Sincrotro de Barcelona
c/o Institute of Physics
8000-C Arhus
DENMARK

Phone: 45-8942-3697
Fax: 45-8612-0740
E-Mail: labrador@dfi.aau.dk

SHARON L. LACKEY
Accelerator Division/Controls
Fermi National Accelerator Laboratory
P.O. Box 500
Batavia, IL 60510

Phone: 630/840-4453
Fax: 630/840-8590
E-Mail: slackey@fnal.gov

FRANK R. LENKSZUS
Accelerator Systems Division/APS
Argonne National Laboratory
Building 401
9700 South Cass Avenue
Argonne, IL 60439

Phone: 630/252-6972
Fax: 630/252-6123
E-Mail: frl@aps.anl.gov

HUNG-CHI LIHN
Wafer Inspection Division
Tencor Instruments, Inc.
2400 Charleston Road
Mountain View, CA 94043

Phone: 415/943-6176
Fax: 415/961-0513
E-Mail: hlihn@tencor.com

ALEX H. LUMPKIN
Accelerator Systems Division/APS
Argonne National Laboratory
Building 401
9700 South Cass Avenue
Argonne, IL 60439

Phone: 630/252-4879
Fax: 630/252-4732
E-Mail: lumpkin@aps.anl.gov

HENGJIE MA
Accelerator Division/MA
Fermi National Accelerator Laboratory
P.O. Box 500, MS 340
Batavia, IL 60510

Phone: 630/840-4490
Fax: 630/840-2677
E-Mail: mahengjie@fnalv.fnal.gov

GEORGE H. MACKENZIE
TRIUMF
4004 Wesbrook Mall
Vancouver B.C. V6T 2A3
CANADA

Phone: 604/222-7342
Fax: 604/222-1074
E-Mail: ghm@triumf.ca

FABIO MARCELLINI
INFN Laboratori Nazionali di Frascati
via Enrico Fermi 40
00044 Frascati
ITALY

Phone: 39-6-94032
Fax: 39-6-94032256
E-Mail: marcel@lnf.infn.it

FELIX MARTI
Cyclotron Laboratory
Michigan State University
East Lansing, MI 48824

Phone: 517/333-6360
Fax: 517/353-5967
E-Mail: marti@nscl.msu.edu

DOUGLAS MCCORMICK
Accelerator Department
SLAC
4349 Sand Hill Road
Menlo Park, CA 94025

Phone: 415/926-2470
Fax: 415/926-2407
E-Mail: domac@slac.stanford.edu

DAVID P. MCGINNIS
Accelerator Division/Proton Source
Fermi National Accelerator Laboratory
P.O. Box 500
Batavia, IL 60510

Phone: 630/840-2789
Fax: 630/840-8737
E-Mail: mcginnis@admail.fnal.gov

RICHARD A. MEADOWCROFT
Accelerator Division/RFI
Fermi National Accelerator Laboratory
P.O. Box 500
Batavia, IL 60510

Phone: 630/840-2127
Fax: 630/840-3754
E-Mail: meadowcroft@
           admail.fnal.gov

KEITH G. MEISNER
Accelerator Division/RFI
Fermi National Accelerator Laboratory
P.O. Box 500
Batavia, IL 60510

Phone: 630/840-4807
Fax: 630/840-3754
E-Mail: meisner@fnal.gov

KARL H. MESS
Department FEB/FE
DESY
Notkestrasse 85
D22603 Hamburg
GERMANY

Phone: 49-40-8998-3055
Fax: 49-40-8998-4445
E-Mail: mess@desy.de

GERHARD MROTZEK
Department SRA
CCLRC Daresbury Laboratory
Keckwick Lane
Warrington, Cheshire WA4 4AD
UNITED KINGDOM

Phone: 44-1925603587
Fax:
E-Mail: gmrotzek@dl.ac.uk

ROMAN J. NAWROCKY
NSLS
Brookhaven National Laboratory
Building 725B
Upton, NY 11973

Phone: 516/344-4449
Fax: 516/344-4745
E-Mail: nawrocky@bnl.gov

JOHNNY S. T. NG
FDET Department
DESY-Hamburg
85 Notkestrasse
22603 Hamburg
GERMANY

Phone: 040-8998-3030
Fax: 040-8998-3094
E-Mail: ng@desy.de

TAKASHI OBINA
Accelerator Department
National Laboratory for
High Energy Physics
1-1 Oho
Tsukuba, Ibaraki 305
JAPAN

Phone: 81-298-64-5283
Fax: 81-298-64-3182
E-Mail: obina@kekvax.kek.jp

MARVIN A. OLSON
Accelerator Division/RFI
Fermi National Accelerator Laboratory
P.O. Box 500
Batavia, IL 60510

Phone: 630/840-4445
Fax: 630/840-3754
E-Mail: molson@fnal.gov

RALPH S. PASQUINELLI
Accelerator Division/Antiproton Source
Fermi National Accelerator Laboratory
P.O. Box 500, MS 341
Batavia, IL 60510

Phone: 630/840-4724
Fax: 630/840-8737
E-Mail: pasquin@fnal.gov

DONALD R. PATTERSON
Advanced Photon Source
Argonne National Laboratory
Building 401
9700 South Cass Avenue
Argonne, IL 60439

Phone: 630/252-6951
Fax: 630/252-4732
E-Mail: drp@aps.anl.gov

WILLIAM PELLICO
Booster/Linac
Fermi National Accelerator Laboratory
Box 500
Batavia, IL 60510

Phone: 630/840-8368
Fax: 630/840-8737
E-Mail: pellico@fnal.gov

DAVID W. PETERSON
Accelerator Division/Antiproton Source
Fermi National Accelerator Laboratory
P.O. Box 500, MS 341
Batavia, IL 60510

Phone: 630/840-3073
Fax: 630/840-8737
E-Mail: peterson@fnal.gov

MAURICE PILLER
Accelerator Development Department
CEBAF
12000 Jefferson Avenue
MS 58B
Newport News, VA 23606

Phone: 804/249-7534
Fax: 804/249-7658
E-Mail: piller@cebaf.gov

PHILIPPE PIOT
Accelerator Development Group
CEBAF
MS 58B
12000 Jefferson Avenue
Newport News, VA 23606

Phone: 804/249-5032
Fax: 804/249-7658
E-Mail: piot@cebaf.gov

MASSIMO PLACIDI
Department SL/BI
CERN
CH-1211 Geneva 23
SWITZERLAND

Phone: 011-4122-767-6638
Fax: 011-4122-782-2850
E-Mail: massimo@cernvm.cern.ch

MICHAEL A. PLUM
Department AOT-2
Los Alamos National Laboratory
MS H838
Los Alamos, NM 87545

Phone: 505/667-7547
Fax: 505/665-2509
E-Mail: plum@lanl.gov

JOHN F. POWER
Department AOT-1
Los Alamos National Laboratory
P.O. Box 1663
MS H808
Los Alamos, NM 87545

Phone: 505/667-7045
Fax: 505/665-2904
E-Mail: jpower@lanl.gov

TOM POWERS
Accelerator Development Department
CEBAF
MS 58B
12000 Jefferson Avenue
Newport News, VA 23606

Phone: 804/249-7660
Fax: 804/249-7658
E-Mail: powers@cebaf.gov

MARCO PULLIA
TERA
CERN
1211 Geneve 23 CH
SWITZERLAND

Phone: 41-22-767-2879
Fax: 41-22-767-9145
E-Mail: pulliam@vxcern.cern.ch

PATRICK M. PUZO
SL/BI Division
CERN
CH-1211 Geneva 23
SWITZERLAND

Phone: 41-22-767-5462
Fax: 41-27-782-2850
E-Mail: patrick.puzo@cern.ch

WILLI J. RADLOFF
Department MKI
DESY
Notkestrasse
22603 Hamburg
GERMANY

Phone: 040-8998-3355
Fax: 040-8994-4303
E-Mail: radloff@desy.de

ULRICH RAICH
Proton Synchrotron Division
CERN
CH 1211 Geneve 23
SWITZERLAND

Phone: 0041-22-767-2632
Fax:
E-Mail: uli@dxcern.cern.ch

WILLIAM R. RAWNSLEY
Cyclotron Department
TRIUMF
4004 Wesbrook Mall
Vancouver, BC V6T 2A3
CANADA

Phone: 604/222-1047
Fax: 604/222-1074
E-Mail: rawnsley@triumf.ca

LUIGI REZZONICO
ARF Department
Paul Scherrer Institute
Würenlingen & Villigen
CH 5232 Villigen PSI
SWITZERLAND

Phone: 41-56-3103377
Fax: 41-56-310-33-83
E-Mail: rezzonico@psi.ch

MARC C. ROSS
Accelerator Department
SLAC
4349 Sand Hill Road
Menlo Park, CA 94025

Phone: 415/926-3526
Fax: 415/926-2407
E-Mail: mcrel@slac.stanford.edu

ALBERTO ROVELLI
Laboratorio Nazionale del Sud (LNS)
INFN
Via S. Sofia 44/A
95125 Catania
ITALIA

Phone: 39-95-542256
Fax: 39-95-542300
E-Mail: rovelli@vaxlns.lns.infn.it

DON RULE
Code 682
Naval Surface Weapons Center (USN)
10901 New Hampshire Avenue
Silver Spring, MD 20903-5640

Phone: 301/394-2260
Fax: 301/394-5135
E-Mail: rule@oasys.dt.navy.mil

FERNANDO SANNIBALE
INFN Laboratori Nazionali di Frascati
via Enrico Fermi 40
00044 Frascati
ITALY

Phone: 39-6-9403 2213
Fax: 39-6-9403 225-6
E-Mail: sannibale@lnf.infn.it

SURAJIT SARKAR
Dept. of Controls & Instrumentation
MIT Bates Linac Laboratory
21 Manning Avenue
Middleton, MA 01949

Phone: 617/253-9207
Fax: 617/253-9799
E-Mail: sarkar@aesir.mit.edu

JAMES J. SEBEK
SSRL
SLAC
2575 Sand Hill Road
Menlo Park, CA 94025

Phone: 415/926-94025
Fax: 415/926-3600
E-Mail: sebek@slac.stanford.edu

WILLIAM C. SELLYEY
Physics Division
Argonne National Laboratory
Building 203
9700 South Cass Avenue
Argonne, IL 60439

Phone: 630/252-4037
Fax: 630/252-9647
E-Mail: sellyey@
          anlphy.phy.anl.gov

NICK SERENO
Accelerator Systems Division/APS
Argonne National Laboratory
Building 401
9700 South Cass Avenue
Argonne, IL 60439

Phone: 630/252-6867
Fax: 630/252-5703
E-Mail: sereno@aps.anl.gov

ROBERT E. SHAFER
Accelerator Operations &
Technology Division
Los Alamos National Laboratory
MS H808
Los Alamos, NM 87545

Phone: 505/667-5877
Fax: 505/665-2904
E-Mail: rshafer@lanl.gov

THOMAS J. SHEA
RHIC Department
Brookhaven National Laboratory
Building 1005-4
Upton, NY 11973

Phone: 516/344-2435
Fax: 516/344-2588
E-Mail: shea@bnlvx1.bnl.gov

VLADIMIR D. SHILTSEV
Accelerator Division/Physics
Fermi National Accelerator Laboratory
P.O. Box 500
Batavia, IL 60510

Phone: 630/840-5241
Fax: 630/840-4552
E-Mail: shiltsev@fnal.gov

TSUMORU SHINTAKE
Accelerator Department
National Laboratory for
High Energy Physics (KEK)
1-1 Oho, Tsukuba-Shi
Ibaraki-Ken, 305
JAPAN

Phone: 81-298-64-5256
Fax: 81-298-64-3182
E-Mail: shintake@jpnkekvx

ROBERT H. SIEMANN
SLAC
P.O. Box 4349, MS 26
Stanford, CA 94309

Phone: 415/926-3892
Fax: 415/926-4999
E-Mail: siemann@slac.stanford.edu

OM V. SINGH
National Synchrotron Light Source
Brookhaven National Laboratory
MS 725B
Upton, NY 11973

Phone: 516/344-5332
Fax: 516/344-4745
E-Mail: singhi@bnl.gov

GARY A. SMITH
Accelerator Department
Brookhaven National Laboratory
Building 911C
Upton, NY 11973

Phone: 516/344-3473
Fax: 516/344-5954
E-Mail: smith1@bnldag.bnl.gov

ROBERT J. SMITH
Synchrotron Radiation Department
CCLRC Daresbury Laboratory
MS B74
Keckwick Lane, Daresbury
Warrington, Cheshire WA4 4AD
UNITED KINGDOM

Phone: 44-1925-603341
Fax: 44-1925-603124
E-Mail: r.j.smith@dl.ac.uk

STEPHEN R. SMITH
Controls Department
SLAC
P.O. Box 4349, MS 50
Stanford, CA 94309

Phone: 415/926-3916
Fax: 415/926-3800
E-Mail: ssmith@slac.stanford.edu

VERNON R. SMITH
Controls Department
SLAC
2575 Sandhill Road
Menlo Park, CA 94025

Phone: 415/926-3519
Fax: 415/926-3800
E-Mail: vrs@slac.stanford.edu

HELMUTH SPIELER
Physics Division
Lawrence Berkeley National Laboratory
MS 50B-6208
1 Cyclotron Road
Berkeley, CA 94720

Phone: 510/486-6643
Fax: 510/486-5401
E-Mail: spieler@lbl.gov

JAMES M. STEIMEL
Accelerator Division/Proton Source
Fermi National Accelerator Laboratory
P.O. Box 500
Batavia, IL 60510

Phone: 630/840-4826
Fax: 630/840-8737
E-Mail: jsteimel@adcalc.fnal.gov

ARNOLD STILLMAN
AGS Department
Brookhaven National Laboratory
Building 911B
Upton, NY 11973

Phone: 516/344-4944
Fax: 516/344-5954
E-Mail: stillman@bnl.gov

GREGORY D. STOVER
Advanced Light Source
Electronics Engineering
Lawrence Berkeley National Laboratory
1 Cyclotron Road
Berkeley, CA 94720

Phone: 510/486-7706
Fax: 510/486-5775
E-Mail: gdstover@lbl.gov

TSUYOSHI SUWADA
Photon Factory, Linac Division
National Laboratory for
High Energy Physics (KEK)
1-1 Oho, Tsukuba
Ibaraki 305
JAPAN

Phone: 81-298-64-5589
Fax: 81-298-64-2801
E-Mail: suwada@kekvax.kek.jp

THOMAS N. TALLERICO
AGS Department
Brookhaven National Laboratory
Building 911B
Upton, NY 11973

Phone: 516/282-4642
Fax: 516/282-5954
E-Mail: tallerico@
              bnldag.ags.bnl.gov

KAZUHIRO TAMURA
Department of Materials Science
Hiroshima University
1-3-1 Kagamiyama
Higashi-Hiroshima, Hiroshima 739
JAPAN

Phone: 81-298-64-1171 ext. 3880
Fax: 81-298-64-2801
E-Mail: tamurak@kekvax.kek.jp

GIANNI R. TASSOTTO
Research Division/EED
Fermi National Accelerator Laboratory
P.O. Box 500
Batavia, IL 60510

Phone: 630/840-4325
Fax: 630/840-2950
E-Mail: tassotto@fnal.gov

DMITRY TEYTELMAN
SLAC
2575 Sand Hill Road
Menlo Park, CA 94025

Phone: 415/926-8532
Fax: 415/926-8533
E-Mail: dim@slac.stanford.edu

MAKOTO TOBIYAMA
Accelerator Department
National Laboratory for
High Energy Physics (KEK)
1-1 Oho
Tsukuba, Ibaraki 305
JAPAN

Phone: 81-298-64-5236
Fax: 81-298-64-3182
E-Mail: tobiyama@kekvax.kek.jp

KLAUS B. UNSER
Department of Development
Bergoz Beam Instrumentation
01170 Crozet
FRANCE

Phone: 33-50-41-0089
Fax: 33-50-41-0199
E-Mail: bergoz@bergoz.com

ROK URSIC
Accelerator Development Department
CEBAF
MS 58B
12000 Jefferson Avenue
Newport News, VA 23608

Phone: 804/249-5004
Fax: 804/249-7658
E-Mail: rok@cebaf.gov

CRISTINA VACCAREZZA
INFN Laboratori Nazionali di Frascati
via Enrico Fermi 40
00044 Frascati
ITALY

Phone: 39-6-9403 2537
Fax: 39-6-9403 2256
E-Mail: varese@lnf.infn.it

WILLEM K. VAN ASSELT
AGS Department
Brookhaven National Laboratory
Building 911B
Upton, NY 11973

Phone: 516/344-7778
Fax:
E-Mail: vanasselt@
         bnldag.ags.bnl.gov

GREGORY L. VOGEL
Accelerator Division/RFI
Fermi National Accelerator Laboratory
P.O. Box 500, MS 308
Batavia, IL 60510

Phone: 630/840-4942
Fax: 630/840-3754
E-Mail: vogel@fnal.gov

DEFA WANG
MIT-Bates
21 Manning Road
Middleton, MA 01949

Phone: 617/253-9574
Fax: 617/253-9599
E-Mail: dwang@aesir.mit.edu

XIJIE WANG
NSLS
Brookhaven Accelerator Test Facility
Upton, NY 11973

Phone: 516/344-5791
Fax: 516/344-3115
E-Mail: xwang@bnl.gov

ROBERT C. WEBBER
Accelerator Division/Proton Source
Fermi National Accelerator Laboratory
P.O. Box 500, MS 341
Batavia, IL 60510

Phone: 630/840-5415
Fax: 630/840-8737
E-Mail: webber@fnal.gov

DAVID W. WILDMAN
Accelerator Division/RFI
Fermi National Accelerator Laboratory
P.O. Box 500
Batavia, IL 60510

Phone: 630/840-4619
Fax: 630/840-4552
E-Mail: wildman@admail.fnal.gov

RICHARD L. WITKOVER
AGS Department
Brookhaven National Laboratory
Building 911B
Upton, NY 11973

Phone: 516/344-4607
Fax: 516/344-5954
E-Mail: witkover@
        bnldag.ags.bnl.gov

BINGXIN YANG
Accelerator Systems Division/APS
Argonne National Laboratory
Building 401
9700 South Cass Avenue
Argonne, IL 60439

Phone: 630/252-9821
Fax: 630/252-4732
E-Mail: bxyang@aps.anl.gov

YAN YIN
TRIUMF
4004 Wesbrook Mall
Vancouver, BC V6T 2A3
CANADA

Phone: 604/222-7360
Fax: 604/222-1074
E-Mail: yanyin@triumf.ca

REUBEN YOTAM
SSRL
SLAC
P.O. Box 4349, MS 69
Stanford, CA 94309-0210

Phone: 415/926-3167
Fax: 415/926-4100
E-Mail: yotam@slac.stanford.edu

JAMES R. ZAGEL
Accelerator Division/RFI
Fermi National Accelerator Laboratory
P.O. Box 500, MS 308
Batavia, IL 60510

Phone: 630/840-4076
Fax: 630/840-3754
E-Mail: zagel@almond.fnal.gov

ALEX ZALTSMAN
Department AGS
Brookhaven National Laboratory
Bldg. 911-B
Upton, NY 11973

Phone: 516/344-2967
Fax: 516/344-5954
E-Mail:

List of Vendors

W. L. COLLINS
MicroGraphics Inc.
P. O. Box 125
Waterloo, Wisconsin 53594

Phone: (414) 478-2889
Fax: (414) 478-3689
E-mail: wlc@MicroG.com

TONY COPLEY
Tektronix, Inc.
5350 Keystone Court
Rolling Meadows, Illinois 60008

Phone: (847) 259-7580
Fax: (847) 259-8388
E-Mail:

DAN DALEY
Hewlett-Packard
1200 E. Diehl Road
Naperville, Illinois 60566

Phone: (630) 245-3696
Fax: (630) 245-3885
E-Mail:

HANK D. GERWERS
Princeton Scientific Corp.
P. O. Box 143
Princeton, New Jersey 08542

Phone: (609) 924-3011
Fax: (609) 924-3018
E-Mail: princescie@aol.com

JIM GOEING
National Instruments
P. O. Box 99
Plainfield, Illinois 60544

Phone: (815) 436-7775 x 4824
Fax: (815) 436-6336
E-Mail: jim.goeing@natinst.com

GARY F. HANCOCK
Princeton Instruments, Inc.
3660 Quakerbridge Road
Trenton, New Jersey 08619

Phone: (609) 587-9797
Fax: (609) 587-1970
E-Mail: prinst@pipeline.com

IAN WALKER
GMW Associates
P. O. Box 2578
Redwood City, California 94064

Phone: (415) 802-8292 x15
Fax: (415) 802-8298
E-Mail: ian@gmw.com

AIP Conference Proceedings

	Title	L.C. Number	ISBN
No. 311	Physics of High Energy Particles in Toroidal Systems (Irvine, CA 1993)	94-72098	1-56396-364-7
No. 312	Molecules and Grains in Space (Mont Sainte-Odile, France 1993)	94-72615	1-56396-355-8
No. 313	The Soft X-Ray Cosmos ROSAT Science Symposium (College Park, MD 1993)	94-72499	1-56396-327-2
No. 314	Advances in Plasma Physics Thomas H. Stix Symposium (Princeton, NJ 1992)	94-72721	1-56396-372-8
No. 315	Orbit Correction and Analysis in Circular Accelerators (Upton, NY 1993)	94-72257	1-56396-373-6
No. 316	Thirteenth International Conference on Thermoelectrics (Kansas City, Missouri 1994)	95-75634	1-56396-444-9
No. 317	Fifth Mexican School of Particles and Fields (Guanajuato, Mexico 1992)	94-72720	1-56396-378-7
No. 318	Laser Interaction and Related Plasma Phenomena 11th International Workshop (Monterey, CA 1993)	94-78097	1-56396-324-8
No. 319	Beam Instrumentation Workshop (Santa Fe, NM 1993)	94-78279	1-56396-389-2
No. 320	Basic Space Science (Lagos, Nigeria 1993)	94-79350	1-56396-328-0
No. 321	The First NREL Conference on Thermophotovoltaic Generation of Electricity (Copper Mountain, CO 1994)	94-72792	1-56396-353-1
No. 322	Atomic Processes in Plasmas Ninth APS Topical Conference (San Antonio, TX)	94-72923	1-56396-411-2
No. 323	Atomic Physics 14 Fourteenth International Conference on Atomic Physics (Boulder, CO 1994)	94-73219	1-56396-348-5
No. 324	Twelfth Symposium on Space Nuclear Power and Propulsion (Albuquerque, NM 1995)	94-73603	1-56396-427-9
No. 325	Conference on NASA Centers for Commercial Development of Space (Albuquerque, NM 1995)	94-73604	1-56396-431-7
No. 326	Accelerator Physics at the Superconducting Super Collider (Dallas, TX 1992-1993)	94-73609	1-56396-354-X

	Title	L.C. Number	ISBN
No. 327	Nuclei in the Cosmos III Third International Symposium on Nuclear Astrophysics (Assergi, Italy 1994)	95-75492	1-56396-436-8
No. 328	Spectral Line Shapes, Volume 8 12th ICSLS (Toronto, Canada 1994)	94-74309	1-56396-326-4
No. 329	Resonance Ionization Spectroscopy 1994 Seventh International Symposium (Bernkastel-Kues, Germany 1994)	95-75077	1-56396-437-6
No. 330	E.C.C.C. 1 Computational Chemistry F.E.C.S. Conference (Nancy, France 1994)	95-75843	1-56396-457-0
No. 331	Non-Neutral Plasma Physics II (Berkeley, CA 1994)	95-79630	1-56396-441-4
No. 332	X-Ray Lasers 1994 Fourth International Colloquium (Williamsburg, VA 1994)	95-76067	1-56396-375-2
No. 333	Beam Instrumentation Workshop (Vancouver, B. C., Canada 1994)	95-79635	1-56396-352-3
No. 334	Few-Body Problems in Physics (Williamsburg, VA 1994)	95-76481	1-56396-325-6
No. 335	Advanced Accelerator Concepts (Fontana, WI 1994)	95-78225	1-56396-476-7 (Set) 1-56396-474-0 (Book) 1-56396-475-9 (CD-Rom)
No. 336	Dark Matter (College Park, MD 1994)	95-76538	1-56396-438-4
No. 337	Pulsed RF Sources for Linear Colliders (Montauk, NY 1994)	95-76814	1-56396-408-2
No. 338	Intersections Between Particle and Nuclear Physics 5th Conference (St. Petersburg, FL 1994)	95-77076	1-56396-335-3
No. 339	Polarization Phenomena in Nuclear Physics Eighth International Symposium (Bloomington, IN 1994)	95-77216	1-56396-482-1
No. 340	Strangeness in Hadronic Matter (Tucson, AZ 1995)	95-77477	1-56396-489-9
No. 341	Volatiles in the Earth and Solar System (Pasadena, CA 1994)	95-77911	1-56396-409-0

	Title	L.C. Number	ISBN
No. 376	Chaos and the Changing Nature of Science and Medicine: An Introduction (Mobile, AL 1995)	96-85220	1-56396-442-2
No. 377	Space Charge Dominated Beams and Applications of High Brightness Beams (Bloomington, IN 1995)	96-85165	1-56396-625-7
No. 378	Surfaces, Vacuum, and Their Applications (Cancun, Mexico 1994)	96-85594	1-56396-418-X
No. 379	Physical Origin of Homochirality in Life (Santa Monica, CA 1995)	96-86631	1-56396-507-0
No. 380	Production and Neutralization of Negative Ions and Beams / Production and Application of Light Negative Ions (Upton, NY 1995)	96-86435	1-56396-565-8
No. 381	Atomic Processes in Plasmas (San Francisco, CA 1996)	96-86304	1-56396-552-6
No. 382	Solar Wind Eight (Dana Point, CA 1995)	96-86447	1-56396-551-8
No. 383	Workshop on the Earth's Trapped Particle Environment (Taos, NM 1994)	96-86619	1-56396-540-2
No. 384	Gamma-Ray Bursts (Huntsville, AL 1995)	96-79458	1-56396-685-9
No. 385	Robotic Exploration Close to the Sun: Scientific Basis (Marlboro, MA 1996)	96-79560	1-56396-618-2
No. 386	Spectral Line Shapes, Volume 9 13th ICSLS (Firenze, Italy 1996)		1-56396-656-5
No. 387	Space Technology and Applications International Forum (Albuquerque, NM 1997)	96-80254	1-56396-679-4 (Case set) 1-56396-691-3 (Paper set)
No. 388	Resonance Ionization Spectroscopy 1996 Eighth International Symposium (State College, PA 1996)	96-80324	1-56396-611-5
No. 389	X-Ray and Inner-Shell Processes 17th International Conference (Hamburg, Germany 1996)	96-80388	1-56396-563-1
No. 390	Beam Instrumentation Proceedings of the Seventh Workshop (Argonne, IL 1996)	97-70568	1-56396-612-3
No. 391	Computational Accelerator Physics (Williamsburg, VA 1996)	97-70181	1-56396-671-9